VECTORS

$$\mathbf{u} = (a, b), \qquad \mathbf{v} = (c, d)$$
$$\mathbf{u} + \mathbf{v} = (a + c, b + d)$$
$$\mathbf{u} \cdot \mathbf{v} = ac + bd$$
$$\|\mathbf{u}\| = \sqrt{a^2 + b^2}$$

POLAR COORDINATES

$$\begin{cases} x = r \cos \theta \\ y = r \sin \theta \end{cases}$$

$$\begin{cases} x^2 + y^2 = r^2 \\ \cos \theta = \dfrac{x}{r}, \quad \sin \theta = \dfrac{y}{r} \end{cases}$$

COMPLEX NUMBERS

$$(a + bi)(c + di) = (ac - bd) + (ad + bc)i$$
$$\overline{a + bi} = a - bi$$
$$|a + bi|^2 = a^2 + b^2$$
$$(\cos \theta + i \sin \theta)^n = \cos n\theta + i \sin n\theta$$

ALGEBRA AND TRIGONOMETRY

ALGEBRA
AND
TRIGONOMETRY

Harley Flanders
Tel Aviv University

Justin J. Price
Purdue University

ACADEMIC PRESS   *New York   San Francisco   London*

A Subsidiary of Harcourt Brace Jovanovich, Publishers

ACADEMIC PRESS, INC.
111 Fifth Avenue, New York, New York 10003

United Kingdom Edition published by
ACADEMIC PRESS, INC. (LONDON) LTD.
24/28 Oval Road, London NW1

Library of Congress Cataloging in Publication Data

Flanders, Harley.
 Algebra and trigonometry.

 Bibliography: p.
 Includes index.
 1. Algebra. 2. Trigonometry. I. Price, Justin J.,
joint author. II. Title.
QA152.2.F55 512′.13 74-17980
ISBN 0–12–259665–X

To Daveed, Michal, Tal, and Zvi

CONTENTS

PREFACE   *xi*

1   BASIC ALGEBRA

1. The Nature of Algebra   *1*
2. Algebraic Notation   *3*
3. The Rules of Algebra   *5*
4. Integers and Rationals   *12*
5. Integral Exponents   *18*
6. Roots and Radicals   *23*
7. Rational Exponents   *28*
8. Polynomials   *31*
9. Polynomials in Several Variables   *35*
10. Factoring   *38*
11. Rational Expressions   *42*
12. Common Errors in Algebra   *46*
    Tests   *48*

2   EQUATIONS AND INEQUALITIES

1. Equations and Identities   *49*
2. Linear Equations   *52*
3. Quadratic Equations   *57*
4. Other Types of Equations   *63*
5. Applications   *66*
6. Order   *72*
7. Absolute Values   *78*

8. Inequalities   *83*
Tests   *88*

3   FUNCTIONS AND GRAPHS

1. Introduction   *89*
2. Coordinates in the Plane   *90*
3. Functions   *93*
4. Construction of Functions   *99*
5. Linear Functions   *101*
6. Quadratic Functions   *109*
7. Tips on Graphing   *116*
Tests   *121*

4   POLYNOMIAL AND RATIONAL FUNCTIONS

1. Introduction   *123*
2. Graphs of Polynomials   *123*
3. Graphs of Factored Polynomials   *130*
4. Rational Functions   *134*
5. Graphs of Rational Functions   *137*
6. Factored Rational Functions   *142*
Tests   *148*

5   EXPONENTIALS AND LOGARITHMS

1. Exponential Functions   *149*
2. Logarithm Functions   *153*
3. Power Functions   *160*
4. Accuracy and Round-off   *167*
5. Tables and Interpolation   *169*
6. Computations with Logarithms (optional)   *172*
7. Computations with a Small Calculator
(optional)   *177*
8. Applications   *183*
9. Business Applications (optional)   *186*
Tests   *192*

6   TRIGONOMETRIC FUNCTIONS

1. Introduction   *193*
2. Distances and Angles   *194*

3. Sine and Cosine   *199*
4. Other Trigonometric Functions   *205*
5. Graphs of Sine and Cosine   *209*
6. Graphs of the Other Functions   *215*
   Tests   *221*

7 IDENTITIES AND INVERSE FUNCTIONS

1. Basic Identities   *223*
2. The Addition Laws   *228*
3. Further Identities   *231*
4. Inverse Functions   *236*
5. Applications   *243*
   Tests   *250*

8 TRIGONOMETRY

1. Right Triangles   *251*
2. Oblique Triangles   *256*
3. Numerical Solution   *260*
4. Applications to Geometry   *267*
5. Vectors (optional)   *272*
6. Length and Inner Product (optional)   *278*
   Tests   *284*

9 TOPICS IN ALGEBRA

1. Division of Polynomials   *285*
2. Zeros of Polynomials   *291*
3. Partial Fractions   *299*
4. Systems of Linear Equations   *305*
5. Determinants   *312*
   Tests   *319*

10 COMPLEX NUMBERS

1. Complex Arithmetic   *321*
2. The Complex Plane   *326*
3. Zeros of Polynomials   *333*
4. De Moivre's Theorem and Roots of Unity   *337*
   Tests   *344*

11 DISCRETE ALGEBRA

1. Sequences *345*
2. Permutations and Combinations *350*
3. Binomial Theorem *354*
4. Summation *357*
5. Mathematical Induction *364*
   Tests *369*

ANSWERS TO ODD-NUMBERED EXERCISES *371*

TABLES

1. 4-place Logarithms *404*
2. 4-place Antilogarithms *406*
3. Powers and Roots *408*
4. Trigonometric (Degrees) *410*
5. Trigonometric (Radians) *414*
6. Log-Trig (Degrees) *416*
7. Trigonometric (π Radians) *420*

INDEX *421*

In this text we present the essentials of algebra and trigonometry with some applications. Our aim is to provide students with a solid working knowledge, which they will be able to apply in other courses and in their occupations. To this end we emphasize practical skills, problem solving, and computational techniques. We always try to justify theory by down-to-earth applications.

Our presentation is informal. We believe that a definition–theorem–proof style quickly deadens the interest of most students at this level. While we include some proofs, we do so only when they give insight into the subject matter.

The topics covered in this text fall into several categories. Let us look briefly at each.

FUNDAMENTALS OF ALGEBRA Chapter 1 contains the absolute essentials of algebra. However, it is much more than a whirlwind review and list of formulas. Each topic is explained, illustrated, and accompanied by an ample set of exercises. We stress the proper use of algebraic notation and practical manipulative skills such as factoring, using exponents and radicals, and simplifying rational expressions. In the last section we pinpoint the most common mistakes in algebra.

EQUATIONS AND INEQUALITIES Chapter 2 treats the solution of linear, quadratic, and other types of equations and systems of equations. It also treats the order relation for real numbers, absolute values, and the solution of inequalities.

FUNCTIONS Chapters 3—6 aim at a practical presentation of the most basic functions: polynomial, rational, exponential, logarithm, and trigonometric. We emphasize the graphs of these functions and facility in graphing generally.

TRIGONOMETRY The trigonometric functions, their basic properties, and their graphs are introduced in Chapter 6. The inverse trigonometric functions and identities are presented in Chapter 7, and the solution of triangles in Chapter 8.

FURTHER ALGEBRA TOPICS Chapter 9 discusses zeros of polynomials, division, partial fractions, linear systems and their matrices, and determinants. Chapter 10 presents the complex number system as a natural extension of the real number system, and discusses the zeros of complex polynomials. The polar form of complex numbers provides an important connection with trigonometric functions. Chapter 11 covers progressions, permutations, combinations, and the binomial theorem. Later sections take up the Σ notation for sums, techniques for computing sums, and mathematical induction.

APPLICATIONS Useful, realistic applications are scattered through the book. For example in Chapter 2, in addition to the usual problems on work, rates, and mixtures, we include supply–demand, investment, and the use of inequalities in making estimates. In Chapter 3 we have some maximum and minimum problems. In Chapter 5 we have a section on physical and biological applications of exponential growth and decay, and a section covering a short course on the mathematics of finance. Besides practical triangle solving, our chapters on trigonometry include applications to polar coordinates, the geometry of the triangle and its related circles, vectors in the plane, and the inner product. Our applications of mathematical induction include inductive definitions and far more than the usual list of summation formulas. We believe our set of exercises is truly representative of the various ways induction is actually used.

COMPUTATION In addition to practical computation via logarithms, we offer in Chapter 5 an optional introduction to scientific pocket calculators. This option is continued in Chapter 8 with some material on solving triangles on a calculator.

AN OMITTED TOPIC Just as every fairy tale must begin with "Once upon a time," so it has become almost mandatory that every mathematics text must begin with the theory of sets. We feel that this obsession with sets has been a disservice to students because it has diverted their attentions from mastering important skills to learning set formalism and jargon. We are more interested in the *solutions* of a quadratic equation than in its solution *set*. We are not alone in this view; it now appears that the "new math" is losing its initial sets-appeal.

FEATURES

There are numerous worked examples and numerous figures, both an essential part of the text. The 2300 exercises are graded in difficulty, and harder ones are marked with an asterisk. Answers to the odd-numbered exercises are provided at the end. Each chapter ends with two sample tests typical of what a student can expect at the end of a unit.

We include some basic numerical tables that are adequate for all of the exercises in this text. For greater accuracy, we recommend the *C.R.C. Standard Mathematical Tables,* a handbook of useful tables and formulas.

ACKNOWLEDGMENTS

We acknowledge with pleasure the fast, accurate work of our typist, Sara Marcus, and the constructive criticisms of Johnny W. Duvall, Mountain View College, Anthony A. Patricelli, Northeastern Illinois University, and Herbert B. Perlman, Wilbur Wright College. We are particularly grateful for the thorough and meaningful suggestions given by Thomas Butts, Michigan State University, and Calvin Lathan, Monroe Community College, and for the high quality graphics of Vantage Art, Inc. and the outstanding editing job of Academic Press.

HARLEY FLANDERS

JUSTIN J. PRICE

ALGEBRA AND TRIGONOMETRY

BASIC ALGEBRA

Algebra is the systematic study of the operations of arithmetic and relations between numbers expressed by these operations. In this section, we discuss two outstanding features of algebra: brevity and generality.

Algebraic notation is the shorthand of mathematics. It makes relations between numbers as short and as clear as possible. As an example, consider this statement: "The area of a rectangle is equal to its length multiplied by its width". This assertion requires 14 words, 57 letters. Certainly not all of them are needed to convey the meaning. The statement can be shortened in various ways, for instance:

Area equals length times width.

Area = length times width.

Area = length · width.

$A = L \cdot W.$

$A = LW.$

The final version, $A = LW$, expresses the given statement boiled down to its bare bones.

Here is another example. To convert a centigrade temperature to Fahrenheit, multiply by $\frac{9}{5}$, then add 32. In algebraic notation, this 13-word statement is abbreviated by the formula

$$F = \tfrac{9}{5}C + 32.$$

One further example: Choose any number. Take the number that is 2 larger than it and the number that is 2 smaller than it. If you multiply these numbers, then you get a number that is 4 smaller than the square of the number you started with.

This long-winded statement can be drastically shortened by translating it into algebraic language. The basic step in translation is just common sense: if you are going to refer to a number several times, give it a name and use that name each time. So call the chosen number x. The number 2 larger than x is $x + 2$; the number 2 smaller than x is $x - 2$; the number 4 less than the square of x is $x^2 - 4$. Now the original clumsy statement becomes a simple algebraic relation,

$$(x + 2)(x - 2) = x^2 - 4.$$

Everything unnecessary has been stripped away, allowing you to see at a glance just what is going on.

Generality

Their ability to abbreviate complicated ideas is an important feature of algebraic formulas. But even more important is their generality; formulas can express many facts in one statement. For example, the formula $A = LW$ is a statement about *all* rectangles, not just one. The formula $A = \pi r^2$ tells the area of every circle; it contains many particular statements such as

the area of a circle of radius 3 is $\pi \cdot 3^2$;

the area of a circle of radius 17 is $\pi \cdot 17^2$;

the area of a circle of radius 9.08 is $\pi \cdot (9.08)^2$;

and so on. We could spend a lifetime writing out the information contained in this one formula and never finish.

The moral is clear: we can achieve both brevity and generality by using formulas with numbers represented by letters. Even though $20 = 4 \times 5$ expresses the area of a 4 by 5 rectangle, it is better to know the short formula $A = LW$, which gives the area of every rectangle.

A Typical Application

In working specific problems, we exploit the brevity and generality of algebra. Here is a simple example. Suppose we wanted to know whether there are any temperatures where Fahrenheit and centigrade readings are the same. Now the formula

$$F = \tfrac{9}{5}C + 32$$

expresses the relation between the two scales in all cases. We are asking if ever $F = C$. Therefore in algebraic language our question is whether the equation

$$C = \tfrac{9}{5}C + 32$$

can ever hold.

The battle is 90% won; we can solve this simple equation by standard algebraic techniques. It has exactly one solution, $C = F = -40°$. By using the generality of algebra, we are sure to have considered all possibilities. Conclusion: $-40°$F and $-40°$C represent the same temperature, and there is no other temperature where this happens.

2. ALGEBRAIC NOTATION

The basis of all computations are the four operations of arithmetic. If x and y represent real numbers, we denote their sum, difference, product, and quotient by

$$x + y, \quad x - y, \quad x \cdot y \text{ or } (x)(y) \text{ or } xy, \quad \frac{x}{y} \text{ or } x/y \text{ or } x \div y.$$

Mixtures of numbers and letters are allowed; for instance

$$x + 4, \quad 5 - 2x, \quad \pi x, \quad \frac{y}{10}$$

are all legitimate. Note that $y/10$ and $\frac{1}{10}y$ are the same.

■ *Example 2.1*

Write a formula for

(a)   one-third of a number,

(b)   the average of two numbers,

(c)   7 less than 5 times a number.

SOLUTION   (a)   If the number is x, then one-third of x is $\frac{1}{3}x$.

(b)   To find the average of two numbers, you add them, then divide by 2. Call the two numbers x and y. Their sum is $x + y$. Divide by 2: their average is $\frac{1}{2}(x + y)$.

(c)   Call the number x. Then 5 times x is $5x$ and 7 less than $5x$ is $5x - 7$.

Answer   (a) $\frac{1}{3}x$ (b) $\frac{1}{2}(x + y)$ (c) $5x - 7$.

Algebraic notation is meant to be absolutely precise. We must always be alert to avoid any possible misunderstanding. For example, the expression $3x + y$ is at first ambiguous. Does it mean "y added to 3 times x" or "y added to x and the sum multiplied by 3"?

We make a convention: multiplications and divisions precede additions and subtractions. So to compute $3x + y$, we first multiply x by 3, then add y. For added insurance we could write

$$3x + y = (3x) + y.$$

We shall not do this, however, because everybody follows the convention.

Parentheses

Sometimes parentheses are essential to avoid ambiguity. If you want to subtract from x the number that is 1 more than y, you must write $x - (y + 1)$, not $x - y + 1$. For instance, if $x = 5$ and $y = 2$, then

$$x - (y + 1) = 5 - (2 + 1) = 5 - 3 = 2, \quad \text{the correct answer,}$$

but

$$x - y + 1 = 5 - 2 + 1 = 4, \quad \text{which is wrong.}$$

The average of x and y is $(x + y)/2$. The parentheses cannot be omitted, because $x + y/2$ means $x + \frac{1}{2}y$, which is different. Similarly, the average can be written as $\frac{1}{2}(x + y)$, which is not the same as $\frac{1}{2}x + y$. For instance, if $x = 2$ and $y = -2$, then

$$\tfrac{1}{2}(x + y) = \frac{x + y}{2} = 0, \qquad x + y/2 = x + \tfrac{1}{2}y = 1, \qquad \tfrac{1}{2}x + y = -1,$$

all different!

Be careful with expressions like $x/y + 1$. By convention this means $(x/y) + 1$, not $x/(y + 1)$. If $x = 2$ and $y = 1$, then

$$x/y + 1 = \frac{x}{y} + 1 = 3, \qquad \text{but} \quad x/(y + 1) = \frac{x}{y + 1} = 1.$$

It is good policy to insert parentheses in case of doubt. Usually no harm can be done, but lots of harm can be done by omitting them.

Some Abbreviations

It is perfectly alright to write $x + x$, but simpler to write $2x$. Similarly,

$$x + x + x = 3x, \qquad x + x + x + x = 4x, \qquad \text{etc.}$$

There are corresponding abbreviations for multiplication. Even though xx makes sense, it is shorter to write x^2. Similarly, we write

$$x^3 \quad \text{for} \quad xxx, \qquad x^4 \quad \text{for} \quad xxxx, \qquad \text{etc.}$$

With this convention $x = x^1$.

The notation applies to repeated multiplication of any algebraic expression. For example,

$$(4x - 7y)(4x - 7y)(4x - 7y) = (4x - 7y)^3.$$

■ *Example 2.2*

The side of a cube has length s. Write formulas for

(a) the volume,   (b) the surface area.

SOLUTION (a) The volume of any rectangular box is the product of its length, width, and height. For the cube, each of these dimensions is s. Hence

$$\text{Volume} = V = s \cdot s \cdot s = s^3.$$

(b)   The cube has 6 faces, each a square of side s, hence area $s \cdot s = s^2$. The total surface area is

$$S = s^2 + s^2 + s^2 + s^2 + s^2 + s^2 = 6s^2.$$

Answer   (a) $V = s^3$ (b) $S = 6s^2$.

Express in algebraic notation:

1.   one number added to one-half of another number
2.   half of one number subtracted from twice another number
3.   double the sum of a and b
4.   the average of x and 1
5.   three percent of a number
6.   5% of the amount by which a number exceeds 100
7.   the time it takes to go 350 miles at r miles per hour
8.   the area of a triangle
9.   the area of a triangle whose height is $\frac{1}{3}$ the base
10.   the area of a circle of radius r
11.   the area of a circle of radius $r + 1$
12.   the area of a circular ring bounded by two concentric circles
13.   the sum of two numbers multiplied by their difference
14.   the surface area of a rectangular box.

15.   It is said that for a perfect wedding the age of the groom should be 17 years more than half the age of the bride. Express this as an algebraic formula.

Compute the value for (a) $x = 4$, $y = 3$ and (b) $x = -4$, $y = 3$:

16. $5x - 2y$
17. $(x + y)/7$
18. $x + y/7$
19. $x^2/4 + 3$
20. $(x - 3y)^2$
21. $(x + 1)(y - 1)$
22. $x/(y + 3)$
23. $x/y + 3$
24. $(x^2 - 1)(x + y)$.

Substitute numbers to show that the expressions are different:

25. $(x + 2)y$ and $x + 2y$
26. $3x(y + 1)$ and $3xy + 1$
27. $(x + y)^2$ and $x + y^2$
28. $1/(x + y)$ and $1/x + 1/y$
29. $4 - 2(x - y)$ and $2x - 2y$
30. $5 - (3x - 5)$ and $-3x$.

3. THE RULES OF ALGEBRA

Real Numbers

In this book we shall deal mostly with *real numbers*. We are accustomed to thinking of the real number system R geometrically as the number line (Fig. 3.1), like a scale

Fig. 3.1   The number line

on a thermometer that goes endlessly in both directions. We'll have a lot more to say about this geometric point of view in Chapter 2.

When a is to the left of b we say "a is less than b" and write

$$a < b.$$

When a is to the left of b or equal to b we say "a is less than or equal to b" and write

$$a \leq b.$$

These same two statements are also written $b > a$ (b is greater than a) and $b \geq a$ (b is greater than or equal to a), respectively. Learn to use this notation for comparison of numbers. We shall use it a great deal when we study inequalities in the next chapter.

Examples:

$$0 < 105, \quad 0 > -3, \quad -4 \leq 1, \quad -9 \geq -10, \quad 7.98 < 7.985, \quad -1.042 \geq -1.043.$$

A basic fact, which we shall use again and again, is:

> If a is any real number, then $a^2 \geq 0$.

Commutative and Associative Laws

The real numbers obey the laws of arithmetic. Therefore symbols representing real numbers obey the same laws. The simplest are the commutative laws: addition or multiplication of two numbers can be done in either order.

> **Commutative Laws**
> If x and y are real numbers, then
> $$x + y = y + x, \qquad xy = yx.$$

It follows that any two expressions representing real numbers may be added or multiplied in either order.

Examples:

$$(\tfrac{3}{4}x - 5y) + (7z + 12) = (7z + 12) + (\tfrac{3}{4}x - 5y),$$
$$(x^2 + y)(3x - 1) = (3x - 1)(x^2 + y).$$

We remark that $3x$ and $x3$ mean exactly the same thing, but it is customary to write $3x$.

We add or multiply numbers two at a time, so what do $x + y + z$ and xyz mean? The sum might mean $(x + y) + z$ or $x + (y + z)$ and the product might mean $(xy)z$ or $x(yz)$. The question is settled by the associative laws, which say that the results are the same.

Associative Laws

If x, y, z are real numbers, then

$$(x + y) + z = x + (y + z), \qquad (xy)z = x(yz).$$

The effect of the associative laws is that $x + y + z$ and xyz make sense without parentheses; each has only one possible meaning.

Not only do $x + y + z$ and xyz have unique meanings, but they can be written in any order:

$$x + y + z = x + z + y = z + y + x, \quad \text{etc,}$$

$$xyz = xzy = zxy = zyx, \quad \text{etc.}$$

All of these assertions can be proved by fiddling around with the associative and commutative laws.

Examples:

$$x + y + z = (x + y) + z = z + (x + y)$$
$$= z + (y + x) = z + y + x,$$

$$xyz = (xy)z = z(xy) = z(yx) = zyx.$$

These principles extend to longer sums and products as well. Thus $x + y + z + w$ and $xyzw$ can be written in any order or grouped in any way by parentheses. This can be helpful in simplifying algebraic expressions.

Examples:

$$x + y + z + y + z + y + y + x = x + x + y + y + y + y + z + z$$
$$= 2x + 4y + 2z,$$

$$(xy)(xz)(xyw) = xxxyyzw = x^3y^2zw.$$

The Distributive Law

The operations of addition and multiplication are not independent of each other. They are connected by the distributive law.

Distributive Law

If x, y, z are real numbers, then

$$x(y + z) = xy + xz, \qquad (y + z)x = yx + zx.$$

The two forms of the distributive law are equivalent because of the commutative law for multiplication.

We use the distributive law all the time in ordinary arithmetic without realizing it.

Examples:

$$4 \times 12 = 4(10 + 2) = 4 \times 10 + 4 \times 2 = 40 + 8 = 48,$$
$$15 \times 49 = 15(50 - 1) = 15 \times 50 - 15 \times 1 = 750 - 15 = 735.$$

■ Example 3.1

Show how the formulas follow from the distributive law:

(1) $x(y + z + w) = xy + xz + xw,$
(2) $(a + b)(x + y) = ax + bx + ay + by.$

SOLUTION   (1)   Write $y + z + w$ as $(y + z) + w$ and use the distributive law:

$$x(y + z + w) = x[(y + z) + w] = x(y + z) + xw.$$

Use the distributive law again, then the associative law:

$$x(y + z) + xw = (xy + xz) + xw = xy + xz + xw.$$

Therefore

$$x(y + z + w) = xy + xz + xw.$$

(2)   Think of $(a + b)$ as a single number and $(x + y)$ as a sum. By the distributive law, applied twice,

$$(a + b)(x + y) = (a + b)x + (a + b)y = (ax + bx) + (ay + by)$$
$$= ax + bx + ay + by.$$

Formula (1) of Example 3.1 is a direct extension of the distributive law. A similar formula applies as well to any number of terms, for example,

$$x(u + v + w + y + z) = xu + xv + xw + xy + xz.$$

Formula (2) of Example 3.1 is an important rule: to multiply $(a + b)$ by $(x + y)$, multiply each term in the first parentheses by each term in the second, then add the four products.

Example:

$$(45)(23) = (40 + 5)(20 + 3) = (40)(20) + (5)(20) + (40)(3) + (5)(3)$$
$$= 800 + 100 + 120 + 15 = 1035.$$

Formula (2) extends in two ways. First, the parentheses may contain more than two terms. For example,

$$(a + b)(x + y + z) = ax + ay + az + bx + by + bz.$$

Again you form all six possible products, taking one factor from each parentheses, then add them. Second, there may be more than two expressions multiplied together. For example,

$$(a + b)(r + s)(x + y) = arx + ary + asx + asy + brx + bry + bsx + bsy.$$

This time you get eight products, each taking one factor from each of the three parentheses.

Here is one more important use of the distributive law. Look again at the formula

$$x(y + z) = xy + xz,$$

but read it from right to left. On the right side, the two terms have a common factor x. On the left side, x is "factored out". Taking out common factors from a sum is an important technique; it often simplifies algebraic expressions.

Examples:

$$35 + 21 = 7 \cdot 5 + 7 \cdot 3 = 7(5 + 3) = 7 \cdot 8 = 56,$$

$$xyz + 4xzw = (xz)y + (xz)(4w) = xz(y + 4w),$$

$$4x - 12xy - 10x^2z = 2x(2) + 2x(-6y) + 2x(-5xz)$$
$$= 2x(2 - 6y - 5xz).$$

Identity and Inverse Laws

The identity laws of algebra single out the special roles of 0 and 1 with respect to addition and multiplication, respectively.

> **Identity Laws**
> For all real numbers a,
>
> $$a + 0 = 0 + a = a, \qquad a \cdot 1 = 1 \cdot a = a.$$

Thus 0 is the identity element with respect to addition of the system of all real numbers, and 1 is the identity element with respect to multiplication: zero added to a leaves a unchanged; one multiplied by a leaves a unchanged.

Also 0 is something special with respect to multiplication. Multiplying a number by 0 completely wipes out that number:

> For each real number a,
> $$a \cdot 0 = 0 \cdot a = 0.$$

The inverse laws say that each real number has an inverse (negative) with respect to addition, and each non-zero real number has an inverse with respect to multiplication.

> **Inverse Laws**
>
> To each real number a there is a real number $-a$ such that
>
> $$a + (-a) = (-a) + a = 0.$$
>
> To each real number $a \neq 0$ there is a real number a^{-1} such that
>
> $$aa^{-1} = a^{-1}a = 1.$$

The number a^{-1} is called the **reciprocal** of a and is usually written $\dfrac{1}{a}$.

Examples:

$$-0 = 0, \quad -(3) = -3, \quad -(-10) = 10, \quad -(-4.5) = 4.5,$$
$$1^{-1} = 1, \quad 5^{-1} = \tfrac{1}{5}, \quad (-\tfrac{1}{2})^{-1} = -2, \quad (\tfrac{1}{3})^{-1} = 3, \quad (-\tfrac{4}{3})^{-1} = -\tfrac{3}{4}.$$

The inverse laws have some useful consequences.

> **Rules for Inverses**
>
> (1) $-(-a) = a.$ (2) $(a^{-1})^{-1} = a$ if $a \neq 0.$
>
> (3) $-(a + b) = (-a) + (-b).$
>
> (4) If $a \neq 0$ and $b \neq 0$, then $(ab)^{-1} = a^{-1}b^{-1}.$

Properties (1) and (2) say that the inverse of the inverse is the original number. Properties (3) and (4) say that the inverse of a sum or a product is the sum or product, respectively, of the inverses. We leave the proofs of these properties as exercises.

Property (4) has an important consequence:

> The product of non-zero real numbers is always non-zero.

For if $a \neq 0$ and $b \neq 0$, then (4) says that ab has an inverse, $(ab)^{-1}$. Thus $(ab)(ab)^{-1} = 1$. It follows that $ab \neq 0$; otherwise $(ab)(ab)^{-1}$ would equal $0 \cdot (ab)^{-1} = 0$.

This simple fact about products is important in solving equations, as we shall see later. It is often expressed in this form: If $ab = 0$, then either $a = 0$ or $b = 0$ or both.

The **quotient** a/b is defined to be the product ab^{-1}:

$$\frac{a}{b} = ab^{-1}, \quad b \neq 0.$$

The usual rules for computations with quotients follow from the properties of inverses:

Rules for Quotients

$$\left(\frac{a}{b}\right)\left(\frac{c}{d}\right) = \frac{ac}{bd}, \qquad \frac{a}{b} + \frac{c}{d} = \frac{ad + bc}{bd},$$

$$\frac{a}{b} = \frac{ka}{kb} \quad (k \neq 0), \qquad \frac{\dfrac{a}{b}}{\dfrac{c}{d}} = \frac{ad}{bc}.$$

Let us prove the fourth formula and leave the other three as exercises. It says that you reduce a four-story fraction by inverting the denominator and multiplying.

We use two steps. First

$$\left(\frac{c}{d}\right)^{-1} = \frac{d}{c}$$

because

$$\left(\frac{c}{d}\right)\left(\frac{d}{c}\right) = (cd^{-1})(dc^{-1}) = c(d^{-1}d)c^{-1} = c \cdot 1 \cdot c^{-1} = cc^{-1} = 1.$$

Second,

$$\frac{\dfrac{a}{b}}{\dfrac{c}{d}} = \frac{a}{b}\left(\frac{c}{d}\right)^{-1} = \left(\frac{a}{b}\right)\left(\frac{d}{c}\right) = (ab^{-1})(dc^{-1})$$

$$= ab^{-1}dc^{-1} = (ad)(b^{-1}c^{-1}) = (ad)(bc)^{-1} = \frac{ad}{bc}.$$

The Names of the Rules

The rules we have given are a system of axioms for much of what we do in algebra. Two things are important: (1) what each rule is and what it says, (2) that the other rules we use for algebraic manipulation are logical consequences of these basic rules.

The names of the rules are not important at all. Mathematics is very different from zoology, where attaching a label to each member of the animal kingdom is important. Knowing a bunch of *names* of rules never solved a single problem. But knowing the *meaning* of the rules is essential for your work.

EXERCISES

Multiply, using the distributive law:

1. $11 \cdot 1492$

2. $16 \cdot 1003$

3. $19 \cdot 211$

4. $18 \cdot 407.$

Simplify by removing common factors:

5. $3 \cdot 4 \cdot 5 \cdot 6 - 2 \cdot 3 \cdot 4 \cdot 5$

6. $55 - 22 + 77 - 66$

7. $12x - 8y + 20$

8. $(5x)(6y) - 10u$

9. $2xyz - 6xy$

10. $x^3y - xy^2$

11. $3x^2 - 9x^2y + 12x^3$

12. $6xyz + x^2y^2z$

13. $2(x + y)^2 + 3v(x + y)$

14. $x + 2y + (x + 2y)^2.$

Multiply:

15. $(x + 1)(y + 1)$

16. $(x + 2)(y + 3)$

17. $(xy + 1)(x - 2)$

18. $(x + 3)(x + 4)$

19. $(2x + 1)(x - 1)$

20. $(2x + 3)(x + y)$

21. $(2x - 5)(2y - 5)$

22. $(x + y)(x - y)$

23. $(x + 2y)(3x - 4y)$

24. $(x + 1)^2$

25. $(y + 2)^2$

26. $(z - 3)^2$

27. $(x + 1)(x + y + 1)$

28. $(x - 2)(2x - 3y - 1)$

29. $(2x + y)(x + y + z)$

30. $(a + b)(u + v + x + y)$

31. $(a + b + 1)(u + v + 2)$

32. $(a + b + c)(x + y + z).$

Prove:

33. $-(-a) = a$

34. $(a^{-1})^{-1} = a \quad (a \neq 0)$

35. $-(a + b) = (-a) + (-b)$

36. $(ab)^{-1} = a^{-1}b^{-1} \quad$ if $\quad a \neq 0, b \neq 0$

37. $\left(\dfrac{a}{b}\right)\left(\dfrac{c}{d}\right) = \dfrac{ac}{bd}$

38. $\dfrac{a}{b} + \dfrac{c}{d} = \dfrac{ad + bc}{bd}$

39. $\dfrac{a}{b} = \dfrac{ka}{kb} \quad (k \neq 0)$

40. $xy(x^{-1} - y^{-1}) = y - x$

41. $(x + y)(x^{-1} - y^{-1}) = yx^{-1} - xy^{-1}$

42. $x^4(1 + x^{-1})^2 = (x^2 + x)^2$

43. $x^{-1}y^{-1}(x^2 + y^2) = xy^{-1} + yx^{-1}$

44. $x^{-1}(x + y)y^{-1} = x^{-1} + y^{-1}.$

4. INTEGERS AND RATIONALS

The most basic of all numbers are the positive integers $1, 2, 3, \cdots$. Together with 0 and the negative integers $-1, -2, -3, \cdots$, they form a number system called the system of integers. The sum, difference, and product of integers are again integers; we say that the system is **closed** with respect to addition, subtraction, and multiplication. It is not closed with respect to division because usually, when you divide one integer by another, the quotient is not an integer.

The integers obey the commutative, associative, and distributive laws. For the positive integers especially these are easy to visualize since each law is a principle of counting. Fig. 4.1 illustrates some simple cases.

Integers are usually denoted by letters from the middle of the alphabet, especially $i, j, k, m,$ and n. This is not compulsory, just customary.

Each integer is either odd or even. An integer is **even** if it is divisible by 2, that is, if it is twice some integer. Therefore, a general expression for an even integer is $2n$, where n is any integer. An integer is **odd** if it is one more than an even integer. So the typical odd integer is $2n + 1$. Since $2n + 1 = 2(n + 1) - 1$, the typical odd integer can also be written $2m - 1$.

■ *Example 4.1*

Show that   (a)   the sum of two even integers is even,

(b)   the product of two odd integers is odd.

SOLUTION (a) Two even integers are $2k$ and $2n$. Their sum is $2k + 2n = 2(k + n)$. This is twice the integer $k + n$, hence even.

(b) Two odd integers are $2k + 1$ and $2n + 1$. Their product is

$$(2k + 1)(2n + 1) = 2k \cdot 2n + 1 \cdot 2n + 2k \cdot 1 + 1 \cdot 1$$
$$= (4kn + 2n + 2k) + 1 = 2(2kn + n + k) + 1.$$

Now $2kn + n + k = m$, an integer. Hence the product is $2m + 1$, odd.

(a) example of commutative law: $5 \times 3 = 3 \times 5$

(b) example of distributive law:
$(3 \times 2) + (3 \times 4) = 3 \times (2 + 4)$

Fig. 4.1 Simple counting principles

The positive integers are often used as labels for various quantities. For example, in expressions involving many quantities, it may be convenient to use indexed symbols such as $x_1, x_2, x_3, \cdots$. Then you never have to worry about running out of letters.

Examples:

The average of 25 test scores is

$$\tfrac{1}{25}(t_1 + t_2 + t_3 + \cdots + t_{25}).$$

The perimeter of a pentagon is $s_1 + s_2 + s_3 + s_4 + s_5$, where s_1 is the length of the first side, s_2 the length of the second side, etc.

The product

$$(x_1 + x_2 + x_3 + x_4)(y_1 + y_2 + y_3 + y_4 + y_5)$$

is the sum of all 20 products of the form $x_i y_j$, where $i = 1, 2, 3,$ or 4 and $j = 1, 2, 3, 4,$ or 5.

Sums Involving Integers

Let us denote the sum of the positive integers $1, 2, 3, \cdots, n$ by S_n. Thus

$$S_n = 1 + 2 + 3 + \cdots + n.$$

This formula involves a new twist; the number of terms on the right is different for different choices of n:

$$S_1 = 1$$
$$S_2 = 1 + 2 = 3$$
$$S_3 = 1 + 2 + 3 = 6$$
$$S_4 = 1 + 2 + 3 + 4 = 10$$
$$\vdots$$
$$S_{10} = 1 + 2 + 3 + 4 + \cdots + 10 = 55, \quad \text{etc.}$$

As another example, take

$$T_n = (n + 1) + (n + 2) + (n + 3) + \cdots + 3n.$$

It is important to be able to interpret such a formula. In words, T_n is the sum of consecutive integers starting with $n + 1$ and ending with $3n$. For example T_4 starts with $4 + 1 = 5$ and ends with $3 \cdot 4 = 12$:

$$T_4 = 5 + 6 + 7 + 8 + 9 + 10 + 11 + 12 = 68.$$

■ *Example 4.2*

If $S_n = 1 + 2 + 3 + \cdots + n$, show that $S_n = \frac{1}{2}n(n + 1)$.

SOLUTION Write S_n twice, the second time in reverse order:

$$S_n = 1 + \quad 2 \quad + \quad 3 \quad + \cdots + (n - 1) + n,$$
$$S_n = n + (n - 1) + (n - 2) + \cdots + \quad 2 \quad + 1.$$

The sum of each column on the right is $n + 1$, and there are n columns. Add:

$$2S_n = (n + 1) + (n + 1) + (n + 1) + \cdots + (n + 1) + (n + 1)$$
$$= n(n + 1).$$

Hence $S_n = \frac{1}{2}n(n + 1)$.

■ *Example 4.3*

Find a formula for $E_n = 2 + 4 + 6 + \cdots + 2n$, the sum of the first n even integers.

SOLUTION Remove the common factor 2:

$$E_n = 2 + 4 + 6 + \cdots + 2n = 2(1 + 2 + 3 + \cdots + n).$$

The quantity in parentheses is S_n of Example 4.2, and we have a formula for S_n:

$$E_n = 2S_n = 2[\tfrac{1}{2}n(n + 1)] = n(n + 1).$$

Answer $E_n = n(n + 1).$

. . . .

■ *Example 4.4*

Find a formula for $T_n = (n + 1) + (n + 2) + \cdots + 3n.$

SOLUTION Use the result of Example 4.2. Since

$$S_{3n} = 1 + 2 + 3 + \cdots + n + (n + 1) + (n + 2) + \cdots + 3n,$$

it follows that

$$T_n = S_{3n} - (1 + 2 + 3 + \cdots + n) = S_{3n} - S_n.$$

A formula for S_n is given by Example 4.2. A formula for S_{3n} follows on replacing n by $3n$. Hence

$$\begin{aligned}
T_n = S_{3n} - S_n &= \tfrac{1}{2}(3n)(3n + 1) - \tfrac{1}{2}n(n + 1) \\
&= \tfrac{1}{2}n[3(3n + 1) - (n + 1)] = \tfrac{1}{2}n[9n + 3 - n - 1] \\
&= \tfrac{1}{2}n(8n + 2) = \tfrac{1}{2}n \cdot 2(4n + 1) = n(4n + 1).
\end{aligned}$$

Answer $T_n = n(4n + 1).$

. . . .

Remark: For $n = 4$, this formula says that $T_4 = 4(4 \cdot 4 + 1) = 4(16 + 1) = 4 \cdot 17 = 68$, which agrees with the value we computed earlier.

When we needed a formula for S_{3n} in the solution of Example 4.4, we replaced n by $3n$ in the formula for S_n. Why is that allowed?

Recall that S_n is the sum of consecutive integers from 1 to n. Then S_{3n} is the sum from 1 to $3n$, so $3n$ plays the same role here as n does in S_n. Hence the formula for S_{3n} is the same as that for S_n, with n replaced by $3n$.

Another way of looking at it is this. Write the formula as

$$S_k = \tfrac{1}{2}k(k + 1).$$

Then take $k = 3n$.

Sometimes it helps to avoid subscripts and use words. In S_n, the subscript n denotes the number of terms in the sum. You might write:

$$S_{(\text{number of terms})} = \tfrac{1}{2}(\text{number of terms})[(\text{number of terms}) + 1].$$

Whatever the number of terms is, whether it is k or n or $n + 1$ or $3n$, you plug that into the formula.

Rationals

A **rational number** (or **fraction**) is a real number that is equal to the quotient m/n of two integers, $n \neq 0$. The same rational number is equal to many quotients of integers, for example

$$\frac{2}{3} = \frac{4}{6} = \frac{6}{9} = \frac{-10}{-15} = \frac{-2}{-3} = \frac{2000}{3000}, \quad \text{etc.}$$

The simplest of these is $\frac{2}{3}$; it is the *only* quotient in this chain whose numerator and denominator have no common integer factor greater than 1, and whose denominator is positive. A fraction with this property is in **lowest terms** (also called **simplest form**).

The equivalence of the various forms of a rational number can be expressed as a rule: You may cancel any common integer factor from the numerator and denominator. Alternatively, you may multiply numerator and denominator by the same non-zero integer:

$$\frac{a}{b} = \frac{ka}{kb} \quad (k \neq 0).$$

Adding fractions with the same denominator is easy:

$$\frac{k}{p} + \frac{n}{p} = \frac{k + n}{p}.$$

It works the same way for more than two fractions as well. For example,

$$\frac{k}{p} + \frac{n}{p} + \frac{m}{p} = \frac{k + n + m}{p}.$$

Now read the last equation from right to left. It says you can break up a fraction whose numerator is a sum into a sum of fractions, one for each term in the numerator.

To add or subtract fractions with different denominators, convert them to equivalent forms with equal denominators. For example, to add $\frac{1}{3}$ and $\frac{4}{7}$, convert each to a fraction with denominator $3 \cdot 7 = 21$:

$$\frac{1}{3} + \frac{4}{7} = \frac{1 \cdot 7}{3 \cdot 7} + \frac{4 \cdot 3}{7 \cdot 3} = \frac{7}{21} + \frac{12}{21} = \frac{19}{21}.$$

The product of the denominators always serves as a **common denominator.** Sometimes there is a smaller common denominator, for instance,

$$\frac{7}{15} + \frac{1}{6} = \frac{14}{30} + \frac{5}{30} = \frac{19}{30};$$

it is not necessary to use $15 \times 6 = 90$ as common denominator.

The general rule for adding fractions is

$$\frac{a}{b} + \frac{c}{d} = \frac{ad + bc}{bd}.$$

It covers all cases, but as we have just seen, the denominator bd may be larger than necessary.

Warning: It is not true that

$$\frac{m}{p + q} = \frac{m}{p} + \frac{m}{q}.$$

Repeat: It is NOT TRUE that $m/(p + q) = m/p + m/q$. For instance,

$$\frac{1}{5} = \frac{1}{2 + 3} \neq \frac{1}{2} + \frac{1}{3}, \qquad 1 = \frac{12}{12} = \frac{12}{2 + 10} \neq \frac{12}{2} + \frac{12}{10} = 6 + \frac{6}{5}.$$

Multiplication and division of rational numbers is simple:

$$\left(\frac{m}{n}\right)\left(\frac{p}{q}\right) = \frac{mp}{nq}, \qquad \frac{m}{n} \bigg/ \frac{p}{q} = \left(\frac{m}{n}\right)\left(\frac{q}{p}\right) = \frac{mq}{np}.$$

To multiply, multiply numerators and denominators. To divide, invert the divisor and multiply. Note that

$$m\left(\frac{p}{q}\right) = \frac{mp}{q},$$

which follows from the multiplication rule; just write $m = m/1$.

Remark 1: The rules for combining rational numbers are special cases of the rules for quotients, p. 11.

Remark 2: It is known that not all real numbers are rational. In fact, it can be proved that $\sqrt{3}$, π, and many other specific real numbers are **irrational** (not rational).

EXERCISES

Show that the assertion is true:

1. The sum of two odd integers is even.
2. The sum of an even integer and an odd integer is odd.
3. The product of an even integer and an odd integer is even.
4. The product of two even integers is divisible by 4.

5. If the product of three integers is odd, what can you say about the **parity** (oddness or evenness) of the factors?
6. If the sum of four integers is odd what can you say about the parity of the summands?
7. Using 2-cent, 4-cent, and 6-cent stamps, how can you make exactly 23 cents postage?
8. On a test, three students made a score of n_1, five students made n_2, and one student each made $n_3, n_4, \cdots, n_{12}$. What was the class average?

Compute the sum, using the results and techniques of Examples 4.2, 4.3, and 4.4:

   9. of all even integers from 10 to $2n$
  10. of all odd integers from 1 to $2n - 1$
  11. (cont.) of all odd integers from $2n + 1$ to $4n - 1$
  12. of the first n integers of the form 2, 5, 8, 11,$\cdots$
  13. of the first $2n$ odd integers
  14. $(n + 1) + (n + 2) + \cdots + 2n$
  15. $5 + 10 + 15 + \cdots + 5n$
  16. $4 + 9 + 14 + \cdots + (5n - 1)$.

Express in lowest terms:

  17. $54/90$ **18.** $40/84$ **19.** $112/21$
  20. $-95/152$ **21.** $34/(-5)$ **22.** $333/(-74)$.

Carry out the operations and express the answer in lowest terms:

  23. $\dfrac{14}{5} + \dfrac{1}{3}$

  24. $\dfrac{1}{15} + \dfrac{1}{6}$

  25. $\left(\dfrac{2}{3}\right)\left(\dfrac{30}{50}\right)$

  26. $\dfrac{3}{2}\left(\dfrac{4}{9} - \dfrac{2}{21}\right)$

  27. $\left(\dfrac{1}{2} + \dfrac{1}{3}\right)\left(\dfrac{1}{2} - \dfrac{1}{3}\right)$

  28. $\dfrac{102}{35}\left(2 - \dfrac{3}{17}\right)$

  29. $\dfrac{1}{2} + \dfrac{1}{3} - \dfrac{1}{4}$

  30. $\dfrac{2}{3} - \dfrac{5}{6} + 1$

  31. $\dfrac{24}{77}\left(\dfrac{4}{15} - \dfrac{1}{2} - \dfrac{21}{20}\right)$

  32. $\dfrac{\frac{2}{3} - \frac{1}{4}}{1 + \frac{1}{12}}$

  33. $\dfrac{3}{4}\left(2 - \dfrac{16}{21}\right)$

  34. $\dfrac{\frac{1}{6} + \frac{1}{9} - \frac{1}{18}}{\frac{1}{6} - \frac{1}{9} + \frac{1}{18}}$.

5. INTEGRAL EXPONENTS

In Section 2 we introduced the abbreviations

$$a^1 = a, \quad a^2 = a \cdot a, \quad a^3 = a \cdot a \cdot a, \quad \cdots, \quad a^n = \underbrace{a \cdot a \cdot a \cdot a \cdot a \cdots a}_{n \text{ factors}}.$$

Read a^n as "a to the n-th power" or "a to the n". The number n is an **exponent.** The number a^n is called the n-th **power** of a.

Examples:

$$\left(-\frac{1}{2}\right)^3 = \left(-\frac{1}{2}\right)\left(-\frac{1}{2}\right)\left(-\frac{1}{2}\right) = -\frac{1}{8}, \qquad 10^4 = 10 \cdot 10 \cdot 10 \cdot 10 = 10000.$$

Look what happens when you multiply a^5 by a^3:

$$a^5 a^3 = \underbrace{(aaaaa)}_{5}\underbrace{(aaa)}_{3} = \underbrace{aaaaaaaa}_{8 \text{ factors}} = a^8.$$

Thus $a^5 a^3 = a^8$. Similarly, if you multiply m factors a by n factors, you get $m + n$ factors. Therefore

$$a^m a^n = a^{m+n}.$$

To multiply, add exponents.

Look at what happens when you divide a^5 by a^3:

$$\frac{a^5}{a^3} = \frac{aaaaa}{aaa} = \frac{aaa}{aaa}\frac{aa}{1} = aa = a^2.$$

Thus $a^5/a^3 = a^{5-3}$. You subtract exponents because the three factors in the denominator cancel three of the factors in the numerator. By the same reasoning

$$\frac{a^m}{a^n} = a^{m-n} \qquad m > n.$$

Examples:

$$\frac{2^5}{2^3} = 2^{5-3} = 2^2, \qquad \text{that is,} \qquad \frac{32}{8} = 4,$$

$$\frac{10^6}{10^4} = 10^2, \qquad \text{that is,} \qquad \frac{1,000,000}{10,000} = 100.$$

There is a natural definition for a^0. Since $a/a = 1$, we ought to have $a^{1-1} = a^0 = 1$. So we boldly *define*

$$\boxed{a^0 = 1 \qquad (a \neq 0).}$$

We do this only for $a \neq 0$. The symbol 0^0 has no meaning in mathematics.

We also define negative powers a^{-n}; however, now we must insist that $a \neq 0$. We already know what a^{-1} is; it is the reciprocal (multiplicative inverse) of a, defined by

$$aa^{-1} = a\frac{1}{a} = 1 \qquad (a \neq 0).$$

If n is a positive integer, we define

$$\boxed{a^{-n} = (a^{-1})^n = \left(\frac{1}{a}\right)^n \qquad (a \neq 0).}$$

Examples:

$$10^{-2} = (10^{-1})^2 = \left(\frac{1}{10}\right)^2 = \frac{1}{10^2} = \frac{1}{100},$$

$$2^{-4} = (2^{-1})^4 = \left(\frac{1}{2}\right)^4 = \left(\frac{1}{2}\right)\left(\frac{1}{2}\right)\left(\frac{1}{2}\right)\left(\frac{1}{2}\right) = \frac{1}{2^4} = \frac{1}{16},$$

$$\left(\frac{1}{5}\right)^{-3} = \left[\left(\frac{1}{5}\right)^{-1}\right]^3 = 5^3 = 125.$$

An important property is that a^{-n} is the multiplicative inverse of a^n, that is, $a^n a^{-n} = 1$. Indeed,

$$a^n a^{-n} = a^n (a^{-1})^n = \underbrace{(a \cdot a \cdot a \cdots a)}_{n} \underbrace{(a^{-1} a^{-1} a^{-1} \cdots a^{-1})}_{n}$$

$$= \underbrace{(aa^{-1})(aa^{-1})(aa^{-1}) \cdots (aa^{-1})}_{n \text{ factors}} = 1 \cdot 1 \cdot 1 \cdots 1 = 1.$$

This property can be written $a^{-n} = (a^n)^{-1}$.

In summary, a^n is defined for every integer exponent n, provided $a \neq 0$. Here are the basic rules of manipulation with exponents.

Rules of Exponents

If a and b are non-zero reals and if m and n are integers, then

(1) $a^m a^n = a^{m+n}$,   (2) $\dfrac{a^m}{a^n} = a^{m-n}$,

(3) $(a^m)^n = a^{mn}$,   (4) $(ab)^n = a^n b^n$.

We emphasize that these rules are valid for *all* integers m and n, not just positive ones.

Examples:

Rule (3):

$$(a^4)^3 = a^4 \cdot a^4 \cdot a^4 = a^{4+4+4} = a^{12} = a^{4 \cdot 3},$$

$$(a^4)^{-3} = \left(\frac{1}{a^4}\right)^3 = \frac{1}{(a^4)^3} = \frac{1}{a^{12}} = a^{-12} = a^{4(-3)}.$$

Rule (4):

$$(ab)^3 = (ab)(ab)(ab) = (aaa)(bbb) = a^3 b^3,$$

$$(ab)^{-3} = [(ab)^3]^{-1} = (a^3 b^3)^{-1} = (a^3)^{-1}(b^3)^{-1} = a^{-3} b^{-3}.$$

The rules can be extended in various ways. For instance, by several applications of (3) and (4) we prove

$$(a^m b^n c^p)^r = a^{mr} b^{nr} c^{pr}.$$

■ *Example 5.1*

Use the rules of exponents to simplify

(a) $(xy)^2 (x^2 y^3)^{-1}$ (b) $(x^2 y^{-3})^{-5}$.

SOLUTION (a) $(xy)^2(x^2y^3)^{-1} = (x^2y^2)\dfrac{1}{x^2y^3} = \dfrac{x^2}{x^2}\dfrac{y^2}{y^3} = \dfrac{1}{y} = y^{-1}.$

Alternatively,

$$(xy)^2(x^2y^3)^{-1} = (x^2y^2)(x^{-2}y^{-3}) = x^{2-2}y^{2-3} = x^0y^{-1} = y^{-1}.$$

(b) $(x^2y^{-3})^{-5} = (x^2)^{-5}(y^{-3})^{-5} = x^{2(-5)}y^{(-3)(-5)} = x^{-10}y^{15}.$

Answer (a) $y^{-1} = 1/y$ (b) $x^{-10}y^{15} = y^{15}/x^{10}.$

■ *Example 5.2*

Express $\dfrac{2^{-3}\cdot 8^5}{4^3\cdot 16}$ as a power of 2.

SOLUTION

$$\frac{2^{-3}\cdot 8^5}{4^3\cdot 16} = \frac{2^{-3}(2^3)^5}{(2^2)^3(2^4)} = \frac{2^{-3}\cdot 2^{15}}{2^6\cdot 2^4} = \frac{2^{12}}{2^{10}} = 2^2.$$

Answer $2^2.$

■ *Example 5.3*

Use the rules of exponents to compute $\dfrac{2^6\cdot 5^7}{25\cdot 10^4}.$

SOLUTION

$$\frac{2^6\cdot 5^7}{25\cdot 10^4} = \frac{2^6\cdot 5^7}{5^2\cdot(2\cdot 5)^4} = \frac{2^6\cdot 5^7}{5^2\cdot 2^4\cdot 5^4} = \frac{2^6\cdot 5^7}{2^4\cdot 5^6} = 2^2\cdot 5 = 20.$$

Answer 20.

Scientific Notation

One important practical application of exponents is in computations. In scientific work, we need an efficient way of writing and computing with very large or very small numbers, e.g.,

$$32000000000, \qquad 1876000, \qquad 0.0000000000006.$$

Imagine multiplying such numbers!

The idea of scientific notation is this: express each positive number in the form $c \times 10^n$, where $1 \le c < 10$ and n is an appropriate exponent.

Examples:

$$140 = 1.4 \times 10^2 \qquad\qquad 0.05 = 5 \times 10^{-2}$$
$$2550 = 2.55 \times 10^3 \qquad\qquad 0.0031 = 3.1 \times 10^{-3}$$
$$1876000 = 1.876 \times 10^6 \qquad\qquad 0.000988 = 9.88 \times 10^{-4}$$
$$32000000000 = 3.2 \times 10^{10} \qquad 0.0000000000006 = 6 \times 10^{-13}.$$

Example 5.4

Multiply: $(140)(32000000000)(0.0000000000006)$.

SOLUTION

$$(1.4 \times 10^2)(3.2 \times 10^{10})(6 \times 10^{-13}) = (1.4)(3.2)(6) \times 10^{2+10-13} = 26.88 \times 10^{-1}.$$

Answer   2.688.

Example 5.5

Compute $\dfrac{(14000)(0.00003)(8800000)}{(1100)(0.000002)}$.

SOLUTION

$$\frac{(1.4 \times 10^4)(3 \times 10^{-5})(8.8 \times 10^6)}{(1.1 \times 10^3)(2 \times 10^{-6})} = \frac{(1.4)(3)(8.8)}{(1.1)(2)} \times 10^{4-5+6-3+6} = 16.8 \times 10^8.$$

Answer 1.68×10^9.

Scientific notation is used in the displays of small scientific calculators, which can display only a limited number of digits, usually 8 or 10. Large numbers and very small numbers are displayed in scientific notation. Look at the number displayed in Fig. 5.1. The 67 on the right means $\times 10^{67}$, so the number displayed is 3.9401625×10^{67}.

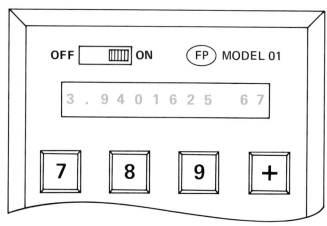

Fig. 5.1   Display of hand-held calculator (electronic slide rule)

Compute:

1. 3^3

2. $(-3)^3$

3. 6^{-2}

4. $\left(\dfrac{1}{5}\right)^2$

5. $\left(\dfrac{1}{2}\right)^{-2}$

6. $\left(\dfrac{1}{3}\right)^0$

7. $\left(-\dfrac{2}{3}\right)^5$

8. $\left(-\dfrac{3}{2}\right)^{-1}$

9. $2^{-3}\left(\dfrac{1}{3}\right)^2$

10. $\dfrac{2^5 \cdot 5^3}{2^6 \cdot 5^{-2}}$

11. $2^4 3^2 6^{-2} 8^{-1}$

12. $10^4 2 5^{-3} 16^{-1}$.

Express as a power of 2:

13. 4^2

14. 8^{-3}

15. $\left(\dfrac{1}{2}\right)^7$

16. $(2^4 \cdot 16^{-2})^3$

17. $2^4 \left(\dfrac{1}{2}\right)^3 8^2 16^{-1}$

18. $2 \cdot 4^2 \cdot 8^3$

19. $\left(\dfrac{1}{2}\right)^{-8}\left(\dfrac{1}{4}\right)^2$

20. $\dfrac{1}{2} \cdot \dfrac{1}{4} \cdot \dfrac{1}{8} \cdot \dfrac{1}{16}$.

Express as simply as possible as a quotient, with no negative exponents:

21. $\dfrac{(xy)^6}{xy^2}$

22. $\dfrac{1}{x^3}(x^2)^3 x^{-4}$

23. $a^2(a^{-1} + a^{-3})$

24. $(aba^{-4})(a^3 b^{-2})^{-1}$

25. $(8a^3 b)^{-4}(2a/b)^{12}$

26. $(-5x^2 y^{-3})^{-20}$

27. $(-xy^2)^3(-2x^2 y^2)^{-4}$

28. $(xy)^{-5}(2xy^2)(3xy)^3$

29. $(4x^3 y^2 z)^2(4x^3 y^2 z)^{-7}$

30. $(2x^2 y^2)^3(-3x^4 y)^{-2}$

31. $\dfrac{x^{-2}}{y^{-2}} + \dfrac{y^2}{x^2}$

32. $\dfrac{(2pqr)^2}{(p^3 q)^2(pr)^{-1}}$

33. $\left[\dfrac{(-x^2 y)^3 y^{-4}}{(xy)^5}\right]^{-2}$

34. $\left(\dfrac{ab^2 c^3 d^4}{a^4 b^3 c^2 d}\right)^2$.

Express in scientific notation:

35. 0.4

36. 0.0081

37. 2.37

38. 0.6008

39. 0.000076

40. 0.0000000059

41. 17

42. 5280

43. 12400

44. 3.005

45. 3832000

46. 452000000000000.

Compute and express the result in scientific notation:

47. $(180)(30000000)(0.00012)$

48. $\dfrac{(20100)(0.006)}{(0.0000002)(402000)}$

49. $(0.002)^3(0.00004)(0.000005)(6000000000)$

50. $\dfrac{1}{(800)(200000)^2(0.00001)^4}$

51. the number of inches in 100 miles

52. the number of cubic centimeters in a cubic kilometer.

6. ROOTS AND RADICALS

Square Roots

For every positive real number a, there is a positive real number r such that $r^2 = a$. We call r the **square root** of a and write $r = \sqrt{a}$. For example, $\sqrt{4} = 2$, $\sqrt{9} = 3$, $\sqrt{100} = 10$, etc. The symbol $\sqrt{}$ is called a **radical.**

Actually, $-r$ is also a square root of a since $(-r)^2 = r^2 = a$. By the symbol $\sqrt{a}$, we shall agree that we always mean the *positive* square root of a. Also, since $0^2 = 0$, we shall write $\sqrt{0} = 0$.

Square roots obey some simple rules.

Rules for square roots

Let a and b be positive real numbers. Then

(1) $(\sqrt{a})^2 = a,$ (2) $\sqrt{a^2} = a,$

(3) $\sqrt{ab} = \sqrt{a}\,\sqrt{b},$ (4) $\sqrt{\dfrac{a}{b}} = \dfrac{\sqrt{a}}{\sqrt{b}}.$

Rules (1) and (2) come directly from the definition of square roots. To check rule (3), let $\sqrt{a} = r$ and $\sqrt{b} = s$. Then $a = r^2$ and $b = s^2$, hence $ab = r^2 s^2 = (rs)^2$. This says that $\sqrt{ab} = rs = \sqrt{a}\,\sqrt{b}$. Rule (4) is checked similarly.

Warning: These rules concern only multiplication and division. Square roots have no simple behavior with respect to addition and subtraction. In particular,

$$\sqrt{a+b} \quad \text{IS NOT} \quad \sqrt{a} + \sqrt{b}\,!$$

For example, $\sqrt{25} = \sqrt{9+16}$ and $\sqrt{25} = 5$, but $5 \neq \sqrt{9} + \sqrt{16} = 3 + 4 = 7$.

Square roots of complicated quantities can often be simplified by rule (3). For instance,

$$\sqrt{48} = \sqrt{16 \times 3} = \sqrt{16}\,\sqrt{3} = 4\sqrt{3}.$$

In general, by rule (3),

$$\sqrt{r^2 b} = \sqrt{r^2}\,\sqrt{b} = r\sqrt{b} \qquad (r > 0, \quad b > 0).$$

Thus a squared factor under the radical can come out in front of the radical, but it loses its square on the way out.

■ *Example 6.1*

Simplify: (a) $\sqrt{36 \times 49}$ (b) $\sqrt{500}$ (c) $\sqrt{\dfrac{75}{16}}.$

SOLUTION (a) By rule (3),

$$\sqrt{36 \times 49} = \sqrt{36}\,\sqrt{49} = 6 \times 7 = 42.$$

(b) By rule (3),

$$\sqrt{500} = \sqrt{100 \times 5} = \sqrt{100}\,\sqrt{5} = 10\sqrt{5}.$$

(c)   Use rule (4), then rule (3):

$$\sqrt{\frac{75}{16}} = \frac{\sqrt{75}}{\sqrt{16}} = \frac{\sqrt{25 \times 3}}{4} = \frac{\sqrt{25}\,\sqrt{3}}{4} = \frac{5\,\sqrt{3}}{4}.$$

Answer   (a)   42   (b) $10\,\sqrt{5}$ (c) $\frac{5}{4}\sqrt{3}$.

. . . .

Note: In the following examples, and in the exercises, all letters represent positive numbers.

◼ *Example 6.2*

Simplify   (a) $\sqrt{16x^2}$ (b) $\sqrt{225p^4q^7}$ (c) $\sqrt{uv}\,\sqrt{\dfrac{8}{u^3v}}$.

SOLUTION   (a)   Use rule (3):

$$\sqrt{16x^2} = \sqrt{16}\,\sqrt{x^2} = 4x.$$

(b)   Use rule (3) several times:

$$\sqrt{225p^4q^7} = \sqrt{225}\,\sqrt{p^4}\,\sqrt{q^7} = 15p^2\,\sqrt{q^6q}$$
$$= 15p^2\,\sqrt{q^6}\,\sqrt{q} = 15p^2q^3\,\sqrt{q}.$$

(c)   Use rule (3), then rule (4):

$$\sqrt{uv}\,\sqrt{\frac{8}{u^3v}} = \sqrt{uv\left(\frac{8}{u^3v}\right)} = \sqrt{\frac{8}{u^2}} = \frac{\sqrt{8}}{\sqrt{u^2}} = \frac{\sqrt{4\cdot 2}}{u} = \frac{2\,\sqrt{2}}{u}.$$

Answer   (a) $4x$ (b) $15p^2q^3\,\sqrt{q}$ (c) $2\,\sqrt{2}/u$.

. . . .

Remark: It is useful to be able to recognize perfect squares of integers, like $15^2 = 225$, $21^2 = 441$, etc. Refer to Table 3, p. 408.

◼ *Example 6.3*

Compute   (a) $(4 + \sqrt{7})(5 - 2\,\sqrt{7})$ (b) $(\sqrt{x} + \sqrt{y})(\sqrt{x} - \sqrt{y})$.

SOLUTION   (a)   Use the distributive law and $(\sqrt{7})^2 = 7$:

$$(4 + \sqrt{7})(5 - 2\,\sqrt{7}) = (4)(5) + (\sqrt{7})(5) - (4)(2\,\sqrt{7}) - (\sqrt{7})(2\,\sqrt{7})$$
$$= 20 + 5\,\sqrt{7} - 8\,\sqrt{7} - 2(\sqrt{7})^2$$
$$= 20 - 3\,\sqrt{7} - (2)(7) = 6 - 3\,\sqrt{7}.$$

(b) $(\sqrt{x} + \sqrt{y})(\sqrt{x} - \sqrt{y}) = (\sqrt{x})^2 - (\sqrt{y})^2 = x - y.$

Answer   (a) $6 - 3\,\sqrt{7}$ (b) $x - y$.

. . . .

Rationalizing the Denominator

Sometimes it is convenient to eliminate square roots from the denominator of a fractional expression. One trick for doing this is shown by these examples:

$$\frac{1}{\sqrt{3}} = \frac{1}{\sqrt{3}} \cdot \frac{\sqrt{3}}{\sqrt{3}} = \frac{\sqrt{3}}{3}, \qquad \frac{4}{\sqrt{5}} = \frac{4}{\sqrt{5}} \cdot \frac{\sqrt{5}}{\sqrt{5}} = \frac{4\sqrt{5}}{5}.$$

Another standard trick is shown in these examples:

$$\frac{1}{\sqrt{3}-1} = \frac{1}{\sqrt{3}-1} \frac{\sqrt{3}+1}{\sqrt{3}+1} = \frac{\sqrt{3}+1}{(\sqrt{3})^2-1^2} = \frac{\sqrt{3}+1}{3-1} = \frac{\sqrt{3}+1}{2},$$

$$\frac{1}{\sqrt{5}+\sqrt{2}} = \frac{1}{\sqrt{5}+\sqrt{2}} \frac{\sqrt{5}-\sqrt{2}}{\sqrt{5}-\sqrt{2}} = \frac{\sqrt{5}-\sqrt{2}}{(\sqrt{5})^2-(\sqrt{2})^2} = \frac{\sqrt{5}-\sqrt{2}}{3}.$$

Higher Order Roots

Suppose n is an integer and $n \geq 2$. To each positive real a, there is a unique *positive* real r for which $r^n = a$. We call r the n-th **root** of a and write $r = \sqrt[n]{a}$. The symbol $\sqrt[n]{}$ is again called a **radical.** (We note that $\sqrt[2]{} = \sqrt{}$.) As with the square root, we shall write $\sqrt[n]{0} = 0$ because $0^n = 0$.

Examples:

$$\sqrt[3]{64} = 4 \qquad \text{because} \qquad 4^3 = 64,$$
$$\sqrt[4]{81} = 3 \qquad \text{because} \qquad 3^4 = 81.$$

Remark 1:   If n is *even,* then $(-r)^n = r^n$, so $-\sqrt[n]{a}$ is also an n-th root of a, a number whose n-th power is a. However, as with square roots, the symbol $\sqrt[n]{a}$ always means the unique *positive* n-th root of a for $a > 0$.

Remark 2:   If n is *odd,* then negative numbers also have n-th roots. For instance $-1 = (-1)^5$, $-8 = (-2)^3$. In fact, if n is odd and a is any real number, positive, negative, or zero, then there is exactly one unique real number r such that $r^n = a$. It is tempting to write $r = \sqrt[n]{a}$ in any case, even $a < 0$. We won't do this yet, until we have mastered the rules for *positive* n-th roots. The rules are not 100% correct without qualifications if negative roots are allowed.

Rules for n-th roots

Let a and b be positive real numbers. Then

(1) $(\sqrt[n]{a})^n = a$ ●

(2) $\sqrt[n]{a^n} = a$

(3) $\sqrt[n]{ab} = \sqrt[n]{a}\,\sqrt[n]{b}$

(4) $\sqrt[n]{\dfrac{a}{b}} = \dfrac{\sqrt[n]{a}}{\sqrt[n]{b}}.$

(5) $\sqrt[m]{\sqrt[n]{a}} = \sqrt[mn]{a}.$

Rules (1)–(4) are just like the corresponding rules for square roots. Rule (5) is new; it corresponds to the rule of exponents $(r^m)^n = r^{mn}$. In fact, if we set

$$r = \sqrt[m]{\sqrt[n]{a}},$$

then by definition r is the positive number whose m-th power is $\sqrt[n]{a}$, that is, $r^m = \sqrt[n]{a}$. But by definition again, r^m is the n-th root of a, that is, $(r^m)^n = a$. Hence $r^{mn} = a$, so r is the mn-th root of a, which proves (5). For example,

$$\sqrt{\sqrt{x}} = \sqrt[4]{x}, \qquad \sqrt[4]{\sqrt[3]{x}} = \sqrt[12]{x}.$$

Example 6.4

Simplify (a) $\sqrt[3]{16x^3y^6}$ (b) $\sqrt[4]{\dfrac{48}{x^{12}y^4}}$.

SOLUTION (a) By rule (3),

$$\sqrt[3]{16x^3y^6} = \sqrt[3]{16}\,\sqrt[3]{x^3}\,\sqrt[3]{(y^2)^3} = \sqrt[3]{16}\,xy^2.$$

But $16 = 2^4 = 2^3 \times 2$, so $\sqrt[3]{16} = \sqrt[3]{2^3}\sqrt[3]{2} = 2\sqrt[3]{2}$. Hence the answer is $2xy^2\sqrt[3]{2}$.

(b) By rules (3) and (4),

$$\sqrt[4]{\frac{48}{x^{12}y^4}} = \frac{\sqrt[4]{48}}{\sqrt[4]{x^{12}y^4}} = \frac{\sqrt[4]{16 \cdot 3}}{\sqrt[4]{x^{12}}\,\sqrt[4]{y^4}} = \frac{\sqrt[4]{16}\,\sqrt[4]{3}}{x^3y} = \frac{2\sqrt[4]{3}}{x^3y}.$$

Answer (a) $2xy^2\sqrt[3]{2}$ (b) $\dfrac{2\sqrt[4]{3}}{x^3y}$.

Remark: Note that we write the answer to (a) in the form $2xy^2\sqrt[3]{2}$ rather than $2\sqrt[3]{2}\,xy^2$. In general we prefer to write radical factors on the right because it is so easy to confuse symbols like $\sqrt[3]{2}x$ and $\sqrt[3]{2x}$. Just draw the bar a bit too long and you aren't sure which one is meant.

EXERCISES

Simplify (assuming all letters represent positive numbers):

1. $\sqrt{81}$

2. $\sqrt{144}$

3. $\sqrt{\dfrac{1}{9}}$

4. $\sqrt{\dfrac{49}{25}}$

5. $\sqrt{9a^6}$

6. $\sqrt{\frac{1}{4}y^2}$

7. $\sqrt{50}$

8. $\sqrt{128}$

9. $\sqrt{\dfrac{1}{12}}$

10. $\sqrt{ab^2c^5}$

11. $\sqrt{\dfrac{18x^3}{(y+z)^4}}$

12. $\dfrac{\sqrt{6}}{\sqrt{3}}$

13. $\sqrt[3]{1000}$

14. $\sqrt[3]{\dfrac{1}{8}}$

15. $\sqrt[3]{\dfrac{27}{64}}$

16. $\sqrt[3]{27 \cdot 64 \cdot 125}$

17. $\sqrt[4]{\dfrac{1}{16}}$ **18.** $\sqrt[4]{81}$ **19.** $\sqrt[4]{32}$ **20.** $\sqrt[4]{162}$

21. $\sqrt[3]{8a^3b^9}$ **22.** $\sqrt[4]{10{,}000u^2}$ **23.** $\sqrt[5]{64u^5v^6}$ **24.** $\sqrt[3]{\dfrac{a^7}{24b^9}}$.

Find an equivalent expression without radicals in the denominator:

25. $\dfrac{1}{\sqrt{6}}$ **26.** $\dfrac{\sqrt{2}}{\sqrt{6}+\sqrt{2}}$ **27.** $\dfrac{\sqrt{x}+\sqrt{y}}{\sqrt{x}-\sqrt{y}}$ **28.** $\dfrac{1}{\sqrt{x}-2\sqrt{y}}$.

Compute and simplify:

29. $(a+b\sqrt{2})(a-b\sqrt{2})$ **30.** $(\sqrt{x}+3\sqrt{y})^2$ **31.** $(\sqrt{x})^4$

32. $\left(\sqrt{x}-\dfrac{1}{\sqrt{x}}\right)^2$ **33.** $\sqrt{\dfrac{5x}{3y}}\,\sqrt{27x^3y^2}$ **34.** $\sqrt{xy}\left(\sqrt{\dfrac{x}{y}}+\sqrt{\dfrac{y}{x}}\right)$.

Find an equivalent expression involving at most one radical sign:

35. $\sqrt{x\sqrt{x^6}}$ **36.** $\sqrt{3}(1+\sqrt{3})$ **37.** $\sqrt{2}\sqrt{xy}\sqrt{10yz}$ **38.** $\sqrt[3]{\dfrac{16xy}{z}}\,\sqrt[3]{4y^2z^7}$.

Verify the formulas:

39. $\sqrt{a^2+x^2}=a\sqrt{1+\left(\dfrac{x}{a}\right)^2}$ **40.** $\sqrt{x^2+1}-\dfrac{1}{\sqrt{x^2+1}}=\dfrac{x^2}{\sqrt{x^2+1}}$

41. $\dfrac{2}{\sqrt{3}+1}-\dfrac{1}{\sqrt{3}-1}=\dfrac{\sqrt{3}-3}{2}$ **42.** $\dfrac{1}{3y}\sqrt{9x^2-y^2}=\sqrt{(x/y)^2-\tfrac{1}{9}}$.

Prove for *n*-th roots:

43. rule (1)   **44.** rule (2)   **45.** rule (3)   **46.** rule (4).

Use the approximations $\sqrt{2}\approx1.4$, $\sqrt{3}\approx1.7$, $\sqrt{5}\approx2.2$ to estimate

47. $\sqrt{6}$ **48.** $\sqrt{10}$ **49.** $\sqrt{3/5}$ **50.** $\sqrt{2/3}$
51. $\sqrt{30}$ **52.** $\sqrt{75}$ **53.** $\sqrt{40}$ **54.** $\sqrt{216}$.

Prove:

55. $\sqrt{5+2\sqrt{6}}=\sqrt{2}+\sqrt{3}$ **56.** $\sqrt{5-2\sqrt{6}}=\sqrt{3}-\sqrt{2}$
57. $\sqrt{9-2\sqrt{14}}=\sqrt{7}-\sqrt{2}$ **58.** $\sqrt{8+2\sqrt{15}}=\sqrt{3}+\sqrt{5}$.

7. RATIONAL EXPONENTS

What meaning can we give to $a^{m/n}$, where m/n is a rational number not necessarily an integer? If we can define such fractional powers, and if the rules of exponents are to hold, then we should have

$$(a^{m/n})^n = a^{(m/n)n} = a^m.$$

This means that $a^{m/n}$ must be the *n*-th root of a^m. Therefore we *define*

$$a^{m/n} = \sqrt[n]{a^m},$$

where m/n is a rational number with $n > 0$.

Examples:

$$8^{2/3} = \sqrt[3]{8^2} = \sqrt[3]{64} = 4, \quad 25^{3/2} = \sqrt{(25)^3} = \sqrt{(5^2)^3} = \sqrt{5^6} = 5^3 = 125,$$

$$9^{-1/2} = \sqrt{9^{-1}} = \sqrt{\tfrac{1}{9}} = \tfrac{1}{3}.$$

According to the definition of $a^{m/n}$, where $m = 1$, we have

$$a^{1/n} = \sqrt[n]{a},$$

so $a^{1/n}$ is just the n-th root of a in new clothes. For example,

$$36^{1/2} = 6, \quad 1000^{1/3} = 10, \quad \left(\frac{1}{32}\right)^{1/5} = \frac{1}{2}.$$

There is a subtle point in the definition of $a^{m/n}$. The same rational number might be expressed in two ways. For instance, suppose $m/n = p/q$. The definition then gives two possibilities,

$$a^{m/n} = \sqrt[n]{a^m}, \quad a^{p/q} = \sqrt[q]{a^p}.$$

These had better be the same, or the definition is plain nonsense. They *are* the same, as is verified by showing that their qn-th powers are equal. We leave the details as an exercise.

Rules of Exponents

The definition of $a^{m/n}$ is a useful one because all the rules of exponents for integer exponents carry over to fractional exponents.

Rules of exponents

If m/n is rational with $n > 0$ and if $a > 0$, then

(1) $a^{m/n} = \sqrt[n]{a^m} = (\sqrt[n]{a})^m$.

If s and t are rationals and if a and b are positive reals, then

(2) $a^s a^t = a^{s+t}$,   (3) $\dfrac{a^s}{a^t} = a^{s-t}$,

(4) $(a^s)^t = a^{st}$,   (5) $a^s b^s = (ab)^s$.

The first equality in (1) is the definition of $a^{m/n}$: take the m-th power of a and then extract its n-th root. The second equality,

$$\sqrt[n]{a^m} = (\sqrt[n]{a})^m,$$

says something new, that you can extract the n-th root first, then take the m-th power, and you get the same thing. To prove it we show that $(\sqrt[n]{a})^m$ has the property that defines $a^{m/n}$; its n-th power is a^m:

$$[(\sqrt[n]{a})^m]^n = (\sqrt[n]{a})^{mn} = [(\sqrt[n]{a})^n]^m = a^m.$$

We prove Rule (2) by writing s and t with a common denominator and using rules for *integer* exponents and radicals. Thus we write

$$s = \frac{m}{n}, \qquad t = \frac{p}{n}, \qquad n > 0.$$

Then

$$a^s a^t = a^{m/n} a^{p/n} = \sqrt[n]{a^m}\, \sqrt[n]{a^p} = \sqrt[n]{a^m a^p} = \sqrt[n]{a^{m+p}}$$
$$= a^{(m+p)/n} = a^{(m/n)+(p/n)} = a^{s+t}.$$

Rule (3) is proved similarly. We shall not give the details of Rules (4) and (5). They are proved by careful use of the rules for integer exponents and radicals.

We remark that the rules for rational exponents include as special cases the rules for integer exponents.

The rule $\sqrt[n]{a}\, \sqrt[n]{b} = \sqrt[n]{ab}$ is just the special case $s = 1/n$ or Rule (5).

Example 7.1

Simplify using rules for exponents:

(a) $(9u^4)^{-3/2}$ (b) $\left(\dfrac{x^3}{8y^{-6}}\right)^{1/3}$ (c) $(16x^4y^8z^{13})^{1/4}$.

SOLUTION

(a) $(9u^4)^{-3/2} = [(3u^2)^2]^{-3/2} = (3u^2)^{2(-3/2)} = (3u^2)^{-3}$

$$= \frac{1}{(3u^2)^3} = \frac{1}{27u^6}.$$

(b) $\left(\dfrac{x^3}{8y^{-6}}\right)^{1/3} = \dfrac{(x^3)^{1/3}}{(8y^{-6})^{1/3}} = \dfrac{x}{8^{1/3}y^{-6/3}} = \dfrac{x}{2y^{-2}} = \dfrac{xy^2}{2}.$

(c) $(16x^4y^8z^{13})^{1/4} = (2^4x^4y^8z^{12}z)^{1/4}$

$$= (2^4)^{1/4}(x^4)^{1/4}(y^8)^{1/4}(z^{12})^{1/4}z^{1/4} = 2xy^2z^3\sqrt[4]{z}.$$

Answer   (a) $\dfrac{1}{27u^6}$ (b) $\dfrac{xy^2}{2}$ (c) $2xy^2z^3\sqrt[4]{z}$.

Example 7.2

Express as a single radical:

(a) $\sqrt[3]{9}\,\sqrt{\tfrac{1}{3}}$ (b) $\dfrac{\sqrt{r^3s^5}}{\sqrt[4]{r^2s}}$.

SOLUTION (a) Convert to fractional exponents:

$$\sqrt[3]{9}\,\sqrt{\tfrac{1}{3}} = 3^{2/3}3^{-1/2} = 3^{2/3-1/2} = 3^{1/6} = \sqrt[6]{3}.$$

(b) $\dfrac{\sqrt{r^3s^5}}{\sqrt[4]{r^2s}} = (r^3s^5)^{1/2}(r^2s)^{-1/4} = r^{3/2}s^{5/2}r^{-1/2}s^{-1/4}$

$$= r^{3/2-1/2}s^{5/2-1/4} = rs^{9/4} = rs^2s^{1/4} = rs^2\,\sqrt[4]{s}.$$

Answer (a) $\sqrt[6]{3}$ (b) $rs^2\,\sqrt[4]{s}$.

Compute:

1. $27^{2/3}$ 2. $64^{3/2}$ 3. $25^{-1/2}$ 4. $16^{-5/4}$

5. $(1000)^{5/3}$ 6. $(1{,}000{,}000)^{5/6}$ 7. $\left(\dfrac{4}{49}\right)^{-3/2}$ 8. $(0.001)^{-5/3}$.

Express as a power of 2:

9. $4\sqrt[3]{2}$ 10. $(\sqrt{2})^{2/3}$ 11. $8(16 \cdot 2^{-7/2})$ 12. $\left(\dfrac{32}{\sqrt[3]{2}}\right)^{1/6}$.

Simplify:

13. $(25x^4)^{-3/2}$ 14. $\left(\dfrac{8}{u^6}\right)^{2/3}$ 15. $\left(\dfrac{u^4}{v^{12}}\right)^{3/4}$

16. $\left(\dfrac{27a^3}{8b^3c^6}\right)^{-4/3}$ 17. $(x^4y^6z^{-8})^{5/2}$ 18. $(x^{-3/2}\sqrt{y})^{-2}$

19. $(x^{4/3}y^{-2/3})^3$ 20. $(xy^2)^{1/3}(x^2y)^{-2/3}$ 21. $(8x\sqrt{x})^{5/3}$

22. $(x^{1/2}+y^{3/2})^2$ 23. $u^{1/3}(2u^{2/3}-u^{-1/6})$ 24. $v^{1/3}(8v^6)^{-2/3}$.

Express in terms of at most a single radical:

25. $\sqrt{2}\cdot\sqrt[3]{2}$ 26. $\sqrt[3]{\sqrt{x^{1/4}}}$ 27. $\sqrt[3]{4}/\sqrt[6]{16}$ 28. $\dfrac{\sqrt{xy}}{\sqrt[4]{x^2y}}$

29. $\left(\dfrac{\sqrt{3a}}{\sqrt[3]{6a^2}}\right)^4$ 30. $\sqrt{b\sqrt{b}}$ 31. $\sqrt{x}\cdot\sqrt{x^2}\cdot\sqrt{x^3}$ 32. $\sqrt{x}\cdot\sqrt[3]{x}\cdot\sqrt[4]{x}$.

Express without radicals, using only positive exponents:

33. $(\sqrt[3]{xy^2})^{-3/5}$ 34. $\sqrt[5]{(xy^2)^{-10/3}/(x^2y)^{-15/7}}$

35. $(\sqrt[4]{x^{14}y^{-21/5}})^{-3/7}$ 36. $(x^{5/6}-x^{-5/6})^2$

37. $\dfrac{x}{(x^{5/6})^{42}(x^{51})^{-2/3}}$ 38. $\left(\dfrac{x}{y}\right)^{1/5}\left(\dfrac{y}{z}\right)^{2/5}\left(\dfrac{z}{w}\right)^{3/5}\left(\dfrac{w}{x}\right)^{4/5}$.

In this section we study properties of the most common algebraic expressions, polynomials. A **polynomial** is an algebraic expression of the type

$$a_0 + a_1x + a_2x^2 + \cdots + a_nx^n,$$

where $a_0, a_1, a_2, \cdots, a_n$ are real numbers called the **coefficients** of the polynomial.

Examples:

$$3 - 2x, \quad 5 + 6x + x^2, \quad -\tfrac{1}{4}\pi + \tfrac{1}{10}x - 2x^3, \quad 1 + 7x^2 + 8x^3 - x^5.$$

The **degree** of a polynomial is the highest exponent occurring with non-zero coefficient. Standard notation is

$$\deg(a_0 + a_1x + \cdots + a_nx^n) = n, \qquad a_n \neq 0.$$

The zero polynomial $f(x) = 0$ is not assigned a degree. The coefficient a_n is called the **leading coefficient,** and the term a_nx^n is called the **leading term.** Polynomials have special names according to degree:

Constant polynomial (zero degree): $f(x) = a_0$.

Linear polynomial (first degree): $f(x) = a_0 + a_1x, \qquad a_1 \neq 0.$

Quadratic polynomial (second degree): $f(x) = a_0 + a_1x + a_2x^2, \quad a_2 \neq 0.$

Cubic polynomial (third degree): $f(x) = a_0 + a_1x + a_2x^2 + a_3x^3, \quad a_3 \neq 0.$

n-th degree polynomial: $f(x) = a_0 + a_1x + a_2x^2 + \cdots + a_{n-1}x^{n-1} + a_nx^n,$
$$a_n \neq 0.$$

Remark: We can write

$$a_nx^n + a_{n-1}x^{n-1} + \cdots + a_1x + a_0$$

just as well as $a_0 + a_1x + \cdots + a_nx^n$, and we shall use both ways of writing polynomials as is convenient.

Polynomial Algebra

To add or subtract polynomials, add or subtract coefficients of like powers. This is not an arbitrary rule, but a consequence of the laws of algebra. Polynomials are sums; add them like any other sums.

■ Example 8.1

Compute:

(a) $(2x^2 + 6x + 5) + (3x^2 - 8x - 1)$
(b) $(x^4 + 4x^2 + x + 12) - (3x^2 + 7x).$

SOLUTION Use the commutative and associative laws to rearrange and regroup the terms. Then use the distributive law to combine like powers of x:

(a) $(2x^2 + 6x + 5) + (3x^2 - 8x - 1) = (2x^2 + 3x^2) + (6x - 8x) + (5 - 1)$
$$= (2 + 3)x^2 + (6 - 8)x + (5 - 1)$$
$$= 5x^2 - 2x + 4.$$

(b) $(x^4 + 4x^2 + x + 12) - (3x^2 + 7x) = x^4 + (4x^2 - 3x^2) + (x - 7x) + 12$
$$= x^4 + (4 - 3)x^2 + (1 - 7)x + 12$$
$$= x^4 + x^2 - 6x + 12.$$

Answer: (a) $5x^2 - 2x + 4$ (b) $x^4 + x^2 - 6x + 12.$

Remark: In practice, nobody adds polynomials with so much fuss. You just combine like powers by inspection.

To multiply polynomials, treat them like any other sums: multiply each term of the first by each term of the second. Simplify the result by using the rule for exponents,

$$x^m \cdot x^n = x^{m+n},$$

then collecting all terms in the same power of x.

■ *Example 8.2*

Compute:

(a) $(3x + 2)(5x - 1)$ (b) $(x - 2)(4x^2 + 5x + 6).$

SOLUTION (a) Multiply each term of $3x + 2$ by each term of $5x - 1$:
$$(3x + 2)(5x - 1) = (3x)(5x) + (2)(5x) + (3x)(-1) + (2)(-1)$$
$$= 15x^2 + 10x - 3x - 2 = 15x^2 + 7x - 2.$$

(b) Same technique, only you get six terms instead of four:
$$(x - 2)(4x^2 + 5x + 6) = (x)(4x^2) - (2)(4x^2)$$
$$+ (x)(5x) - (2)(5x) + (x)(6) - (2)(6)$$
$$= 4x^3 - 8x^2 + 5x^2 - 10x + 6x - 12$$
$$= 4x^3 - 3x^2 - 4x - 12.$$

Answer (a) $15x^2 + 7x - 2$ (b) $4x^3 - 3x^2 - 4x - 12.$

■ *Example 8.3*

Compute: (a) $(x + c)^2$ (b) $(x + c)^3.$

SOLUTION

(a) $(x + c)^2 = (x + c)(x + c) = x \cdot x + c \cdot x + x \cdot c + c \cdot c$
$$= x^2 + cx + cx + c^2 = x^2 + 2cx + c^2.$$

(b) Use the result of part (a):

$$(x + c)^3 = (x + c)(x + c)^2 = (x + c)(x^2 + 2cx + c^2)$$
$$= x \cdot x^2 + c \cdot x^2 + x \cdot 2cx + c \cdot 2cx + x \cdot c^2 + c \cdot c^2$$
$$= x^3 + cx^2 + 2cx^2 + 2c^2x + c^2x + c^3$$
$$= x^3 + 3cx^2 + 3c^2x + c^3.$$

Answer (a) $x^2 + 2cx + c^2$ (b) $x^3 + 3cx^2 + 3c^2x + c^3$.

. . . .

Because of the rules of exponents, the degree of a product of polynomials is the sum of their degrees. For if we multiply $a_m x^m + \cdots + a_0$ by $b_n x^n + \cdots + b_0$, we obtain

$$a_m b_n x^{m+n} + \cdots + a_0 b_0.$$

Thus the degree of the product is $m + n$ since $a_m b_n \neq 0$, being the product of the non-zero leading coefficients of the factors.

> The degree of a product of polynomials is the sum of their degrees.

Compute:

1. $(x + 4) + (5x - 3)$ **2.** $(3x - 2) + (x + 6)$
3. $(8x^2 + 5x + 6) + (3x - 4)$ **4.** $(x^2 + x + 7) - (4x + 1)$
5. $(3x^2 + 7x - 2) - (9x^2 + x + 1)$ **6.** $(15x^2 - 8x - 11) - (-10x^2 + 4x + 5)$
7. $(\frac{1}{3}x^2 - 2x - 6) + (\frac{4}{3}x^2 - \frac{1}{2})$ **8.** $(4.1x^2 + 7.05x) - (2x^2 - 1.34x - 0.02)$
9. $(x^3 + 2x^2 - 4x - 5) + (6x^2 - 12x + 1)$
10. $(x^4 + 7x^3 + 8x^2 + 2x + 1) - (3x^3 + 4x^2 + 2x - 1)$
11. $(-x^5 + x + 9) - (x^4 + x^3 + x^2 + x + 1)$
12. $(x^5 - 4x^4 - 6x^3 - 7) - (x^5 + x^4 + 7x^2 + x)$
13. $(x^8 + 6x^6 + 4x^4 + 2x^2) + (x^6 - 5x^5 - 4x^3)$
14. $(-x^{10} + x + 1) + (x^3 - 4x^2 + 2x + 2)$.

Multiply:

15. $(x + 2)(x - 3)$ **16.** $(x - 4)(x - 7)$
17. $(2x + 5)(3x + 1)$ **18.** $(8x - 3)(x + 2)$
19. $(x + 6)(x^2 + x - 1)$ **20.** $(2x - 1)(x^2 + 3x + 2)$
21. $(x^3 + x)(x^2 + 1)$ **22.** $(2x^3 - 1)(-x^2 + x)$
23. $(x^5 + \frac{1}{2}x^2)(x^2 - x - \frac{1}{2})$ **24.** $(x^4 - 3x)(2x^2 + 5x + 1)$
25. $x(x + 1)(x^2 + x + 1)$ **26.** $(x + 1)(x + 2)(x + 3)$
27. $(x - 1)(x^4 + x^3 + x^2 + x + 1)$ **28.** $(x + 1)(x^4 - x^3 + x^2 - x + 1)$.

Compute:

29. $(1 + x)(1 + x^2)$
30. (cont.) $(1 + x)(1 + x^2)(1 + x^4)$

31. (cont.) Can you see a pattern?
32. $(1 - x)(1 + x)$
33. (cont.) $(1 - x)(1 + x)(1 + x^2)$
34. (cont.) $(1 - x)(1 + x)(1 + x^2)(1 + x^4)$
35. (cont.) Can you see a pattern?
36. $(1 - x^2)(1 + x + x^2 + x^3 + x^4)$
37. $(x^2 + x + 1)(x^2 - x + 1)$
38. $(x^2 + 2x + 2)(x^2 - 2x + 2)$.

Express in the form $a_n x^n + \cdots + a_2 x^2 + a_1 x + a_0$:

39. $(2x + 3)(x^2 + 1) + (x - 1)(x - 2)$
40. $x(x + 1)(x^2 + 1) - (2x + 1)(2x - 1)$
41. $(x + a)(x + b)$
42. $(x + a)(x + b) - (x + a)(x + c)$
43. $(x^2 + 1)(2x - 1) + (x^2 + 1)(x^2 + 2)$
44. $x(3x^2 + x - 1) + (x + 1)(x^2 + 2x)$
45. $(ax + b)(bx + a)$
46. $(x^2 + c)(x^2 - d)$.

Find the coefficient of x^3 in the product:

47. $(x^2 + 3x + 1)(2x - 1)$
48. $(x^3 - 7x + 6)(x^2 + 1)$
49. $x^2(2x - 5)(x + 6)$
50. $(x + 1)(2x - 1)(4x - 1)$
51. $(x^4 - 6x^3 + 2x^2 + 5x + 2)(x^3 + x + 4)$
52. $(x^3 + 2x^2 + 3x + 4)(6x^3 + 7x^2 - x - 5)$
53. $(1 + x^3)(1 + x^4)(1 + x^5)(1 + x^6)$
54. $(1 + 2x)(1 + 3x^2)(1 + 4x^3)(1 + 5x^4)$
55. $x(2x + 1)(3x + 1) - (x^2 + 2)(3x - 1)$
56. $(1 + x + x^2 + x^3)^3$.

9. POLYNOMIALS IN SEVERAL VARIABLES

A **polynomial** in several variables is a sum of terms of the form

$$1, \quad x, \quad y, \quad x^2, \quad xy, \quad y^2, \quad x^3, \quad x^2 y, \quad xyz, \quad x^3 y^2 z, \quad \cdots$$

with various coefficients. A term such as $ax^k y^m z^n$ is called a **monomial.** Its **degree** is the sum $k + m + n$ of its exponents. Thus the degree of $5x^2 y^4$ is 6; the degree of $-8xyz$ is 3. The **degree** of a polynomial is the highest degree of the monomials it contains. For example,

$$x - 5y + 2, \quad x^2 + xy + 4y^2 + 2x, \quad \tfrac{1}{2}x^4 y^2 - x^3 y - 6x^2 y^2 + 3y + 4,$$

are polynomials in two variables of degree 1, 2, and 6, respectively, and

$$x + 3y + 2z + 7, \quad x^2 + y^2 + z^2 + 5xy - 6xz - yz + 8x - 4y + z + 2$$

are polynomials in three variables of degree 1 and 2, respectively.

Polynomials in several variables are sums. They are added, subtracted, and multiplied just like any other sums.

■ *Example 9.1*

Compute:

(a) $(x^2 + 3y^2 + xy + 2x) + (4x^2 - y^2 + 2xy + 1)$

(b) $(4x^3 + x^2y + xy^2 + z^2 + xyz) - (4x^3 - 2x^2y - xyz + 5xz)$.

SOLUTION   (a)   Rearrange and combine similar terms:

$$(x^2 + 3y^2 + xy + 2x) + (4x^2 - y^2 + 2xy + 1)$$
$$= (x^2 + 4x^2) + (3y^2 - y^2) + (xy + 2xy) + 2x + 1$$
$$= 5x^2 + 2y^2 + 3xy + 2x + 1.$$

(b) $(4x^3 + x^2y + xy^2 + z^2 + xyz) - (4x^3 - 2x^2y - xyz + 5xz)$

$$= (4x^3 - 4x^3) + (x^2y + 2x^2y) + xy^2 + z^2 + (xyz + xyz) - 5xz$$
$$= 3x^2y + xy^2 + z^2 + 2xyz - 5xz.$$

Answer   (a) $5x^2 + 2y^2 + 3xy + 2x + 1$

(b) $3x^2y + xy^2 + z^2 + 2xyz - 5xz$.

■ *Example 9.2*

Multiply:

(a) $(2x - y)(x^2 + xy + 3y^2)$ (b) $(xy + z)(xy^2 + y^2z^2 + y)$.

SOLUTION   (a)   Multiply each term in the first parentheses by each term in the second, use rules of exponents, then collect similar terms:

$$(2x - y)(x^2 + xy + 3y^2)$$
$$= 2x \cdot x^2 - y \cdot x^2 + 2x \cdot xy - y \cdot xy + 2x \cdot 3y^2 - y \cdot 3y^2$$
$$= 2x^3 + (-x^2y + 2x^2y) + (-xy^2 + 6xy^2) - 3y^3$$
$$= 2x^3 + x^2y + 5xy^2 - 3y^3.$$

(b) $(xy + z)(xy^2 + y^2z^2 + y)$

$$= xy \cdot xy^2 + z \cdot xy^2 + xy \cdot y^2z^2 + z \cdot y^2z^2 + xy \cdot y + z \cdot y$$
$$= x^2y^3 + xy^2z + xy^3z^2 + y^2z^3 + xy^2 + yz.$$

Answer   (a) $2x^3 + x^2y + 5xy^2 - 3y^3$

(b) $x^2y^3 + xy^2z + xy^3z^2 + y^2z^3 + xy^2 + yz$.

Perfect Squares

If we square $x + y$, we obtain

$$(x + y)^2 = (x + y)(x + y) = x^2 + xy + yx + y^2 = x^2 + 2xy + y^2.$$

The result is an important formula:

$$(x + y)^2 = x^2 + 2xy + y^2.$$

This formula holds for all real numbers x and y. Therefore, it holds just as well if x and y are replaced by any expressions that represent real numbers.

Examples:

$$(a - b)^2 = a^2 - 2ab + b^2, \qquad (2a + 3b)^2 = 4a^2 + 12ab + 9b^2,$$

$$(x + \tfrac{1}{2})^2 = x^2 + x + \tfrac{1}{4}.$$

In practice, it helps to be able to recognize such squares of sums. For example, given the right sides of the equations above, to realize that they are the squares of the expressions on the left.

In hunting for such a perfect square, you need three terms: two squares plus twice the unsquared quantities. In words, the square of a sum of two terms looks like this:

$$(\text{first})^2 + (\text{second})^2 + 2(\text{first})(\text{second}).$$

■ *Example 9.3*

Express as a perfect square:

(a) $x^2 - 10x + 25$ (b) $x^2y^2 + 6xyz^2 + 9z^4$

(c) $36x^2 + 6xy + \tfrac{1}{4}y^2$ (d) $9x^2 + 8xy + 16y^2.$

SOLUTION (a) Does $x^2 - 10x + 25$ fall into the pattern of $a^2 + 2ab + b^2$? Yes, if $a = x$ and $b = -5$:

$$(x - 5)^2 = x^2 + 2x(-5) + (-5)^2 = x^2 - 10x + 25.$$

(b) The logical choice is $a^2 = x^2y^2$ and $b^2 = 9z^4$, that is, $a = xy$ and $b = 3z^2$:

$$(xy + 3z^2)^2 = (xy)^2 + 2(xy)(3z^2) + (3z^2)^2 = x^2y^2 + 6xyz^2 + 9z^4.$$

(c) Try $a = 6x$ and $b = \tfrac{1}{2}y$:

$$(6x + \tfrac{1}{2}y)^2 = (6x)^2 + 2(6x)(\tfrac{1}{2}y) + (\tfrac{1}{2}y)^2 = 36x^2 + 6xy + \tfrac{1}{4}y^2.$$

(d) The only logical choices are $a = 3x$ and $b = 4y$:

$$(3x + 4y)^2 = (3x)^2 + 2(3x)(4y) + (4y)^2 = 9x^2 + 24xy + 16y^2.$$

We obtain $24xy$, not $8xy$; therefore the given expression (d) is not a square.

     Answer (a) $(x - 5)^2$ (b) $(xy + 3z^2)^2$
              (c) $(6x + \tfrac{1}{2}y)^2$ (d) not a square.

Compute:

1. $(x + 2y + 3) + (4x - y - 1)$
2. $(2x - y + 8) + (3x + 2y - 5)$
3. $(-4x + 5y + 6z) - (x - y + z + 2)$
4. $(y + 2z) - (3x + y + z + 1)$
5. $(x^2 + 3y^2) + (x^2 + y^2 - 3y)$
6. $(x^2 + 6xy + 2y^2) - (x^2 + 4xy - y^2 - x - 2)$
7. $(x^2 + y^2 + z^2 + xy + yz + xz) - (3x^2 + y^2 + z^2 + xy + x + y)$
8. $x(3x^2 + xy^3 + yz^4) - y(xz^4 + xy + x^2y^2)$.

Multiply:

9. $(x + y)(2x + 3y)$
10. $(2x - y)(3x - y)$
11. $(5x + 2y)(5x - 2y)$
12. $(\frac{1}{2}x - \frac{1}{3}y)(2x - 3y)$
13. $(4u - 3v)^2$
14. $(uv + 2w)^2$
15. $(x + y)^3$
16. $(x - y)(x + y)(x^2 + y^2)$
17. $(x + y + z)(2x - y)$
18. $(y - z)(3x + 4y + z + 1)$
19. $(x + y - 1)(x^2 - y^2 + x)$
20. $(x + y + z)(x + y - z)$
21. $(x^2 + y^2)(x^2 - 3y^2)$
22. $(x^4 - y^4)(x^4 + y^4)$
23. $(x + y + z)^2$
24. $(ax + by + cz)(cx + by + az)$
25. $(r + 1)(s + 1)(t + 1)$
26. $(r + s)(r - s)^2$
27. $(a + b)(a + 2b)(a + 3b)$
28. $(a + b)(b + c)(c + a)$
29. $(x^2 + 2xy - 2y^2)(x^2 - 2xy + 2y^2)$
30. $(x + y + z)(x^2 + y^2 + z^2 - xy - yz - zx)$.

Is the expression a perfect square? If so, of what?

31. $x^2 + 16x + 64$
32. $y^4 + 2y^2 + 1$
33. $z^6 + 4z^3 + 4$
34. $16a^2 + 8a + 1$
35. $x^2y^2 + 3xy + 9$
36. $9c^2 - 30cd + 25d^2$
37. $(x^2 + 1)^2 + 2(x^2 + 1) + 1$
38. $9u^2 + 9u + 1$
39. $a^2b^8 - 4ab^4c^2 + 4c^4$
40*. $x^2 + 2xy + y^2 + 4xz + 4yz + 4z^2$.

Explain these party games:

41. Take a number from 1 to 10. Square the number one larger and square the number one smaller. Subtract. Divide by the number you started with. Now you have 4.
42. Multiply your number by the number 4 larger. Add 4. Take the square root. Subtract the number you started with. Now you have 2.

Find the coefficient of x^2y in

43. $(2x^2 + 3x - 4)(y^2 + 8y + 1)$
44. $(x + y)(x^2 + y^2)(1 + x)$
45. $(x + 4y)^3$
46. $(1 + x + y + xy + x^2 + y^2)(1 + 2x + 3y + 2xy + x^2)$.

10. FACTORING

Factoring is the process of expressing a polynomial as a product of polynomials of lower degree (factors). It is useful for simplifying algebraic expressions, for solving certain types of equations, and for numerical work. When we write

$$x^2 + 5x + 6 = (x + 2)(x + 3),$$

we are factoring the left-hand side. If a given polynomial has integer coefficients, we generally look for factors with integer coefficients.

A familiar kind of factoring is removing a common polynomial factor.

Examples:

$$x^4 - 6x^3 - 7x^2 = x^2(x^2 - 6x - 7),$$

$$(3x - 1)(x^2 + 4) + 2(3x - 1)(4x + 5) + (3x - 1)^2$$
$$= (3x - 1)[(x^2 + 4) + 2(4x + 5) + (3x - 1)]$$
$$= (3x - 1)(x^2 + 11x + 13).$$

Most factoring depends on recognizing certain polynomial products. In Section 8, we had some practice in recognizing perfect squares. Here are some other useful factored forms:

Factoring Formulas

(1) $x^2 + (c + d)x + cd = (x + c)(x + d)$

(2) $acx^2 + (ad + bc)x + bd = (ax + b)(cx + d)$

(3) $x^2 - y^2 = (x + y)(x - y)$

(4) $x^3 + y^3 = (x + y)(x^2 - xy + y^2)$

(5) $x^3 - y^3 = (x - y)(x^2 + xy + y^2)$

(6) $x^n - y^n = (x - y)(x^{n-1} + x^{n-2}y + x^{n-3}y^2 + \cdots + xy^{n-2} + y^{n-1}).$

Formulas (1)–(5) are easily checked by multiplication. Formula (6) we leave as an exercise.

These formulas hold for all real values of x and y. Therefore, they hold just as well if x and y are replaced by any expressions that represent real numbers. For example, (3) is a formula for the difference of any two squares. From it, follow such formulas as

$$4a^2 - 9b^2 = (2a)^2 - (3b)^2 = (2a + 3b)(2a - 3b),$$

$$x^2 - y^4z^6 = x^2 - (y^2z^3)^2 = (x + y^2z^3)(x - y^2z^3),$$

$$(u + v)^2 - 49 = (u + v)^2 - 7^2 = (u + v + 7)(u + v - 7).$$

▧ *Example 10.1*

Factor   (a) $x^2 + 10x + 9$ (b) $2x^2 - 7x + 3$.

SOLUTION   (a)   Try $(x + c)(x + d)$. By formula (1), we need numbers c and d for which $c + d = 10$ and $cd = 9$. Such a pair is $c = 1$, $d = 9$. So $x^2 + 10x + 9 = (x + 1)(x + 9)$ as is checked by multiplying.

(b)   Try $(ax + b)(cx + d)$. By formula (2), we need $ac = 2$. So try $a = 1$ and $c = 2$, that is, $(x + b)(2x + d)$. Now we need $bd = 3$ and $2b + d = -7$. A little trial and error shows that $b = -3$ and $d = -1$ work, so

$$2x^2 - 7x + 3 = (x - 3)(2x - 1).$$

Answer   (a) $(x + 1)(x + 9)$ (b) $(x - 3)(2x - 1)$.

■ *Example 10.2*

Factor $x^5 - 8x^3 + 16x$.

SOLUTION   First remove the common factor x:

$$x^5 - 8x^3 + 16x = x(x^4 - 8x^2 + 16).$$

The polynomial in parentheses is a perfect square, $(x^2 - 4)^2$. Also, by formula (3), we have $x^2 - 4 = (x + 2)(x - 2)$. Therefore

$$x^5 - 8x^3 + 16x = x(x^2 - 4)^2 = x[(x + 2)(x - 2)]^2.$$

Answer $x(x + 2)^2(x - 2)^2$.

■ *Example 10.3*

Factor   (a) $x^3 + 8$ (b) $27x^3 - 125y^3$.

SOLUTION   (a)   Use formula (4) for the sum of x^3 and 2^3:

$$x^3 + 8 = x^3 + 2^3 = (x + 2)(x^2 - x \cdot 2 + 2^2) = (x + 2)(x^2 - 2x + 4).$$

(b)   Use formula (5) for the difference of cubes:

$$27x^3 - 125y^3 = (3x)^3 - (5y)^3 = (3x - 5y)[(3x)^2 + (3x)(5y) + (5y)^2]$$
$$= (3x - 5y)(9x^2 + 15xy + 25y^2).$$

Answer   (a) $(x + 2)(x^2 - 2x + 4)$

(b) $(3x - 5y)(9x^2 + 15xy + 25y^2)$.

Remark: It's not a bad idea to check your factoring by multiplying out.

■ *Example 10.4*

Without dividing, show that 999,992 is divisible by 98.

SOLUTION   999,992 is the difference of cubes:

$$999{,}992 = 1{,}000{,}000 - 8 = (100)^3 - 2^3$$
$$= (100 - 2)(100^2 + 100 \cdot 2 + 2^2)$$
$$= 98(10000 + 200 + 4)$$
$$= (98)(10204).$$

■ *Example 10.5*

Factor $x^4 - 1$.

SOLUTION   This is a difference of squares:

$$x^4 - 1 = (x^2)^2 - 1^2 = (x^2 + 1)(x^2 - 1).$$

The second factor is also a difference of squares; hence

$$x^4 - 1 = (x^2 + 1)(x + 1)(x - 1).$$

Answer $(x^2 + 1)(x + 1)(x - 1).$

What Is Factoring Really?

All examples in this section have involved integer coefficients, both in the given polynomials and in their factors. There is, however, a big difference between seeking factors with integer coefficients and factors with *any* real coefficients. For example, if we do not require integer coefficients, it is easy to factor $x^2 - 5$:

$$x^2 - 5 = (x - \sqrt{5})(x + \sqrt{5}).$$

But if we insist on integer coefficients, it can't be done, as some trial-and-error will show.

Not every polynomial can be factored, even into factors with real coefficients. Consider the polynomial $x^2 + 1$. Suppose it had factors with *real* coefficients,

$$x^2 + 1 = (x + a)(x + b).$$

(There is no real harm in assuming that the coefficients of x are $+1$.) Set $x = -a$ in this formula:

$$(-a)^2 + 1 = (-a + a)(-a + b) = 0.$$

Hence $(-a)^2 = -1$. But this is impossible because $-1 < 0$ and $(-a)^2 \geq 0$. We conclude that $x^2 + 1$ cannot be factored.

We have seen that not every polynomial can be factored, and whether a given polynomial can might depend on what coefficients are allowed.

There is no sure-fire method for deciding whether a given polynomial can be factored; this may be a hard problem. Certainly it is not obvious that $x^3 + y^5$ *cannot* be factored, no matter what coefficients you allow. Nor is it obvious that $x^4 + 4$

can be factored into two quadratic factors with integer coefficients, $x^2 + 2x + 2$ and $x^2 - 2x + 2$.

EXERCISES

Factor:

1. $x^2 + 4x + 3$	**2.** $x^2 + 11x + 30$	**3.** $x^2 + x - 6$
4. $x^2 - 8x + 15$	**5.** $x^2 - 9x + 14$	**6.** $x^2 + 5x - 36$
7. $x^2 - 6x + 9$	**8.** $x^3 - 4x^2 + 3x$	**9.** $x^4 + x^3 - 20x^2$
10. $3x^2 + 4x + 1$	**11.** $2x^2 + 11x + 12$	**12.** $10x^2 - 17x + 3$
13. $4x^2 + 4x + 1$	**14.** $x^6 + 7x^3 + 12$	**15.** $x^4 - 5x^2 + 6$
16. $x^4 - x^2$	**17.** $4x^2 - 81$	**18.** $9y^2 - 100$
19. $(x + y)^2 - z^2$	**20.** $9a^2 - 4 + (3a - 2)(a + 1)$	**21.** $a^4 - 16b^2$
22. $(u^2 - 1) + 2(u + 1)$	**23.** $(v^4 - 81) + v^3 + 9v$	**24.** $b^3 - 27$
25. $8c^3 + 27$	**26.** $x^4 - 8x$	**27.** $8(x + y)^3 + 27x^3$
28. $(2a + 3b)^2 - (a - b)^2$	**29.** $x^5 - 1$	**30.** $x^5 - 32$
31. $(x^5 - 1) - (x^2 - 1)$	**32.** $x^6 - 64$	**33.** $x^8 - 1$
34. $(x^2 - 1)^2 - 6(x^2 - 1) + 9$	**35.** $8x^3y^6 + 1000$	
36. $(16x^4 - 1) + (4x^4 - 3x^2 - 1)$.		

37. Work Example 10.5 using formula (6) and show that your answer agrees with the answer in the text.

38. Write out formula (6) for $n = 5$ and verify it by multiplication.

39. Without dividing, show that 1027 is divisible by 13.

40. Factor $x^2 + 2xy - 8y^2$ by writing $-8y^2 = y^2 - 9y^2$.

11. RATIONAL EXPRESSIONS

Quotients of polynomials are called **rational expressions.**

Examples:

$$\frac{x}{4x^2 + 3}, \quad \frac{y}{x}, \quad \frac{x + 4}{2y - 7}, \quad \frac{xy}{z^3 + z + 1}, \quad \text{etc.}$$

Since rational expressions represent real numbers, we combine them by the usual rules of algebra for sums, differences, products, and quotients, using polynomial algebra to simplify the results.

A rational expression is in **lowest terms** if numerator and denominator have no common polynomial factors.

■ *Example 11.1*

Express in lowest terms:

(a) $\dfrac{x + y}{(x + y)(2x + y)}$ 　　(b) $\dfrac{3x - 6}{x^2 - 4x + 4}$ 　　(c) $\dfrac{x^2 - 1}{x^3 - 1}$.

SOLUTION   (a)   Simply cancel the common factor $x + y$:

$$\frac{x + y}{(x + y)(2x + y)} = \frac{1}{2x + y}.$$

(b)   Factor the numerator and denominator, then cancel the common factor:

$$\frac{3x - 6}{x^2 - 4x + 4} = \frac{3(x - 2)}{(x - 2)^2} = \frac{3}{x - 2}.$$

(c) $\dfrac{x^2 - 1}{x^3 - 1} = \dfrac{(x - 1)(x + 1)}{(x - 1)(x^2 + x + 1)} = \dfrac{x + 1}{x^2 + x + 1}.$

Answer   (a) $\dfrac{1}{2x + y}$ (b) $\dfrac{3}{x - 2}$ (c) $\dfrac{x + 1}{x^2 + x + 1}.$

Suppose we wish to add two rational expressions, for instance,

$$\frac{1}{x + y} \quad \text{and} \quad \frac{3x - 1}{x - 1}.$$

It is not immediately clear that their sum is again a rational expression, a *single* quotient of polynomials. However, guided by our experience with rational *numbers,* we express the summands as quotients with a *common denominator,* and then add the numerators.

■ *Example 11.2*

Compute:

(a) $\dfrac{x}{3x - y} + \dfrac{y}{3x + y}$ (b) $\dfrac{2x + 1}{x^2 - 4} + \dfrac{2}{x + 2} - \dfrac{1}{x - 2}.$

SOLUTION   (a)   A common denominator is $(3x - y)(3x + y)$:

$$\frac{x}{3x - y} + \frac{y}{3x + y} = \frac{x(3x + y)}{(3x - y)(3x + y)} + \frac{(3x - y)y}{(3x - y)(3x + y)}$$

$$= \frac{x(3x + y) + (3x - y)y}{(3x - y)(3x + y)} = \frac{3x^2 + 4xy - y^2}{9x^2 - y^2}.$$

(b)   A common denominator is $x^2 - 4 = (x + 2)(x - 2)$:

$$\frac{2x + 1}{x^2 - 4} + \frac{2}{x + 2} - \frac{1}{x - 2} = \frac{2x + 1}{x^2 - 4} + \frac{2(x - 2)}{x^2 - 4} - \frac{x + 2}{x^2 - 4}$$

$$= \frac{(2x + 1) + 2(x - 2) - (x + 2)}{x^2 - 4} = \frac{3x - 5}{x^2 - 4}.$$

Answer (a) $\dfrac{3x^2 + 4xy - y^2}{9x^2 - y^2}$ (b) $\dfrac{3x - 5}{x^2 - 4}.$

∎ *Example 11.3*

Compute and express in lowest terms:

(a) $\left(\dfrac{xy}{(x - 1)^2}\right)\left(\dfrac{x^2 - 1}{y^2 + y}\right)$ (b) $\dfrac{x + 1}{x^2 + 1} \bigg/ \dfrac{x}{x + 3}.$

SOLUTION (a) Multiply numerators and denominators, then cancel common factors:

$$\left(\dfrac{xy}{(x - 1)^2}\right)\left(\dfrac{x^2 - 1}{y^2 + y}\right) = \dfrac{xy(x^2 - 1)}{(x - 1)^2(y^2 + y)} = \dfrac{xy(x - 1)(x + 1)}{(x - 1)^2 y(y + 1)} = \dfrac{x(x + 1)}{(x - 1)(y + 1)}.$$

(b) Invert the divisor and multiply:

$$\dfrac{x + 1}{x^2 + 1} \bigg/ \dfrac{x}{x + 3} = \left(\dfrac{x + 1}{x^2 + 1}\right)\left(\dfrac{x + 3}{x}\right) = \dfrac{(x + 1)(x + 3)}{x(x^2 + 1)}.$$

There are no common factors; the product is already in lowest terms.

Answer (a) $\dfrac{x(x + 1)}{(x - 1)(y + 1)}$ (b) $\dfrac{(x + 1)(x + 3)}{x(x^2 + 1)}.$

∎ *Example 11.4*

Simplify $\dfrac{\dfrac{1}{x} - \dfrac{1}{x + 1}}{1 + \dfrac{1}{x}}.$

SOLUTION One way to handle a 4-story fraction like this is to compute the numerator and denominator separately, then to divide the two fractions. A second way, sometimes shorter, is to multiply numerator and denominator by a common denominator. In this case multiply by $x(x + 1)$:

$$\dfrac{\dfrac{1}{x} - \dfrac{1}{x + 1}}{1 + \dfrac{1}{x}} = \dfrac{x(x + 1)\left(\dfrac{1}{x} - \dfrac{1}{x + 1}\right)}{x(x + 1)\left(1 + \dfrac{1}{x}\right)} = \dfrac{(x + 1) - x}{x(x + 1) + (x + 1)} = \dfrac{1}{(x + 1)^2}.$$

Answer $\dfrac{1}{(x + 1)^2}.$

Express in lowest terms:

1. $\dfrac{x - 4}{x^2 - 16}$

2. $\dfrac{x^4 + 3x^2}{x}$

3. $\dfrac{x + 1}{x^3 + 1}$

4. $\dfrac{x^2 - x + 30}{x^2 - 25}$

5. $\dfrac{x^2 - y^2}{x^3 - y^3}$

6. $\dfrac{(x + y)(x^2 - xy + y^2)}{x^4 + xy^3}$

7. $\dfrac{x + y - 2z}{(x + y)^2 - 4z^2}$

8. $\dfrac{2xyz + 3xz^2}{4xy^2 + 12xyz + 9xz^2}$.

Compute:

9. $\dfrac{1}{x - 1} - \dfrac{1}{x}$

10. $\dfrac{3}{x - 1} + \dfrac{4}{x + 1}$

11. $\dfrac{1}{2x + 3} - \dfrac{1}{2x + 5}$

12. $\dfrac{1}{x} - \dfrac{1}{2x + 1}$

13. $\dfrac{x}{x^2 - 4} + \dfrac{1}{x + 2} + \dfrac{3}{x - 2}$

14. $\dfrac{3x}{x^2 - 4} + \dfrac{1}{(x - 2)^2}$

15. $\dfrac{x}{y} + \dfrac{y}{x}$

16. $\dfrac{1}{x} + \dfrac{1}{xy} + \dfrac{1}{xy^2}$

17. $\dfrac{x + y}{x + 2y} - \dfrac{y}{x + 3y}$

18. $1 - \dfrac{y^2}{x^2 + y^2}$.

Multiply and express in lowest terms:

19. $\left(\dfrac{x}{x - 1}\right)\left(\dfrac{x^2 - 1}{x^2}\right)$

20. $\left(\dfrac{2x + 3}{x^2 + 1}\right)\left(\dfrac{x}{x^2 - 1}\right)$

21. $\left(\dfrac{x}{y}\right)\left(\dfrac{x + 1}{y + 1}\right)$

22. $\left(\dfrac{x^2 + xy}{x + 3y}\right)\left(\dfrac{2y}{x^2 + y^2}\right)$

23. $\left(\dfrac{1}{x + y}\right)\left(\dfrac{1}{x} + \dfrac{1}{y}\right)$

24. $\left(\dfrac{1}{x - y}\right)\left(\dfrac{y}{x} - \dfrac{x}{y}\right)$

25. $xyz\left(\dfrac{1}{x} + \dfrac{2}{y} - \dfrac{3}{z}\right)$

26. $\left(\dfrac{x + y}{x - 2y}\right)^2\left(\dfrac{x^2 - 3xy + 2y^2}{(x^2 - y^2)^2}\right)$.

Compute the quotient:

27. $\dfrac{x + 3}{x + 4} \bigg/ \dfrac{2x}{x + 4}$

28. $\dfrac{x + y}{x + 2y} \bigg/ \left(\dfrac{x + y}{xy}\right)^2$

29. $\dfrac{a^2 - 1}{a^3 + a} \bigg/ \dfrac{a + 1}{a^2 + 1}$

30. $(a + b)^2 \bigg/ \dfrac{a^2 - b^2}{ab}$.

Simplify:

31. $\dfrac{\dfrac{u^2}{v^2} - \dfrac{v^2}{u^2}}{\dfrac{u}{v} + \dfrac{v}{u}}$

32. $\dfrac{\dfrac{1}{t - 3} + \dfrac{4}{t^2 - 9}}{2 + \dfrac{1}{t + 3}}$

33. $\left(\dfrac{u^2}{v^2} - \dfrac{v^2}{u^2}\right)\left(\dfrac{u^4}{v^4} + 1 + \dfrac{v^4}{u^4}\right)$

34. $\dfrac{x^4 - \dfrac{1}{y^4}}{x^2y^2 - 1}$

35. $\dfrac{(xu - yv)^2 + (xv + yu)^2}{x^2 + y^2}$

36. $\dfrac{x^2 + \dfrac{x}{y} - \dfrac{6}{y^2}}{x^2 + 4\dfrac{x}{y} + \dfrac{3}{y^2}}.$

12. COMMON ERRORS IN ALGEBRA

The basic rules of algebra discussed in this chapter are indispensable in this course and all other mathematics courses you will take. It is worth the effort to master them now so you will avoid silly mistakes in the future. Let us list some of the most common types of errors in algebra.

1. Incorrect use of rules of real numbers:

WRONG	RIGHT
$x/(y + 2) = x/y + 2$	Keep the parentheses: $x/(y + 2)$
$(x + 3)(x + 4) = x + 3 \cdot x + 4$	$(x + 3)(x + 4) = x^2 + 7x + 12$
$(x + y)^2 = x + y^2$	Do not remove the parentheses: $(x + y)^2.$

2. Poor use of parentheses:

WRONG	RIGHT
$3(x + 1) = 3x + 1$	$3(x + 1) = 3x + 3$
$(\frac{1}{2}x)(\frac{1}{2}y) = \frac{1}{2}xy$	$(\frac{1}{2}x)(\frac{1}{2}y) = \frac{1}{4}xy$
$x - (2y + 1) = x - 2y + 1$	$x - (2y + 1) = x - 2y - 1.$

3. Mistakes with exponents and radicals:

WRONG	RIGHT
$(2x)^3 = 2x^3$	$(2x)^3 = 8x^3$
$(x^3)^4 = x^7$	$(x^3)^4 = x^{12}$
$x^3x^4 = x^{12}$	$x^3x^4 = x^7$
$\sqrt{3x} = 3\sqrt{x}$	$\sqrt{3x} = \sqrt{3}\sqrt{x}.$

4. *Incorrect use of formulas:*

WRONG	RIGHT
$\dfrac{1}{x} + \dfrac{1}{y} = \dfrac{1}{x+y}$	$\dfrac{1}{x} + \dfrac{1}{y} = \dfrac{x+y}{xy}$
$\sqrt{x+y} = \sqrt{x} + \sqrt{y}$	Leave as $\sqrt{x+y}$.
$(x+y)^2 = x^2 + y^2$	$(x+y)^2 = x^2 + 2xy + y^2$.

5. *Sloppy writing:*

$\sqrt{x} + 1$	Is it $\sqrt{x+1}$ or $\sqrt{x} + 1$?
$\dfrac{1}{x+y} + 3$	Is it $\dfrac{1}{x+y+3}$ or $\dfrac{1}{x+y} + 3$?
$x - 5$	Is it x^{-5} or $x - 5$?
$\dfrac{1}{2}x$	Is it $\dfrac{1}{2}x$ or $\dfrac{1}{2x}$?

These examples illustrate some, but not all, of the common mistakes made in algebra. Get into the habit of checking your work. Read it over and see if it says what you mean. Are your formulas ambiguous? Are they legible?

EXERCISES

Find the mistake in each formula and correct it.

1. $\left(\dfrac{x}{5}\right)^2 = \dfrac{x^2}{5}$

2. $(3x + 1)^2 = 9x^2 + 1$

3. $\sqrt{4x + 4} = 4\sqrt{x + 1}$

4. $(-x)^4 = -x^4$

5. $\sqrt{2^9} = 2^3$

6. $\dfrac{x^4}{x^2 + x^4} = x^2 + 1$

7. $\sqrt[3]{\sqrt[3]{x}} = \sqrt[6]{x}$

8. $(x - y)(x + y) = x^2 + y^2$

9. $\sqrt{x^2 + 1} = x + 1$

10. $\left(\dfrac{1}{\sqrt{x} + \sqrt{y}}\right)^2 = \dfrac{1}{x + y}$

11. $2a(a^2 + 4a + 1) = 2a^3 + 8a^2 + 1$

12. $\sqrt{3x + 5} + \sqrt{2x + 1} = \sqrt{5x + 6}$

13. $a^3 + b^3 = (a + b)^3$

14. $\sqrt[3]{a^2} = a^{-2/3}$

15. $a^{-2}a^{-2} = a^4$

16. $(x/y^2)^2 = x/y^4$

17. $\sqrt{x^3 + 2x^2 + x} = x(x + 1)$

18. $x + x + x + x = x^4$

19. $(25x)(4x) = 100x$

20. $\sqrt{x} + \sqrt{2x} = \sqrt{3x}$

21. $\dfrac{x + y}{x + z} = 1 + \dfrac{y}{z}$

22. $x^{-2} = \dfrac{-2}{x}$

23. $(x + 1)(y + 1)(z + 1)$
$= xyz + x + y + z + 1$

24. $1 + \dfrac{1}{x} = \dfrac{2}{1 + x}$

25. $\dfrac{4x^3 - 7x^2 + 1}{x} = 4x^2 - 7x + 1$

26. $\dfrac{1/x}{1/y} = \dfrac{1}{xy}$

27. $(x + y)^3 = x^3 + 3xy + y^3$

28. $x^{p/q} = x^p/x^q$

29. $\dfrac{x^2 + 4x + 8}{x^2 + 2x + 2} = 1 + 2 + 4 = 7$

30. $\sqrt[3]{8 + 27} = 2 + 3 = 5$

31. $-2^4 = 2^{-4}$

32. $(-2)^{-4} = -2^{-4}.$

Test 1

1. Expand:

(a) $(x - z)(2x - 3y - 2z)$ (b) $(x - y)(x + z).$

2. Express in algebraic notation:
(a) 15% of the sum of two numbers times the quotient of two other numbers;
(b) The area between a circle and a square inside the circle.

3. Factor:

(a) $16x^2 - 24xy + 9y^2$ (b) $z^4 + 8z.$

4. Express as simply as possible, without radicals or negative exponents:

(a) $\dfrac{(x^2 y^{-1})^3 z^4}{(xy^{-2})^2 y^5 z^{-1}}$ (b) $\sqrt[3]{\dfrac{z}{16xy}} \sqrt[3]{\dfrac{1}{4y^2 z^7}}.$

Test 2

1. Compute the sum:

$$5 + 12 + 19 + 26 + \cdots + (7n - 2).$$

2. Prove $a(1 - ab)^{-1}b = (1 - ba)^{-1} - 1.$

3. Express in scientific notation:

(a) $\frac{1}{5} \cdot 2^{10}$ (b) $0.002000351.$

4. Simplify $\left(\sqrt{\dfrac{3x}{7y}}\right)(\sqrt{27x^2 y^3})^{-1}.$

5. Multiply and express in lowest terms:

$$\left(\dfrac{a^2}{b^2} - \dfrac{b^2}{a^2}\right)\left(\dfrac{b^2}{a - b}\right).$$

EQUATIONS AND INEQUALITIES

Can you see a fundamental difference between these equations?

$$(1)\quad x^2 - 1 = (x + 1)(x - 1), \qquad (2)\quad x^2 - 1 = 99.$$

Equation (1) is a true statement for every real number x, but equation (2) is true only if $x = 10$ or $x = -10$.

An equation that holds for all real values of the variables is called an **identity.** Thus $x^2 - 1$ and $(x + 1)(x - 1)$ are exactly the same thing, only expressed in different ways. Similarly,

$$(x + y)^2 = x^2 + 2xy + y^2$$

is a true statement because $(x + y)^2$ and $x^2 + 2xy + y^2$ are equal no matter what real numbers are substituted for x and y.

An identity, then, is like a law of the universe, a fundamental truth. This is definitely not the case for the equation $x^2 - 1 = 99$, which is false for most values of x. An equation that is not an identity is called a **conditional equation.** This kind of equation is a question, asking what values of x make it a true statement. Each such value is called a **solution** or **root** of the equation. To solve a conditional equation is to find its set of solutions.

Remark: In practice, both identities and conditional equations are loosely referred to as equations.

Sometimes identities are called formulas. But usually the word **formula** is used for an equation in which the letters involved denote physical or geometrical quantities. For instance,

$$d = rt \qquad \text{(distance)} = \text{(rate)(time)}$$

$$A = \tfrac{1}{2}bh \qquad \text{(area)} = \tfrac{1}{2}\text{(base)(height)}$$

$$E = mc^2 \qquad \text{(energy)} = \text{(mass)(speed of light)}^2$$

$$1 + 2 + 3 + \cdots + n = \tfrac{1}{2}n(n + 1) \qquad \text{(sum of first } n \text{ positive integers)}$$
$$= \tfrac{1}{2}n(n + 1)$$

are formulas expressing laws of nature when the letters have the indicated meanings.

Solution of Equations

The easiest way to solve an equation is by inspection—if the equation is simple enough.

Examples:

$$x + 1 = 3 \qquad \text{solution:} \quad 2$$

$$4x = 4 \qquad \text{solution:} \quad 1$$

$$x^2 = 25 \qquad \text{solutions:} \quad 5 \quad \text{and} \quad -5.$$

For solving harder equations, the usual strategy is to keep modifying the equation until you arrive at some equation you can solve. Take, for example,

$$x^2 - x = 25 - x.$$

If x satisfies this equation, it also satisfies

$$x^2 = 25,$$

and conversely. The latter equation is simpler; $x = 5$ or $x = -5$.
    If x satisfies

$$2x + 1 = 7,$$

then x also satisfies

$$2x = 6,$$

from which $x = 3$.
    In the first example, we added x to both sides of the equation; in the second we subtracted 1 from both sides. We solved $2x = 6$ by multiplying each side by $\tfrac{1}{2}$, or equivalently, by dividing each side by 2. These examples illustrate some important principles in action:

Adding or subtracting the same quantity from both sides of an equation does not affect its solutions.
    The same is true for multiplication by a non-zero quantity and division by a non-zero quantity.

Using these principles, you can often change an equation into one that is easier to solve *and* whose solutions are the same as the solutions of the harder equation you started with.

Note that multiplication by 0 is not allowed. Take the equation $x = 3$, for example. It holds only for one value of x. But if you multiply both sides by 0, you get the equation $0 \cdot x = 0$, which holds for all values of x.

■ *Example 1.1*

Solve $4x - 3 = 9$.

SOLUTION   Add 3 to both sides:
$$(4x - 3) + 3 = 9 + 3, \qquad 4x = 12.$$
Divide both sides by 4:
$$\frac{4x}{4} = \frac{12}{4}, \qquad x = 3.$$

Answer   3.

■ *Example 1.2*

Solve $x^2 + x = 24 - x$.

SOLUTION   Add x to both sides:
$$x^2 + 2x = 24.$$
Now add 1:
$$x^2 + 2x + 1 = 25.$$
The left side is $(x + 1)^2$, so the equation is
$$(x + 1)^2 = 25.$$
By inspection, $x + 1 = 5$ or $x + 1 = -5$. The first possibility yields $x = 4$, the second, $x = -6$.

Answer -4, 6.

Remark 1: In the solution we replaced $x^2 + 2x + 1$ by $(x + 1)^2$. You are always allowed to replace any expression in an equation by an identical expression. If, for example, $x^2 - 1$ appears somewhere, you may replace it by $(x + 1)(x - 1)$ without having to apologize.

Remark 2: It makes no sense to solve an *identity* since all values are solutions. For instance, nobody would want to *solve* the identity $x^2 - 4 = (x + 2)(x - 2)$. However, sometimes a complicated identity may be disguised and look like a conditional equation. If you try to solve, you will probably end up with something like $0 = 0$ or $x = x$. So if this ever happens, look carefully at the equation you started to solve. Probably it is an identity.

We have had examples of equations with one solution and with two solutions. Other possibilities exist as well. For example, the equation

$$(x - 1)(x - 2)(x - 3)(x - 4)(x - 5) = 0$$

has five solutions: 1, 2, 3, 4, 5. Some equations have no solutions, e.g.,

$$x^2 + 1 = 0.$$

The left side can never be 0 since x^2 is non-negative for every real number x. Learn to be suspicious; just because an equation is written down does not guarantee that it has any solutions.

EXERCISES

State whether the equation is an identity or a conditional equation:

1. $5x + 4 = 2x$
2. $\dfrac{1}{x + 1} = 2$
3. $4(2x - 3) + 1 = 6(x - 1) + 2x - 5$
4. $x^2 + 2x = x(x + 1) + x$
5. $x^3 + 2x = 1$
6. $x^2(x - 3) = 0$
7. $1 + \dfrac{1}{x} = \dfrac{x + 1}{x}$
8. $\sqrt{1 + x^2} = x + 3$
9. $4x^2 - 7x = 2x + 1$
10. $(x + 1)^2 - x^2 = 2(x + 1) - 1.$

Find all solutions:

11. $4x = 24$
12. $\frac{1}{3}x = 2$
13. $x^2 + 4x + 4 = 9$
14. $3x + 7 = 4(x + 1) - x$
15. $x(x - 2)(x - 7) = 0$
16. $2x^2 + 1 = 3x^2 - 8$
17. $\sqrt{x - 2} = 1$
18. $1/x^2 = 9$
19. $(x^2 - 4)(x^2 - 9) = 0$
20. $x^2 = x.$

2. LINEAR EQUATIONS

In this section and the next, we study two of the simplest and most common types of equations, linear and quadratic equations.

A **linear equation** is one of the form

$$ax + b = 0 \qquad (a \neq 0),$$

or one that can be brought into that form by moving all terms to one side of the equation. For example,

$$2x + 1 = -7, \qquad x - \tfrac{1}{2} = 2x + 1$$

are linear equations. They can be brought into the forms

$$2x + 8 = 0, \qquad x + \tfrac{3}{2} = 0$$

by shifting terms. Other examples:

$$\frac{x - 1}{3} = 4x + 5, \qquad 8(4x + 1) + 7(3x - 5) = 4(2x + 1).$$

To solve the equation $ax + b = 0$, we move b to the right side and divide by a:

$$ax + b = 0, \qquad ax = -b, \qquad x = -b/a.$$

This short calculation shows that the equation has at most one solution, $x = -b/a$. But $-b/a$ is a solution, since if $x = -b/a$, then

$$ax + b = a(-b/a) + b = -b + b = 0.$$

> A linear equation
> $$ax + b = 0 \qquad (a \neq 0)$$
> has exactly one solution, $x = -b/a$.

It is often convenient to write a linear equation in the form

$$ax = b \qquad (a \neq b),$$

which is equivalent to

$$ax + (-b) = 0 \qquad (a \neq 0),$$

and the solution is simply $x = b/a$.

■ *Example 2.1*

Solve   (a) $3x + 1 = 8$ (b) $4(x - 2) = 2x + 5$.

SOLUTION   (a)   Subtract 1 from both sides:

$$3x = 7.$$

Now multiply by $\frac{1}{3}$:

$$\tfrac{1}{3}(3x) = \tfrac{1}{3} \cdot 7, \qquad x = \tfrac{7}{3}.$$

(b)   Expand the left side:

$$4x - 8 = 2x + 5.$$

Add 8 to both sides, then add $-2x$:

$$4x = 2x + 13, \qquad 2x = 13.$$

Hence $x = 13/2$.

Answer   (a)   7/3   (b)   13/2.

Some equations can be brought into linear form by various algebraic manipulations.

■ *Example 2.2*

Solve: $\dfrac{3}{x-6} = \dfrac{1}{2x-4}$.

SOLUTION The equation makes no sense for $x = 2$ and $x = 6$. So just by writing it down, we are agreeing to exclude these values. To eliminate fractions, we multiply both sides by $(x-6)(2x-4)$. This quantity is non-zero since $x \neq 2$ and $x \neq 6$. Therefore the multiplication will yield a new equation with the same solutions as the given one, unless $x = 2$ or $x = 6$ happens to be a solution of the new equation.

$$\left(\frac{3}{x-6}\right)(x-6)(2x-4) = \left(\frac{1}{2x-4}\right)(x-6)(2x-4),$$

$$3(2x-4) = x-6.$$

This equation is linear. We solve it in the usual way:

$$6x - 12 = x - 6, \qquad 5x = 12 - 6 = 6, \qquad x = \tfrac{6}{5}.$$

Check: Substitute $x = 6/5$:

$$\frac{3}{x-6} = \frac{3}{\frac{6}{5}-6} = \frac{3 \cdot 5}{6 - 6 \cdot 5} = \frac{15}{6-30} = -\frac{15}{24} = -\frac{5}{8};$$

$$\frac{1}{2x-4} = \frac{1}{2(\frac{6}{5})-4} = \frac{5}{2 \cdot 6 - 4 \cdot 5} = \frac{5}{12-20} = -\frac{5}{8}.$$

Answer 6/5.

■ ■

Some equations involve several variables. Take

$$x^2 y + 3y = 4x^3 - xy - 5$$

for example. This equation looks complicated, but is actually a linear equation as far as y is concerned. With a little rearrangement of terms, it can be written as

$$(x^2 + x + 3)y = 4x^3 - 5,$$

which is of the form $ay = b$. It follows that

$$y = \frac{4x^3 - 5}{x^2 + x + 3}.$$

We say that we have solved for y **in terms of** x, or expressed y in terms of x.

■ *Example 2.3*

The relation between Fahrenheit and centigrade temperatures is $F = \tfrac{9}{5}C + 32$. Express C in terms of F.

SOLUTION The equation is linear in F and also linear in C. Solve for C:

$$F - 32 = \tfrac{9}{5}C, \qquad \tfrac{5}{9}(F - 32) = C.$$

Answer $C = \tfrac{5}{9}(F - 32)$.

- -

■ *Example 2.4*

Solve for x: $\dfrac{1}{x} - \dfrac{1}{y} = 10$.

SOLUTION Assume $xy \neq 0$ and multiply both sides by xy:

$$y - x = 10xy,$$
$$y = 10xy + x = (10y + 1)x,$$
$$\frac{y}{10y + 1} = x.$$

Answer $x = \dfrac{y}{10y + 1}$.

- -

Systems of Linear Equations

So far we have been solving equations in one unknown. Now let us look at equations in two unknowns.

Most often, two unknowns occur in a system of equations such as

$$\begin{cases} x + y = 3 \\ x - y = 1. \end{cases}$$

A **solution** is a pair (x, y) of real numbers that makes both equations correct statements at the same time, that is, *simultaneously*. A solution of the preceding system is $x = 2$, $y = 1$. Let us show that there is no other solution.

Suppose (x, y) is any solution. Then both equations are true statements. Add them and you will get another true statement:

$$\begin{aligned} x + y &= 3 \\ \underline{x - y} &= \underline{1} \\ 2x \quad\;\; &= 4. \end{aligned}$$

The y drops out, $2x = 4$, so $x = 2$. Therefore, if (x, y) is a solution, then $x = 2$. Now $x + y = 3$, $2 + y = 3$, so $y = 1$. Conclusion: the only possible solution is $(2, 1)$.

It was convenient that y disappeared when we added the equations. Note that x disappears if we subtract.

■ *Example 2.5*

Solve $\begin{cases} 5x + 2y = 5 \\ 4x + 3y = 6. \end{cases}$

SOLUTION 1   To eliminate y we multiply the first equation by 3 and the second by -2, then add:

$$
\begin{array}{c|cc}
3 & 5x + 2y = 5 & 15x + 6y = 15 \\
-2 & 4x + 3y = 6 & -8x - 6y = -12 \\
\hline
 & & 7x = 3.
\end{array}
$$

Hence $x = \frac{3}{7}$. From the first equation,

$$5\left(\frac{3}{7}\right) + 2y = 5, \qquad 2y = 5 - \frac{15}{7} = \frac{20}{7}, \qquad y = \frac{10}{7}.$$

SOLUTION 2   Another idea is to solve one equation for x in terms of y, substitute the result into the second equation, and then solve the second equation. First,

$$5x + 2y = 5, \qquad x = \tfrac{1}{5}(5 - 2y) = 1 - \tfrac{2}{5}y.$$

Now substitute:

$$4x + 3y = 6, \qquad 4(1 - \tfrac{2}{5}y) + 3y = 6,$$
$$4 - \tfrac{8}{5}y + 3y = 6, \qquad \qquad \tfrac{7}{5}y = 2.$$

Finally solve for y. We obtain $y = \frac{10}{7}$. To find x, use the expression for x in terms of y:

$$x = 1 - \frac{2}{5}y = 1 - \frac{2}{5}\left(\frac{10}{7}\right) = 1 - \frac{4}{7} = \frac{3}{7}.$$

Answer $x = 3/7, \quad y = 10/7.$

· ·

Remark 1:   Not every system has a solution. For example

$$\begin{cases} x + 2y = 1 \\ x + 2y = 3 \end{cases}$$

has no solution. For if it did, we would be forced to the false conclusion that $1 = 3$. A system with no solution is called **inconsistent.**

Remark 2:   A system of two equations may really be a single equation in disguise. For example, in the system

$$\begin{cases} 3x + 6y = 12 \\ x + 2y = 4, \end{cases}$$

the first equation is three times the second. Any attempt to eliminate one unknown from the system will simultaneously eliminate the other, so the result will be $0 = 0$. Such a system, called **underdetermined,** is inadequate to determine a single answer.

EXERCISES

Solve:

1. $4x + 1 = 25$
2. $\frac{1}{2}x - 4 = 3$
3. $2x + 3 = 6x - 5$
4. $-3(2x + 1) = 4(4x - 3)$
5. $2(\frac{1}{3}x + 1) + 3(\frac{2}{3}x + 8) = -4$
6. $5x + 2 = \frac{3}{2}x - 11$
7. $\dfrac{2x - 1}{2x + 1} = \dfrac{1}{4}$
8. $\dfrac{2x - 1}{2x + 1} = \dfrac{x + 2}{x + 4}$
9. $\dfrac{1}{x} = \dfrac{4}{3x + 1}$
10. $2x - \dfrac{x + 5}{4} = \dfrac{3x - 1}{2} + 1$
11. $(x + 1)^2 = (x + 2)^2 + 5$
12. $x^3 + 5x + 3 = x^2(x - 2) + 2(x^2 + 4x + 1)$
13. $ax + b = cx + d, \quad a \neq c$
14. $(ax + 1)(bx + 1) = (ax + 3)(bx - 2), \quad 3a \neq 2b.$

Solve for x in terms of y:

15. $2x + 3y - 4 = 0$
16. $x(y + 3) = 4x + 5y + 1$
17. $\dfrac{1}{x} + \dfrac{2}{y + 1} = 3$
18. $\dfrac{x + y}{x - y} = 4$
19. $y = \dfrac{2x + 1}{3x + 5}$
20. $\dfrac{x}{y} = \dfrac{x + y + 1}{2y - 3}.$

Solve the system

21. $\begin{cases} x + 3y = 4 \\ 2x - y = 1 \end{cases}$
22. $\begin{cases} x - y = 4 \\ x + 2y = 1 \end{cases}$
23. $\begin{cases} x - 2y = -2 \\ 2x - y = 5 \end{cases}$

24. $\begin{cases} 2x + y = 2 \\ -3x + y = -3 \end{cases}$
25. $\begin{cases} x - 6y = -17 \\ x + 2y = 7 \end{cases}$
26. $\begin{cases} x - \frac{1}{2}y = 4 \\ 2x + y = 6 \end{cases}$

27. $\begin{cases} 2x - 7y = -1 \\ 3x + 4y = 2 \end{cases}$
28. $\begin{cases} 6x + 5y = -3 \\ 5x - 6y = \frac{23}{3} \end{cases}$
29. $\begin{cases} 2x - 3y = 1 \\ 6x - 9y = 2 \end{cases}$

30. $\begin{cases} x + 4y = 7 \\ 2x + 8y = 10 \end{cases}$
31. $\begin{cases} x^2 + y^2 = 5 \\ x^2 - y^2 = 3 \end{cases}$
32. $\begin{cases} x + y^2 = 10 \\ x + 3y^2 = 28. \end{cases}$

3. QUADRATIC EQUATIONS

A **quadratic equation** is an equation of the form

$$ax^2 + bx + c = 0 \qquad (a \neq 0),$$

or one that can be transformed into this type by rearranging terms, e.g.,

$$3x^2 - 5x = 6x + 1, \qquad (x + 3)(2x - 9) = x^2 - 4x,$$
$$\tfrac{1}{2}x^2 + 8 = -x^2 + 2x + \tfrac{1}{4}.$$

Simple quadratic equations such as

$$x^2 + 1 = 0, \qquad x^2 = 0, \qquad x^2 - 1 = 0$$

show that there may be no real solutions, or one solution, or two solutions. We shall see in this chapter that there are no other possibilities.

Solution by Factoring

This method depends on a basic property of real numbers:

> If r and s are real numbers such that $rs = 0$, then either $r = 0$, or $s = 0$, or both.

Put in different words, the product of non-zero numbers cannot be zero. We exploit this property to solve quadratic equations $ax^2 + bx + c = 0$ when we are able to factor the left-hand side.

■ *Example 3.1*

Solve   (a) $x^2 - 2x = 0$ (b) $x^2 - 3x - 4 = 0$.

SOLUTION   (a)   Factor the left side:

$$x(x - 2) = 0.$$

The product is 0 only if either $x = 0$ or $x - 2 = 0$, that is, $x = 2$. Thus 0 and 2 are the only possible solutions. Furthermore, each is a solution since each value makes the left side 0.
   (b)   Factor the left side:

$$(x - 4)(x + 1) = 0.$$

The product is 0 only if $x - 4 = 0$ or $x + 1 = 0$, that is, only if $x = 4$ or $x = -1$.

Answer   (a) $0, 2$ (b) $4, -1$.

. . . .

Completing the Square

The quadratic equation

$$(x - 2)^2 = 9$$

is easy to solve because the left-hand side is a perfect square. By inspection, either $x - 2 = 3$ or $x - 2 = -3$. Hence $x = 5$ and $x = -1$ are solutions.
   Now look at the equation

$$x^2 + 6x + 4 = 0.$$

The left side is not a perfect square, but something similar is: $x^2 + 6x + 9$. So add

5 to both sides:

$$x^2 + 6x + 9 = 5,$$
$$(x + 3)^2 = 5.$$

Therefore $x + 3$ must be one of the two square roots of 5, that is, $x + 3 = \pm\sqrt{5}$. Hence

$$x = -3 + \sqrt{5} \quad \text{or} \quad x = -3 - \sqrt{5}.$$

In general, each expression $x^2 + bx$ is part of a perfect square,

$$\left(x + \frac{b}{2}\right)^2 = (x^2 + bx) + \frac{b^2}{4}.$$

Examples:

$$x^2 - 10x \quad \text{is part of} \quad (x - 5)^2 = x^2 - 10x + 25,$$
$$x^2 + 3x \quad \text{is part of} \quad (x + \tfrac{3}{2})^2 = x^2 + 3x + \tfrac{9}{4}.$$

Given $x^2 - 10x$, we complete the square by adding 25, and given $x^2 + 3x$, we complete the square by adding $\frac{9}{4}$.

By completing the square, any expression $x^2 + bx + c$ can be written in the form

$$(x + p)^2 + q.$$

Just add and subtract $b^2/4$:

$$x^2 + bx + c = x^2 + bx + \left(\frac{b}{2}\right)^2 + c - \left(\frac{b}{2}\right)^2$$

$$= \left(x + \frac{b}{2}\right)^2 + \left(c - \frac{b^2}{4}\right).$$

Examples:

$$x^2 - 10x + 3 = x^2 - 10x + 25 + (3 - 25) = (x - 5)^2 - 22,$$
$$x^2 + 3x - 4 = x^2 + 3x + \left(\frac{3}{2}\right)^2 + \left[-4 - \left(\frac{3}{2}\right)^2\right] = \left(x + \frac{3}{2}\right)^2 - \frac{25}{4}.$$

■ *Example 3.2*

Solve by completing the square:

$$\text{(a)} \quad x^2 - 10x + 17 = 0, \quad \text{(b)} \quad x^2 - 2x + 8 = 0.$$

SOLUTION   (a)   Write

$$x^2 - 10x + 17 = x^2 - 10x + 25 + 17 - 25 = (x - 5)^2 - 8.$$

The equation is equivalent to

$$(x - 5)^2 - 8 = 0, \quad (x - 5)^2 = 8.$$

Hence $x - 5 = \sqrt{8}$ or $x - 5 = -\sqrt{8}$, so $x = 5 + \sqrt{8}$ or $x = 5 - \sqrt{8}$.

(b) Complete the square by adding 1 to $x^2 - 2x$:

$$x^2 - 2x + 8 = 0, \qquad x^2 - 2x + 1 + 7 = 0, \qquad (x - 1)^2 + 7 = 0,$$

so

$$(x - 1)^2 = -7.$$

This is impossible since the square of a real number cannot be negative. Therefore there are no solutions.

Answer   (a) $5 \pm 2\sqrt{2}$ (b) no solutions.

The Quadratic Formula

The method of completing the square will work for every quadratic equation. But instead of completing the square each time, let us do it once and for all. We take the most general quadratic equation:

$$ax^2 + bx + c = 0 \qquad (a \neq 0).$$

First we divide by a:

$$x^2 + \frac{b}{a}x + \frac{c}{a} = 0.$$

Next we complete the square, adding $(b/2a)^2$ to each side of the equation:

$$x^2 + \frac{b}{a}x + \left(\frac{b}{2a}\right)^2 + \frac{c}{a} = \left(\frac{b}{2a}\right)^2.$$

We subtract c/a from both sides, then put the right-hand side over a common denominator:

$$\left(x + \frac{b}{2a}\right)^2 = \left(\frac{b}{2a}\right)^2 - \frac{c}{a}$$

$$= \frac{b^2}{4a^2} - \frac{c}{a} = \frac{b^2 - 4ac}{4a^2},$$

$$\left(x + \frac{b}{2a}\right)^2 = \frac{b^2 - 4ac}{4a^2}.$$

Let us rewrite this result in the form

$$\left(x + \frac{b}{2a}\right)^2 = \frac{D}{4a^2}, \qquad \text{where} \quad D = b^2 - 4ac.$$

The sign of D is crucial. If $D < 0$, then the right side is negative, it has no real square roots, and *there are no solutions*. If $D = 0$, the equation reduces to

$$\left(x + \frac{b}{2a}\right)^2 = 0 \qquad \text{or} \qquad x + \frac{b}{2a} = 0$$

so *there is exactly one solution,* $x = -b/2a$.

If $D > 0$, then $D/4a^2 > 0$ and there are two distinct square roots,

$$x + \frac{b}{2a} = \frac{\sqrt{b^2 - 4ac}}{2a} \quad \text{or} \quad x + \frac{b}{2a} = -\frac{\sqrt{b^2 - 4ac}}{2a},$$

so *there are two solutions:*

$$x = \frac{-b + \sqrt{b^2 - 4ac}}{2a} \quad \text{and} \quad x = \frac{-b - \sqrt{b^2 - 4ac}}{2a}.$$

We abbreviate the two possibilities by use of the plus-minus sign:

$$x = \frac{-b \pm \sqrt{b^2 - 4ac}}{2a}.$$

The Quadratic Formula

$ax^2 + bx + c = 0, \qquad a \neq 0, \qquad D = b^2 - 4ac:$

(1) $D < 0$ No real solutions.

(2) $D = 0$ One real solution, $x = \dfrac{-b}{2a}.$

(3) $D > 0$ Two real solutions,

$$x = \frac{-b + \sqrt{D}}{2a} \quad \text{and} \quad x = \frac{-b - \sqrt{D}}{2a}.$$

The quantity D is called the **discriminant** of the quadratic polynomial. The sign of D gives instant information about the solution of $ax^2 + bx + c = 0$.

Suggestion: You will probably memorize the quadratic formula,

$$x = \frac{-b \pm \sqrt{b^2 - 4ac}}{2a}.$$

However, there is some danger of becoming too dependent on the letters a, b, c. For instance, could you solve $bx^2 + cx + a = 0$ with the formula?

To avoid trouble, identify the roles of a, b, c; they are the coefficient of x^2, the coefficient of x, and the constant term, or if you prefer, the first, middle, and last coefficients. Then you can remember the quadratic formula as

$$x = \frac{-(\text{middle}) \pm \sqrt{(\text{middle})^2 - 4(\text{first})(\text{last})}}{2(\text{first})}.$$

■ *Example 3.3*

Solve for x by the quadratic formula:

(a) $3x^2 - 10x + 4 = 0,$ (b) $x^2 y^3 + 5xy - 50 = 0.$

SOLUTION (a) Use the quadratic formula with $a = 3$, $b = -10$, and $c = 4$:

$$x = \frac{-(-10) \pm \sqrt{(-10)^2 - 4 \cdot 3 \cdot 4}}{2 \cdot 3} = \frac{10 \pm \sqrt{52}}{2 \cdot 3} = \frac{5 \pm \sqrt{13}}{3}.$$

(b) Treat this as a quadratic equation in x:

$$(y^3)x^2 + (5y)x - 50 = 0.$$

By the quadratic formula with $a = y^3$, $b = 5y$, and $c = -50$,

$$x = \frac{-5y \pm \sqrt{(5y)^2 - 4y^3(-50)}}{2y^3} = \frac{-5y \pm \sqrt{25y^2 + 25(8y^3)}}{2y^3}$$

$$= \frac{-5y \pm 5y\sqrt{1 + 8y}}{2y^3} = \frac{-5 \pm 5\sqrt{1 + 8y}}{2y^2}.$$

Answer (a) $\dfrac{5 \pm \sqrt{13}}{3}$ (b) $\dfrac{-5 \pm 5\sqrt{1 + 8y}}{2y^2}.$

EXERCISES

Solve by factoring:

1. $3x^2 - 4x = 0$
2. $x(x + 1) + 2x = 0$
3. $x^2 - 5x - 14 = 0$
4. $x^2 + 7x + 6 = 0$
5. $6x^2 - 5x + 1 = 0$
6. $4x^2 - 12x + 9 = 0$
7. $5x^2 + 17x - 12 = 0$
8. $12x^2 + 17x + 6 = 0.$

Express each polynomial in the form $a(x + p)^2 + q$:

9. $x^2 - 8x$
10. $x^2 + 4x$
11. $x^2 + 6x + 1$
12. $x^2 - 3x - 6$
13. $x^2 + 2kx$
14. $x^2 + 2bx + c$
15. $4x^2 + 12x$
16. $4x^2 - 10x$
17. $2x^2 - \frac{1}{2}x$
18. $2x^2 + 3x$
19. $-x^2 + 4x$
20. $-3x^2 + 4x$
21. $3x^2 - 5x + 1$
22. $-5x^2 + 2x$
23. $-3x^2 + x + 5$
24. $-6x^2 - 5x + 1.$

Solve by any method:

25. $x^2 + 4x + 2 = 0$
26. $x^2 + 2x - 5 = 0$
27. $2x^2 + 4x + 1 = 0$
28. $9x^2 - 6x + 1 = 0$
29. $x^2 + 6x + 11 = 0$
30. $3x^2 + x + 4 = 0$
31. $6(x^2 + x) + 7x = 3x - 1$
32. $3x^2 + x = 5x + 1$
33. $(2x + 3)^2 = x - 8$
34. $(x + 1)(2x - 3) = x^2 + x + 7$
35. $\frac{1}{2}x^2 - 4x - \frac{9}{2} = 0$
36. $x + 3 = -\frac{1}{2}x^2$
37. $x + \dfrac{1}{x} = 2$
38. $\dfrac{x}{x^2 + 1} = \dfrac{1}{3}$
39. $\dfrac{1}{x + 1} + \dfrac{3}{x - 1} = 5$
40. $\dfrac{x - 2}{3x - 8} = \dfrac{x + 5}{5x - 2}.$

Without solving the equation find how many real solutions it has:

41. $9x^2 - 13x + 5 = 0$
42. $x^2 + 14x + 31 = 0$

43. $ax^2 - 2x + \dfrac{1}{a} = 0$ **44.** $2x^2 - 8x - 3 = 0$

45. $41x^2 + 10,000x + 57 = 0$ **46.** $3x^2 + 13x + 10,000 = 0.$

Solve for x:

47. $x^2 + 6xy^2 = y$ **48.** $x(x + y + 2) = y(2x + 1)$

49. $\dfrac{x + y}{x - y} = \dfrac{x}{3x + y}$ **50.** $\dfrac{x}{y} + \dfrac{y}{x} = 2y$

51. $6x^2 - 5xy + y^2 = 0$ **52.** $x^2 + 2xy - 5y^2 = 0.$

4. OTHER TYPES OF EQUATIONS

Solution by Factoring

Solving a polynomial equation

$$a_n x^n + a_{n-1} x^{n-1} + \cdots + a_1 x + a_0 = 0$$

is usually difficult when its degree n is greater than 2. However, there are certain favorable circumstances when the equation can be solved, for instance, when you are lucky or skillful enough to factor the left-hand side.

▣ *Example 4.1*

Solve $x^5 - 16x = 0.$

SOLUTION Factor the left side:

$$x^5 - 16x = x(x^4 - 16) = x(x^2 + 4)(x^2 - 4)$$
$$= x(x^2 + 4)(x + 2)(x - 2) = 0.$$

A product is 0 precisely when one of its factors is 0. The factor $x^2 + 4$ is never 0, but each of the other three factors can be 0: for the values 0, -2, and 2, respectively.

    Answer: 0, -2, 2.

▣ *Example 4.2*

Solve $(x^3 + 1)(2x^2 + 7x + 1) = 0.$

SOLUTION The left side is already a product; it can be 0 only if one of the factors is 0. Therefore, solving the given equation amounts to solving both

$$x^3 + 1 = 0 \quad \text{and} \quad 2x^2 + 7x + 1 = 0.$$

Solve the first equation by factoring:

$$x^3 + 1 = (x + 1)(x^2 - x + 1) = 0.$$

Set both factors equal to 0. The first yields $x = -1$; the second yields the equation

$$x^2 - x + 1 = 0.$$

This quadratic has no real solution because its discriminant is negative:

$$b^2 - 4ac = 1 - 4 = -3 < 0.$$

Solve the second equation, $2x^2 + 7x + 1 = 0$, by the quadratic formula:

$$x = \frac{-7 \pm \sqrt{49 - 8}}{4} = \frac{-7 \pm \sqrt{41}}{4}.$$

Answer: $-1, \frac{1}{4}(-7 \pm \sqrt{41})$.

. . . .   ■

Equations of Quadratic Type

An equation of the form

$$ax^4 + bx^2 + c = 0$$

is really a disguised quadratic since it can be expressed as

$$a(x^2)^2 + b(x^2) + c = 0.$$

We can find x^2 by the quadratic formula, then find x. Similarly the equations

$$ax^6 + bx^3 + c = 0 \quad \text{and} \quad d(x^2 - 3x)^2 + e(x^2 - 3x) + f = 0$$

are quadratic relative to the quantities x^3 and $(x^2 - 3x)$.

■ *Example 4.3*

Solve $x^4 - 2x^2 - 2 = 0$.

SOLUTION   Write the equation as

$$(x^2)^2 - 2(x^2) - 2 = 0.$$

Solve for x^2 by the quadratic formula

$$x^2 = \frac{2 \pm \sqrt{4 + 8}}{2} = \frac{2 \pm \sqrt{12}}{2} = \frac{2 \pm 2\sqrt{3}}{2} = 1 \pm \sqrt{3}.$$

Hence

$$x^2 = 1 + \sqrt{3} \quad \text{or} \quad x^2 = 1 - \sqrt{3}.$$

The first equation above has the solutions

$$x = \pm \sqrt{1 + \sqrt{3}}.$$

The second equation has no solutions since $1 - \sqrt{3} < 0$.

Answer: $\pm \sqrt{1 + \sqrt{3}}$.

. . . .   ■

■ *Example 4.4*

Solve $x - 10\sqrt{x} + 23 = 0$, where $\sqrt{x}$ denotes the non-negative square root of x.

SOLUTION The equation is quadratic in $\sqrt{x}$:

$$(\sqrt{x})^2 - 10(\sqrt{x}) + 23 = 0.$$

From the quadratic formula,

$$\sqrt{x} = 5 + \sqrt{2} \quad \text{or} \quad \sqrt{x} = 5 - \sqrt{2}.$$

Both $5 + \sqrt{2}$ and $5 - \sqrt{2}$ are positive, hence acceptable as the square root of x. Square:

$$x = (5 + \sqrt{2})^2 = 27 + 10\sqrt{2} \quad \text{or} \quad x = (5 - \sqrt{2})^2 = 27 - 10\sqrt{2}.$$

Check: If $x = 27 \pm 10\sqrt{2}$, then

$$x - 10\sqrt{x} + 23 = (27 \pm 10\sqrt{2}) - 10(5 \pm \sqrt{2}) + 23$$
$$= (27 - 50 + 23) \pm 10\sqrt{2} \mp 10\sqrt{2} = 0.$$

Answer: $27 \pm 10\sqrt{2}$.

Remark: When the dual signs, $\pm$ and $\mp$ appear in an expression, you either take both top signs or both bottom signs. For example,

$$(\pm a) + (\pm b) = a + b \quad \text{or} \quad -a - b,$$
$$(\pm a) + (\mp b) = a - b \quad \text{or} \quad -a + b,$$
$$(\pm a)(\pm b) = ab.$$

Simultaneous Quadratic and Linear Equations

■ *Example 4.5*

Solve the system $\begin{cases} x^2 + y^2 = 1 \\ x + 2y = 1. \end{cases}$

SOLUTION   The simplest method is to solve the linear equation for one unknown in terms of the other, substitute into the quadratic equation, and then solve the resulting quadratic in one unknown. The steps are first

$$x + 2y = 1, \quad x = 1 - 2y.$$

Second,

$$x^2 + y^2 = 1, \quad (1 - 2y)^2 + y^2 = 1,$$
$$1 - 4y + 4y^2 + y^2 = 1, \quad 5y^2 - 4y = 0.$$

The left side factors: $y(5y - 4) = 0$, so the solutions are $y = 0$, $y = \frac{4}{5}$. The corresponding values of x are

$$x = 1 - 2y = 1 - 2 \cdot 0 = 1; \qquad x = 1 - 2y = 1 - 2 \cdot \tfrac{4}{5} = -\tfrac{3}{5}.$$

Answer $x = 1, y = 0; \quad x = -\frac{3}{5}, y = \frac{4}{5}$.

EXERCISES

Solve:

1. $x^4 + 5x^3 + 6x^2 = 0$
2. $x^6 = x^3$
3. $(2x + 1)^3 = (2x + 1)(x^2 + 9x + 3)$
4. $x^2(x - 1) = 4(x - 1)$
5. $(x^2 - 4x + 4)(x^2 + 3x + 1) = 0$
6. $(x - 3)(2x - 3)(5x + 3) = 0$
7. $(x^2 - 9)(4x^2 - 1) = 0$
8. $(x^2 - 16)(25x^2 + 9) = 0$
9. $x^7 + 5x^6 + x + 5 = 0$
10. $x^4 - (2x - 1)^2 = 0$
11. $x^8 - 1 = 0$
12. $x^8 - x^5 = 0$
13. $x^4 + 7x^2 + 6 = 0$
14. $x^4 - 5x^2 - 14 = 0$
15. $10x^4 - 7x^2 + 1 = 0$
16. $\frac{1}{2}x^4 = 3x^2 - 1$
17. $x - 5\sqrt{x} + 6 = 0$
18. $x - 6\sqrt{x} + 9 = 0$
19. $(2x - 5)(x^6 + x^4 - 7x^2) = 0$
20. $x^8 - 15x^4 - 16 = 0$

21. $\dfrac{1}{x^4} = \dfrac{6}{x^2} - 8$

22. $\left(\dfrac{x}{4x + 1}\right)^2 = \dfrac{x}{4x + 1} + 2$

23. $\dfrac{x^2 + 2}{x^2 + 5} = \dfrac{1}{x^2 - 2}$

24. $1 - x^2 = 1 + \dfrac{1}{x^2}$.

Solve the system:

25. $\begin{cases} x + y = 16 \\ \quad xy = 63 \end{cases}$

26. $\begin{cases} x^2 - y^2 = 25 \\ -x - 2y = 11 \end{cases}$

27. $\begin{cases} x^2 + xy + y^2 = 3 \\ \quad x + y = 1 \end{cases}$

28. $\begin{cases} x^2 + 2xy + y^2 = 9 \\ \quad x - y = 1 \end{cases}$

29. $\begin{cases} 2x^2 + xy = 1 \\ 3x - \frac{1}{2}y = 1 \end{cases}$

30. $\begin{cases} \frac{1}{9}x^2 + \frac{1}{4}y^2 = 1 \\ -2x + y = 6 \end{cases}$

31. $\begin{cases} \frac{1}{9}x^2 - \frac{1}{25}y^2 = 1 \\ \frac{1}{3}x + \frac{2}{5}y = 1 \end{cases}$

32. $\begin{cases} x^4 + y^4 = 1 \\ x^2 + 2y^2 = 1. \end{cases}$

5. APPLICATIONS

Algebra can be used to solve a variety of practical problems from the natural sciences, the social sciences, technology, etc. The solution of a problem involves two steps.

(1) Set up. Identifying precisely what is known (given) and what is unknown (to be found). Assigning letters to various quantities in the statement of the problem and translating the data into an equation or system of equations.

(2) Solution. Solving the equation(s).

The first step is often the harder one. It requires practice to translate data into equations, to "say it in math". Here are two examples of step (1) only.

■ *Example 5.1*

Find two numbers whose product is 5 and the sum of whose cubes is 30. (Set up the problem only.)

SOLUTION Two numbers are unknown. Call one of them x, the other y. "Whose product is 5" translates to $xy = 5$, and "the sum of whose cubes is 30" translates to $x^3 + y^3 = 30$. Thus the algebra problem is: Solve the system

$$\begin{cases} xy = 5 \\ x^3 + y^3 = 30. \end{cases}$$

(The solution happens to be $\sqrt[3]{5}$, $\sqrt[3]{25}$, but that is another story.)

■ *Example 5.2*

A tinsmith wants to cut a square from each corner of a 3-ft by 6-ft rectangular sheet and fold up the edges to make an open rectangular box of volume 4.5 ft³. Find the dimensions of the box. (Set up the problem only.)

SOLUTION A drawing helps enormously. First we show the rectangle with the squares cut out (Fig. 5.1a), then the folded up box (Fig. 5.1b).

We choose x for the side of the square and show all dimensions accordingly. The sides of the box are x, $3 - 2x$, and $6 - 2x$, and their product is the given volume 4.5. Hence the algebra problem is: Solve the equation

$$x(3 - 2x)(6 - 2x) = 4.5.$$

(An approximation, good enough for cutting sheet metal, is $x \approx 0.9076$ ft $\approx 10\frac{7}{8}$ in., $3 - 2x \approx 14\frac{3}{16}$ in., $6 - 2x \approx 50\frac{3}{16}$ in.)

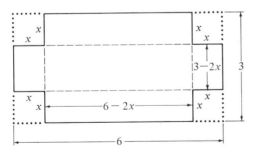

Fig. 5.1   (a) sheet with corners cut out   (b) open rectangular box

Suggestion: Take your time in setting up a problem. Read it several times to be sure you know what is given and what is to be found. Look for an equation disguised in the statement of the problem. Such expressions as "the same as" or "as much as" indicate an equation lurking somewhere. If you get stuck, check to see that you have used *all* of the data.

Rate Problems

If a factory produces goods at a rate of R units per hour, then the total amount it produces in t hours is $A = Rt$. This simple idea applies to a variety of situations. For example, $D = Rt$ (distance D in time t at speed R), $W = rt$ (work done in time t at rate r), $I = Prt$ (total interest earned by principal amount P in time t at interest rate r), etc.

■ *Example 5.3*

Suppose A can do a certain job in 8 hours and B can do the same job in 5 hours. How long will it take them working together to do the job?

SOLUTION We call the rate at which A works r_A and the rate at which B works r_B. Then in time t, worker A does

$$W_A = r_A t$$

work. We are given $W_A = 1$ job when $t = 8$ hr, hence

$$1 = r_A \cdot 8, \qquad r_A = \tfrac{1}{8}.$$

Similarly $r_B = \tfrac{1}{5}$, so

$$W_A = \tfrac{1}{8}t \qquad \text{and} \qquad W_B = \tfrac{1}{5}t.$$

Working together, in time t they accomplish

$$W = W_A + W_B = \tfrac{1}{8}t + \tfrac{1}{5}t = \tfrac{13}{40}t.$$

Consequently $W = 1$ job for

$$1 = \tfrac{13}{40}t, \qquad t = \tfrac{40}{13} = 3\tfrac{1}{13}.$$

Answer $3\tfrac{1}{13}$ hr.

■ *Example 5.4*

I borrow \$200 at 6% simple interest and another \$300 at 7%. What is the effective interest rate on my \$500 loan?

SOLUTION If I borrow the principal amount P at the simple interest rate r, then the amount of interest I due after time t is

$$I = Prt.$$

The interest due on the first loan is

$$I_1 = (200)(0.06)t = 12t,$$

and that due on the second is

$$I_2 = (300)(0.07)t = 21t.$$

The total interest due is

$$I = I_1 + I_2 = 12t + 21t = 33t.$$

But $I = 500rt$, hence

$$500r = 33, \qquad r = \tfrac{33}{500} = 0.066.$$

Answer   6.6%.

. . . .

Certain practical problems involve mixtures of various substances. Usually solving them requires more common sense than algebra. A typical example: Suppose 50 grams (g) of a 12% salt solution is mixed with 30 g of an 18% salt solution. What is the strength of the resulting mixture?

The mixture contains $50 + 30 = 80$ g total solution of which

$$(0.12)(50) + (0.18)(30) = 6.0 + 5.4 = 11.4 \text{ g}$$

is salt. Consequently the proportion of salt in the mixture is

$$\frac{11.4}{80} = 0.1425,$$

so the answer is 14.25%.

Example 5.5

How many quarts of 75% alcohol must be mixed with 15 quarts of 45% alcohol to produce 70% alcohol.

SOLUTION   Let x denote the number of quarts of 75% alcohol. Then the mixture contains $15 + x$ quarts of solution including $0.75x + (0.45)(15)$ quarts of pure alcohol. We want this number to be 70% of $15 + x$, that is

$$0.75x + (0.45)(15) = (0.70)(15 + x).$$

Solve:

$$0.05x = (0.70)(15) - (0.45)(15) = (0.25)(15),$$

$$x = (0.25)(15)/(0.05) = 75.$$

Answer   75 qts.

. . . .

Example 5.6

I want to invest $5000, part in a high risk mining stock that grows at rate 8% annually and part in a relatively safe steel stock that grows at 3% annually. How much of each should I buy to achieve a 5% growth rate?

SOLUTION Let x be the amount in the mining stock and y the amount in the steel stock. Then $x + y = 5000$.

Investment x grows by $0.08x$ and investment y by $0.03y$ annually, and these must total $(0.05)(5000)$. Thus we have the system

$$\begin{cases} x + y = 5000 \\ 0.08x + 0.03y = (0.05)(5000). \end{cases}$$

Multiply the first equation by -0.03 and add to the second. The result is

$$0.05x = (0.02)(5000), \qquad x = 2000.$$

Answer $2000 in mining, $3000 in steel.

. . . .

The economics of supply and demand provides some interesting examples. Generally, demand for a product is high when its price is low, and decreases as the price increases. Generally, the supply of a product increases as the price increases. The **equilibrium price** is the price at which supply equals demand.

■ *Example 5.7*

For a certain product, the quantity demanded (Q_d units) is related to the unit price (P dollars) by

$$Q_d = 10^5(100 - P).$$

The quantity supplied (Q_s units) is related to the price by

$$Q_s = 10^5(P - 10).$$

Find the equilibrium price and the quantity sold at that price.

SOLUTION We must find the value of P for which $Q_s = Q_d$ and the value $Q = Q_s = Q_d$. Therefore we must solve the system

$$\begin{cases} Q = 10^5(100 - P) \\ Q = 10^5(P - 10) \end{cases} \quad \text{that is,} \quad \begin{cases} Q + 10^5P = 10^7 \\ Q - 10^5P = -10^6. \end{cases}$$

Add these equations, then subtract them:

$$2Q = 10^7 - 10^6 \qquad 2 \times 10^5P = 10^7 + 10^6$$
$$2Q = 9 \times 10^6 \qquad 2P = 110$$
$$Q = 4.5 \times 10^6 \qquad P = 55.$$

Answer $P = \$55$, $Q = 4\frac{1}{2}$ million units.

Remark: Note the meaning of the given equations. At price 0, the demand is for 10^7 units, at price $10 the supply is 0. At price $100, the demand is 0 and the supply is 9 million units.

. . . .

■ *Example 5.8*

Find all right triangles whose hypotenuse is two units longer than one leg and one unit longer than the other leg.

SOLUTION The sides are x, $x + 1$, and hypotenuse $x + 2$. By the Pythagorean Theorem,

$$(x + 2)^2 = x^2 + (x + 1)^2.$$

Hence

$$x^2 + 4x + 4 = x^2 + x^2 + 2x + 1,$$

$$x^2 - 2x - 3 = 0.$$

Factor: $(x + 1)(x - 3) = 0$. The only positive solution is $x = 3$.

Answer The sides are 3, 4, 5.

■ *Example 5.9*

Find two numbers whose sum is 19 and whose product is 100.

SOLUTION Call the numbers x and y. Then

$$\begin{cases} x + y = 19 \\ \quad xy = 100. \end{cases}$$

From the first equation $y = 19 - x$. Substitute into the second:

$$x(19 - x) = 100, \qquad x^2 - 19x + 100 = 0.$$

The discriminant is $D = 19^2 - 400 = -39 < 0$, hence there is no solution.

Answer No solution.

1. I borrow $100 from a friend, no interest, and I borrow $200 from the bank at 8% interest. What is the effective interest rate on the total?
2. I borrow $340 at 4% and $150 at $7\frac{1}{2}$%. What is the effective rate on the total?
3. I borrow $500 on my insurance policy at 4% and lend a friend $200 at 6%. What is the effective rate on the $300 I am actually holding?
4. Suppose $1\frac{1}{2}$ men can do $1\frac{1}{2}$ jobs in $1\frac{1}{2}$ days. How long does it take one man to do one job?
5. Suppose an average professor can do a certain job in 9 hours and an average student can do the same job in 6 hours. If 2 professors and 5 students work together, how long will the job take?
6. If A can plow a certain field in 5 hours, B in 4 hours, and C in 3 hours, then working together, how long will it take them to plow the field?

7. Suppose one pump can fill a certain tank in h_1 hours and a second pump can fill the tank in h_2 hours. How long will it take them together to fill 2 such tanks?

8. Suppose one pump can fill a tank in 8 hours and a second pump can empty the (full) tank in 12 hours. Suppose the tank is half full and both pumps are started. When will the tank be full?

9. If your midterm grades are 85, 62, 73, what do you need on the final to average 75?

10. (cont.) What if the final carries double weight?

11. How much pure water must be added to 100 g of 30% salt solution to produce an 18% solution?

12. How much pure salt must be added to 200 g of 5% salt solution to triple its strength?

13. How many grams of 60–40 solder (60% tin, 40% lead) must be mixed with 40–60 solder to produce 500 g of 55–45 solder?

14. How many gallons of 100 proof whiskey (50% alcohol) must be mixed with 80 proof whiskey to produce 50 gal of 95 proof whiskey?

15. Find two integers whose sum is odd and whose difference is even.

16. Find two integers whose difference is even and such that 3 times one plus 5 times the other is odd.

17. Risky stock X pays 10% dividend and safe bond B pays 2% dividend. How should I invest $10,000 to have an income of $450?

18. How should $20,000 be split between a 4% savings account and a 7% investment fund to yield a 6% return?

19. Suppose demand and supply in units and price in dollars of a certain toy are related by

$$Q_d = 2 \times 10^6(1 - \tfrac{1}{15}P), \qquad Q_s = 5 \times 10^4(P - 5).$$

Find the equilibrium price and quantity sold.

20. Suppose demand and supply per week and price per egg in cents are related by

$$\frac{Q_d}{10^9} + \frac{P}{10} = 1, \qquad Q_s = \tfrac{1}{3} \times 10^9(P - 4).$$

Find the equilibrium price of eggs and production per week.

21. Team A has 11 wins and 9 losses. Team B has 20 wins and 8 losses. How many consecutive games must A win from B so the teams will have the same winning percentages?

22. A falling body falls $16t^2$ ft from rest in t sec. Object A is dropped from a window 400 ft up and one second later object B is dropped from a window 300 ft up. When does A "catch up" to B?

23. Demand, supply, and price in dollars of a certain sports car are related by

$$Q_s = 2(P - 16000), \qquad Q_d P = 5.25 \times 10^7.$$

Find the equilibrium price and quantity produced at this price.

24. Express x and y in terms of t, assuming $x^2 + y^2 = 1$, $y = t(x + 1)$, and $x \neq -1$.

25. Find two numbers whose product is twice their sum, and such that the product of the two numbers two less than each is 4.

26*. Find two numbers such that twice their product equals three times their sum, and the sum of whose squares is 40.

6. ORDER

An important relation in the real number system is called **order.** This relation asserts that one number is less than, or greater than, another. Recall the notation

$a < b$	a is less than b
$a \le b$	a is less than or equal to b
$a > b$	a is greater than b
$a \ge b$	a is greater than or equal to b.

Picturing the real number system as a number line (Fig. 6.1) helps us to understand order.

Fig. 6.1   The number line

The numbers greater than 0 are the **positive** numbers, and the numbers less than 0 are the **negative** numbers. The relation $a < b$ means that the difference $b - a$ is positive:

$a < b$	if and only if $b - a > 0$
$a \le b$	if and only if $b - a \ge 0$.

The notation is mathematical shorthand for saying that a is to the left of b (left of or equal to b in the second case).

The set of all numbers between two fixed numbers is a *interval* **segment** on the number line. A segment may include one or both of its end points, or neither, depending on how it is specified. For example, $-2 \le x \le 1$ describes the segment of all numbers between -2 and 1, *including* the end points, while $3 < x < 7$ describes the segment *strictly* between 3 and 7, that is, *excluding* the end points (Fig. 6.2).

Fig. 6.2   Examples of segments: (a) $-2 \le x \le 1$,   (b) $3 < x < 7$

Properties of Order

Two properties are obvious from the number line. (1) If a is to the left of b and b is to the left of c, then a is to the left of c. In other words, $a < b$ and $b < c$ imply $a < c$. (2) If a is to the left of b and both points are shifted c units to the left or right, the shifted points are in the same order. In other words, if $a < b$ and c is any real number, then $a + c < b + c$. These two properties can be verified without geometry:

(1)   Suppose $a < b$ and $b < c$. This means $b - a > 0$ and $c - b > 0$. But the sum of positive numbers is positive, hence

$$c - a = (c - b) + (b - a) > 0.$$

Therefore $a < c$.

(2)   If c is any real number, then $a \pm c < b \pm c$ because

$$(b \pm c) - (a \pm c) = b - a > 0.$$

The second property says that you can add (or subtract) the same number to both sides of an inequality and get a valid inequality:

$$\begin{array}{c} a < b \\ c = c \\ \hline a \pm c < b \pm c. \end{array}$$

Question: Can you multiply both sides of an inequality by the same number? Answer: Yes, provided the number is positive. If it is negative, you reverse the inequality. For example, suppose we start with $2 < 3$. Then

$$\tfrac{1}{17} \cdot 2 < \tfrac{1}{17} \cdot 3, \qquad 5 \cdot 2 < 5 \cdot 3, \qquad (8.07)2 < (8.07)3, \qquad \text{etc.,}$$

but

$$(-\tfrac{1}{17})2 > (-\tfrac{1}{17})3, \qquad (-5)2 > (-5)3, \qquad (-8.07)2 > (-8.07)3, \qquad \text{etc.}$$

In general, we have properties (3) if $a < b$ and $c > 0$, then $ac < bc$ and (4) if $a < b$ and $c < 0$, then $ac > bc$. The proofs depend on the fact that the product of positive numbers is positive:

(3) $b - a > 0$ and $c > 0$, so $bc - ac = (b - a)c > 0$, hence $ac < bc$.

(4) $b - a > 0$ and $c < 0$, so $-c > 0$ and $ac - bc = (b - a)(-c) > 0$, that is, $bc < ac$.

The same rule holds for division of both sides of an inequality by a non-zero number c. For division by c is the same as multiplication by $1/c$, and $c > 0$ if and only if $1/c > 0$.

Finally we observe that taking reciprocals *reverses* an inequality, provided the numbers involved have the same sign. For example,

$$0 < 2 < 3 \quad \text{but} \quad \frac{1}{2} > \frac{1}{3}; \qquad -4 < -3 < 0 \quad \text{but} \quad \frac{1}{-4} > \frac{1}{-3}.$$

In general, we have property (5)   if $0 < a < b$ or $a < b < 0$, then $1/b < 1/a$. The proof is left as an exercise.

Rules for Order

(1)   If $a < b$ and $b < c$,   then $a < c$.

(2)   If $a < b$ and if c is any real number, then $a \pm c < b \pm c$.

(3)   If $a < b$ and $c > 0$,   then $ac < bc$.

(4)   If $a < b$ and $c < 0$,   then $bc < ac$.

(5)   If $0 < a < b$ or $a < b < 0$,   then $\dfrac{1}{b} < \dfrac{1}{a}$.

These rules apply just as well with $\le$ in place of $<$, except in (5) you must have

$$0 < a \le b \quad \text{or} \quad a \le b < 0.$$

■ *Example 6.1*

Given $1.4 < \sqrt{2} < 1.5$, show that

(a) $2 + \sqrt{2} < 3.5$, (b) $8.4 < 6\sqrt{2} < 9$, (c) $\dfrac{1}{3 - 2\sqrt{2}} > 5$.

SOLUTION (a) Start with $\sqrt{2} < 1.5$. By Rule (2), you may add 2 to both sides:

$$2 + \sqrt{2} < 2 + 1.5 = 3.5.$$

(b) Start with the two inequalities

$$1.4 < \sqrt{2}, \qquad \sqrt{2} < 1.5.$$

By Rule (3), you may multiply both sides of each by 6:

$$8.4 = 6(1.4) < 6\sqrt{2}, \qquad 6\sqrt{2} < 6(1.5) = 9.$$

Combine: $8.4 < 6\sqrt{2} < 9$.

(c) One way to show that a number is big is to show that its reciprocal is small. By Rule (3), $2.8 < 2\sqrt{2} < 3$, and it follows by Rule (4) that $-3 < -2\sqrt{2} < -2.8$. Add 3:

$$0 < 3 - 2\sqrt{2} < 0.2$$

by Rule (2). Finally, by Rule (5),

$$\frac{1}{3 - 2\sqrt{2}} > \frac{1}{0.2} = 5.$$

. . . .

Several natural extensions of the rules of order are convenient in applications.

> (6) If $a < b$, then $-b < -a$.
>
> (7) If $a < A$ and $b < B$, then $a + b < A + B$.
>
> (8) If $0 < a < A$ and $0 < b < B$, then $ab < AB$.

The proofs are left as exercises. Note that Rules (7) and (8) are easily extended to three or more inequalities. For instance, if $a < A$, $b < B$, and $c < C$, then

$$a + b + c < A + B + C.$$

Estimates

The rules of order are useful in making estimates with a minimum of computation.

■ *Example 6.2*

Verify the estimate, using as little arithmetic as possible:

(a) $6 \cdot 8 \cdot 9 \cdot 11 < 5000,$

(b) $\pi^2 < 10,$

(c) $\frac{5}{9} < \frac{1}{5} + \frac{1}{6} + \frac{1}{7} + \frac{1}{8} + \frac{1}{9} < 1.$

SOLUTION   (a)   Use Rule (8):

$$6 \cdot 8 \cdot 9 \cdot 11 = 48 \cdot 99 < 50 \cdot 100 = 5000.$$

(b)   Use the fact that $\pi < 3.1415 < 3.15 = 63/20$ and Rule (8):

$$\pi^2 = \pi \cdot \pi < \frac{63}{20} \cdot \frac{63}{20} = \frac{3969}{400} < \frac{4000}{400} = 10.$$

(c)   Each term after the first is less than $1/5$. Hence, by Rule (7),

$$\frac{1}{5} + \frac{1}{6} + \frac{1}{7} + \frac{1}{8} + \frac{1}{9} < \frac{1}{5} + \frac{1}{5} + \frac{1}{5} + \frac{1}{5} + \frac{1}{5} = 1.$$

Similarly, each term but the last is greater than $1/9$. Hence

$$\frac{1}{5} + \frac{1}{6} + \frac{1}{7} + \frac{1}{8} + \frac{1}{9} > \frac{1}{9} + \frac{1}{9} + \frac{1}{9} + \frac{1}{9} + \frac{1}{9} = \frac{5}{9}.$$

■ *Example 6.3*

Find the largest possible error in computing the product of two numbers, $a = 8.4$ and $b = 12.3$, determined by an experiment in which the last digit might be off by 1.

SOLUTION   We have

$$8.3 \leq a \leq 8.5 \qquad \text{and} \qquad 12.2 \leq b \leq 12.4.$$

By Rule (8),

$$(8.3)(12.2) \leq ab \leq (8.5)(12.4),$$

that is, $101.26 \leq ab \leq 105.40$. Since $(8.4)(12.3) = 103.32$, the larger of

$$105.40 - 103.32 = 2.08 \qquad \text{and} \qquad 103.32 - 101.26 = 2.06$$

is the largest possible error.

Answer   2.08

Abbreviate, using as few symbols as possible:

1. x is strictly between 4 and 9.
2. x is between 3 and 5, with 3 and 5 allowed.
3. x is non-negative.
4. x is positive but not more than 10.
5. x is less than y and y is at most z.
6. x is strictly between 1 and 2 and y is strictly between 2 and 3.

Indicate the portion of the number line defined by the inequality:

7. $x \leq -2$ **8.** $x > 4$ **9.** $0 < x < 1$
10. $2 \leq x < 5$ **11.** $x^2 > 9$ **12.** $x + 1 > 0$.

Verify the inequality, using as little computation as possible:

13. $5667 - 4128 < 6093 - 4128$
14. $33 \cdot 89 < 36 \cdot 89$
15. $1/\pi > 1/4$
16. $11/5 < 14/5$
17. $9 \cdot 11 \cdot 13 < 1300$
18. $97 \cdot 98 \cdot 99 < 1{,}000{,}000$
19. $\frac{1}{3}(8561 + 8774 + 8819) > 8561$
20. $100\sqrt{2} > 140$
21. $\pi^6 < 1000$
22. $6.28 < 2\pi < 6.30$
23. $9 + 99 + 999 + 9999 < 11110$
24. $\frac{1}{2} < \frac{1}{11} + \frac{1}{12} + \frac{1}{13} + \cdots + \frac{1}{20} < \frac{10}{11}$
25. $\frac{4}{3} < \frac{1}{\sqrt{5}} + \frac{1}{\sqrt{6}} + \frac{1}{\sqrt{7}} + \frac{1}{\sqrt{8}} < 2$
26. $1 < \frac{1}{4}(1 + \pi) < 1.04$.

27. If $0 < a < b$, show that $a^2 < b^2$.
28. (cont.) Show that the conclusion is not necessarily true if a and b are not both positive.
29. If a and b have opposite signs and $a < b$, find the relation between $1/a$ and $1/b$.
30. If $a < b$, prove that $c - b < c - a$.

Prove:

31. Rule (5)   **32.** Rule (6)   **33.** Rule (7)   **34.** Rule (8).
35. Suppose $1.2 < a < 1.3$ and $10.0 < b < 10.5$. Find the set of possible values of ab.
36. Suppose $-1 < a < 1$ and $-3 < b < 5$. Find the set of possible values of ab.
37. I measure five numbers between 12 and 13. Estimate their sum S.
38. If $3 \leq a \leq 4$ and $5 \leq b \leq 6$, find the set of possible values of $1/a + 1/b$.

Which is larger?

39. $(67)(69)$, 68^2
40. $(1047)(1051)$, 1049^2
41. $(121)(122)(124)(125)$, 123^4
42. $(a - 1)(a + 1)$, a^2.

Which is the best and which the worst buy:

43. toothpaste; *giant* $5\frac{3}{4}$ oz size at 59¢, *large economy* $6\frac{7}{9}$ oz size at 67¢, or *family* $7\frac{1}{3}$ oz size at 75¢?
44. corn flakes; *brand X* of 14 oz at 38¢, *brand Y* of 18 oz at 49¢, or *brand Z* of 1 lb at 44¢?
45. detergent: *mammoth* $5\frac{1}{2}$ lb size at 89¢, *super* $7\frac{3}{4}$ lb size at \$1.29, or *extra savings* $10\frac{1}{4}$ lb size at \$1.69?
46. peanut butter; *small* 4 oz jar at 23¢, *medium* 6.6 oz jar at 36¢, or *large* 12.4 oz jar at 65¢?

47. Suppose $x > 0$ and

$$y = \cfrac{1}{1 + \cfrac{1}{1 + \cfrac{1}{x}}}.$$

If x increases, does y increase or decrease?

48*. Brain teaser. Ten people sit at a round table. Each has an amount of money equal to the average of the amounts held by his two neighbors. All together they have \$100. Find all possible ways the \$100 can be distributed.

7. ABSOLUTE VALUES

Often we need a measure of the size of a number, regardless of its sign. In some sense, -7 seems larger than 2, but we cannot write $-7 > 2$. We introduce absolute values in order to express this feeling precisely. We define the **absolute value** of a real number a, written $|a|$, as follows:

$$|a| = \begin{cases} a & \text{if } a \geq 0 \\ -a & \text{if } a < 0. \end{cases}$$

For example, $|-10| = 10$, $|-5| = 5$, $|3| = 3$, $|0| = 0$.

No matter what sign a has, we can say that either $|a| = a$ or $|a| = -a$, and $|a| \geq 0$. Thus we can write

$$|a| = \pm a \qquad \text{and} \qquad |a| \geq 0,$$

and these relations completely determine $|a|$. Obviously

$$a \leq |a| \qquad \text{and} \qquad -a \leq |a|.$$

Every real number except 0 has a positive absolute value which measures its "size", or "magnitude". With absolute values, we can say correctly what we were trying to express before, $|-7| > |2|$.

Absolute values satisfy some simple rules:

Rules for Absolute Values

(1) $|a| = \pm a$, $|a| > 0$ if $a \neq 0$, $|0| = 0$

(2) $|-a| = |a|$

(3) $|ab| = |a| \cdot |b|$

(4) $\left| \dfrac{a}{b} \right| = \dfrac{|a|}{|b|}$ $(b \neq 0)$

(5) $|a + b| \leq |a| + |b|$ (the triangle inequality).

We have discussed (1), and (2) is obvious by the definition. To prove (3), note that $|ab| = \pm ab$. Also $|a| = \pm a$ and $|b| = \pm b$, so whatever the signs, $|a| \cdot |b| = \pm ab$. Hence $|ab| = |a| \cdot |b|$, each being equal to whichever of $\pm ab$ is non-negative.

Rule (4) follows from (3) applied to $b(a/b) = a$:

$$\left| b\left(\frac{a}{b}\right) \right| = |a|, \qquad |b|\left|\frac{a}{b}\right| = |a|, \qquad \left|\frac{a}{b}\right| = \frac{|a|}{|b|}.$$

Rule (5) is proved in several steps. First,

$$a \leq |a| \quad \text{and} \quad b \leq |b|, \quad \text{so} \quad a + b \leq |a| + |b|.$$

Likewise

$$-a \leq |a| \quad \text{and} \quad -b \leq |b|, \quad \text{so} \quad -(a + b) \leq |a| + |b|.$$

Hence $\pm(a + b) \leq |a| + |b|$. But for one of the choices of sign, $\pm(a + b) = |a + b|$. Therefore

$$|a + b| \leq |a| + |b|.$$

Examples:

Rule (3): $\quad |4 \cdot 5| = |20| = 20 = |4| \cdot |5|$

$\qquad\qquad |7 \cdot (-4)| = |-28| = 28 = |7| \cdot |-4|$

$\qquad\qquad |(-6)(-5)| = |30| = 30 = |-6| \cdot |-5|.$

Rule (5): $\quad |3 + 4| = |7| = 7 = |3| + |4|$

$\qquad\qquad |-3 + (-4)| = |-7| = 7 = |-3| + |-4|$

$\qquad\qquad |5 - 3| = |2| = 2 < 8 = |5| + |-3|$

$\qquad\qquad |-5 + 3| = |-2| = 2 < 8 = |-5| + |3|.$

Remark 1: Rules (3) and (5) extend easily to three or more numbers; for instance,

$$|abc| = |a| \cdot |b| \cdot |c|, \qquad |a + b + c| \leq |a| + |b| + |c|.$$

Remark 2: Rule (5) has two useful consequences:

$$\boxed{\begin{array}{ll} (6) & |a - b| \leq |a| + |b| \\ (7) & |a| - |b| \leq |a - b|. \end{array}}$$

To prove (6), replace b by $-b$ in (5). To prove (7), apply (5) to the sum $(a - b) + b = a$:

$$|a| = |(a - b) + b| \leq |a - b| + |b|, \qquad \text{hence} \quad |a| - |b| \leq |a - b|.$$

■ ***Example 7.1***

Compute

$$\text{(a)} \quad |7 - 30|, \qquad \text{(b)} \quad |1 - \sqrt{2}|, \qquad \text{(c)} \quad |(-2)^5|.$$

SOLUTION   (a) $|7 - 30| = |-23| = 23.$

(b) $\sqrt{2} > 1.4$ so $1 - \sqrt{2} < 0$. Therefore

$$|1 - \sqrt{2}| = -(1 - \sqrt{2}) = \sqrt{2} - 1.$$

(c) $|(-2)^5| = |-32| = 32.$ Alternatively, $|(-2)^5| = |-2|^5 = 2^5 = 32.$

Answer   (a)   23   (b) $\sqrt{2} - 1$ (c)   32.

· ·

■ *Example 7.2*

Find all numbers x that satisfy

(a) $|x - 4| = 1,$ (b) $|x| \leq 10,$ (c) $|x - 1| < 5.$

SOLUTION   (a)   If $|x - 4| = 1$, then either $x - 4 = 1$ or $x - 4 = -1$. Hence $x = 5$ or $x = 3$.

(b)   If $x \geq 0$, then $|x| = x$, and $|x| \leq 10$ means $0 \leq x \leq 10$. If $x < 0$, then $|x| = -x$, and $|x| \leq 10$ means $0 \leq -x \leq 10$, that is, $-10 \leq x \leq 0$. In all cases then, $-10 \leq x \leq 10$.

(c)   By the same reasoning as in (b), we see that $|x - 1| < 5$ means $-5 < x - 1 < 5$. Adding 1 to these inequalities, we obtain $-4 < x < 6$.

Answer   (a) $x = 3, 5$ (b) $-10 \leq x \leq 10$

(c) $-4 < x < 6.$

· ·

Geometric Interpretation

Absolute values have an important geometric meaning on the number line. By its very definition, $|a|$ is the distance of the point a from the point 0. Thus the "size" of a is measured by its *distance* from 0, a positive quantity no matter whether a is right or left of 0.

Given distinct points a and b on the number line, the distance between them is $|a - b|$. See Fig. 7.1. The distance between 5 and 9 is $|5 - 9| = |-4| = 4$. The distance between -2 and 3 is $|-2 - 3| = |-5| = 5$. Since $|a - b| = |b - a|$, it doesn't matter which you subtract from which; the answer comes out the same.

Fig. 7.1   Length of a segment

Using absolute values and inequalities, we can often express geometric facts algebraically, or interpret algebraic relations geometrically. Here are a few examples:

ARITHMETIC STATEMENT	GEOMETRIC STATEMENT
a is positive.	The point a lies to the right of the point 0.
$a > b$.	a lies to the right of b.
$a - b = c > 0$.	a lies c units to the right of b.
$a < b < c$.	b lies between a and c.
$|3 - a| < \frac{1}{2}$.	The point a is within $\frac{1}{2}$ unit of the point 3.
$|a| < |b|$.	The point a is closer to 0 than the point b is.

This close relationship between algebra and geometry is extremely important. It can provide two ways of looking at a problem, hence can increase the chances of solving it. Sometimes we are able to attack geometric problems algebraically, sometimes vice versa.

Example 7.3

Express in algebraic notation using absolute values:

(a) The distance from x to -3 is 4.

(b) x is closer to 4 than y is to -6.

(c) x is at most 2 units from 5.

SOLUTION (a) $|x - (-3)| = 4$, that is, $|x + 3| = 4$.

(b) $|x - 4| < |y - (-6)|$, that is, $|x - 4| < |y + 6|$.

(c) The distance from x to 5 is less than or equal to 2, hence $|x - 5| \leq 2$.

Answer (a) $|x + 3| = 4$ (b) $|x - 4| < |y + 6|$
(c) $|x - 5| \leq 2$.

Example 7.4

Find all points on the number line that are twice as far from 0 as from 6.

SOLUTION The distance from x to 0 is $|x|$ and the distance from x to 6 is $|x - 6|$. We must find points x for which

$$|x| = 2|x - 6|.$$

Since $|x| = \pm x$ and $|x - 6| = \pm(x - 6)$, there are two possibilities, the same signs or opposite signs:

$$x = 2(x - 6) \qquad\qquad x = -2(x - 6)$$
$$x = 2x - 12 \qquad\qquad x = -2x + 12$$
$$x = 12 \qquad\qquad 3x = 12, \quad x = 4.$$

Answer 4 and 12.

EXERCISES

Find a simpler expression, not involving absolute values:

1. $|-4|$ **2.** $-|-6|$ **3.** $|4 - 11|$

4. $|8 - 3|$ **5.** $|-2| + |-3|$ **6.** $|(-2)(-3)|$

7. $|-5| \cdot |-6|$ **8.** $|-5|/5$ **9.** $|-4/9|$

10. $||-2| - |-3||$ **11.** $|-6|^2$ **12.** $|(-6)^2|$.

Express without absolute values all numbers x that satisfy the condition:

13. $|x| = 0$ **14.** $|x| = -1$ **15.** $|3x| = 9$

16. $|\frac{1}{2}x| = 3$ **17.** $|x - 4| = 1$ **18.** $|x + 4| = 1$

19. $|x| \leq \frac{1}{2}$ **20.** $|x| > \frac{2}{3}$ **21.** $|2x| < 8$

22. $|-4x| > 6$ **23.** $|x - 1| < 3$ **24.** $|2 - x| < 5$.

Express in algebraic notation using absolute values:

25. x is either 2 or -2
26. x is farther from 0 than from y
27. x is closer to 1 than to 5
28. x lies either to the left of -2 or to the right of 2
29. x is within 3 units of 7
30. x is between 6 and 8.

31. Find all points that are 3 times as far from 5 as from 1.
32. Find all points that are 10 times as far from 5 as from 1.
33. Explain how you can tell at a glance that there is no x for which both $|x - 1| < 2$ and $|x - 12| < 3$.
34. If $a \leq b$ and $-a \leq b$, show that $|a| \leq b$.
35. Explain why $|a| + |b| + |c| > 0$ is algebraic shorthand for "at least one of the numbers a, b, c is different from 0".
36. (cont.) Find an analogous shorthand version of "a, b, c are all different from 0".
37. Suppose $|x - a| < 10^{-6}$. Show that $|7x - 7a| < 10^{-5}$.
38. Suppose $|x - 7| < 10^{-6}$ and $|y - 5| < 10^{-6}$. Show that $|(x + y) - 12| < 10^{-5}$. [Hint: Use the triangle inequality.]
39. Suppose $|x - 7| < \frac{1}{10}$ and $|y - 5| < \frac{1}{10}$. Prove that $|xy - 35| < 1.5$. [Hint: $xy - 35 = x(y - 7) + 7(x - 5)$.]
40. (cont.) Suppose $|x - 7| < 10^{-6}$ and $|y - 5| < 10^{-6}$. Prove that $|xy - 35| < 2 \times 10^{-5}$.
41. Suppose $|x - 3| < 10^{-6}$. Prove that $|x^2 - 9| < 10^{-5}$. [Hint: Factor $x^2 - 9$.]
42*. (cont.) Prove $|x^3 - 27| < 5 \times 10^{-5}$.

Let us look at inequalities involving a variable. Statements such as $x^2 + 1 > 0$, which are true for all real numbers x, are called **absolute inequalities.** They correspond to identities in the case of equations. Statements such as $x + 1 > 10$, which are not true for all real numbers x, are called **conditional inequalities.** They correspond to conditional equations.

Solving a conditional inequality means finding all real numbers for which the statement is true. For example, the solution of $x + 1 > 10$ is the set of all real numbers greater than 9.

Some techniques for solving conditional inequalities are similar to techniques for solving equations. For example, by (2) on p. 74, you may add or subtract the same quantity from both sides of an inequality. By (3), you may multiply both sides by the same *positive* quantity. If you multiply by a negative quantity, you must *reverse* the inequality.

■ *Example 8.1*

Solve $3x - 7 > x + 1$.

SOLUTION   Add 7 to both sides:

$$3x > x + 8.$$

Subtract x from both sides:

$$2x > 8.$$

Now divide by 2, a positive number:

$$x > 4.$$

Answer $x > 4$.

■ *Example 8.2*

Solve $\dfrac{x - 3}{2x + 1} < 0$.

SOLUTION   The quotient is negative if and only if numerator and denominator have opposite signs. To see where this happens, we make a diagram (Fig. 8.1).

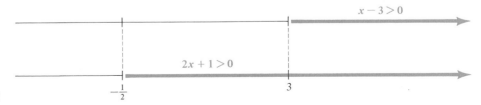

Fig. 8.1

The signs are opposite only on the segment $-\frac{1}{2} < x < 3$.

Answer $-\frac{1}{2} < x < 3$.

$\blacksquare$ *Example 8.3*

Solve $\dfrac{13x + 3}{2x + 1} < 6$.

SOLUTION 1   If this were an equation, we would multiply both sides by $2x + 1$. We can do the same with this inequality *provided* we distinguish two cases, $2x + 1 > 0$ and $2x + 1 < 0$.

Case 1: $2x + 1 > 0$, that is, $x > -\frac{1}{2}$. Multiply both sides by the *positive* quantity $2x + 1$:

$$(2x + 1)\left(\frac{13x + 3}{2x + 1}\right) < 6(2x + 1), \qquad 13x + 3 < 6(2x + 1).$$

Solve this linear inequality:

$$13x + 3 < 12x + 6, \qquad x < 3.$$

The solution in Case 1 is $-\frac{1}{2} < x < 3$.

Case 2: $2x + 1 < 0$, that is, $x < -\frac{1}{2}$. Multiply both sides by the *negative* quantity $2x + 1$, *reversing* the *inequality:*

$$13x + 3 > 6(2x + 1),$$
$$13x + 3 > 12x + 6, \qquad x > 3.$$

Both $x < -\frac{1}{2}$ and $x > 3$ together are impossible, so there are no solutions in Case 2.

SOLUTION 2   We can avoid cases by subtracting 6 from both sides of the conditional inequality and using the method of the previous example:

$$\frac{13x + 3}{2x + 1} - 6 < 0.$$

Express the left side as a quotient:

$$\frac{13x + 3 - 6(2x + 1)}{2x + 1} < 0, \qquad \frac{x - 3}{2x + 1} < 0.$$

This is the conditional inequality solved in Example 8.2.

Answer $-\frac{1}{2} < x < 3$.

■ *Example 8.4*

Solve $x^2 + 3 \geq 4x$.

SOLUTION   Quadratic inequalities are best solved by transposing all terms to one side and factoring:

$$x^2 - 4x + 3 \geq 0,$$

$$(x - 1)(x - 3) \geq 0.$$

The product is positive only if both factors have the same sign; the product is 0 only if one of the factors is 0. It is convenient to make a diagram (Fig. 8.2).

Fig. 8.2

The signs agree (or the product is 0) for $x \leq 1$ or $x \geq 3$.

> *Answer* $x \leq 1$ or $x \geq 3$.

. . . .

■ *Example 8.5*

Solve $|5x - 4| < 20$.

SOLUTION   This inequality is equivalent to two inequalities without absolute values:

$$-20 < 5x - 4 < 20.$$

Add 4 to each term, then divide by 5:

$$-16 < 5x < 24, \qquad \frac{-16}{5} < x < \frac{24}{5}.$$

> *Answer* $-16/5 < x < 24/5$.

. . . .

The next example is a typical application of inequalities.

■ *Example 8.6*

The assets of an established bank are $200 million and increasing $10 million per year. A new bank starts with $100 million and, under aggressive management, its assets increase $12 million per year. When will its assets exceed 75% of those of the older bank?

SOLUTION For convenience, take as the unit of money the megabuck, $1 million. In x years, the assets of the older bank will increase by $10x$. Hence its total assets will be $200 + 10x$. Similarly, those of the new bank will be $100 + 12x$. The problem asks for what value of x is

$$100 + 12x > \tfrac{3}{4}(200 + 10x).$$

Multiply both sides by 4 and solve:

$$400 + 48x > 3(200 + 10x) = 600 + 30x,$$

$$18x > 200, \qquad x > \frac{200}{18} = \frac{100}{9} = 11\tfrac{1}{9}.$$

Answer After $11\tfrac{1}{9}$ years.

Applications of $a^2 \geq 0$

The absolute inequality $a^2 \geq 0$ has many applications. We give two.

■ *Example 8.7*

Let a, b, c be constants with $a > 0$. Prove that

$$ax^2 + bx + c \geq \frac{4ac - b^2}{4a}$$

is an absolute inequality (true for all x).

SOLUTION Complete the square:

$$ax^2 + bx + c = a\left(x + \frac{b}{2a}\right)^2 + \frac{4ac - b^2}{4a}.$$

Since $a > 0$ and $(x + b/2a)^2 \geq 0$ we have $a(x + b/2a)^2 \geq 0$, so

$$ax^2 + bx + c = a\left(x + \frac{b}{2a}\right)^2 + \frac{4ac - b^2}{4a} \geq \frac{4ac - b^2}{4a}.$$

■ *Example 8.8*

Prove that the inequality

$$\sqrt{ab} \leq \frac{a + b}{2}$$

is true for all $a \geq 0$ and $b \geq 0$.

SOLUTION We start with $(\sqrt{b} - \sqrt{a})^2 \geq 0$ and expand:

$$b - 2\sqrt{ab} + a \geq 0, \qquad a + b \geq 2\sqrt{ab}.$$

The result follows easily.

Remark: The result is called the **arithmetic-geometric mean** inequality. For other versions see Exercises 35–36.

Solve:

1. $6x - 1 < 3$

2. $2x + 10 < 7$

3. $4x - 5 > 8x + 1$

4. $3(x - 4) > \frac{1}{2}x - 6$

5. $-4 < 2x + 9 < 16$

6. $6 < \frac{1}{2}(x + 3) < 10$

7. $(2x - 1)^2 \leq 4x^2 + 2$

8. $(x + 1)(x + 2)$
$\geq (x - 3)(x + 4)$

9. $\dfrac{1}{x + 5} > \dfrac{1}{8}$

10. $\dfrac{1}{2} > \dfrac{4}{3 - x}$

11. $x^2 > 6x - 7$

12. $x^2 + 4x \leq 12$

13. $\dfrac{2x - 3}{x^2} < 1$

14. $(3x - 5)^2 \geq 0$

15. $\dfrac{x^2}{x^2 + 8} < 1$

16. $|x - 4| < 5$

17. $|2x - 1| < 4$

18. $|3x + 1| > 1$

19. $|x^2 - 1| < 3$

20. $|x| < 2x + 3$

21. $|x| < |x + 5|$

22. $\dfrac{2x + 1}{4x + 1} < 1$

23. $\dfrac{1}{x + 1} > \dfrac{2}{x}$

24. $\dfrac{x}{x - 3} < 0$

25. $\dfrac{x}{8x - 3} > 0$

26. $(x - 1)(x - 2)(x - 3) > 0$

27. $(x - 2)(x^2 + 1) > 0$

28. $(x - 2)^2(x^2 + 1)(x - 3) > 0$.

29. A projectile is fired from level ground. After t sec, its height is $320t - 16t^2$ ft above the ground. During what period is the projectile higher than 1200 ft?

30. Company A will rent a car for $10 per day. Company B will rent a car for $6 per day plus an initial fee of $60. When is it cheaper to rent from Company B?

31. Prove $x(1 - x) \leq \frac{1}{4}$ for $0 \leq x \leq 1$.

32. (cont.) Find the largest possible area of a rectangle of perimeter 4.

33. Find all values of c for which it is possible to find two numbers whose sum is c and whose product is 100. [Hint: Reduce to a quadratic equation; check the sign of its discriminant.]

34. Find all values of c for which it is possible to find two numbers, the sum of whose squares is c, and whose sum is 10.

35. Prove

$$\frac{x^2 + y^2}{2} \geq \left(\frac{x + y}{2}\right)^2$$

for all x and y.

36. Prove $(xz + yw)^2 \leq (x^2 + y^2)(z^2 + w^2)$ for all x, y, z, w. [Hint: $(xw - yz)^2 \geq 0$.]

37. Prove $\frac{1}{4}(x + y + z + w) \geq \sqrt[4]{xyzw}$ for all $x \geq 0$, $y \geq 0$, $z \geq 0$, $w \geq 0$. [Hint: Use the result of Example 8.8 several times.]

38*. (cont.) Prove $\frac{1}{3}(x + y + z) \geq \sqrt[3]{xyz}$ for all $x \geq 0$, $y \geq 0$, $z \geq 0$. [Hint: Set $w = \frac{1}{3}(x + y + z)$ in Ex. 37.]

Prove the absolute inequalities

39. $x^4 + x^2 - 2x + 3 > 0$

40*. $x^4 - 4x^3 + 5x^2 - 2x + 1 > 0$.

Test 1

1. Solve

    (a) $3 + \dfrac{2x}{x+1} = \dfrac{x-1}{x+1}$ (b) $\begin{cases} x + 9y = 5 \\ x + 8y = -3. \end{cases}$

2. Solve $x^2 - x - 30 = 0$

    (a) by factoring   (b) by completing the square.

3. How much 97% pure aluminum must be mixed with 99.5% pure aluminum to produce one ton of 98% pure metal?

4. Solve   (a) $|2 - x| < 2$ (b) $\dfrac{36 - x^2}{5x} < 1.$

Test 2

1. Solve   (a) $(x - 2)(x + 3) = (x + 4)^2$

    (b) $\begin{cases} x - 3y = 1 \\ -2x + 6y = 2. \end{cases}$

2. Solve the system $\begin{cases} x + 2y = 1 \\ xy = -21. \end{cases}$

3. If A can do a job in 8 hours and A and B together can do the same job in 5 hours, how long does it take B to do the job?

4. Solve $|x| + |x - 2| < 3.$

3

FUNCTIONS AND GRAPHS

Everyone is familiar with the use of graphs to summarize data (Fig. 1.1). The figure shows three typical graphs. There are many others; one sees graphs concerning length, time, speed, voltage, blood pressure, supply, demand, etc.

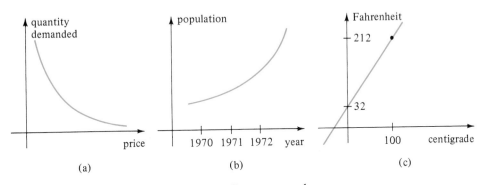

Fig. 1.1 Common graphs

All graphs have an essential common feature; they illustrate visually the way one numerical quantity depends on (or varies with) another. In Fig. 1.1, (a) shows how the demand for a certain item depends on its price, (b) shows how the population depends on (varies with) time, and (c) shows how Fahrenheit readings depend on (are related to) centigrade readings.

Graphs are pictures of **functions.** Roughly speaking, a function describes the way one quantity depends on another or the way one quantity varies with another. We say, for instance, that pressure is a function of temperature, or that population is a function of time, etc. Functions lurk everywhere; they are the basic idea in almost every application of mathematics. Therefore, a great deal of study is devoted to their nature and properties.

As Fig. 1.1 illustrates, a graph is an excellent tool in understanding the nature of a function. For it is a kind of "life history" of a function, to be seen at a glance. That is why there is much emphasis on graphs in this book.

2. COORDINATES IN THE PLANE

When the points of a line are specified by real numbers, we say that the line is **coordinatized:** each point has a label or **coordinate.** It is possible also to label, or coordinatize, the points of a plane.

Draw two perpendicular lines in the plane. Mark their intersection O and coordinatize each line as shown in Fig. 2.1. By convention, call one line horizontal and name it the x-axis; call the other line vertical and name it the y-axis.

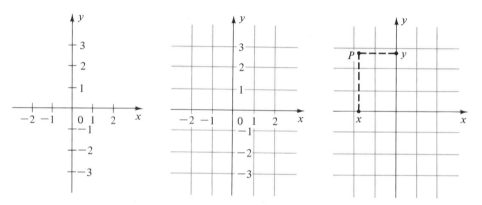

Fig. 2.1 Coordinate axes in the plane   **Fig. 2.2** Rectangular grid   **Fig. 2.3** Coordinates of a point

Consider all lines parallel to the x-axis and all lines parallel to the y-axis (Fig. 2.2). These two systems of parallel lines impose a rectangular grid on the whole plane. We use this grid to coordinatize the points of the plane.

Take any point P of the plane. Through P pass one vertical line and one horizontal line (Fig. 2.3). They meet the axes in points x and y respectively. Associate with P the ordered pair (x, y); it completely describes the location of P.

Conversely, take any ordered pair (x, y) of real numbers. The vertical line through x on the x-axis and the horizontal line through y on the y-axis meet in a point P whose coordinates are precisely (x, y). Thus there is a one-to-one correspondence,

$$P \longleftrightarrow (x, y),$$

between the set of points of the plane and the set of all ordered pairs of real numbers. The numbers x and y are the x-**coordinate** and y-**coordinate** of P. The point $(0, 0)$

is called the **origin.** Such a coordinate system in the plane is called a **rectangular** or **cartesian** coordinate system.

Remark 1: The pair (x, y) is also sometimes called (ungrammatically) the **coordinates** of *P.*

Remark 2: Some writers refer to the *x*-coordinate of a point as its **abscissa** and the *y*-coordinate as its **ordinate.** This language is old-fashioned.

The coordinate axes divide the plane into four quadrants which are numbered as in Fig. 2.4.

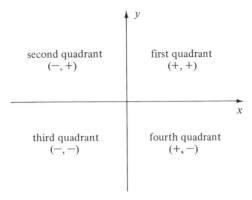

Fig. 2.4 The four quadrants

Sometimes the two coordinate axes are used to represent incompatible quantities. When this is the case, there is no reason whatsoever for choosing equal unit lengths on the two axes. On the contrary, it may be best to take different unit lengths, or scales.

Figure 2.5 shows the distance y in miles covered by a car in t seconds moving in city traffic. If we are interested in the car's progress for about one minute, a

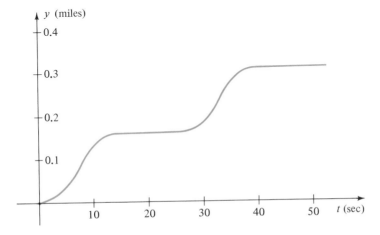

Fig. 2.5 Example of different scales on the axes

reasonable choice of unit on the t-axis is 10 sec. Since we expect the car's speed to be at most 40 mph (about 0.1 mi per 10 sec), a reasonable choice for the unit on the y-axis is 0.1 mi. If we choose 1 sec and 1 mi for units, the graph will be silly and impractical. Try it!

If, however, we wish to plot the car's progress for 10 or 15 min, then 10 sec would probably be too small as a unit of time. A more practical choice might be 1 min as the time unit and 0.5 mi as the distance unit.

Subsets of the Plane

It is often possible to describe a subset of the coordinate plane by an inequality or a system of inequalities. An example will illustrate the idea.

■ *Example 2.1*

Sketch the set of all points (x, y) for which $y \leq 0$ and $|x| \geq 1$.

SOLUTION   First we shade (Fig. 2.6) the points where $y \leq 0$, that is, the region *below* the x-axis. Then we shade the region where $|x| \geq 1$, which consists of two parts; the part to the right of the vertical line $x = 1$ where $x \geq 1$, and the part to the left of the vertical line $x = -1$ where $x \leq -1$. The region shaded twice (dark) is the desired subset.

Fig. 2.6   The set where $y \leq 0$ *and* $|x| \geq 1$ is shaded dark.

EXERCISES

Plot and label the points on one graph:

1. $(-4, 1)$, $(3, 2)$, $(5, -3)$, $(1, 4)$
2. $(0, -2)$, $(3, 0)$, $(-2, 2)$, $(1, -3)$
3. $(0.2, -0.5)$, $(-0.3, 0)$, $(-1.0, -0.1)$
4. $(75, -10)$, $(-15, 60)$, $(95, 40)$.

Choose suitable scales on the axes and label the points:

5. (150, 0.3), (50, 0.6)
6. (−0.02, 5), (0.03, 12)
7. (0.1, −0.003), (−0.3, 0.007)
8. (−0.02, 35), (0.00, −60).

Indicate on a suitable diagram all points (x, y) in the plane for which:

9. $x = -3$
10. $y = 2$
11. x and y are positive
12. either x or y (or both) is zero
13. $1 \leq x \leq 3$
14. $-1 \leq y \leq 2$
15. $-2 \leq x \leq 2$ and $-2 \leq y \leq 2$
16. $x > 2$ and $y < 3$
17. both x and y are integers
18. $x^2 > 4$
19. $|x| \geq 1$ and $|y| \leq 2$
20. $|x| \geq 2$ and $|y| \geq 2$
21. $xy > 0$ and $|x| \leq 3$
22. $|x| + |y| > 0$.

Write the coordinates (x, y) of the

23. vertices of a square centered at $(0, 0)$, sides of length 2 and parallel to the axes
24. vertices of a square centered at $(1, 3)$, sides of length 2, at $45°$ angles with the axes
25. vertices of a 3-4-5 right triangle in the first quadrant, right angle at $(0, 0)$, hypotenuse of length 15
26. vertices of an equilateral triangle, sides of length 2, base on the x-axis, vertex on the positive y-axis.

3. FUNCTIONS

A basic mathematical idea that we shall study in this book is the idea of a function. We have indicated that in some way a function describes the dependence of one quantity on another. It is time to make this more precise.

Let the symbol x represent a real number, taken from a certain set D of real numbers. Suppose there is a rule that associates with each such x a real number y. Then this rule is called a **function** whose **domain** is D.

For instance, suppose that to each real x is assigned a number y by the rule $y = x^2$. Then this assignment is a function whose domain is the set of all real numbers.

As another example, take the assignment of $+\sqrt{x}$ to each real number x which has a square root. This assignment is a function whose domain is the set of non-negative numbers.

Notation: The symbol x used to denote a typical real number in the domain of a function is sometimes called the **independent variable.** The symbol y used to denote the real number assigned to x is called the **dependent variable.**

Generally, but not always, variables are denoted by lowercase letters such as t, x, y, z. Functions are denoted by f, g, h and by capital letters.

If f denotes a function, x the independent variable, and y the dependent variable, then it is common practice to write "$y = f(x)$", read "y equals f of x" or "y equals f at x". The notation means that the function f assigns to each x in its domain a number $f(x)$ which is abbreviated by y.

There are several common variations of this notation. For instance, if f is the function that assigns to each real number its square, then we write "$f(x) = x^2$" or "$y = x^2$".

Warning 1: It is logically incorrect to say "the function $f(x)$", or "the function x^2", or the function "$y = f(x)$". The symbols "$f(x)$", "y", "x^2" represent numbers, the numbers assigned by the function f to the numbers x. A function is not a number, but an assignment of a number y or $f(x)$ to each number x in a certain domain. Nevertheless, these slight inaccuracies are so universal, we shall not try to avoid them.

Warning 2: A function is not a formula, and need not be specified by a formula. It is true that in practice most functions are indeed *computed* by formulas. For instance, f may assign to each real number x the real number y computed by formulas such as $y = x^2$, or $y = (\sqrt{x^2 + 1})/(1 + 7x^4)$, etc. Yet there are perfectly good functions not given by formulas. Here are a few examples:

(a) $f(x) = $ the largest integer (whole number) y for which $y \leq x$.

(b) $f(x) = \begin{cases} 1 & \text{if } x > 0 \\ 0 & \text{if } x = 0 \\ -1 & \text{if } x < 0. \end{cases}$

(c) $f(x) = 1$ if x is an integer, $f(x) = -1$ if x is not an integer.

(d) $f(x) = $ number of letters in the English spelling of the rational number x in lowest terms. For example, $f(\frac{1}{2}) = 7$, $f(3) = 5$.

More on notation: Keep in mind that $f(x)$ is the *number* assigned to x by the function f. If, for instance, $f(x) = x^2 + 3$, then $f(1) = 4$, $f(2) = 7$, $f(3) = 12$. By the same token $f(x + 1) = (x + 1)^2 + 3$, $f(x^2) = (x^2)^2 + 3 = x^4 + 3$, etc. For this particular function, you must boldly square and add 3 to whatever appears in the window, no matter what it is called:

$$f(x + y) = (x + y)^2 + 3, \qquad f\left(\frac{1}{x}\right) = \left(\frac{1}{x}\right)^2 + 3,$$

$$f[f(x)] = [f(x)]^2 + 3 = (x^2 + 3)^2 + 3 = x^4 + 6x^2 + 12.$$

On domains of functions: Most functions arising in practice have simple domains. The most common domains are the whole line, an interval (segment) $a \leq x \leq b$, a "half-line" such as $x \geq 0$ or $x < 2$ or some simple combination of these. Examples:

FUNCTION	DOMAIN
$f(x) = 2x + 1$	all real x (the whole line)
$f(x) = \sqrt{x + 2}$	$x \geq -2$ (half line)
$f(x) = \sqrt{1 - x^2}$	$-1 \leq x \leq 1$ (interval)
$f(x) = \dfrac{1}{x}$	all x except $x = 0$ (union of two half-lines)

Given a real-valued function f, we construct its graph, a geometric picture of the function. Here is how we do it: for each real number x in the domain of f, we find the associated number $y = f(x)$ and plot the point (x, y). The locus (totality) of all such points is called the **graph** of $f(x)$.

■ *Example 3.1*

Graph the (constant) function $f(x) = 2$.

SOLUTION The function is extremely simple since it assigns to each real number the same number, 2. The graph consists of all points in the plane of the form $(x, 2)$. Since it extends indefinitely in both directions, we only show part of it (Fig. 3.1).

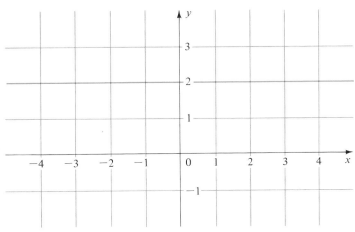

Fig. 3.1   Graph of $f(x) = 2$

When we construct a graph, we indicate all points in the plane of a certain special type. Think of it this way: Imagine at each point of the plane a flag bearing the coordinates of the point (Fig. 3.2). Now to graph $f(x)$ knock down all of the flags except those that show (x, y) where $y = f(x)$. What remains is a curve of flags standing over the graph of $f(x)$. See next page.

■ *Example 3.2*

Graph the function $f(x) = x$.

SOLUTION For each x the corresponding y is $y = f(x) = x$. Thus if $x = 0$ then $y = 0$, so $(0, 0)$ is on the graph. Likewise $(2.5, 2.5)$, $(1, 1)$, $(-1.5, -1.5)$ are on the graph. We must knock down all flags except those whose two numbers are equal. The result (Fig. 3.3) is a straight line through the origin $(0, 0)$ at an angle $45°$ with the positive x-axis. See next page.

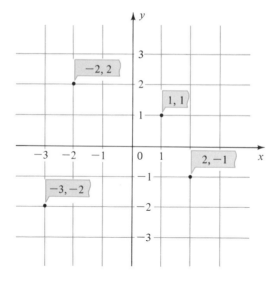

Fig. 3.2 Flag at each point

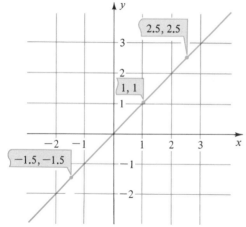

Fig. 3.3 Graph of $f(x) = x$ **Fig. 3.4** Graph of $f(x) = x + 1$

■ *Example 3.3*

Graph the function $f(x) = x + 1$.

SOLUTION For each x, we have $y = f(x) = x + 1$, so the corresponding point on the graph is $(x, x + 1)$. This point is one unit higher than the point (x, x) on the graph of the function $g(x) = x$. We therefore start with the graph in Fig. 3.3, and move each point up one unit. The result is Fig. 3.4.

■ ■

Example 3.3 illustrates an important point. Adding a positive constant c to a function shifts its graph upwards c units. Similarly subtracting c shifts the graph downwards c units (Fig. 3.5).

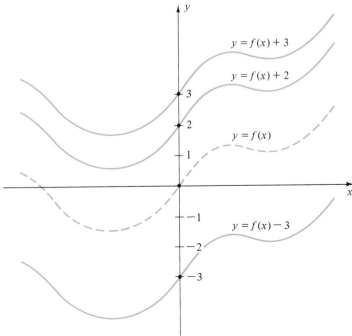

Fig. 3.5 Graph of
$y = f(x) + c$

Remark: The following notation is convenient and common: Instead of referring to the graph of $f(x) = x + 1$, we often refer to "the graph of $y = x + 1$". Thus we may say "graph $y = f(x)$" instead of "graph the function $f(x)$".

Not only does each function have a graph, but each graph defines a function. By a graph, we mean here a collection of points (x, y) in the plane such that no two of the points have the same first coordinate (only one point can lie above a point on the x-axis). Such a graph automatically defines a function $f(x)$: to each x that occurs as a first coordinate of a point (x, y), it assigns the second coordinate y. Thus, $f(x)$ is the "height" of the graph above x.

The graphical definition of functions is standard procedure in science. For instance a scientific instrument recording temperature or blood pressure on a graph is defining a function of time. There is hardly ever an explicit formula for such a function.

We close this section with graphs of two such functions that lack explicit formulas. The first is the "nearest integer" function (Fig. 3.6a, next page), defined by

$$f(x) = \begin{cases} \text{nearest integer to } x \text{ if there is } one \text{ such} \\ x \text{ if there are } two \text{ such.} \end{cases}$$

The second is the "saw-tooth" function (Fig. 3.6b), used sometimes in electronics.

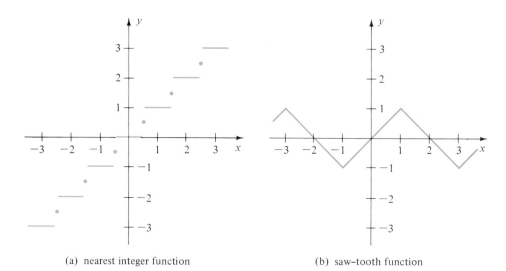

(a) nearest integer function   (b) saw-tooth function

Fig. 3.6   Functions without explicit formulas

1. Let $f(x) = 2x + 5$. Compute:
   (a) $f(0)$ (b) $f(2)$ (c) $f(\frac{1}{2})$ (d) $f(1/x)$ (e) $f(x - 3)$.
2. Let $f(x) = x^2 + x + 1$. Compute:
   (a) $f(0)$ (b) $f(-x)$ (c) $f(x^2)$ (d) $f(\sqrt{x})$ (e) $f(x + h) - f(x)$.

Graph:

3. $f(x) = x + 2$ 4. $f(x) = x - 1$ 5. $f(x) = -x$
6. $f(x) = -x + 1$ 7. $f(x) = -17$ 8. $f(x) = 0.03$
9. $f(x) = x + 0.01$ 10. $f(x) = -x - 2.5$ 11. $f(x) = |x|$

12. $f(x) = |x^2 - 1|$ 13. $f(x) = \begin{cases} 0, & x \le 0 \\ 2x, & x > 0 \end{cases}$ 14. $f(x) = \begin{cases} x - 1, & x \le 3 \\ 2, & x > 3 \end{cases}$

15. $f(x) = \begin{cases} 1, & x > 0 \\ 0, & x = 0 \\ -1, & x < 0 \end{cases}$ 16. $f(x) = \begin{cases} 1 & \text{if } x \text{ is an integer} \\ -1 & \text{if } x \text{ is not an integer.} \end{cases}$

Find the domain of $f(x)$:

17. $f(x) = 3x - 2$ 18. $f(x) = -7x + 6$ 19. $f(x) = 4x - 5$
20. $f(x) = 7 - x$ 21. $f(x) = 1/(2x - 3)$ 22. $f(x) = x/(x + 2)$
23. $f(x) = x/(3x - 5)$ 24. $f(x) = x/(x - 1)(x - 3)$ 25. $f(x) = \sqrt{x - 6}$
26. $f(x) = \sqrt{5 - 2x}$ 27. $f(x) = \sqrt{4 - 9x^2}$ 28. $f(x) = \sqrt{15x^2 + 11}$

29. $f(x) = \sqrt{2x - 3}$ 30. $f(x) = \dfrac{1}{\sqrt{x + 4}}$ 31. $f(x) = \sqrt{\frac{1}{4} - x^2}$

32. $f(x) = \sqrt{x^2 - 1}$ **33.** $f(x) = \sqrt{(x - 1)(x - 4)}$ **34.** $f(x) = \sqrt{x^3 + 1}$.

35. Graph the function that gives the first class postage on a letter as a function of its weight.
36. Graph $f(x)$, the distance from the real number x to the nearest integer.

4. CONSTRUCTION OF FUNCTIONS

There are several standard methods for building new functions out of old ones. We shall list the most common of these constructions.

1. *Addition of functions.* If f and g are functions of x defined on the same domain, then their **sum** $f + g$ is a function defined on the same domain by

$$[f + g](x) = f(x) + g(x).$$

For example, let $f(x) = 2x - 3$ and $g(x) = x^2 - x - 1$. Then

$$[f + g](x) = (2x - 3) + (x^2 - x - 1) = x^2 + x - 4.$$

2. *Multiplication of a function by a constant.* If c is a constant and f is a function, the function cf is defined by

$$[cf](x) = cf(x).$$

For example, if $f(x) = x^2 - 2x - 1$ and $c = -5$, then

$$[-5f](x) = (-5)(x^2 - 2x - 1) = -5x^2 + 10x + 5.$$

3. *Multiplication of functions.* If f and g are functions of x defined on the same domain, then their **product** fg is defined by

$$[fg](x) = f(x)g(x).$$

For example, if $f(x) = 2x - 1$ and $g(x) = 3x + 4$, then

$$[fg](x) = (2x - 1)(3x + 4) = 6x^2 + 5x - 4.$$

4. *Composition of functions.* If g is a function whose values lie in the domain of a second function f, then the **composite** $f \circ g$ of f and g is defined by the formula

$$[f \circ g](x) = f[g(x)].$$

Think of substituting one function into the other, or replacing the variable of f by the function g. Here are some examples:

1. $f(x) = x^2 + 2x, \quad g(x) = -3x.$

$$[f \circ g](x) = f[g(x)] = [g(x)]^2 + 2[g(x)]$$
$$= (-3x)^2 + 2(-3x)$$
$$= 9x^2 - 6x.$$

Note that the domain of f is all real numbers, hence the values of g certainly lie in the domain of f.

2. $f(x) = 3x - 4$, $g(x) = 2x^2 - x + 1$.

$$[f \circ g](x) = f[g(x)] = 3g(x) - 4$$
$$= 3(2x^2 - x + 1) - 4$$
$$= 6x^2 - 3x - 1.$$

Again the domain of f is all real numbers.

3. $f(x) = \sqrt{x - 1}$, $g(x) = -x^2$.

The domain of f is the set of real numbers x with $x \geq 1$. But $g(x) \leq 0$. Therefore the composition $f[g(x)]$ is not defined. Stated briefly, $\sqrt{-x^2 - 1}$ makes no sense. If, however, $g(x) = 4x^2$, then $g(x) \geq 1$ provided $|x| \geq \frac{1}{2}$. Hence

$$[f \circ g](x) = \sqrt{4x^2 - 1}$$

is defined for $|x| \geq \frac{1}{2}$.

EXERCISES

Find $[f + g](x)$, and $[fg](x)$, where

1. $f(x) = 3x + 1$, $g(x) = -2$
2. $f(x) = 2x - 1$, $g(x) = 2x + 3$
3. $f(x) = x^2$, $g(x) = -2x + 1$
4. $f(x) = x^2 + 1$, $g(x) = -x^2 + x$.

5. Does it make sense to add the functions $y = \sqrt{1 - x}$ and $y = \sqrt{x - 2}$?
6. A function f is called **strictly increasing** if whenever $x_1 < x_2$, then $f(x_1) < f(x_2)$). Show that the sum of two strictly increasing functions is strictly increasing.

Find $f \circ g$ and $g \circ f$, where

7. $f(x) = 3x + 1$, $g(x) = x - 2$
8. $f(x) = 2x - 1$, $g(x) = -x^2 + 3x$
9. $f(x) = 2x^2$, $g(x) = -x - 1$
10. $f(x) = x + 1$, $g(x) = -x + 1$
11. $f(x) = 2x$, $g(x) = -2x$
12. $f(x) = x + 3$, $g(x) = -x + 1$
13. $f(x) = x^2$, $g(x) = 3$
14. $f(x) = \pi x^2$, $g(x) = 2x + 5$.

15. If $f(x) = x$ and $g(x)$ is any function, find $f \circ g$.
16. If $g(x) = x$ and $f(x)$ is any function, find $f \circ g$.
17. Let $f(x) = 1 - x$. Compute $[f \circ f](x)$.
18. Let $f(x) = 1/x$ for $x \neq 0$. Compute $[f \circ f](x)$.
19. Find an example of a function $f(x)$ such that $f(x^2) \neq [f(x)]^2$.
20. Find an example of a function $f(x)$ such that $f(1/x) \neq 1/f(x)$.
21. Does it make sense to form $f \circ g$ if $f(x) = \sqrt{2x - 5}$ and $g(x) = 1 - x^2$?
22. Prove that if $f(x) = 3x - 5$, then

$$f\left(\frac{x_0 + x_1}{2}\right) = \frac{f(x_0) + f(x_1)}{2}.$$

23. (cont.) Is the same true for $f(x) = ax + b$?

24. (cont.) Is the same true for $f(x) = x^2$?

25. If $f(x) = 1/x$, show that

$$f\left(\frac{x_0 + x_1}{2}\right) = 2f(x_0 + x_1).$$

26. If $f(x) = 1/x^2$, show that

$$f(x_0 x_1) = f(x_0)f(x_1).$$

5. LINEAR FUNCTIONS

A function $f(x)$ is called **linear** if

$$f(x) = ax + b$$

for all real values of x, where a and b are constants. If $a = 0$, then $f(x) = b$ is a constant function; thus the class of linear functions includes the class of constant functions. Here are two basic facts about linear functions and their graphs:

> The graph of each linear function $y = ax + b$ is a non-vertical straight line.
>
> Conversely, each non-vertical straight line is the graph of a linear function.

(The word "linear" is used because the graph of a linear function is a straight line.)

In order to prove the two assertions we shall first take the special case $b = 0$. We shall prove: (1) the graph of $y = ax$ is a non-vertical straight line through the origin, and (2) each non-vertical straight line through the origin is the graph of a linear function $y = ax$.

(1) Consider the graph of $y = ax$ for $a > 0$. The points $(0, 0)$ and $(1, a)$ are on the graph. The line L through these points lies in the first and third quadrants (Fig. 5.1).

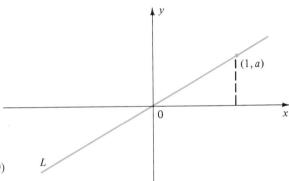

Fig. 5.1 Line L through $(0, 0)$ and $(1, a)$

Each point (x, y) on this line L has the form (x, ax), and each point (x, ax) is on L. Why? Because the right triangles in Fig. 5.2 are all similar, hence the ratios of their corresponding legs are equal:

$$\frac{|y|}{|x|} = \frac{a}{1} = a.$$

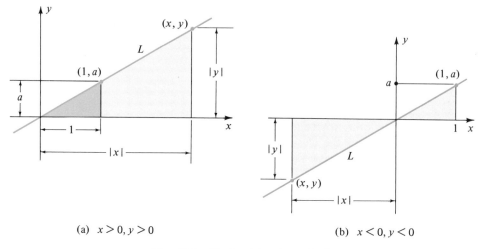

(a) $x > 0, y > 0$ (b) $x < 0, y < 0$

Fig. 5.2 Similar triangles: $a > 0$

Since (x, y) is in the first or third quadrant, x and y have the same sign. Hence $|y|/|x| = y/x$,

$$\frac{y}{x} = a, \qquad y = ax.$$

If $a < 0$, a similar argument applies, but we must be careful with signs (Fig. 5.3). This time L lies in the second and fourth quadrants, hence $|y|/|x| = -y/x$. Also

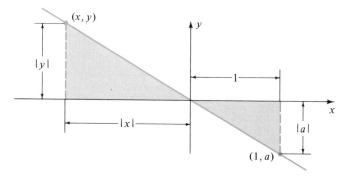

Fig. 5.3 Similar triangles: $a < 0$

$a < 0$, so $|a| = -a$. Therefore

$$\frac{|y|}{|x|} = \frac{|a|}{1}, \qquad -\frac{y}{x} = -\frac{a}{1}, \qquad \frac{y}{x} = a, \qquad y = ax.$$

If $a = 0$, the graph of $y = ax = 0$ is the x-axis. Thus in all cases the graph of $y = ax$ is a non-vertical straight line through $(0, 0)$.

(2)   Conversely, let L be any non-vertical line through $(0, 0)$. Then L passes through a point $(1, a)$, and the same reasoning shows that each point (x, y) on L satisfies $y = ax$.

For the general linear function $y = ax + b$, with $b \neq 0$, the graph is just the graph of $y = ax$ moved up or down $|b|$ units, hence a non-vertical straight line. Conversely each non-vertical straight line is parallel to a non-vertical straight line through $(0, 0)$. Hence it is the graph of $y = ax + b$ for a suitable constant b. See Fig. 5.4.

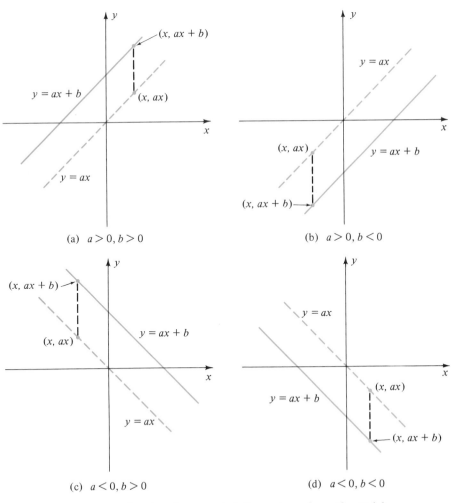

(a) $a > 0, b > 0$

(b) $a > 0, b < 0$

(c) $a < 0, b > 0$

(d) $a < 0, b < 0$

Fig. 5.4   Graphs of $y = ax + b$ for various signs of a and b

Knowing that the graph of a linear function is a line makes it easy to plot the graph. We simply find any two points on the graph and then draw the straight line through them.

■ *Example 5.1*

Graph $y = 2x$ for $-1 \leq x \leq 1$.

SOLUTION If $x = -1$, then $y = 2(-1) = -2$, hence $(-1, -2)$ is a point on the graph. Similarly $(1, 2)$ is another point on the graph. Plot these two points, then join them by a line segment (Fig. 5.5).

Fig. 5.5 Graph of $y = 2x$ for $-1 \leq x \leq 1$

Fig. 5.6 Graph of $y = \frac{1}{2}x - 1$ for $-1 \leq x \leq 3$

■ *Example 5.2*

Graph $y = \frac{1}{2}x - 1$ for $-1 \leq x \leq 3$.

SOLUTION The values $x = -1$ and $x = 3$ yield the points $(-1, -\frac{3}{2})$ and $(3, \frac{1}{2})$ of the graph. Plot and connect them by a straight line (Fig. 5.6).

■ *Example 5.3*

Graph $y = -x + 9$ for $8 \leq x \leq 10$.

SOLUTION The points $(8, 1)$ and $(10, -1)$ are on the graph. Plot and join them by a straight line (Fig. 5.7).

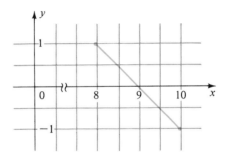

Fig. 5.7   Graph of $y = -x + 9$ for $8 \le x \le 10$

Slope

   Examine the four lines in Fig. 5.8. The arrows indicate the direction of increasing x. Line C moves upwards steeply; line D moves upwards gently; line B moves downwards steeply; line A moves downwards gently. We associate with each non-vertical line in the coordinate plane a measure of its steepness of climb or descent called its **slope.** More precisely, slope is a measure of the amount y changes relative to a change in x.

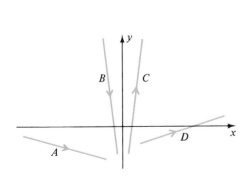

Fig. 5.8   Various degrees of steepness

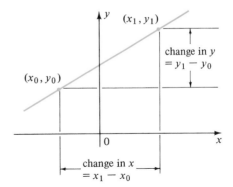

Fig. 5.9   Changes in x and y

   In Fig. 5.9, choose any two points on the line, (x_0, y_0) and (x_1, y_1). As x advances from x_0 to x_1, the variable y changes from y_0 to y_1, so the change in y is $y_1 - y_0$ while the change in x is $x_1 - x_0$. The slope is the ratio of the change in y to the change in x:

$$\text{Slope} = \frac{y_1 - y_0}{x_1 - x_0}.$$

If the line rises steeply, the change in y is greater than the change in x. Hence the slope is a large number. If the line rises slowly, the slope is a small number.
   Notice that if the line falls as x increases, $y_1 - y_0$ is *negative* while $x_1 - x_0$ is positive. Hence the slope is negative.

Remark: The formula

$$\text{slope} = \frac{y_1 - y_0}{x_1 - x_0}$$

is valid whether $x_1 > x_0$ or $x_1 < x_0$, i.e., whether (x_1, y_1) is to the right of (x_0, y_0) or to the left of (x_0, y_0). That is because

$$\frac{y_1 - y_0}{x_1 - x_0} = \frac{y_0 - y_1}{x_0 - x_1}.$$

Of course, $x_1 = x_0$ is strictly forbidden!

To compute the slope of the line $y = ax + b$, use *any* two points on the line:

$$\frac{y_1 - y_0}{x_1 - x_0} = \frac{(ax_1 + b) - (ax_0 + b)}{x_1 - x_0} = \frac{a(x_1 - x_0)}{x_1 - x_0} = a.$$

This simple calculation proves an important fact:

> The slope of the line $y = ax + b$ is a.

The line $y = ax + b$ meets the y-axis at $(0, b)$ as we see by setting $x = 0$. The number b is called the y-**intercept** of the graph.

> The y-intercept of the line $y = ax + b$ is b.

We can now tell by inspection that the graph of $y = ax + b$ is a straight line with slope a and y-intercept b. Conversely, given the slope and y-intercept of a line, we can write its equation immediately.

■ *Example 5.4*

Find the equation of the line with slope a through the point (x_0, y_0).

SOLUTION   The equation is of the form $y = ax + b$, where b is the y-intercept. To compute b, use the fact that (x_0, y_0) satisfies the equation of the line:

$$y_0 = ax_0 + b, \qquad b = y_0 - ax_0.$$

Hence

$$y = ax + b = ax + y_0 - ax_0,$$

that is,

$$y - y_0 = a(x - x_0).$$

Answer $y - y_0 = a(x - x_0)$,   that is, $y = ax + (y_0 - ax_0)$.

■ *Example 5.5*

Find the equation of the straight line passing through two given points (x_0, y_0) and (x_1, y_1). Assume $x_0 \neq x_1$.

SOLUTION The slope of the line is

$$a = \frac{y_1 - y_0}{x_1 - x_0}.$$

Therefore, by the answer to Example 5.4, the equation is

$$y - y_0 = a(x - x_0) = \frac{y_1 - y_0}{x_1 - x_0}(x - x_0),$$

that is,

$$\frac{y - y_0}{x - x_0} = \frac{y_1 - y_0}{x_1 - x_0}.$$

Answer $y - y_0 = \left(\dfrac{y_1 - y_0}{x_1 - x_0}\right)(x - x_0)$, that is,

$$y = ax + b, \text{ where } a = \frac{y_1 - y_0}{x_1 - x_0} \quad \text{and} \quad b = y_0 - ax_0.$$

■ ■

A particular case of Example 5.5 is worth noting: when the two points are $(a, 0)$ and $(0, b)$, that is, when we are given the *x*-intercept *a* and the *y*-intercept *b*.

■ *Example 5.6*

Find the equation of the line through $(a, 0)$ and $(0, b)$. Assume $a \neq 0$ and $b \neq 0$.

SOLUTION Apply the last example with $(x_0, y_0) = (a, 0)$ and $(x, y) = (0, b)$:

$$y - y_0 = \left(\frac{y_1 - y_0}{x_1 - x_0}\right)(x - x_0), \qquad y - 0 = \left(\frac{b - 0}{0 - a}\right)(x - a),$$

$$y = -\frac{b}{a}(x - a) = -\frac{b}{a}x + b.$$

Divide by *b* and rearrange terms to obtain a convenient form of this equation:

$$\frac{x}{a} + \frac{y}{b} = 1.$$

Answer $\dfrac{x}{a} + \dfrac{y}{b} = 1.$

■ ■

Remark: This equation is particularly easy to remember, and it obviously *is* the right answer. For just look at it. If $y = 0$, then $x/a = 1$, $x = a$, so $(a, 0)$ is on its graph. If $x = 0$, then $y/b = 1$, $y = b$, so $(0, b)$ is on its graph.

We have derived four useful formulas for a straight line:

Slope–intercept form:	$y = ax + b$
Point–slope form:	$y - y_0 = a(x - x_0)$
Two-point form:	$y - y_0 = \dfrac{y_1 - y_0}{x_1 - x_0}(x - x_0)$
Two-intercept form:	$\dfrac{x}{a} + \dfrac{y}{b} = 1.$

These formulas are used to obtain the equation of a given line. Conversely, given an equation in one of these forms, it represents a line that can be easily identified. For example, $y - 3 = 2(x - 1)$ is the equation of the line through $(1, 3)$ with slope 2.

Bear in mind that an equation $cx + dy + f = 0$ can be put in the slope–intercept form if $d \neq 0$. For example, $3x - 2y + 2 = 0$ can be written

$$y = \tfrac{3}{2}x + 1,$$

hence it represents a line of slope $\tfrac{3}{2}$ and y-intercept 1.

Remark: Each horizontal line has the equation $y = b$, where b is its y-intercept. We have excluded *vertical* lines from this discussion because a vertical line cannot be presented in the form $y = f(x)$. However, each vertical line has the equation $x = a$, where a is its x-intercept.

EXERCISES

Graph:

1. $y = 2x - 3, 0 \leq x \leq 4$
3. $y = 2x + 9, 1 \leq x \leq 2$
5. $y = -3x + 1, -5 \leq x \leq 5$
7. $y = 3x + 40, 25 \leq x \leq 50$
9. $y = 0.1x + 1.5, 2 \leq x \leq 3$

2. $y = 2x - 3, -2 \leq x \leq 0$
4. $y = -3x + 1, 0 \leq x \leq 1$
6. $y = -2x + 1, -20 \leq x \leq -10$
8. $y = 9x - 50, 100 \leq x \leq 200$
10. $y = -0.3x + 0.2, -1 \leq x \leq 1.$

Graph; t in seconds, x in feet:

11. $x = 0.2t - 1, 0 \leq t \leq 5$
13. $x = 9t - 9, 1 \leq t \leq 2$
15. $x = -t + 10, 25 \leq t \leq 50$

12. $x = 25t + 15, 50 \leq t \leq 100$
14. $x = -100t + 20, -1 \leq t \leq 1$
16. $x = 40t + 40, 0 \leq t \leq 100.$

Find the slope of the line through the given points:

17. $(0, 0), (3, 4)$
20. $(-1, 2), (1, 0)$

18. $(0, 0), (2, 6)$
21. $(0, 1), (1, 2)$

19. $(-1, 2), (1, 2)$
22. $(0, -1), (1, 2)$

23. $(-1, -1), (1, 2)$ 24. $(-1, 2), (2, -1)$ 25. $(-3, 1), (-2, 2)$
26. $(-2, -2), (3, -4)$.

Find the equation and y-intercept of the line with given slope a and passing through the given point:

27. $a = 1, (1, 2)$ 28. $a = -1, (2, -1)$ 29. $a = 0, (4, 3)$
30. $a = 2, (1, 3)$ 31. $a = \frac{1}{2}, (2, -2)$ 32. $a = \frac{2}{3}, (-1, 1)$.

Find the equation of the line through the two given points:

33. $(0, 0), (1, 2)$ 34. $(1, 0), (3, 0)$ 35. $(-1, 0), (2, 4)$
36. $(-1, -1), (2, 6)$ 37. $(\frac{1}{2}, 1), (\frac{3}{2}, 2)$ 38. $(-2, 0), (-\frac{1}{2}, -1)$
39. $(0.1, 3.0), (0.3, 2.0)$ 40. $(-2.01, 4.10), (-2.00, 4.00)$
41. $(0, 3), (0, -3)$ 42. $(-2, 5), (-2, 8)$.

Find the slope and y-intercept:

43. $3x - y - 7 = 0$ 44. $x + 2y + 6 = 0$
45. $3(x - 2) + y + 5 = 2(x + 3)$ 46. $2(x + y + 1) = 3x - 5$.

Find both intercepts and write the equation of the line in two-intercept form:

47. $\dfrac{x}{2} + \dfrac{y}{3} = 1$ 48. $x - y = 2$
49. $2x + 3y = 1$ 50. $3x - 5y = 15$.

6. QUADRATIC FUNCTIONS

A function $f(x)$ is called **quadratic** if

$$f(x) = ax^2 + bx + c,$$

where a, b, and c are constants. A quadratic function is defined for all values of the independent variable x because $ax^2 + bx + c$ is a real number for each real number x. If $a = 0$, then $f(x) = bx + c$ is a linear function; thus the class of quadratic functions includes the class of linear functions.

Quadratic functions occur frequently in applications. For example, if a projectile is shot upwards with muzzle velocity v_0, then its height y at time t is given by

$$y = -\tfrac{1}{2}gt^2 + v_0 t,$$

where g is the constant of gravity.

We begin our study of quadratic functions by graphing $y = x^2$. First we consider only $x \geq 0$. When x is small, x^2 is very small. For example $(0.1)^2 = 0.01$, and $(0.001)^2 = 0.000001$. Therefore, as x increases starting from 0, the graph of $y = x^2$ rises very slowly from 0. See Fig. 6.1, next page.

On the other hand, when x is large, x^2 is very large. For example $(10)^2 = 100$, $(1000)^2 = 1000000$. Therefore, as x gets larger and larger, the graph of $y = x^2$ rises very steeply (Fig. 6.2, next page).

For $x < 0$, we use the fact that $(-x)^2 = x^2$. That means the value of y at $-x$ is the same as at x. So whenever (x, y) is on the graph, so is $(-x, y)$. Hence the graph of $y = x^2$ for $x < 0$ is the mirror image in the y-axis of the graph for $x > 0$. See Fig. 6.3, page 111.

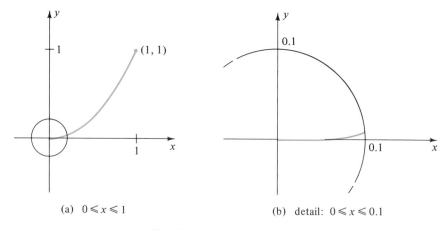

(a) $0 \leqslant x \leqslant 1$ (b) detail: $0 \leqslant x \leqslant 0.1$

Fig. 6.1   Graph of $y = x^2$

(a) $0 \leqslant x \leqslant 2$ (b) $0 \leqslant x \leqslant 10$ **Fig. 6.2**   Graph of $y = x^2$

Next we graph $y = ax^2$, assuming first that $a > 0$. The graph of $y = ax^2$ can be obtained from the graph of $y = x^2$ in a simple way. For if (x_0, y_0) is any point on the graph of $y = x^2$, then (x_0, ay_0) is on the graph of $y = ax^2$ because if $y_0 = x_0^2$, then $ay_0 = ax_0^2$. Therefore, if we stretch the graph of $y = x^2$ by the factor a in the y-direction, we obtain precisely the graph of $y = ax^2$. See Fig. 6.4, next page.

If $a < 0$, then $-a > 0$, and the graph of $y = ax^2$ is obtained from the graph of $y = (-a)x^2$ by changing each y to $-y$, that is, by forming a mirror image in the x-axis (Fig. 6.5, page 112).

Note that $(0, 0)$ is the lowest point on the graph of $y = ax^2$ if $a > 0$, and is the highest point on the graph if $a < 0$.

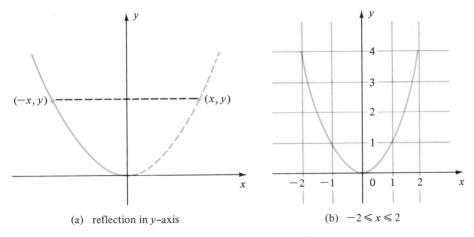

(a) reflection in y-axis

(b) $-2 \leqslant x \leqslant 2$

Fig. 6.3   Graph of $y = x^2$

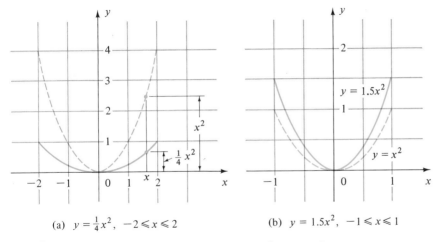

(a) $y = \frac{1}{4}x^2$, $-2 \leqslant x \leqslant 2$

(b) $y = 1.5x^2$, $-1 \leqslant x \leqslant 1$

Fig. 6.4   Graphs of $y = ax^2$ for $a > 0$

The graph of $y = ax^2 + c$ is obtained by shifting the graph of $y = ax^2$ up or down by $|c|$ units (Fig. 6.6, next page).

The General Quadratic

To graph the most general quadratic function $y = ax^2 + bx + c$, we shall complete the square. (We can suppose $a \neq 0$, otherwise the function is linear.) We obtain

$$y = ax^2 + bx + c = a\left(x + \frac{b}{2a}\right)^2 + \left(\frac{4ac - b^2}{4a}\right),$$

that is,

$$y = a(x - h)^2 + d,$$

where

$$h = \frac{-b}{2a} \quad \text{and} \quad d = \frac{4ac - b^2}{4a}.$$

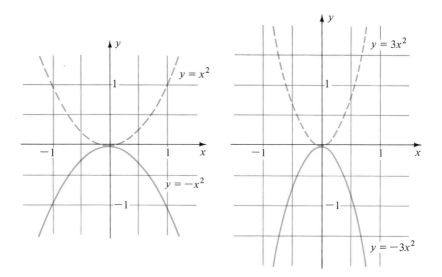

Fig. 6.5   Graphs of $y = ax^2$ for $a < 0$

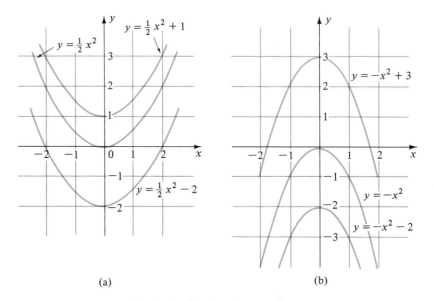

(a)   (b)

Fig. 6.6   Graphs of $y = ax^2 + c$

Thus the graph of $y = ax^2 + bx + c$, where $a \neq 0$, is the same as the graph of $y = a(x - h)^2 + d$. Clearly, if (x_0, y_0) satisfies this latter equation, then $(x_0 - h, y_0)$ satisfies the equation $y = ax^2 + d$. Therefore the graph of $y = ax^2 + bx + c$ is the graph of $y = ax^2 + d$ shifted horizontally h units; shifted right if $h > 0$, shifted left if $h < 0$. See Fig. 6.7.

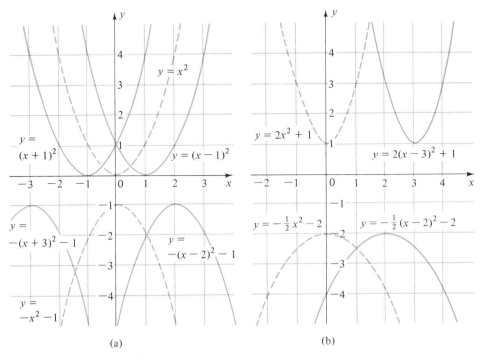

(a)   (b)

Fig. 6.7   Graphs of $y = a(x - h)^2 + d$

Note that (h, d) is the lowest point on the graph of $y = a(x - h)^2 + d$ if $a > 0$, and the highest point if $a < 0$.

■ *Example 6.1*

Graph $y = x^2 + 2x + 4$.

SOLUTION   Complete the square:

$$y = x^2 + 2x + 4 = (x^2 + 2x + 1) + 3 = (x + 1)^2 + 3.$$

The graph is obtained by shifting the graph of $y = x^2$ one unit to the left, then three units up (Fig. 6.8, next page).

■ *Example 6.2*

Graph $y = x^2 - 6x$. Find the lowest point on the curve.

SOLUTION   Complete the square:

$$y = x^2 - 6x = x^2 - 6x + 9 - 9 = (x - 3)^2 - 9.$$

If $x = 3$, then $y = 9$. If $x \neq 3$, then $(x - 3)^2 > 0$, so $y > -9$. Hence the lowest point is $(3, -9)$.

Answer $(3, -9)$. See Fig. 6.9.

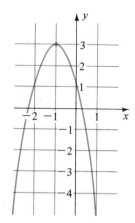

Fig. 6.8   Graph of
$y = x^2 + 2x + 4$

Fig. 6.9   Graph of
$y = x^2 - 6x$

Fig. 6.10   Graph of
$y = -2x^2 - 4x + 1$

■ *Example 6.3*

Graph $y = -2x^2 - 4x + 1$ and find the highest point on the curve.

SOLUTION   Complete the square:

$$\begin{aligned}
y = -2x^2 - 4x + 1 &= -2(x^2 + 2x) + 1 \\
&= -2(x^2 + 2x + 1) + 2 + 1 = -2(x + 1)^2 + 3.
\end{aligned}$$

If $x = -1$, then $y = 3$. If $x \neq -1$, then $(x + 1)^2 > 0$, hence $-2(x + 1)^2 < 0$, so $y < 3$. The highest point is $(-1, 3)$.

Answer $(-1, 3)$. See Fig. 6.10.

Maximum Problems

Problems involving the maximum value of a quadratic function can be solved by completing the square.

■ *Example 6.4*

What is the area of the largest rectangular field that can be enclosed with 2000 ft of fencing?

SOLUTION   Let x and y denote the length and width of the field. Then

$$\text{perimeter} = 2x + 2y = 2000,$$
$$y = 1000 - x.$$

Therefore the area is

$$A = xy = x(1000 - x) = -x^2 + 1000x.$$

The problem is equivalent to finding the highest point (x, A) on the graph of $A = -x^2 + 1000x$. Complete the square:

$$A = -(x^2 - 1000x + 500^2) + 500^2 = -(x - 500)^2 + 500^2.$$

Because the first term is negative or zero, the largest possible value of A is 500^2; it occurs only for $x = 500$. Then $y = 1000 - x = 500$, so the rectangle is a square.

Answer 500^2 ft^2; the rectangle is a square.

. . . .

■ *Example 6.5*

A projectile is fired at time $t = 0$ at a 30 degree angle to the ground. Its height above the ground after t seconds is shown in physics to be $y = \frac{1}{2}v_0 t - 16t^2$ ft, where v_0 is the muzzle velocity (air resistance is neglected). If $v_0 = 1000$ ft/sec, what is the greatest height the projectile reaches?

SOLUTION   Complete the square:

$$y = 500t - 16t^2 = -16(t^2 - \tfrac{125}{4}t)$$
$$= -16[t^2 - \tfrac{125}{4}t + (\tfrac{125}{8})^2] + 16(\tfrac{125}{8})^2$$
$$= -16(t - \tfrac{125}{8})^2 + \tfrac{125^2}{4}.$$

Therefore the largest possible value of y is $125^2/4$.

Answer $125^2/4 = 3906.25$ ft.

EXERCISES

Graph:

1. $y = 2x^2$ 2. $y = -2x^2$ 3. $y = -\frac{1}{2}x^2$
4. $y = \frac{1}{2}x^2$ 5. $y = x^2 + 3$ 6. $y = -x^2 - 3$
7. $y = 2x^2 - 1$ 8. $y = -2x^2 - 1$ 9. $y = -\frac{1}{4}x^2 + 2$
                                         10. $y = \frac{1}{4}x^2 - 2$.

Graph on the indicated range (use different scales on the axes if necessary):

11. $y = 0.1x^2, 0 \leq x \leq 100$ **12.** $y = -x^2, -0.1 \leq x \leq 0.$

Graph and find the highest (or lowest) point on each graph:

13. $y = x^2 - 4x + 1$ **14.** $y = x^2 + 2x - 5$ **15.** $y = x^2 + x + 1$
16. $y = x^2 - x + 1$ **17.** $y = -x^2 - 2x$ **18.** $y = -x^2 + 2x$
19. $y = -x^2 - 4x - 3$ **20.** $y = -x^2 + 4x + 1$ **21.** $y = 2x^2 - 6x + 1$
22. $y = 2x^2 + 4x$ **23.** $y = 3x^2 + 12x - 8$ **24.** $y = -3x^2 + 12x - 8$
25. $y = -2x^2 + 8x - 10$ **26.** $y = -2x^2 + 12x$ **27.** $y = -4x^2 + x$
28. $y = 2x^2 + 2x + 2$ **29.** $y = 2x^2 - 3x$ **30.** $y = x^2 - 6x + 2$
31. $y = x^2 + x - 4$ **32.** $y = 3x^2 + 3x$ **33.** $y = -x^2 + x - 2$
34. $y = -x^2 - 2x$ **35.** $y = -2x^2 + x$ **36.** $y = -2x^2 - 6x + 1.$

37. Show that the graph of $y = ax^2 + bx$ passes through the origin for all choices of a and b. *by making x, y value equal zeroes,*

38*. For what value of c does the lowest point of the graph of $y = x^2 + 6x + c$ fall on the x-axis?

39. Under what conditions is the lowest point of the graph of $y = x^2 + bx + c$ on the y-axis?

40*. What is the relation between the graph of $y = ax^2 + bx + c$ and that of $y = ax^2 - bx + c$?

41. Show that for $0 \leq x \leq 1$, the product $x(1 - x)$ never exceeds $\frac{1}{4}$.

42. A farmer will make a rectangular pen with 100 ft of fencing, using part of a wall of his barn for one side of the pen. What is the largest area he can enclose?

43. A 4-ft line is drawn across a corner of a rectangular room, cutting off a triangular region. Show that its area cannot exceed 4 ft². [Hint: Use the Pythagorean theorem and work with A^2.]

44. A rectangular solid has a square base, and the sum of its 12 edges is 4 ft. Show that its total surface area (sum of the areas of its 6 faces) is largest if the solid is a cube.

45*. Suppose a projectile fired as in Example 6.5 reaches a maximum height of 2500 ft. What was its muzzle velocity?

46*. (cont.) In general, if the muzzle velocity is doubled, is the maximum height of the trajectory also doubled? If not, what does happen?

7. TIPS ON GRAPHING

This section is kind of a lazy man's guide to graphs, featuring techniques that can reduce the work in graphing.

Symmetry

Consider the graphs of $y = x^2$ and $y = x^3$ shown in Fig. 7.1. (The graph of $y = x^3$ will be discussed in Chapter 4.) These graphs possess certain symmetries. The one on the left is symmetric in the y-axis. The one on the right is symmetric in the origin, i.e., to each point of the graph corresponds an opposite point as seen through a peephole in the origin. In either case we need plot the curve only for $x \geq 0$; we obtain the rest by symmetry. Thus the work is "cut in half".

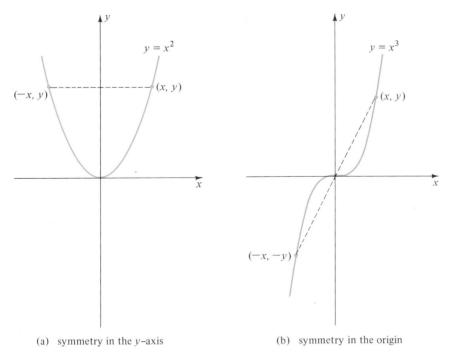

(a) symmetry in the y-axis   (b) symmetry in the origin

Fig. 7.1 Symmetry

When we plot $y = f(x)$, how can we recognize symmetry in advance? Look at Fig. 7.1a. The curve $y = f(x)$ is **symmetric in the y-axis** if for each x, the value of y at $-x$ is the same as at x; in mathematical notation, $f(-x) = f(x)$. If $f(x)$ satisfies this condition, it is called an even function.

Look at Fig. 7.1b. The curve $y = f(x)$ is **symmetric in the origin** if for each x, the value of y at $-x$ is the negative of the value at x, that is, $f(-x) = -f(x)$. If $f(x)$ satisfies this condition, it is called an odd function.

> An **even** function $f(x)$ is one for which $f(-x) = f(x)$.
>   The graph of an even function is symmetric in the y-axis.
>
> An **odd** function $f(x)$ is one for which $f(-x) = -f(x)$.
>   The graph of an odd function is symmetric in the origin.

Important

Vertical and Horizontal Shifts

We know that adding or subtracting a positive constant c to $f(x)$ shifts the graph of $y = f(x)$ up or down c units. Now let us consider horizontal shifts. How can we shift the graph of $y = f(x)$ three units to the right? More precisely how can we find a function $g(x)$ for which the graph of $y = g(x)$ is precisely that of $y = f(x)$ shifted three units to the right?

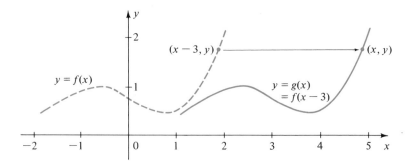

Fig. 7.2 Horizontal shift

Consider Fig. 7.2. For each point (x, y) on the curve $y = g(x)$, there corresponds a point $(x - 3, y)$ on the curve $y = f(x)$. The values of y are the same. But on the first curve $y = g(x)$, on the second, $y = f(x - 3)$. Conclusion: $g(x) = f(x - 3)$. This makes sense. If x represents time, then the value of g "now" is the same as the value of f three seconds ago.

The same reasoning shows the graph of $y = f(x + 3)$ is the graph of $y = f(x)$ shifted three units to the left.

> Let $c > 0$. The graph of
>
> $$\left.\begin{array}{l} y = f(x) + c \\ y = f(x) - c \\ y = f(x - c) \\ y = f(x + c) \end{array}\right\} \text{ is the graph of } y = f(x) \text{ shifted } c \text{ units} \left\{\begin{array}{l} \text{upward} \\ \text{downward} \\ \text{to the right} \\ \text{to the left.} \end{array}\right.$$

Stretching and Reflecting

If $c > 0$, the graph of $y = cf(x)$ is obtained from that of $y = f(x)$ by stretching by a factor of c in the y-direction. Each point (x, y) is replaced by (x, cy). Note: "stretching" by a factor less than one is interpreted as shrinking (Fig. 7.3).

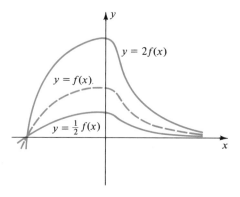

Fig. 7.3 Stretching in the y-direction

The graph of $y = -f(x)$ is obtained by reflecting the graph of $y = f(x)$ in the x-axis (turning it upside down). That is because each point (x, y) is replaced by the point $(x, -y)$. See Fig. 7.4.

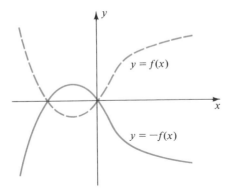

Fig. 7.4   Reflection in the x-axis

■ *Example 7.1*

Graph $y = -\frac{1}{2}(x - 5)^2$.

SOLUTION   We know the graph of $y = x^2$. We plot successively $y = x^2$, then $y = \frac{1}{2}x^2$, then $y = -\frac{1}{2}x^2$, then $y = -\frac{1}{2}(x - 5)^2$. See Fig. 7.5.

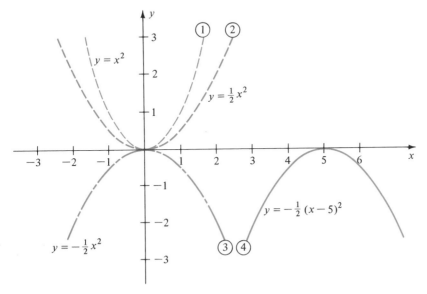

Fig. 7.5   Steps in graphing $y = -\frac{1}{2}(x - 5)^2$

Free Information

Very often you can get valuable information about a graph for free, just by looking at the equation involved. It's good practice not to start right off plotting points, but to take a minute to think. You should look for symmetry, shifts, and stretching and reflecting. Here are some other things to look for.

Domain. If you are graphing $y = f(x)$, is y defined for all real x or is there some restriction on x? For example $y = \sqrt{1 - x^2}$ is defined only for $|x| \leq 1$ and $y = 1/(x - 1)(x - 4)$ is not defined for $x = 1$ or $x = 4$.

Extent. Is there some limitation on y? For example, if $y = 1/(1 + x^2)$, then by inspection $0 < y \leq 1$. The graph does not extend above the level $y = 1$ or below the level $y = 0$.

Sign of y. Can you tell where $y > 0$ or $y < 0$? For example $y = 1/x$ is positive for $x > 0$ and negative for $x < 0$. Also y is never 0.

Increasing or decreasing? For example, $y = 1/x$ decreases as x increases through positive values.

You will not always be able to check all these points. At least try to see what you can find easily.

■ *Example 7.2*

Plot $y = \dfrac{1}{1 + x^4}$.

SOLUTION The graph is defined for all x. Since $1 + x^4 \geq 1$, we have $0 < y \leq 1$. In fact $y = 1$ only at $x = 0$, so the highest point of the curve is $(0, 1)$. As x increases through positive values, y decreases towards 0. The curve is symmetric in the y-axis since the function is even. This free information is enough for a fairly good idea of the curve. Plotting a few points helps fix the shape (Fig. 7.6).

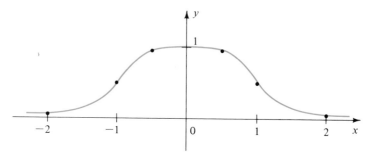

Fig. 7.6 Rough graph of $y = 1/(1 + x^4)$

1. Which of the functions are even?

$$x^2, \quad x^3, \quad x^4, \quad \frac{1}{x^2 + 1}, \quad \frac{1}{x^3 + 1}, \quad \frac{x}{x^2 + 1}, \quad \left(\frac{1}{x^3 + x}\right)^5, \quad x^3 + x^2 + 1.$$

2. (cont.) Which are odd?

3. Plot on the same graph:

$$y = x, \quad y = 3x, \quad y = x - 1,$$

$$y = 3(x - 1), \quad y = x + 4, \quad y = -3(x + 4).$$

4. Plot on the same graph

$$y = x^2, \quad y = 2x^2,$$

$$y = 2(x - 1)^2, \quad y = 2(x + 1)^2, \quad y = -2(x - 3)^2.$$

5. Plot (See Fig. 7.1):

$$y = \tfrac{1}{2}x^3, \quad y = \tfrac{1}{2}(x - 1)^3 + 3, \quad y = -\tfrac{1}{2}(x + 2)^3 + 1.$$

6. Complete the square, then plot

$$y = x^2 + x, \quad y = \tfrac{1}{2}x^2 - \tfrac{1}{8}x + 1, \quad y = 3x^2 - 2x - 5.$$

7. Compute $\tfrac{1}{2}[f(x) + f(-x)]$ if $f(x)$ is

$$x^3 + 1, \quad \frac{1}{x - 3}, \quad \frac{x^2}{x + 1}.$$

Show in each case that the answer is an even function.

8. (cont.) Prove that for any function $f(x)$, the function $g(x) = \tfrac{1}{2}[f(x) + f(-x)]$ is even.

9. (cont.) Prove that for any function $f(x)$, the function $g(x) = \tfrac{1}{2}[f(x) - f(-x)]$ is odd.

10. (cont.) Prove that any function can be expressed as the sum of an odd function and an even function.

Test 1

1. Plot the points $(2, 3)$, $(2, -3)$, $(-2, 3)$, and $(-2, -3)$. Show that they are the vertices of a rectangle. Find the coordinates of the midpoint of the top side.

2. Given $f(x) = 3x + 1$:
   (a) Compute $f(0)$, $f(-2)$, $f[f(x)]$.
   (b) Show that $f(a + b) = f(a) + f(b) - 1$.

3. Are the points $(0, 0)$, $(2, 5)$, $(3, 8)$ collinear or not? Explain.

4. Find a linear function whose graph passes through $(0, 3)$ and $(1, 5)$.

5. Graph and find the lowest point:

$$y = 2x^2 - 12x + 14.$$

Test 2

1. Plot all points (x, y) in the plane for which (a) $y > 0$, (b) $0 \leq x \leq 1$, and y is an integer.
2. If $f(x) = \sqrt{x}$ and $g(x) = 3 - x$, compute

   (a) $[f \circ g](x)$,   (b) $[g \circ f](x)$.

   In each case state the domain of the function.
3. Find a linear function $f(x) = ax + b$ whose graph passes through $(0, 6)$ and is parallel to the graph of $y = -x$.
4. For what numbers b is the value of $x^2 + bx + 1$ positive regardless of the choice of x?
5. Plot on the same set of coordinate axes:

   (a) $y = (x - 2)^2$ (b) $y = -3(x - 2)^2$.

4

POLYNOMIAL AND RATIONAL FUNCTIONS

In this chapter we shall study **polynomial** functions

$$f(x) = a_m x^m + a_{m-1} x^{m-1} + \cdots + a_2 x^2 + a_1 x + a_0$$

and **rational** functions

$$f(x) = \frac{a_m x^m + \cdots + a_1 x + a_0}{b_n x^n + \cdots + b_1 x + b_0}.$$

Such functions play a central role in mathematics for a number of reasons. (1) They arise naturally in many applications. (2) Their values can be computed using only the simplest operations of arithmetic: addition, subtraction, multiplication, division. (3) They can be used to approximate highly complicated functions. (4) Because of (2), numerical work with these functions is perfectly suited to high speed computers; because of (3), such calculations are indispensable in applied mathematics.

In this section we shall graph certain special, but important, polynomials.

We start with $y = x^3$, an important cubic. First we study its graph for $x \geq 0$. Since $0^3 = 0$, the point $(0, 0)$ is on the graph. As x increases, starting from 0, its cube x^3 also increases, so the graph always goes up. We check some values near 0:

x	0.0	0.1	0.2	0.3	0.4	0.5	0.6	0.7	0.8	0.9	1.0
x^3	0.0	0.001	0.008	0.027	0.064	0.125	0.216	0.343	0.512	0.729	1.0

This gives us a pretty good idea of the graph for $0 \le x \le 1$. See Fig. 2.1. The graph is quite flat near $x = 0$, much flatter even than the graph of $y = x^2$. (See Fig. 2.5, next page.)

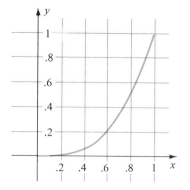

Fig. 2.1 Graph of $y = x^3$ for $0 \le x \le 1$

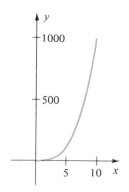

Fig. 2.2 Graph of $y = x^3$ for $0 \le x \le 10$ (Note the different scales.)

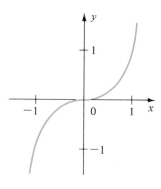

Fig. 2.3 Graph of $y = x^3$

Now consider some larger values of x:

x	0	1	2	3	4	5	6	7	8	9	10
x^3	0	1	8	27	64	125	216	343	512	729	1000

The graph rises very fast as x grows (Fig. 2.2).

For $x < 0$ we observe that $f(x) = x^3$ is an odd function, hence the graph is symmetric in the origin. (Even and odd functions were discussed in Section 3.7, p. 117.) We use this symmetry to complete the graph (Fig. 2.3). For x negative and small, x^3 is negative and very small; for x negative and large, x^3 is negative and very large.

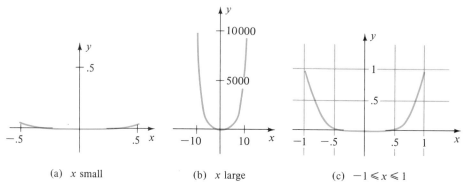

(a) x small

(b) x large

(c) $-1 \le x \le 1$

Fig. 2.4 Graph of $y = x^4$

We obtain the graph of $y = x^4$ similarly. If x is small, then y is very very small; if x is large, then y is very very large. Since $f(x) = x^4$ is an even function, its graph is symmetric in the y-axis. The result is Fig. 2.4.

It is interesting to compare the graphs of x^2, x^3, and x^4 for small x and for large x. When x is small, x^2 is very small, x^3 is even smaller, and x^4 is even smaller yet (Fig. 2.5a). But when x is large, x^2 is very large, x^3 is even larger, and x^4 is even larger yet (Fig. 2.5b). The accompanying Tables 2.1, 2.2, and graphs (Fig. 2.5) show this clearly.

Table 2.1 *Powers of small x*

x	0.0	0.1	0.2	0.3	0.4	0.5	0.6	0.7	0.8	0.9	1.0
x^2	0.00	0.01	0.04	0.09	0.16	0.25	0.36	0.49	0.64	0.81	1.00
x^3	0.000	0.001	0.008	0.027	0.064	0.125	0.216	0.343	0.512	0.729	1.000
x^4	0.0000	0.0001	0.0016	0.0081	0.0256	0.0625	0.1296	0.2401	0.4096	0.6561	1.0000

Table 2.2 *Powers of large x*

x	0	1	2	3	4	5	6	7	8	9	10
x^2	0	1	4	9	16	25	36	49	64	81	100
x^3	0	1	8	27	64	125	216	343	512	729	1000
x^4	0	1	16	81	256	625	1296	2401	4096	6561	10000

(a) $0 \leqslant x \leqslant 1$

(b) $0 \leqslant x \leqslant 10$
(Note the scales.)

Fig. 2.5 Graphs of $y = x^2$, $y = x^3$, $y = x^4$

The graphs of $y = x^5$, $y = x^7$, $y = x^9$, $\cdots$, $y = x^{2n+1}$, $\cdots$, where the exponent is odd, are all more or less like the graph of $y = x^3$. They are increasingly flat near $x = 0$ and grow increasingly rapidly for x large. They are all symmetric in the origin (Fig. 2.6). Similar remarks apply to the graphs of $y = x^4$, $y = x^6$, $y = x^8$, $y = x^{10}$, $\cdots$, $y = x^{2n}$, $\cdots$, where the exponent is even (Fig. 2.7). See next page.

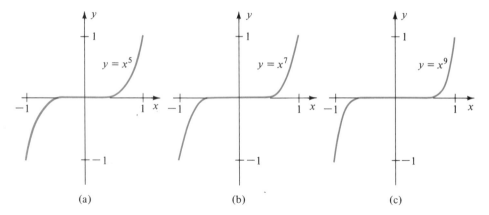

Fig. 2.6 Graphs of $y = x^{2n+1}$

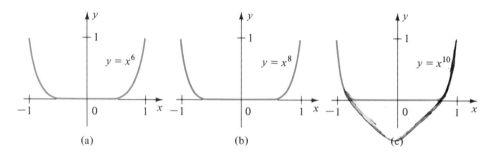

Fig. 2.7 Graphs of $y = x^{2n}$

It is important to remember that the graph of $y = x^n$ crosses the x-axis at $x = 0$. if n is odd, but does not cross if n is even. Algebraically this is obvious: if n is odd, x^n changes sign as x changes from negative to positive; if n is even, x^n is positive everywhere except at $x = 0$. Note that for $n = 1$, the graph of $y = x^n = x$ crosses the axis sharply, but for $n = 3, 5, 7, \cdots$ the graph of $y = x^n$ slithers across.

The graph of $y = c(x - r)^n$ is similar to that of $y = x^n$, except it touches the x-axis at r instead of 0, and is stretched or contracted by a factor $|c|$ in the y-direction (and reflected in the x-axis if $c < 0$). See Fig. 2.8.

Notation: The graph of $y = x^n$ rises indefinitely (without limit) as x increases without limit. This behavior is commonly abbreviated by writing $x^n \longrightarrow \infty$ as $x \longrightarrow \infty$.

If n is even, the graph also rises indefinitely as x becomes more and more negative without limit. This is abbreviated by writing $x^n \longrightarrow \infty$ as $x \longrightarrow -\infty$.

If n is odd, the graph falls indefinitely as x becomes more and more negative. This is abbreviated by writing $x^n \longrightarrow -\infty$ as $x \longrightarrow -\infty$.

We read "x approaches infinity" for $x \longrightarrow \infty$. If a is a real number, we shall also say "x approaches a" and write $x \longrightarrow a$ when x takes values nearer and nearer to a.

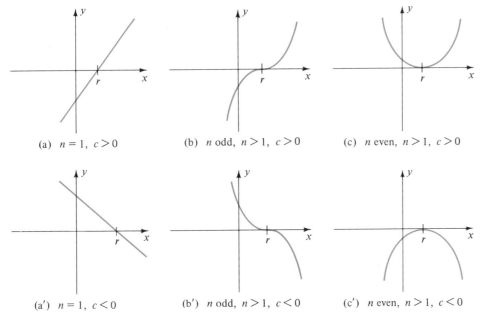

(a) $n = 1, \ c > 0$ (b) n odd, $n > 1, \ c > 0$ (c) n even, $n > 1, \ c > 0$

(a') $n = 1, \ c < 0$ (b') n odd, $n > 1, \ c < 0$ (c') n even, $n > 1, \ c < 0$

Fig. 2.8 Graphs of $y = c(x - r)^n$

Cubics

We know how to graph $y = x^3$. Does this help us graph the general cubic function

$$y = ax^3 + bx^2 + cx + d\,?$$

Unfortunately, no. It does help us graph special cases like $y = ax^3$, $y = ax^3 + d$, $y = a(x - h)^3 + d$. Each of these graphs has the same general shape as the graph of $y = x^3$. A positive factor a stretches in the y-direction; a negative factor a stretches by amount $|a|$ in the y-direction and forms a mirror image in the x-axis. The summand d moves the graph up or down; the replacement of x by $x - h$ moves the graph left or right. For some examples see Fig. 2.9, next page.

A complete discussion of cubic graphs requires differential calculus. It is proved that the possible shapes are those shown in Fig. 2.10. If you must graph a cubic, plot a large number of points and fit a smooth curve. Try to determine which of the shapes in Fig. 2.10 (on p. 129) applies.

Here are four properties of the graph of $y = ax^3 + bx^2 + cx + d$ from which the basic shapes are derived.

1. If $a > 0$, then $y \longrightarrow \infty$ as $x \longrightarrow \infty$, and $y \longrightarrow -\infty$ as $x \longrightarrow -\infty$.

2. If $a < 0$, then $y \longrightarrow -\infty$ as $x \longrightarrow \infty$, and $y \longrightarrow \infty$ as $x \longrightarrow -\infty$.

3. The graph intersects each horizontal line at least once and at most three times.

4. The graph either rises steadily, falls steadily, or has one peak and one pit.

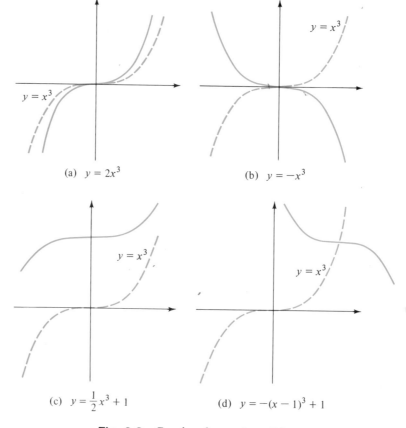

(a) $y = 2x^3$

(b) $y = -x^3$

(c) $y = \frac{1}{2}x^3 + 1$

(d) $y = -(x-1)^3 + 1$

Fig. 2.9   Graphs of $y = a(x - h)^3 + d$

We do not intend to prove these properties here. One of them, property 4, requires calculus. Properties 1 and 2 are more elementary. They are true because for large values of $|x|$, the quantity $ax^3 + bx^2 + cx + d$ is about the size of the term ax^3. To see why, write

$$ax^3 + bx^2 + cx + d = ax^3 \left(1 + \frac{b}{ax} + \frac{c}{ax^2} + \frac{d}{ax^3} \right).$$

When $|x|$ is very large, the quantity in brackets is very close to 1; hence the graph of $y = ax^3 + bx^2 + cx + d$ behaves very much like the graph of $y = ax^3$.

Property 3 follows from the fact, shown in the next section, that a cubic equation has at least one and at most three real solutions. The graph of $y = ax^3 + bx^2 + cx + d$ intersects the horizontal line $y = k$ whenever

$$ax^3 + bx^2 + cx + d = k, \quad \text{that is,} \quad ax^3 + bx^2 + cx + (d - k) = 0.$$

This cubic equation has at least one and at most three real solutions, so property 3 follows.

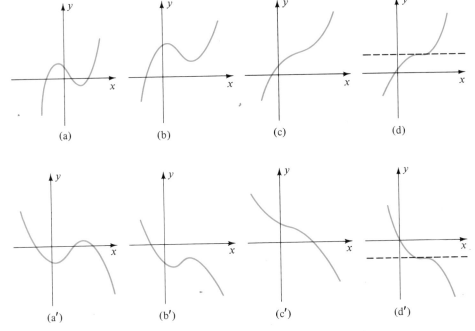

Fig. 2.10   Graphs of cubics showing all possible shapes

Graph:

1. $y = \frac{1}{4}x^4$ **2.** $y = -x^4$ **3.** $y = \frac{1}{8}x^3$

4. $y = -x^3 + 1$ **5.** $y = -x^6$ **6.** $y = x^3 - 2$

7. $y = x^4 - 1$ **8.** $y = x^5 + 1$ **9.** $y = -x^5$

10. $y = \frac{1}{16}x^5$ ✓   **11.** $y = (x + 1)^3$ **12.** $y = (x - 1)^4$ ✓

13. $y = -(x + 1)^3$ **14.** $y = -(x - 1)^4$ **15.** $y = (x - 1)^3 + 1$

16. $y = -(x + 1)^4 - 1$ **17.** $y = \frac{1}{3}(x + 2)^4$ **18.** $y = -\frac{1}{2}(x + 2)^4 - 4$

19. $y = (x - \frac{1}{2})^5 - 1$ **20.** $y = (x + \frac{1}{2})^6 - 2$.

Graph accurately:

21. $y = (x - 2)^3, 1 \le x \le 3$ **22.** $y = -(x + 1)^3, -2 \le x \le 0$

23. $y = -(x - 1)^4, 0 \le x \le 2$ **24.** $y = \frac{1}{2}(x - 2)^4, 1 \le x \le 3$. ✓

Graph roughly; plot for the given values of x:

25. $y = x^3 - 4x, x = 0, \pm 1, \pm 2, \pm 3$

26. $y = x^3 + 2x^2 - x - 2, x = -3, -2, -1, 0, 1, 2$

27. $y = \frac{1}{3}x^3 - \frac{1}{2}x^2 - 6x + 1, x = -3, -2, -1, 0, 1, 2, 3, 4$ ✓

28. $y = x^3 - 3x + 3, x = -3, -2, -1, 0, 1, 2$. ✓

3. GRAPHS OF FACTORED POLYNOMIALS

It is pretty hard to say much about the graph of a general polynomial

$$f(x) = a_n x^n + a_{n-1} x^{n-1} + \cdots + a_1 x + a_0, \qquad a_n \neq 0.$$

Given one, about the most we can do is plot a bunch of points and hope for the best. In calculus it is shown that the graph of $f(x)$ changes direction at most $n - 1$ times, which is a little help. This means for instance that if Fig. 3.1 represents the graph of a polynomial, then its degree must be seven or more.

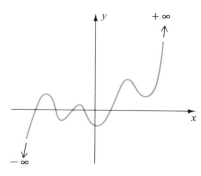

Fig. 3.1   The graph changes direction six times.

We can, however, say something about the behavior of the graph as $x \longrightarrow \infty$ or $x \longrightarrow -\infty$. We write

$$f(x) = a_n x^n + a_{n-1} x^{n-1} + \cdots + a_1 x + a_0$$

$$= a_n x^n \left(1 + \frac{a_{n-1}}{a_n x} + \cdots + \frac{a_1}{a_n x^{n-1}} + \frac{a_0}{a_n x^n} \right).$$

For $|x|$ very large, the quantity in parentheses is very close to 1, so $f(x)$ is about the size of $a_n x^n$. Hence for $|x|$ very large, the graph of $y = f(x)$ is like the graph of $y = a_n x^n$. As $x \longrightarrow \infty$ or $x \longrightarrow -\infty$, it either zooms up or down, depending on the sign of a_n and (for $x \longrightarrow -\infty$) whether n is even or odd.

Factored Polynomials

Polynomials of the form

$$f(x) = (x - r_1)(x - r_2) \cdots (x - r_n)$$

are particularly easy to graph. Each r_i is a **zero** of $f(x)$, that is, a solution of the equation $f(x) = 0$. For instance,

$$f(r_1) = (r_1 - r_1)(r_1 - r_2) \cdots (r_1 - r_n)$$
$$= 0 \cdot (r_1 - r_2) \cdots (r_1 - r_n) = 0.$$

There are no other zeros of $f(x)$, for the product

$$(x - r_1)(x - r_2) \cdots (x - r_n)$$

can be 0 only if one of the factors is 0, that is, only if x is one of the numbers $r_1, r_2, \cdots, r_n$.

We begin with a detailed study of factored cubic polynomials ($n = 3$).

Case 1. $f(x) = (x - r)^3$. This is the simplest cubic; its graph is like that of $f(x) = x^3$, except shifted horizontally so that it crosses the x-axis at r, not at 0. (See Fig. 3.2.)

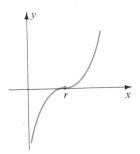

Fig. 3.2 $y = (x - r)^3$

Fig. 3.3 Signs of
$f(x) = (x - r)^2(x - s)$,
where $r < s$

Case 2. $f(x) = (x - r)^2(x - s)$, where $r < s$. First of all, $f(x)$ is zero at $x = r$ and $x = s$. Now suppose $x \neq r$. Then $(x - r)^2 > 0$, so the sign of $f(x)$ is the same as the sign of $(x - s)$, negative for $x < s$, positive for $x > s$. We conclude that $f(x)$ changes from negative to positive as x passes through s. As x passes through r, the cubic $f(x)$ does not change sign (Fig. 3.3). Therefore the graph touches the x-axis at r without crossing. Also $f(x) \longrightarrow -\infty$ as $x \longrightarrow -\infty$ and $f(x) \longrightarrow \infty$ as $x \longrightarrow \infty$. The result is Fig. 3.4.

For $r > s$ the situation is similar, and the graph is shown in Fig. 3.5.

Fig. 3.4 Rough graph of
$y = (x - r)^2(x - s)$,
where $r < s$

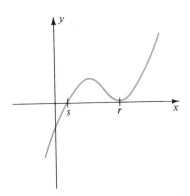

Fig. 3.5 Rough graph of
$y = (x - r)^2(x - s)$,
where $r > s$

Case 3. $f(x) = (x - r)(x - s)(x - t)$ with $r < s < t$. As x passes through each zero, the sign of f changes (Fig. 3.6). As before $f(x) \longrightarrow \infty$ as $x \longrightarrow \infty$ and $f(x) \longrightarrow -\infty$ as $x \longrightarrow -\infty$. The rough graph is now evident (Fig. 3.7).

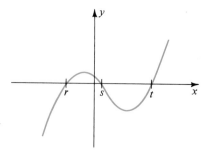

Fig. 3.6 Signs of
$f(x) = (x - r)(x - s)(x - t)$,
where $r < s < t$

Fig. 3.7 Rough graph of
$y = (x - r)(x - s)(x - t)$,
where $r < s < t$

We are ready to examine the general factored polynomial of the form

$$f(x) = (x - r_1) \cdots (x - r_n).$$

It is important how many times each factor $(x - r_i)$ occurs, so we write

$$f(x) = (x - r_1)^{m_1} \cdots (x - r_k)^{m_k}, \qquad r_1 < r_2 < \cdots < r_k,$$

to show clearly each zero r_i with its **multiplicity** m_i.

Let us choose one of the zeros r of multiplicity m and study the graph of $y = f(x)$ near $x = r$. We write $f(x) = (x - r)^m h(x)$, lumping all the other factors together in $h(x)$. Now $h(r) \neq 0$ as can be seen from the factored form above, and near r the function $h(x)$ does not change sign. Thus, for x near r, the graph of $y = f(x)$ looks pretty much like the graph of $y = h(r)(x - r)^m$; it must be close to one of the six types shown in Fig. 2.8, p. 127. The graph crosses the x-axis if m is odd, but does not cross if m is even.

In calculus it is shown that there is a single peak (or pit) between successive zeros of a completely factored polynomial.

■ *Example 3.1*

Sketch the graph of $y = f(x) = (x + 1)^3 x^2 (x - 2)(x - 3)^5$.

SOLUTION Since the degree of $f(x)$ is 11, an odd number, $y \longrightarrow -\infty$ as $x \longrightarrow -\infty$ and $y \longrightarrow \infty$ as $x \longrightarrow \infty$. Taking the multiplicities into account, we find the various signs and sign changes (Fig. 3.8). Next we look closely at $f(x)$ near each zero (Fig. 3.9). Now we can sketch the rough shape of the graph (Fig. 3.10).

Answer Fig. 3.10.

Fig. 3.8 Signs of $f(x) = (x + 1)^3 x^2(x - 2)(x - 3)^5$

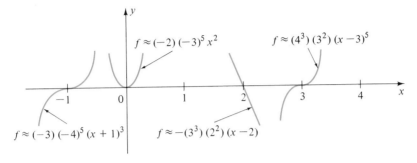

Fig. 3.9 Behavior of $f(x) = (x + 1)^3 x^2(x - 2)(x - 3)^5$ near its zeros

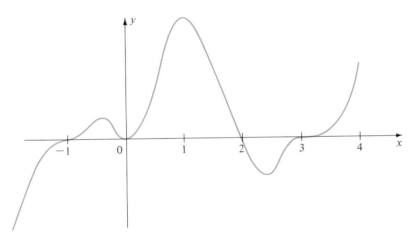

Fig. 3.10 Rough graph of $y = (x + 1)^3 x^2(x - 2)(x - 3)^5$

EXERCISES

Draw a line graph (like Fig. 3.8) of the signs and sign changes; then sketch the graph roughly:

1. $f(x) = x(x - 1)(x - 2)$
2. $f(x) = (x + 2)x(x - 2)$
3. $f(x) = (x + 2)(x + 1)(x - 1)$
4. $f(x) = (x + 2)(x - 1)(x - 2)$
5. $f(x) = x^2(x - 1)$
6. $f(x) = (x - 1)(x - 2)^2$
7. $f(x) = -(x + 1)(x - 1)^2$
8. $f(x) = -x^2(x - 1)$
9. $f(x) = \frac{1}{6}(x - 1)(x - 2)(x - 3)(x - 4)$
10. $f(x) = \frac{1}{24}x(x - 2)(x - 3)(x - 4)$
11. $f(x) = x^2(x^2 - 1)$
12. $f(x) = \frac{1}{4}(x - 1)^2(x^2 - 4)$
13. $f(x) = -\frac{1}{4}(x + 1)^2(x^2 - 4)$
14. $f(x) = -\frac{1}{12}x^2(x - 2)(x - 3)$
15. $f(x) = x(x - 1)^3$
16. $f(x) = \frac{1}{8}(x^2 - 1)^2$.

Solve the inequality:

17. $(x - 3)(x - 5)(x - 8) > 0$ **18.** $(x + 1)(x - 2)^2 < 0$
19. $x^4 - 5x^2 + 4 > 0$ **20.** $(x - 3)^2(x - 4)^5(x - 5)^6 < 0$.

4. RATIONAL FUNCTIONS

A **rational function** $r(x)$ is the quotient of two polynomials:

$$r(x) = \frac{f(x)}{g(x)},$$

where $f(x)$ and $g(x)$ are polynomials and $g(x) \neq 0$.

Obviously each polynomial is itself a rational function because

$$f(x) = \frac{f(x)}{1}.$$

Therefore, the set of rational functions includes the set of polynomials. It also includes many functions that are not polynomials, for instance

$$\frac{1}{x}, \qquad \frac{1}{x^2 + 1}, \qquad \frac{3x}{x^2 + 1}, \qquad \frac{x^3 - 2x^2 - 1}{x}, \qquad \frac{x + 2}{(x + 1)(x + 3)}.$$

A rational function can be written as a quotient of polynomials in many ways, for example,

$$\frac{1}{x + 1} = \frac{x}{x^2 + x} = \frac{x^{10}}{x^{11} + x^{10}} = \frac{x^2 + 1}{(x + 1)(x^2 + 1)} = \frac{-3}{-3x - 3} = \cdots.$$

We say that

$$\frac{1}{x + 1}, \qquad \frac{x}{x^2 + x}, \qquad \frac{x^{10}}{x^{11} + x^{10}}, \qquad \frac{x^2 + 1}{(x + 1)(x^2 + 1)}, \cdots,$$

are different *expressions* for the same rational function. Of all these, the simplest is $1/(x + 1)$ because its numerator and denominator have no common polynomial factors, whereas

$$\frac{x}{x^2 + x} = \frac{1 \cdot x}{(x + 1)x} \qquad \text{(common factor } x\text{)},$$

$$\frac{x^2 + 1}{(x + 1)(x^2 + 1)} = \frac{1 \cdot (x^2 + 1)}{(x + 1)(x^2 + 1)} \qquad \text{(common factor } x^2 + 1\text{)}, \quad \text{etc.}$$

We say that $1/(x + 1)$ expresses the rational function in **lowest terms.**

Each rational number and each rational function possess expressions in lowest terms. This fact seems fairly obvious, but its precise proof is a bit technical; we shall omit the proof. We remark that the lowest terms expression is unique except for a constant factor, e.g.,

$$\frac{x - 3}{x^2}, \qquad \frac{3x - 9}{3x^2}, \qquad \text{and} \qquad \frac{-10x + 30}{-10x^2}$$

can all be considered as lowest terms expressions for the same rational function.

Domain of a Rational Function

Suppose $r(x)$ is a rational function and

$$\frac{f(x)}{g(x)}$$

is one of its expressions as a quotient of polynomials. Each real number a for which $g(a) \neq 0$ can be substituted for x, yielding the value

$$r(a) = \frac{f(a)}{g(a)}.$$

Now suppose $g(a) = 0$, that is, a is a zero of $g(x)$. Does that mean $r(a)$ is not defined? Maybe yes and maybe no. Let us consider some examples before answering the question in all cases.

Example:

$r(x) = x/x$, that is, $r(x) = f(x)/g(x)$, where $f(x) = g(x) = x$. The only zero of $g(x)$ is 0. Thus if $a \neq 0$,

$$r(a) = \frac{f(a)}{g(a)} = \frac{a}{a} = 1.$$

Look at the corresponding graph (Fig. 4.1a). It is the graph of $y = 1$ except for the single missing point $(0, 1)$ on the y-axis. To fill in the missing point take another, simpler, representation of the same rational function:

$$r(x) = \frac{x}{x} = \frac{1}{1} = 1.$$

Thus the obvious value of $r(x)$ at $x = 0$ is 1. See Fig. 4.1b.

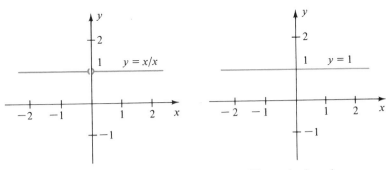

(a) graph of $y = x/x$,
 defined for $x \neq 0$

(b) graph of $y = 1$,
 defined for all x

Fig. 4.1

Example:

$$r(x) = \frac{x - 3}{(x^2 + 1)(x - 3)}.$$

The only zero of the denominator is 3. Hence $r(x)$ is defined for each real number x, except perhaps for $x = 3$. But a simpler representation of the same rational function is

$$r(x) = \frac{1}{x^2 + 1},$$

which is perfectly well defined for $x = 3$.

In these examples, the presence of a common linear factor in the numerator and denominator creates an *artificial* difficulty, which we avoid by expressing $r(x)$ in lowest terms.

In the general case, we first express a rational function $r(x)$ in lowest terms. Then it can be shown (by the Factor Theorem, Chapter 9, p. 290) that any factor $x - a$ of the denominator cannot cancel out, and that there is simply no way to define $r(a)$. For example, suppose

$$r(x) = \frac{x(x + 2)}{(x - 1)(x^2 - 4)}.$$

We reduce $r(x)$ to lowest terms by canceling the common factor $x + 2$:

$$r(x) = \frac{x}{(x - 1)(x - 2)}.$$

Now the difficulties at $x = 1$ and at $x = 2$ are not artificial; we are stuck with them. The domain of $r(x)$ consists of all x except 1 and 2, the zeros of the denominator.

> Represent a rational function $r(x)$ as the quotient
>
> $$\frac{f(x)}{g(x)}$$
>
> of polynomials in lowest terms, i.e., with no common factor. Then the domain of $r(x)$ consists of all real numbers except for the real zeros of $g(x)$.

EXERCISES

Find the domain of

1. $\dfrac{x + 1}{x - 1}$

2. $\dfrac{x^2 + 2}{3x^2 + 2}$

3. $\dfrac{x^2 + 2x + 1}{x + 1}$

4. $\dfrac{x^2 + 2x + 1}{x + 2}$

5. $\dfrac{x^2 + 3x + 2}{x^2 - 4}$

6. $\dfrac{x^2 - 1}{x^2 + 3x + 2}$

7. $\dfrac{x^2}{x^6 - 8x^3}$

8. $\dfrac{x^2 - 1}{x^3 - 1}$

9. $\dfrac{x + 1}{x^3 - x^2 + x - 1}$

10. $\dfrac{5x + 8}{x^4 - 3x^2 + 2}$

11. $\dfrac{1}{x^4 - 2x^2 + 1}$

12. $\dfrac{x + 1}{x^3 + 125}$

13. $\dfrac{x^2 - 1}{3x^2 - 5x - 3}$

14. $\dfrac{(x + 1)(x - 2)}{x^6 + 2x^3 + 4}.$

Compute $r(5)$, $r(-3)$, and $r(1)$:

15. $r(x) = \dfrac{(x^2 + 2x + 1)(x^2 - 5x + 4)}{x(x + 1)(x - 4)}$

16. $r(x) = \dfrac{(x^2 - 1)^3}{(x - 1)^2(x + 1)^3}.$

5. GRAPHS OF RATIONAL FUNCTIONS

When we studied graphs of polynomials, we began with powers $y = x^n$; these were the basic components in the system. Now we shall study the graphs of rational functions beginning with reciprocals of powers, $y = 1/x^n$.

Notation: When we write $x \longrightarrow a+$, we shall mean that x gets closer and closer to a, but only $x > a$ is allowed. (Read "x approaches a from above".) When we write $x \longrightarrow a-$, we shall mean x gets closer and closer to a, but only $x < a$ is allowed. (Read "x approaches a from below".)

Example 5.1

Graph   (a) $y = \dfrac{1}{x}$,   (b) $y = \dfrac{1}{x^2}$,   (c) $y = \dfrac{1}{x^n}$.

SOLUTION (a) The graph is symmetric in the origin since the function $f(x) = 1/x$ is odd. Therefore we plot the graph only for $x > 0$ and obtain the rest by symmetry.

Assuming $x > 0$, we see that $y = 1/x > 0$ and that y decreases as x increases; in fact $y \longrightarrow 0+$ as $x \longrightarrow \infty$. We see also that $y \longrightarrow \infty$ as $x \longrightarrow 0+$. This information together with a few plotted points gives a reasonable idea of the graph (Fig. 5.1).

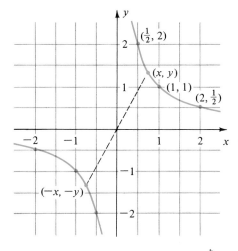

Fig. 5.1   Graph of $y = 1/x$

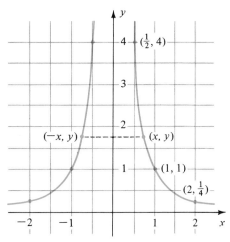

Fig. 5.2   Graph of $y = 1/x^2$

We find the graph of $y = 1/x^2$ similarly. The main differences are that y is always positive and that the graph is symmetric in the y-axis because $1/x^2$ is an even function (Fig. 5.2). The graphs in Figs. 5.1 and 5.2 are quite typical of the graphs of $y = 1/x^n$. See Fig. 5.3.

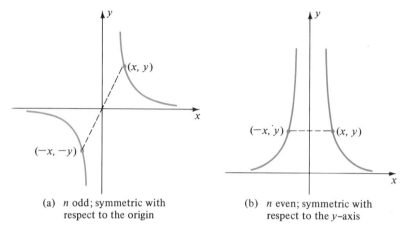

(a) n odd; symmetric with
respect to the origin

(b) n even; symmetric with
respect to the y-axis

Fig. 5.3 Graph of $y = 1/x^n$

Remark: The larger the exponent n, the faster the curve $y = 1/x^n$ approaches 0 as $x \longrightarrow \infty$, and the faster it zooms up to ∞ as $x \longrightarrow 0+$.

■ *Example 5.2*

Graph   (a) $y = \dfrac{1}{x - 1}$,   (b) $y = \dfrac{-1}{(x + 2)^2}$.

SOLUTION (a) This is the graph of $y = f(x - 1)$, where $f(x) = 1/x$. Just shift the graph of $y = 1/x$ one unit to the right (Fig. 5.4a).
    (b) This is the graph of $y = f(x + 2)$ where $f(x) = -1/x^2$. Just turn the graph of $y = 1/x^2$ upside down and then shift it two units to the left (Fig. 5.4b).

The reasoning of Example 5.2 shows that the graph of $y = 1/(x - a)^n$ looks just like that of $y = 1/x^n$, except that it is centered about the vertical axis $x = a$ instead of the y-axis (Fig. 5.5).

■ *Example 5.3*

Graph $y = \dfrac{x + 1}{x}$.

SOLUTION Rewrite the function as $y = 1 + 1/x$. The graph is now easy: just shift the graph of $y = 1/x$ up by one unit (Fig. 5.6).

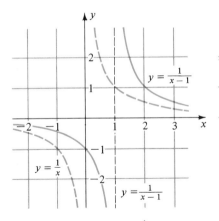

(a) graph of $y = \dfrac{1}{x-1}$

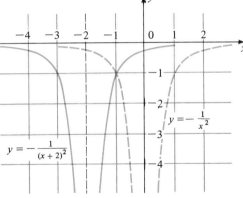

(b) graph of $y = \dfrac{-1}{(x+2)^2}$

Fig. 5.4

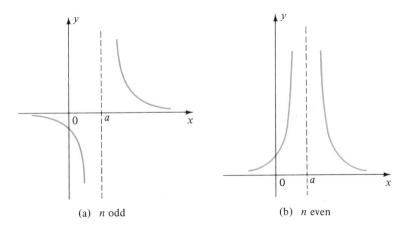

(a) n odd

(b) n even

Fig. 5.5 Graph of $y = 1/(x-a)^n$

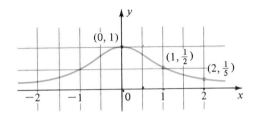

Fig. 5.6 Graph of $y = (x+1)/x$

Fig. 5.7 Graph of $y = 1/(x^2+1)$

■ *Example 5.4*

Graph $y = \dfrac{1}{x^2 + 1}$.

SOLUTION The function $1/(x^2 + 1)$ is even, so the graph is symmetric about the y-axis. Since $x^2 + 1 \geq 1$, we see that $0 < y \leq 1$ for all values of x. If x starts at 0 and increases, then y starts at 1 and decreases; if $x \longrightarrow \infty$, then $y \longrightarrow 0+$. This information plus a few plotted points yields a rough graph (Fig. 5.7, previous page).

■ *Example 5.5*

Graph $y = \dfrac{x}{x^2 + 1}$.

SOLUTION The graph is symmetric with respect to the origin; we need only plot it for $x \geq 0$. Certainly $(0, 0)$ is on the curve, and $y > 0$ when $x > 0$.

Near $x = 0$, the denominator is close to 1, hence the graph is like that of $y = x$. To study the curve for large values of x, divide numerator and denominator by x^2:

$$y = \frac{x}{x^2 + 1} = \frac{1}{x} \cdot \frac{1}{1 + (1/x^2)}.$$

For large values of x, the second factor is near 1, so the curve behaves like $y = 1/x$. Hence $y \longrightarrow 0$ as $x \longrightarrow \infty$. See Fig. 5.8a.

The clues shown in Fig. 5.8a suggest that the curve starts upward from the origin at about $45°$, but soon starts to decline, ultimately dying out towards zero. Plotting a few points confirms this (Fig. 5.8).

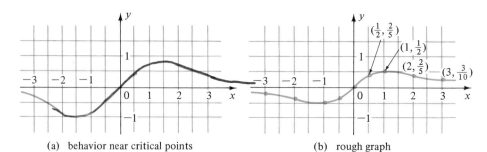

(a) behavior near critical points   (b) rough graph

Fig. 5.8 Graph of $y = x/(x^2 + 1)$

There is a simple and useful criterion for the behavior of a rational function as $|x| \longrightarrow \infty$.

Suppose

$$r(x) = \frac{a_m x^m + a_{m-1} x^{m-1} + \cdots + a_1 x + a_0}{b_n x^n + b_{n-1} x^{n-1} + \cdots + b_1 x + b_0}, \qquad a_m \neq 0, \quad b_n \neq 0.$$

Then as $|x| \longrightarrow \infty$,

$$|r(x)| \longrightarrow \infty \qquad \text{if} \quad m > n,$$
$$r(x) \longrightarrow 0 \qquad \text{if} \quad m < n,$$
$$r(x) \longrightarrow \frac{a_m}{b_n} \qquad \text{if} \quad m = n.$$

Thus, if the degree of the numerator exceeds the degree of the denominator (top-heavy case), then $|r(x)| \longrightarrow \infty$ as $|x| \longrightarrow \infty$. In the opposite (bottom-heavy) case, $r(x) \longrightarrow 0$ as $|x| \longrightarrow \infty$. If the degrees of the numerator and denominator are equal, then $r(x)$ tends to a finite non-zero number, the quotient of the leading coefficients.

The proof of these assertions is easy, but a bit tedious to write out in full generality. The trick is to divide everything in sight by the highest power of x that occurs. We shall merely illustrate with three typical examples.

Example 1:

$$r(x) = \frac{x^4 + 3x}{x^2 + 12} \qquad (m = 4, \quad n = 2).$$

Divide numerator and denominator by x^2:

$$r(x) = \frac{x^2 + \dfrac{3}{x}}{1 + \dfrac{12}{x^2}}.$$

For large values of $|x|$, the denominator is approximately 1, the numerator approximately x^2. Hence $r(x)$ behaves like $y = x^2$ as $|x| \longrightarrow \infty$. Therefore $|r(x)| \longrightarrow \infty$.

Example 2:

$$r(x) = \frac{x^2 + 5x - 3}{x^3 + x^2 + 1} \qquad (m = 2, \quad n = 3).$$

Divide numerator and denominator by x^3:

$$r(s) = \frac{\dfrac{1}{x} + \dfrac{5}{x^2} - \dfrac{3}{x^3}}{1 + \dfrac{1}{x} + \dfrac{1}{x^3}}.$$

For large values of $|x|$, the numerator is near 0, the denominator near 1. Therefore $r(x) \longrightarrow 0$ as $|x| \longrightarrow \infty$.

Example 3:

$$r(x) = \frac{2x^3 + 1}{5x^3 + x - 4} \qquad (m = n = 3, \quad a_m = 2, \quad b_n = 5).$$

Divide by x^3:

$$r(x) = \frac{2 + \dfrac{1}{x^3}}{5 + \dfrac{1}{x^2} - \dfrac{4}{x^3}}.$$

For large values of $|x|$, the numerator is approximately 2, the denominator approximately 5, so $r(x) \longrightarrow \frac{2}{5} = a_m/b_n$ as $|x| \longrightarrow \infty$.

EXERCISES

Graph:

1. $y = \dfrac{1}{x^2} - 1$ 2. $y = \dfrac{1}{x} + 3$ 3. $y = -\dfrac{1}{x} + 1$

4. $y = -\dfrac{1}{x^2} - 2$ 5. $y = \dfrac{x}{5x - 3}$ 6. $y = \dfrac{2}{1 + x^4}$

7. $y = \dfrac{-1}{4 + x^2}$ 8. $y = \dfrac{-x}{3x + 7}$ 9. $y = \dfrac{x^2}{x + 1}$

10. $y = \dfrac{x - 1}{x^2}$.

Describe the behavior of $r(x)$ as $x \longrightarrow \infty$:

11. $r(x) = \dfrac{1}{x + 1} - \dfrac{1}{x - 1}$ 12. $r(x) = x^2 - \dfrac{1}{x^2}$.

6. FACTORED RATIONAL FUNCTIONS

It is fairly easy to sketch the graph of a rational function that is completely factored into linear factors:

$$y = a \frac{(x - r_1)^{m_1}(x - r_2)^{m_2} \cdots (x - r_h)^{m_h}}{(x - s_1)^{n_1}(x - s_2)^{n_2} \cdots (x - s_k)^{n_k}}.$$

We assume this expression is in lowest terms, hence none of the numbers r_i is the same as any of the numbers s_j.

Suppose r is one of the zeros of the numerator. Write

$$y = g(x)(x - r)^m,$$

where $g(x)$ is composed of all the other factors of the numerator and denominator lumped together. Note that $g(r) \neq 0$. If $g(r) = c$, then near $x = r$ the graph is like that of $y = c(x - r)^m$. Similarly, near a zero s of the denominator, the graph is like that of $y = d/(x - s)^n$.

We have further information too: we can find the behavior of y as $x \longrightarrow \pm\infty$, and we know that y changes sign at r_i or s_j if the corresponding exponent m_i or n_j is odd.

■ *Example 6.1*

Graph $y = \dfrac{(x - 2)^2}{x + 1}$.

SOLUTION   There is one zero of the numerator, at $x = 2$, and one zero of the denominator, at $x = -1$.

To study the graph near $x = 2$, write

$$y = g(x)(x - 2)^2,$$

where $g(x) = 1/(x + 1)$. Since $g(2) = \frac{1}{3}$, the graph behaves like that of

$$y = \tfrac{1}{3}(x - 2)^2$$

near $x = 2$.

To study the graph near $x = -1$, write

$$y = g(x)\frac{1}{x + 1},$$

where $g(x) = (x - 2)^2$. Since $g(-1) = 9$, the graph behaves like

$$y = \frac{9}{x + 1}$$

near $x = -1$.

Further information: the given rational function is top-heavy, hence $|y| \longrightarrow \infty$ as $|x| \longrightarrow \infty$. Check signs:

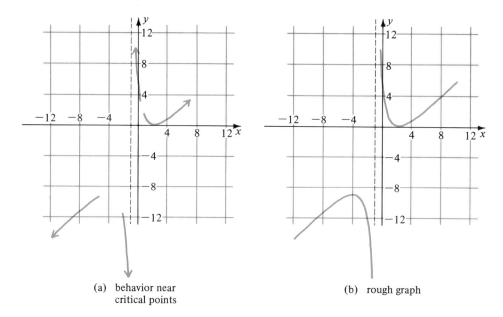

(a) behavior near
     critical points

(b) rough graph

Fig. 6.1   Graph of $y = (x - 2)^2/(x + 1)$

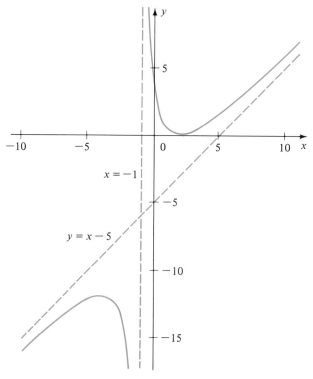

Fig. 6.2   More accurate graph of $y = \dfrac{(x - 2)^2}{(x + 1)}$

and sketch these facts (Fig. 6.1a). Then fill in the rough graph (Fig. 6.1b).

With a little skill, the graph can be improved considerably. By long division

$$y = \frac{(x-2)^2}{x+1} = x - 5 + \frac{9}{x+1}.$$

Note that $9/(x+1) \longrightarrow 0+$ as $x \longrightarrow \infty$, and $9/(x+1) \longrightarrow 0-$ as $x \longrightarrow -\infty$. Therefore, the graph is very slightly above the line $y = x - 5$ as $x \longrightarrow \infty$ and very slightly below this line as $x \longrightarrow -\infty$. See Fig. 6.2.

■ *Example 6.2*

Graph $y = \dfrac{x-1}{x^2}$.

SOLUTION Near $x = 0$, the curve is like $y = -1/x^2$. Near $x = 1$, it is like $y = x - 1$. As $x \longrightarrow -\infty$ we have $y \longrightarrow 0-$, and as $x \longrightarrow \infty$ we have $y \longrightarrow 0+$. This information is shown in Fig. 6.3a, and the rough graph in Fig. 6.3b.

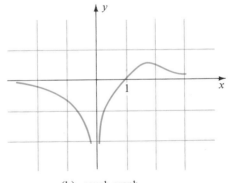

(a) behavior near critical points

(b) rough graph

Fig. 6.3   Graph of $y = (x-1)/x^2$

■ *Example 6.3*

Graph $y = \dfrac{x^2}{(x+2)(x-1)}$.

SOLUTION Near $x = 0$, the curve is like

$$y = \frac{x^2}{(0+2)(0-1)} = -\frac{x^2}{2}.$$

Near $x = -2$, it resembles

$$y = \frac{(-2)^2}{(x+2)(-2-1)} = -\frac{4}{3}\left(\frac{1}{x+2}\right).$$

Near $x = 1$, it resembles

$$y = \frac{(1)^2}{(1 + 2)(x - 1)} = \frac{1}{3}\left(\frac{1}{x - 1}\right).$$

Further information: y changes sign at $x = -2$ and at $x = 1$. Also since

$$y = \frac{x^2}{x^2 + x - 2}$$

our criterion shows that $y \longrightarrow 1$ as $x \longrightarrow \pm\infty$.

One last point: does $y \longrightarrow 1$ from above or below? To decide just check whether $y > 1$ or $y < 1$ for large positive and negative values of x. If x is large and positive then $x^2 + (x - 2) > x^2$, so

$$y = \frac{x^2}{x^2 + x - 2} < 1.$$

If x is large and negative, then $x^2 > x^2 + x - 2 > 0$, so $y > 1$. It follows that $y \longrightarrow 1+$ as $x \longrightarrow -\infty$ and $y \longrightarrow 1-$ as $x \longrightarrow \infty$.

We show all of these clues in Fig. 6.4a, from which we obtain the graph in Fig. 6.4b.

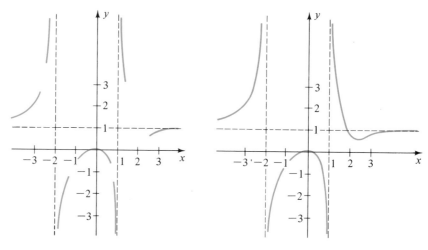

(a) behavior near critical points   (b) rough graph

Fig. 6.4   Graph of $y = x^2/(x + 2)(x - 1)$

Two more rational functions are graphed in Fig. 6.5. A glance at Fig. 6.5a shows that the lines $x = 1$, $x = -1$, and $y = 1$ play a special role. These lines are called asymptotes of the graph. In general the line $x = a$ is called a **vertical asymptote** of the graph $y = f(x)$ if $|f(x)| \longrightarrow \infty$ as $x \longrightarrow a$. A non-vertical line L is called

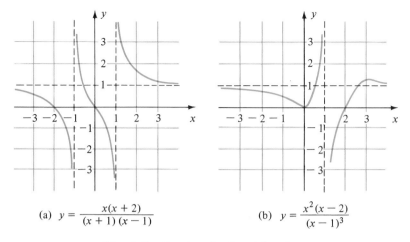

(a) $y = \dfrac{x(x+2)}{(x+1)(x-1)}$ (b) $y = \dfrac{x^2(x-2)}{(x-1)^3}$

Fig. 6.5 Graphs of rational functions

an **asymptote** of the graph $y = f(x)$ if the vertical distance between the line and the graph approaches zero as $x \longrightarrow \infty$ or $x \longrightarrow -\infty$ (or both). An example of an asymptote that is neither vertical nor horizontal is the line $y = x - 5$ in Fig. 6.2.

EXERCISES

Graph:

1. $y = \dfrac{(x+1)(x-1)}{x^3}$ **2.** $y = \dfrac{x^3}{(x+1)(x-1)}$ **3.** $y = \dfrac{(x+2)^2}{x^3}$

4. $y = \dfrac{x^3}{(x-1)^2}$ **5.** $y = \dfrac{-x^2}{(x+1)^2}$ **6.** $y = \dfrac{x^3}{(x+1)^3}$

7. $y = \dfrac{x}{x+3} + 2$ **8.** $y = \dfrac{3x^2}{(x+1)^2} - 2$ **9.** $y = \dfrac{x(x-1)}{(x+1)(x-2)}$

10. $y = \dfrac{(x+2)(x-3)}{(x+1)(x-2)}$ **11.** $y = \dfrac{(x+2)x(x-2)}{(x+1)(x-1)}$ **12.** $y = \dfrac{(x+1)(x-1)}{(x+2)x(x-2)}$

13. $y = \dfrac{x^4+2}{x(x-1)(2x^2+5)}$ **14.** $y = \dfrac{x^2+2}{(x-1)(x+1)(x^2+1)}$

15. $y = \dfrac{x^2+1}{x(x+1)(x^2+4)}$ **16.** $y = \dfrac{x^2+4}{x(x^2+1)}$.

17*. Suppose $r(x) = f(x)/g(x)$ is expressed in lowest terms and $\deg f(x) = 1 + \deg g(x)$. Why does the graph of $y = r(x)$ have an oblique asymptote?

18*. Under what circumstances does the graph of a rational function have a horizontal asymptote? vertical asymptote?

Test 1

1. Graph $y = x^3 - 3x$.
2. Graph $y = \frac{1}{4}(x + 2)(x - 1)^3$.

3. Find the domain of $r(x) = \dfrac{1}{(x - 1)(x^2 + 1)} + \dfrac{2x - 3}{x^2 - 1}$.

4. Graph $y = x + \dfrac{1}{x}$.

5. Graph $y = \dfrac{x^3}{(x + 1)^2(x - 2)(x - 3)}$.

Test 2

1. Graph $y = x^3 - 4x$.
2. Find a cubic polynomial whose graph crosses the x-axis at $x = -1$, is tangent to the x-axis at $x = 2$, and crosses the y-axis at $y = 8$.

3. Graph $y = \dfrac{x^2}{(x - 3)^2}$.

4. Graph $y = \dfrac{x(x - 1)}{x^2 - 4}$.

5. Construct a rational function with vertical asymptotes $x = 0$ and $x = 4$, and horizontal asymptote $y = 3$.

5

EXPONENTIALS AND LOGARITHMS

In this chapter we introduce, for the first time, functions that cannot be expressed in terms of the elementary algebraic operations of addition, subtraction, multiplication, division, and root extractions (radicals). First we study exponential functions, then turn these around to obtain logarithm functions. In Chapter 1 we discussed powers a^n where n is an integer and powers a^r where r is rational and $a > 0$. Now we propose to introduce exponential functions $f(x) = a^x$, where $a > 0$ and x takes *all real values*.

Certainly it is not obvious how to define such a number as $a^{\sqrt{2}}$. Rather than attempt the technical definition of exponential functions, let us see what properties they ought to have. For example, if $f(x) = 2^x$ were defined, what would it be like?

If x is an integer n, then 2^x should agree with our former definition of 2^n. We can tabulate some values of the function:

x	0	1	2	3	4	5	6	7	8	9	10
2^x	1	2	4	8	16	32	64	128	256	512	1024

The values increase rapidly! Plot the points (Fig. 1.1, next page) and join with a smooth curve. This should give some idea of the graph of $y = 2^x$.

We expect the exponential 2^x to satisfy the law of exponents: $2^{-x} = 1/2^x$. Assuming this is so, we tabulate the function for some negative values of x, using two-place accuracy. The data suggest the graph shown in Fig. 1.2 (next page).

x	-10	-9	-8	-7	-6	-5	-4	-3	-2	-1	0
$f(x) = 2^x$	0.00	0.00	0.00	0.01	0.02	0.03	0.06	0.12	0.25	0.50	1.00

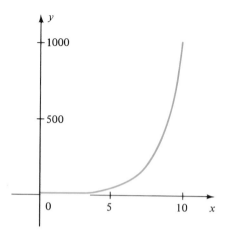

Fig. 1.1 Graph of $y = 2^x$ for $0 \leq x \leq 10$ (Note the scales.)

Fig. 1.2 Graph of $y = 2^x$ for $-10 \leq x \leq 0$ (Note the scales.)

Now let us plot $y = 2^x$, however this time using the same scale on both axes (Fig. 1.3). Several properties are evident from the graph. The curve always rises as x increases. It rises very fast as x increases through positive values, and it dies out towards zero very fast as x decreases through negative values. The same is true for the graph of $y = a^x$ for any $a > 1$, as we can see in a similar manner (Fig. 1.4).

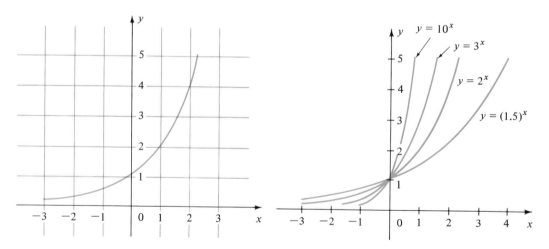

Fig. 1.3 Graph of $y = 2^x$

Fig. 1.4 Graphs of $y = a^x$ for various $a > 1$

Properties of Exponential Functions

We shall leave the actual construction of exponential functions to more advanced courses. We shall simply accept their existence and list their properties, based on experimental evidence.

For each number $a > 0$, there exists an exponential function a^x with the following properties:

(1) a^x is defined for all real x, and $a^x > 0$.

(2) $a^n = \underbrace{a \cdot a \cdot a \cdots a}_{n \text{ factors}}$ for each positive integer n.

(3)   If $a > 1$, then a^x is an increasing function ($a^x < a^y$ whenever $x < y$) and $a^x \longrightarrow \infty$ as $x \longrightarrow \infty$.

(4)   The rules of exponents hold:

$$a^{x+y} = a^x a^y, \qquad a^{x-y} = \frac{a^x}{a^y}, \qquad a^{-x} = \frac{1}{a^x}, \qquad a^0 = 1,$$

$$(a^x)^y = a^{xy}, \qquad a^x b^x = (ab)^x, \qquad 1^x = 1.$$

The number a is called the **base** of the exponential function a^x.

Remark: Tables of exponential functions for various values of a are available, and we shall make use of them as needed.

Graph of $y = a^x$ for $a < 1$

So far we have sketched the graphs of exponential functions $y = a^x$ only for $a > 1$. What does the graph look like if $0 < a < 1$?

Let $b = 1/a$. Then $b > 1$. By the rules of exponents

$$a^x = b^{-x} = \frac{1}{b^x}.$$

Since b^x is an increasing function, a^x is a decreasing function. We can say even more: the graph of $y = a^x$ is the mirror image in the y-axis of the graph of $y = b^x$ because the height of the curve $y = a^x$ at $-x$ is the height of the curve $y = b^x$ at x. For example, the graph of $y = (\frac{1}{2})^x$ is the mirror image of the graph of $y = 2^x$. See Fig 1.5.

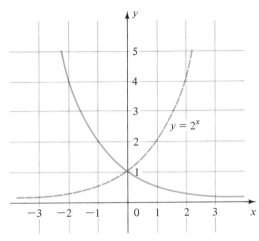

Fig. 1.5   Graph of $y = (\frac{1}{2})^x$

Rate of Growth

The exponential function 2^x grows rapidly as x increases. From the table on p. 149 we see that $2^{10} = 1024 > 10^3$. It follows that

$$2^{20} = (2^{10})^2 > (10^3)^2 = 10^6, \qquad 2^{30} > 10^9, \qquad 2^{300} > 10^{90}, \qquad \text{etc.}$$

This rate of growth as $x \longrightarrow \infty$ is extremely rapid, much more rapid than the growth of any polynomial. For example, let us compare 2^x with the polynomial x^{10}, which itself grows quite fast:

$$x = 100: \qquad 2^x = 2^{100} = (2^{10})^{10} > (10^3)^{10} = 10^{30},$$

$$x^{10} = 100^{10} = (10^2)^{10} = 10^{20},$$

$$\frac{2^x}{x^{10}} = \frac{2^{100}}{100^{10}} > \frac{10^{30}}{10^{20}} = 10^{10}.$$

$$x = 1000: \qquad 2^x = 2^{1000} = (2^{10})^{100} > (10^3)^{100} = 10^{300},$$

$$x^{10} = 1000^{10} = (10^3)^{10} = 10^{30},$$

$$\frac{2^x}{x^{10}} = \frac{2^{1000}}{1000^{10}} > \frac{10^{300}}{10^{30}} = 10^{270}.$$

Thus if $x = 100$, then 2^x is more than 10^{10} times as large as x^{10}, and if $x = 1000$, then 2^x is more than 10^{270} times as large as x^{10}. Even though $2^x < x^{10}$ for small values of x, still 2^x far outdistances x^{10} as $x \longrightarrow \infty$.

The exponential function a^x increases very rapidly for large values of x even when a is only slightly larger than 1. For example, take $a = 1.01$. We find from tables, or by other means, that $(1.01)^{900} > 10$. Therefore

$$(1.01)^{1800} > 10^2, \qquad (1.01)^{2700} > 10^3, \qquad (1.01)^{9000} > 10^{10}, \qquad \text{etc.}$$

Each time x increases by one unit, $(1.01)^x$ increases by a factor of (1.01), i.e., by 1%. At first these increases are small; nevertheless $(1.01)^x$ eventually becomes as big as you like. Once it reaches, say 10^{10}, a 1% increase is enormous. Moral: if you invest a dollar at 1% interest per year and hold it long enough, say 9000 years, you will become fabulously rich.

Note: For a further discussion of this investment see Exercises 53–56, p. 177. Also see Example 3.2 on p. 355 for the estimate $(1.01)^{900} > 10$.

It is important to have a feeling for the rate of decrease of the exponential function a^x as $x \longrightarrow -\infty$ as well as its rate of growth as $x \longrightarrow \infty$ (assuming $a > 1$). The decrease (decay) towards zero is very rapid. The reason is simple: because $a^{-x} = 1/a^x$, the values of a^x for $x < 0$ are the reciprocals of its values for $x > 0$. Since a^x increases very rapidly, $1/a^x$ decreases very rapidly. In fact, $a^x \longrightarrow 0+$

much faster as $x \longrightarrow -\infty$ than any function $1/x^n$. For example, 2^x is more than 10^{270} times as small as $1/x^{10}$ for $x \leq -1000$.

One final observation: even though 2^x and 3^x increase very rapidly as $x \longrightarrow \infty$, the function 3^x far outdistances 2^x. The ratio of the functions is

$$\frac{3^x}{2^x} = \left(\frac{3}{2}\right)^x,$$

which itself increases rapidly. Similarly, if $b > a > 1$, then b^x is much larger than a^x as $x \longrightarrow \infty$.

EXERCISES

Graph:

1. $y = 3^x, -5 \leq x \leq 0$ **2.** $y = 3^x, 0 \leq x \leq 5$ **3.** $y = 3^x, -2 \leq x \leq 2$
4. $y = 10^x, 0 \leq x \leq 6$ **5.** $y = 10^{-x}, -6 \leq x \leq 0$ **6.** $y = (1.5)^x, -3 \leq x \leq 3$
7. $y = 2^{x-1}, 0 \leq x \leq 2$ **8.** $y = \frac{1}{2}(2^x + 2^{-x}), -2 \leq x \leq 2$
9. $y = \frac{1}{2}(2^x - 2^{-x}), -2 \leq x \leq 2$ **10.** $y = 3^x - 2^x, -1 \leq x \leq 1$.

11. Water flows into a tank in such a way that the volume of water is doubled each minute. If it takes 10 minutes to fill the tank, when is the tank half full?
12. Compare the values of 2^{-x} and x^{-2} for $x = 1, 2, 3, \cdots, 10$ by computing their ratio.
13. Find a value of n for which $2^n > 10^{50}$.
14. Find a value of x for which $2^x > x^{100}$.
15. Find a function $f(x)$ for which $f(x_1 + x_2) = f(x_1)f(x_2)$.
16. Find a function $f(x)$ for which $f(2x) = [f(x)]^2$.
17. Show graphically that $2^x = x$ has no solution.
18. Determine graphically the number of solutions to $2^x = x + 3$.
19*. Compare 2^x and x^{100} for $x = 10^3$ and $x = 10^6$ by computing their ratio.
20*. Compare $(1.1)^x$ and x^{10} for $x = 10^3$ and $x = 10^6$ by computing their ratio. Use $2 < (1.1)^{10}$.
21. Show that $(\frac{3}{2})^4 \approx 5$ and use this fact to get a quick estimate of $(\frac{3}{2})^{20}$.
22. Make a table comparing the values of 2^x, 3^x, and 5^x for $x = 1, 2, 3, \cdots, 10$.

2. LOGARITHM FUNCTIONS

Let us consider some properties of the exponential function $y = 10^x$. Its graph is shown in Fig. 2.1 (next page). The function is positive and increasing; $y \longrightarrow \infty$ as $x \longrightarrow \infty$ and $y \longrightarrow 0+$ as $x \longrightarrow -\infty$. Hence its graph crosses each horizontal line $y = c$, where $c > 0$, at exactly one point (x, c). See Fig. 2.2, next page. We state this property of 10^x in other words:

> If $y > 0$, then there is one and only one real number x such that
>
> $$10^x = y.$$
>
> This number x is called the **logarithm** of y, and is written $x = \log y$.

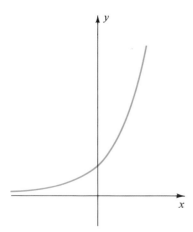

Fig. 2.1   Graph of $y = 10^x$

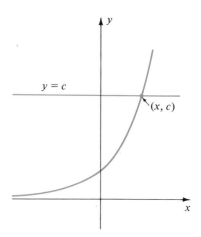

Fig. 2.2   Graphs of $y = 10^x$ and $y = c$, where $c > 0$

We have defined a new function, the logarithm function, whose domain is the set of positive real numbers. Its graph, by very definition, is the graph in Fig. 2.1, *with y interpreted as the independent variable.* To get the usual picture, y a function of x, we interchange x and y. The result is Fig. 2.3. Since the graph of $y = \log x$ is obtained from the graph of $y = 10^x$ by interchanging x and y, it follows that these two graphs are mirror images of each other in the line $y = x$. See Fig. 2.4.

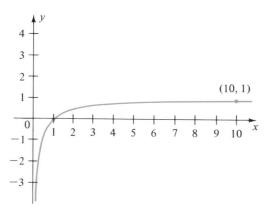

Fig. 2.3   Graph of $y = \log x$

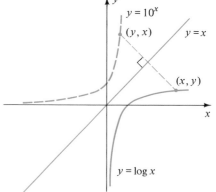

Fig. 2.4   The graphs of $y = \log x$ and $y = 10^x$ are reflections of each other in the line $y = x$.

We note five important properties:

---

(1) The function $y = \log x$ is defined for all $x > 0$.

(2) The function $y = \log x$ increases as x increases:

$$\text{if } x_1 < x_2, \text{ then } \log x_1 < \log x_2.$$

(3) $y \longrightarrow -\infty$ as $x \longrightarrow 0+$.

(4) $y \longrightarrow \infty$ as $x \longrightarrow \infty$.

(5) Each horizontal line $y = c$ meets the graph of $y = \log x$ in exactly one point.

Property (1) states that each positive number has a logarithm, and Property (5) states that each real number is the logarithm of a unique positive number. Therefore, we should be able to identify a positive number by its logarithm. In practice this is indeed what happens; the logarithm function is so well tabulated that any positive number can be identified (except for a relatively tiny error) by its logarithm.

Note: The statements "$y = 10^x$" and "$x = \log y$" mean precisely the same thing.

EXPONENTIAL STATEMENT	EQUIVALENT LOGARITHMIC STATEMENT
$10^0 = 1$	$\log 1 = 0$
$10^1 = 10$	$\log 10 = 1$
$10^2 = 100$	$\log 10^2 = 2$
$10^3 = 1000$	$\log 10^3 = 3$
$10^{-1} = \frac{1}{10}$	$\log 10^{-1} = -1$
$10^{-2} = \frac{1}{100}$	$\log 10^{-2} = -2$
$10^{1/2} = \sqrt{10}$	$\log 10^{1/2} = \frac{1}{2}$.

Note that $\log 1 = 0$, that $\log x > 0$ if $x > 1$, and that $\log x < 0$ if $0 < x < 1$. Like every property of logarithms, this is just a restatement of a property of exponentials: $10^0 = 1$, while $10^x > 1$ if $x > 0$, and $0 < 10^x < 1$ if $x < 0$.

The relation between "logarithm" and "ten to the x" is an inverse one. Each function undoes what the other does:

$$10^{\log x} = x, \qquad \log 10^x = x.$$

Why? Because $\log x$ is that unique number such that "10 to the $\log x$" is x, that is, $10^{\log x} = x$. Also $\log 10^x$ is that unique number y such that $10^y = 10^x$. Hence $y = x$, so $\log 10^x = x$.

Rules of Logarithms

Logarithms satisfy certain rules (algebraic properties) of great importance in theory and in computation:

$$\begin{aligned}
\log(x_1 x_2) &= \log x_1 + \log x_2 \\
\log(x_1/x_2) &= \log x_1 - \log x_2 \\
\log x^b &= b \log x.
\end{aligned}$$

These properties are inherited from corresponding properties of 10^x. Take the first one for example. Suppose

$$y_1 = \log x_1, \quad y_2 = \log x_2, \quad \text{that is,} \quad x_1 = 10^{y_1}, \quad x_2 = 10^{y_2}.$$

Then, by a rule for exponentials,

$$x_1 x_2 = 10^{y_1} 10^{y_2} = 10^{y_1 + y_2},$$

which means

$$\log(x_1 x_2) = y_1 + y_2 = \log x_1 + \log x_2.$$

The other two properties are proved similarly.

The first rule for logarithms converts multiplication problems into much easier addition problems. To multiply x_1 and x_2, add their logarithms (found in a table). Then $x_1 x_2$ is the number whose logarithm is the sum.

The rule $\log(x_1/x_2) = \log x_1 - \log x_2$ applies in a similar way to division. The rule $\log x^b = b \log x$ greatly simplifies the computation of powers and roots. For example, computing the cube root of 1291 to five decimal places can be a nasty job. However, if we write

$$\sqrt[3]{1291} = (1291)^{1/3}, \quad \log(\sqrt[3]{1291}) = \tfrac{1}{3}\log 1291,$$

the job is much easier. We divide log 1291 by 3 and then find the number whose logarithm this is.

The practical techniques of computing with logarithms will be studied in Section 6.

Notation: When we want to indicate that two numbers a and b are approximately equal, we shall write $a \approx b$. An expression like $x \approx 0.358$ will generally imply that 0.358 is the closest we can estimate x on the basis of information at hand.

■ *Example 2.1*

Given $\log 2 \approx 0.3010$ and $\log 5 \approx 0.6990$, estimate $\log[(\tfrac{2}{5})^{1/3}]$.

SOLUTION By the rules for logarithms,

$$\log(\tfrac{2}{5})^{1/3} = \tfrac{1}{3}\log\tfrac{2}{5} = \tfrac{1}{3}(\log 2 - \log 5) \approx \tfrac{1}{3}(0.3010 - 0.6990)$$
$$= -\tfrac{1}{3}(0.3980) \approx -0.1327.$$

Answer -0.1327.

It is possible to define logarithms not only in terms of 10^x, but in terms of other exponential functions as well. Consider the graph (Fig. 2.5) of $y = b^x$ for $b > 1$. This graph, like that of $y = 10^x$, meets each horizontal line $y = c$ for $c > 0$ in a single point. Hence we can define a logarithm function relative to b^x just as we did for 10^x.

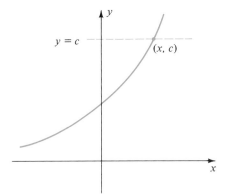

Fig. 2.5   Graph of $y = b^x$ for $b > 1$

Fix a base $b > 1$. If $y > 0$, there is one and only one real number x such that

$$b^x = y.$$

This number x is called the **logarithm of** y **to the base** b and is written

$$x = \log_b y.$$

When $b = 10$, we have the ordinary $\log x = \log_{10} x$, also called the **common logarithm** of x. In this case we shall write "$\log x$" without indicating the base 10. Here are some examples for the base $b = 2$:

EXPONENTIAL RELATION	EQUIVALENT LOGARITHMIC RELATION
$2^0 = 1$	$\log_2 1 = 0$
$2^1 = 2$	$\log_2 2 = 1$
$2^2 = 4$	$\log_2 4 = 2$
$2^{10} = 1024$	$\log_2 1024 = 10$
$2^{-1} = \frac{1}{2}$	$\log_2 \frac{1}{2} = -1$
$2^{-4} = \frac{1}{16}$	$\log_2 \frac{1}{16} = -4$
$2^{1/2} = \sqrt{2}$	$\log_2 \sqrt{2} = \frac{1}{2}.$

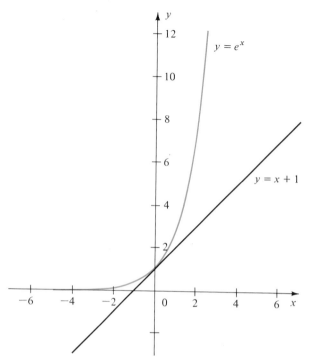

Fig. 2.6 Graph of $y = e^x$. The tangent line at $(0, 1)$ is $y = x + 1$.

The logarithm function to the base b obeys the same rules as the common logarithm function:

$$\log_b(x_1 x_2) = \log_b x_1 + \log_b x_2$$
$$\log_b(x_1/x_2) = \log_b x_1 - \log_b x_2$$
$$\log_b(x^c) = c \log_b x.$$

There is an important relation between the functions $\log x$ and $\log_b x$. Write

$$x = b^{\log_b x}$$

and take logs to base 10 on both sides:

$$\log x = \log(b^{\log_b x}) = (\log_b x)(\log b).$$

Hence

$$\log_b x = \frac{\log x}{\log b}.$$

Therefore the function $\log_b x$ is merely a constant multiple of the function $\log x$. For example, from the approximation $\log 3 \approx 0.4771$, we deduce the approximation

$$\log_3 x \approx \frac{\log x}{0.4771}.$$

The basic principle here is that logs to the base b are proportional to logs to the base 10. It follows that all log tables are proportional. Once you have an accurate table of logs to the base 10, you can approximate logs to any base.

The base 10 is the most used base in practical computations because it goes so well with decimals. However there is another base, called e, that in many respects is the most natural one for theoretical work. To ten places

$$e \approx 2.7182818285.$$

There are several ways to define e, but they all require knowledge usually contained in calculus courses. One definition of e has an easy graphical interpretation (Fig. 2.6). Each exponential graph $y = a^x$ passes through the point $(0, 1)$. The graph of $y = e^x$ is the only one of these tangent to the line $y = x + 1$ of slope 1.

The function $\log_e x$ is called the **natural logarithm** and is often written $\ln x$.

EXERCISES

Find the common logarithm:

1. 10,000
2. 1,000,000
3. 0.01
4. 0.00001
5. $\sqrt{1000}$
6. $\sqrt[3]{0.01}$.

Find the logarithm to the base 2:

7. 8
8. 128
9. 1024
10. $\sqrt[3]{256}$
11. 1/16
12. 1/64
13. $1/2\sqrt{2}$
14. $1/(\sqrt[3]{2})^5$.

Use the approximations $\log 2 \approx 0.301$, $\log 3 \approx 0.477$, and $\log 5 \approx 0.699$ to estimate

15. $\log 6$
16. $\log 48$
17. $\log(9/16)$
18. $\log \sqrt{12}$
19. $\log 45$
20. $\log 225$
21. $\log(\sqrt{5}/96)$
22. $\log \sqrt[3]{36/5}$
23. $\log(3/25)$
24. $\log(6/125)^{1/5}$.

25. Find $10^{\log 17}$.
26. Find $\log_5 5^{12}$.
27*. Simplify $(\log_a b)(\log_b a)$.
28*. Express $\log_b x$ in terms of $\log_a x$.
29. Let $a > 0$ and $b > 0$ and solve for x:

$$\log x = \tfrac{1}{2}(\log a + \log b).$$

30. Find all x such that $-2 < \log x < -1$.
31. From the properties of exponential functions derive the formula $\log(x/y) = \log x - \log y$.
32. (cont.) Do the same for $\log x^c = c \log x$.
33. Does the rule $a^x b^x = (ab)^x$ imply something about logarithms?
34. By considering some numerical values of x show that the function $y = \log x$ increases much more slowly than $y = x$ as $x \longrightarrow \infty$.
35. Suppose $a > 0$, $b > 0$, and $a \neq 1$. Show that a number c exists such that $b^x = a^{cx}$ for all real x.
36. (cont.) What does the result of Ex. 35 imply about the shape of the graphs of $y = a^x$ and $y = b^x$ if $a > 1$ and $b > 1$?

37. Find the domain of the function log log x.
38. Compare log log x and log x for $x = 10^{1000}$ and $x = 10^{10^6}$.
39. Sketch $y = \log_2 x$ and $y = \log x$ on the same graph.
40. Sketch $y = \log 5x$.
41. Sketch $y = \log x^3$.
42. Find the relation between $\log_2 x$ and $\log_8 x$.
43*. Which is larger, $\log_6 5$ or $\log_7 5$?
44*. (cont.) Express the ratio $(\log_6 5)/(\log_7 5)$ in terms of common logs.
45. We have defined $\log_b x$ for base $b > 1$. Show how to define it for $0 < b < 1$.
46. (cont.) Suppose $0 < b < 1$ and let $c = 1/b$. Show that $\log_b x = -\log_c x$.

3. POWER FUNCTIONS

In Chapters 3 and 4 we studied functions $f(x) = x^n$ where n is an integer, either positive or negative. Now we consider more general power functions of the form

$$f(x) = x^a,$$

where a is any real number.

How shall we define x^a if a is not an integer? Recall that

$$x = 10^{\log x}.$$

If n is an integer, then

$$x^n = (10^{\log x})^n = 10^{n \log x}.$$

This is a round-about way of writing x^n, but there is one great advantage: it makes no difference whether or not n is an integer. Therefore we *define*

$$x^a = 10^{a \log x}.$$

For any real number a, this definition makes sense provided $x > 0$. Furthermore, it agrees with our old notion of x^n when $a = n$, an integer.

For each real number a there is a power function
$$f(x) = x^a$$
defined for $x > 0$ by
$$x^a = 10^{a \log x}.$$

Remark: An equivalent definition is $x^a = e^{a \ln x}$ because $x = e^{\ln x}$.

Power functions inherit important algebraic properties from exponential and logarithm functions:

> The power of a product is the product of powers:
> $$(xy)^a = x^a y^a.$$
> The product of power functions is a power function:
> $$x^a x^b = x^{a+b}.$$
> The reciprocal of a power function is a power function:
> $$x^{-a} = \frac{1}{x^a}.$$

We shall prove only the first of these rules. To do so, we start with

$$a \log(xy) = a(\log x + \log y) = a \log x + a \log y.$$

From this,

$$(xy)^a = 10^{a \log(xy)} = 10^{a \log x + a \log y}$$
$$= 10^{a \log x} 10^{a \log y} = x^a y^a.$$

Power functions also inherit growth properties from exponential and logarithm functions:

> Let $a > 0$. Then the power function $y = x^a$ is strictly increasing:
> $$\text{if } x_1 < x_2, \text{ then } x_1^a < x_2^a.$$
> If $x \longrightarrow 0+$, then $x^a \longrightarrow 0+$.
> If $x \longrightarrow \infty$, then $x^a \longrightarrow \infty$.

> Let $a < 0$. Then the power function $y = x^a$ is strictly decreasing:
> $$\text{if } x_1 < x_2, \text{ then } x_1^a > x_2^a.$$
> If $x \longrightarrow 0+$, then $x^a \longrightarrow \infty$.
> If $x \longrightarrow \infty$, then $x^a \longrightarrow 0+$.

Let us prove the statements for positive a; those for negative a follow directly because $x^{-a} = 1/x^a$.

Suppose $a > 0$. Then $a \log x$ increases as x increases, $a \log x \longrightarrow \infty$ as $x \longrightarrow \infty$, and $a \log x \longrightarrow -\infty$ as $x \longrightarrow 0+$. Since

$$x^a = 10^{a \log x},$$

it follows that x^a is strictly increasing, $x^a \longrightarrow \infty$ as $x \longrightarrow \infty$, and $x^a \longrightarrow 0+$ as $x \longrightarrow 0+$. Because of the last assertion, it is logical to *define* $0^a = 0$ provided $a > 0$.

Rational Powers

Let us study the graphs of the power functions $y = x^r$, where r is a rational number. We begin with $r = \frac{1}{2}, \frac{1}{3}, \frac{1}{4}, \cdots$. For the graph of $y = x^{1/n}$ we observe that $y^n = x$ and $y \geq 0$. Therefore we want that part of the graph of $y^n = x$ where $y \geq 0$. But the graph of $y^n = x$ is obtained from that of $x^n = y$ by interchanging x and y, that is, by reflection in the line $y = x$. See Fig. 3.1.

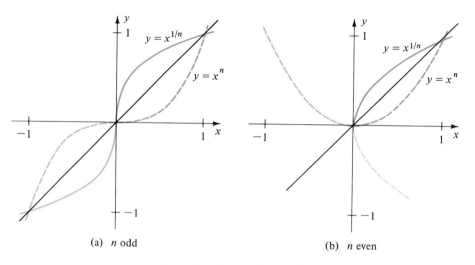

(a) *n* odd   (b) *n* even

Fig. 3.1   Graph of $y = x^{1/n}$

Remark: If n is an odd integer, it is possible to define $x^{1/n}$ also for $x < 0$. We can define $x^{1/n}$ at $-x$ by the corresponding point on the graph of $y^n = x$:

$$(-x)^{1/n} = -x^{1/n}.$$

This cannot be done if n is even. Why?

We use the following relation to graph $y = x^r$ for more general positive rational powers.

If r is a positive rational number of the form m/n, then

$$x^r = (x^{1/n})^m.$$

Once the values of $x^{1/n}$ are known, it is easy to compute the values of $x^{m/n}$. The graphs of $y = x^r$ for various rational values of r are shown in Fig. 3.2.

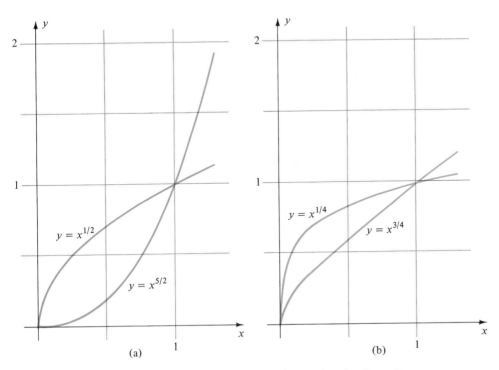

(a)   (b)

Fig. 3.2   Graphs of $y = x^r$ for various rational values of r

The graphs of $y = x^r$ for $r = -\frac{1}{2}, -\frac{1}{3}, -\frac{1}{4}$ are shown in Fig. 3.3.

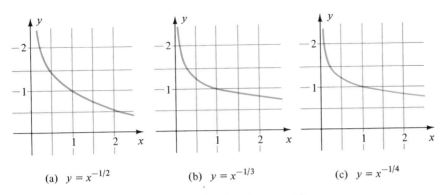

(a) $y = x^{-1/2}$ (b) $y = x^{-1/3}$ (c) $y = x^{-1/4}$

Fig. 3.3   Graphs of $y = x^{-1/n}$

The graphs of $y = x^{-2/3}$ and $y = x^{-3/2}$ are shown in Fig. 3.4 (next page). They are mirror images of each other in the line $y = x$. Why?

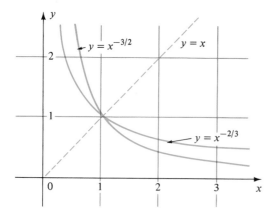

Fig. 3.4 Graphs of $y = x^{-2/3}$ and $y = x^{-3/2}$

Functions Involving Square Roots

The most common power function with non-integer exponent is the square root function $f(x) = x^{1/2} = \sqrt{x}$. Composite functions involving square roots occur often, for example,

$$f(x) = \sqrt{1 - x^2}, \qquad \sqrt{1 + x^2}, \qquad \sqrt{x^2 - 1}.$$

These are defined *only* for values of x which make the **radicand** (the quantity under the radical) non-negative. Thus

$f(x)$	DOMAIN
$\sqrt{1 - x^2}$	$-1 \le x \le 1$
$\sqrt{1 + x^2}$	all x
$\sqrt{x^2 - 1}$	$x \le -1$ or $1 \le x$.

There are two important tricks that often simplify computations involving square roots. The first, rationalizing the denominator, was discussed in Chapter 1, p. 26. It eliminates radicals from the denominator; for example

$$\frac{1}{\sqrt{7} - 1} = \frac{1}{\sqrt{7} - 1} \frac{\sqrt{7} + 1}{\sqrt{7} + 1} = \frac{\sqrt{7} + 1}{7 - 1} = \frac{1}{6}(\sqrt{7} + 1).$$

The second, called **rationalizing the numerator,** eliminates radicals from the numerator. Here is the idea:

$$\sqrt{b} - \sqrt{a} = (\sqrt{b} - \sqrt{a})\frac{\sqrt{b} + \sqrt{a}}{\sqrt{b} + \sqrt{a}} = \frac{b - a}{\sqrt{b} + \sqrt{a}}.$$

We shall look at two applications of this trick.

■ *Example 3.1*

Show that $\sqrt{1 + x^2} - x \longrightarrow 0+$ as $x \longrightarrow \infty$.

SOLUTION Rationalize the numerator:

$$\sqrt{1 + x^2} - x = (\sqrt{1 + x^2} - x)\frac{\sqrt{1 + x^2} + x}{\sqrt{1 + x^2} + x} = \frac{(1 + x^2) - x^2}{\sqrt{1 + x^2} + x}$$

$$= \frac{1}{\sqrt{1 + x^2} + x}.$$

Certainly $\sqrt{1 + x^2} + x \longrightarrow \infty$ as $x \longrightarrow \infty$; therefore

$$\sqrt{1 + x^2} - x \longrightarrow 0+ \quad \text{as } x \longrightarrow \infty.$$

■ *Example 3.2*

Graph $y = \sqrt{1 + x^2}$.

SOLUTION The function is defined for all real x. The graph is symmetric in the y-axis since $\sqrt{1 + (-x)^2} = \sqrt{1 + x^2}$. Therefore it suffices to consider only $x \geq 0$.

Obviously $y = 1$ for $x = 0$, and y increases as x increases. Since $\sqrt{1 + x^2} > \sqrt{x^2} = x$, we have $y > x$; the graph is above the line $y = x$. But this line is actually an asymptote of the graph because $y - x \longrightarrow 0+$ as $x \longrightarrow \infty$ by the result of Example 3.1.

This information is shown in Fig. 3.5a, and a rough graph is indicated in Fig. 3.5b. Here is a subtle point however. In sketching the graph, we *assumed* that the curve would be rounded at its low point $(0, 1)$ and not have a sharp corner there. How can we justify this?

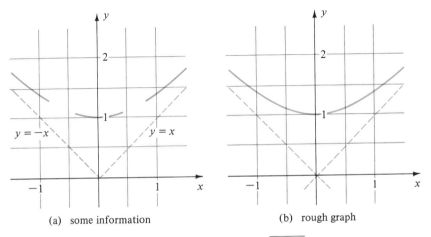

(a) some information   (b) rough graph

Fig. 3.5 Graph of $y = \sqrt{1 + x^2}$

Figure 3.5b suggests that the graph touches the line $y = 1$ very smoothly at $(1, 0)$, just as $y = x^2$ or $y = x^4$ touches the x-axis at $(0, 0)$. Let us confirm this by computing $y - 1$.

$$y - 1 = \sqrt{x^2 + 1} - 1 = (\sqrt{x^2 + 1} - 1) \frac{\sqrt{x^2 + 1} + 1}{\sqrt{x^2 + 1} + 1}$$

$$= \frac{(x^2 + 1) - 1}{\sqrt{x^2 + 1} + 1} = \frac{x^2}{\sqrt{x^2 + 1} + 1}.$$

Near $x = 0$, the denominator is approximately 2. Hence

$$y - 1 \approx \tfrac{1}{2}x^2, \qquad y \approx 1 + \tfrac{1}{2}x^2,$$

and the graph $y = 1 + \tfrac{1}{2}x^2$ is rounded at its low point $(0, 1)$.

EXERCISES

Graph carefully; use tables or slide rule:

1. $y = x^{1/2}, \quad 0 \le x \le 1$
2. $y = x^{1/2}, \quad 0 \le x \le 10$
3. $y = x^{1/3}, \quad 0 \le x \le 1$
4. $y = x^{1/3}, \quad 0 \le x \le 100$
5. $y = x^{1/3}, \quad 0 \le x \le 0.1$
6. $y = x^{1/4}, \quad 0 \le x \le 1$
7. $y = x^{3/2}, \quad 0 \le x \le 2$
8. $y = x^{5/2}, \quad 0 \le x \le 2$
9. $y = x^{-1/2}, \quad 0.2 \le x \le 2$
10. $y = x^{-4/3}, \quad 0.2 \le x \le 1$.

11. Graph $y = \sqrt{1 - x^2}$.
12. Graph $y = \sqrt{x^2 - 1}$. Show that $y = x$ is an asymptote.
13. Graph $y = 1/\sqrt{1 + x^2}$.
14. Graph $y = 1/\sqrt{2x - 3}$.

Rationalize the numerator:

15. $\dfrac{\sqrt{7} - \sqrt{5}}{2}$

16. $\dfrac{\sqrt{5} - \sqrt{2}}{\sqrt{5}}$

17. $\dfrac{\sqrt{x} + 1}{x}$

18. $\dfrac{\sqrt{x^2 + x} + \sqrt{x}}{\sqrt{x}}.$

19. Show that

$$\frac{\sqrt{1 + x} - 1}{x} \longrightarrow \frac{1}{2} \quad \text{as} \quad x \longrightarrow 0.$$

20. Show that $\sqrt{x + 1} - \sqrt{x} \longrightarrow 0+$ as $x \longrightarrow \infty$.
21. Show that $\sqrt{10001} - 100 < 0.005$.
22. Show that $\sqrt{x^2 + 2x} - x \longrightarrow 1$ as $x \longrightarrow \infty$.
23. Show that

$$\frac{1}{\sqrt[3]{b} - \sqrt[3]{a}} = \frac{\sqrt[3]{b^2} + \sqrt[3]{ab} + \sqrt[3]{a^2}}{b - a}$$

24. (cont.) Rationalize the denominator of $1/(\sqrt[3]{2} - 1)$.

In Sections 5 and 6 we shall discuss computations with logarithms—not exact, but approximate computations. For this reason, we must now discuss certain practical questions of accuracy that constantly arise in numerical work.

When we analyze data, we usually decide in advance on a certain degree of accuracy, no more than the accuracy of our measurements. Consider an example: a chemist's analytic balance that weighs anything from 0 to 150 grams with one-milligram accuracy. The read-out always has 3-decimal-place accuracy. That means the maximum error in a reading is $\pm 5 \times 10^{-4}$ gm. For one sample the read-out might be, say, 0.493 gm. That means the sample actually weighs between 0.4925 and 0.4935 gm. For another sample the read-out might be, say, 104.228 gm. This sample actually weighs between 104.2275 and 104.2285 gm.

Compare these two readings, 0.493 and 104.228. The second seems much more accurate than the first, because its possible error is only about 5 parts in one million, whereas the possible error in the first reading is about 5 parts in 5000.

The number of digits (after possible zeros on the left) provides a measure of how accurate the data is. In general, if a number is written with a decimal point, its number of **significant figures** is the number of digits from the left-most non-zero digit to the right-most digit. In our example, the number 0.493 has 3 significant figures; the number 100.223 has 6 significant figures.

Notice that 12.80 implies greater accuracy than 12.8. For a read-out of 12.80 implies an error within $\pm 5 \times 10^{-3}$ whereas a read-out of 12.8 implies an error within $\pm 5 \times 10^{-2}$.

Examples:

NUMBER	SIG. FIGS.	NUMBER	SIG. FIGS.
12.8	3	0.04	1
12.80	4	1.336	4
1500.0	5	4.38×10^{-6}	3
1.5×10^3	2	3.1416	5
10^9	1	3.14159	6

If we say the population of Paris is 11 million, we mean it is between 10.5 and 11.5 million. We should write 1.1×10^7 to indicate clearly that the number is given to 2 significant figures.

The chemist's balance we discussed gives readings to **3-decimal-place accuracy.** This means its readings have three figures to the right of the decimal point—not the same thing as three significant figures.

Round-off

Suppose we have a 5-place table, but we only require 2-place accuracy. Then we must **round off** each 5-place entry to 2 places (and this means decimal places).

Examples:

NUMBER	ROUNDED-OFF NUMBER
0.48265	0.48
0.48701	0.49
0.49013	0.49
0.49501	0.50
0.49500	?

The last entry is a problem. Do we round off to 0.49 or 0.50? The convention varies, but we shall adopt the rule "make the last digit even". So we round off to 0.50 because 0 is even and 9 is odd.

> **Rules for Rounding Off:**
> (1) If the discarded portion is less than 5000 · · · , then drop it.
> (2) If the discarded portion is greater than 5000 · · · , then drop it and add 1 to the last digit kept.
> (3) If the discarded portion is exactly 5000 · · · , then drop it; if the last digit kept is even, do nothing; if the last digit kept is odd, add 1 to it.

Each time we round off, we introduce an error. But we feel these rules are fair and hope that in a series of calculations with round-off, the errors will more or less average out, not pile up.

Example:

Round off 9.86507 to 0, 1, 2, 3, and 4 decimal places. Solution: 10, 9.9, 9.87, 9.865, 9.8651.

Example:

Round off 9.865 to 2 decimal places. Solution: 9.86.

Remark: Note the different results:

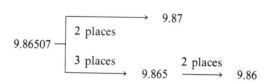

The one-step round-off is more accurate than the two-step procedure.

There is a lesson to be learned here: if you want 4 places it is better to use a 4-place table rather than round off from a 5-place table. The reason is that the entries in the tables are already rounded off by the table-makers; you may lose accuracy in rounding off again. For example, in 4-place and 5-place common log tables we find

$$\log 1.19 \approx 0.0755 \quad \text{and} \quad \log 1.19 \approx 0.07555.$$

If we round off the second entry to 4 places we get 0.0756. But a 6-place table shows $\log 1.19 \approx 0.075547$ so that 0.0755 is more accurate.

EXERCISES

Round off to 2 decimal places:

1. 0.4444, 0.3128, 0.1075, 0.2555

2. 6.411, 10.91, 2.0041, 3.0095.

Round off to 3 decimal places:

3. 0.0005, 0.00049, 16.2445, 3.7855

4. 1.8125, 3.14159265, 0.9997, 0.99946.

5. (a) Compute $1.255 + 0.395 + 2.116 + 1.336$, then round off to 2 places.
(b) Round off each term first, then add. Compare the answers. Which answer is more accurate?

6. (cont.) Do the same for $0.255 + 0.365 + 0.166 + 0.823$.

Round off to 3 significant figures:

7. 1046.0, 55.521, 10.05

8. 9.095, 9.094, 9.0949.

An inexperienced technician uses a voltmeter with 2 significant figure readings. Rewrite accurately his data:

9. 0.4   **10.** 12.0   **11.** 2.3   **12.** 9.

5. TABLES AND INTERPOLATION

In order to compute with logarithms, we shall have to familiarize ourselves with log tables. Recall that any positive number p can be expressed in scientific notation as $p = 10^n x$, where n is an integer and $1 \leq x < 10$. Then

$$\log p = \log(10^n x) = \log 10^n + \log x = n + \log x.$$

Since $1 \leq x < 10$, we have $0 \leq \log x < 1$. It is not $\log p$, but $\log x$ that we find in a table; this number is called the **mantissa** of $\log p$, and the number n is called the **characteristic** of $\log p$. (Mantissas are given in tables without the decimal point.) To find the log of a number that differs from p by a power of 10 (shift of the decimal point) we merely have to add an appropriate integer to $\log p$.

■ *Example 5.1*

Given $\log 2.7 \approx 0.4314$, estimate the logs of

$$27, \quad 27000, \quad 0.000027.$$

SOLUTION

$$27 = 2.7 \times 10, \quad 27000 = 2.7 \times 10^4, \quad 0.000027 = 2.7 \times 10^{-5}.$$

Hence,

$$\log 27 \approx 1.4314, \qquad \log 27000 \approx 4.4314, \qquad \log 0.000027 \approx 0.4314 - 5.$$

Answer   1.4314,   4.4314,   0.4314 − 5.

Remark: The third answer is 0.4314 − 5, which is equal to −4.5686. Do not confuse this with −5.4314.

. . . .

Look at the log table in the back of the book. To each 3-digit number from 100 to 999, the table gives a 4-digit mantissa. For example, corresponding to 534 the table gives 7255. This means

$$\log 5.34 \approx 0.7275.$$

Since $53.4 = 5.34 \times 10$, $0.0534 = 5.34 \times 10^{-2}$, and $5340 = 5.34 \times 10^3$, we have

$$\log 53.4 \approx 1.7275, \quad \log 0.0534 \approx 0.7275 - 2, \quad \log 5340 \approx 3.7275, \quad \text{etc.}$$

Therefore the table gives approximate logs for any positive number with 3 significant digits.

There exist finer tables too; for instance, 5-place tables give mantissas of 5 decimal digits for numbers with four significant digits.

Linear Interpolation

Suppose we want log 3.1517. From a 5-place table we find

$$\log 3.151 \approx 0.49845, \qquad \log 3.152 \approx 0.49859.$$

What we do is pretend that the logarithm function $\log x$ is linear for $3.151 \leq x \leq 3.152$. Now 3.1517 is $\frac{7}{10}$ of the way from 3.151 to 3.152, so log 3.1517 must be about $\frac{7}{10}$ of the way from 0.49845 to 0.49859. See Fig. 5.1. But

$$\tfrac{7}{10}(0.49859 - 0.49845) = \tfrac{7}{10}(0.00014) \approx 0.00010.$$

Therefore

$$\log 3.1517 \approx 0.49845 + 0.00010 = 0.49855.$$

Many logarithm (and other) tables give a list of "Proportional parts" (P.p.) on the side. With this the work is very easy. Look up 3.151 in the table. What is actually listed is the decimal part without the decimal point: 49845. The next entry is 49859; mentally note the difference, 14. In the P.p. table for 14 you find 9.8 opposite 7. So add 10 to 49845 to get 49855, the decimal part of the answer.

Remark: Linear interpolation is not very accurate in the 5-place log table for the range $1.0 \leq x \leq 1.2$. Therefore many books of tables include 6- or 7-place tables for this range.

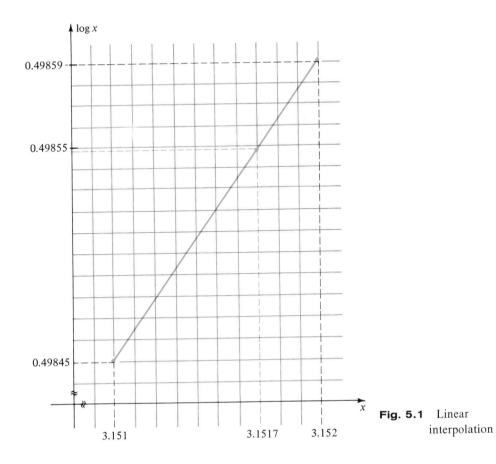

Fig. 5.1 Linear interpolation

When you use logs for a computation you usually end up with the log of the answer, so it is important to know how to find x, given $\log x$. There are two ways: (1) locate $\log x$ in the body of a log table, and see what x it corresponds to; (2) use an antilog table (table of the function 10^x).

■ *Example 5.2*

Given $\log x \approx 2.7423$, estimate x using 4-place log tables and 4-place anti-log tables.

SOLUTION Let us find the number whose log is 0.7423, then multiply by 10^2 to account for the characteristic 2.

Method 1. From the 4-place log table,

$$\left.\begin{array}{l} \log 5.52 \approx 0.7419 \\ \log 5.53 \approx 0.7427 \end{array}\right\rangle \text{difference} = 8.$$

Now 23 is $\frac{4}{8}$-ths of the way from 19 to 27. But $\frac{4}{8} = 0.5$, so by interpolation,

$$\log 5.525 \approx 0.7423.$$

Method 2. By the antilog table,

$$10^{0.742} \approx 5.521$$
$$10^{0.743} \approx 5.534$$

difference = 13.

The difference is 13 and $\frac{3}{10} \times 13 = 3.9 \approx 4.0$, so by interpolation

$$x \approx 10^{0.7423} \approx 5.525.$$

Answer 552.5.

Remark: Note carefully how the decimal point is omitted in the tables, and where it is supposed to be.

EXERCISES

Use 4-place tables to estimate the log of:

1. 1.46	**2.** 902	**3.** 35.5
4. 0.024	**5.** 11.0	**6.** 1.01.

Use 4-place tables and interpolation to estimate the log of:

7. 52.41	**8.** 0.6822	**9.** 300.4
10. 0.004218	**11.** 1.559	**12.** 37270
13. 16.07	**14.** 25.34	**15.** 0.7113
	16. 0.0609.	

Use 4-place tables to estimate the antilog:

17. 0.1482	**18.** 0.9111	**19.** 3.2817
20. 0.9121 − 1	**21.** 5.6789	**22.** 4.0100
23. 0.7813 − 5	**24.** 0.9001 − 10	**25.** 0.0416 − 2
	26. 0.5273 − 3.	

6. COMPUTATIONS WITH LOGARITHMS [optional]

Let us now combine the theory of logarithms with our knowledge of tables to do some computations, the kind that actually arise in scientific work. In the following examples we repeatedly use the rules of logarithms, p. 156.

■ *Example 6.1*

Use 4-place tables to compute $(24.86)(0.01392)(1.787)$.

SOLUTION If the answer is x, then

$$\log x = \log 24.86 + \log 0.01392 + \log 1.787.$$

Look up these logs in the table, add them, then find the antilog:

$$\log 24.86 \approx 1.3955$$
$$\log 0.01392 \approx 0.1436 - 2$$
$$\underline{\log 1.787 \approx 0.2521}$$
$$\log x \approx 0.7912 - 1$$
$$x \approx 0.6183.$$

Answer   0.6183.

. . . .

▪ *Example 6.2*

Use 4-place tables to compute

$$\frac{(24.86)(0.01392)}{1.787}.$$

SOLUTION   If the answer is x, then

$$\log x = \log 24.86 + \log 0.01392 - \log 1.787.$$

In problems like this, it is a good idea to lay out the work in advance:

$$
\begin{array}{r|l}
 & \log 24.86 \approx \\
+ & \log 0.01392 \approx \\
\hline
 & (\text{sum}) \approx \\
- & \log 1.787 \approx \\
\hline
 & \log x \approx \\
 & x \approx
\end{array}
$$

Now fill in the numbers:

$$\log 24.86 \approx 1.3955$$
$$\underline{\log 0.01392 \approx 0.1436 - 2}$$
$$(\text{sum}) \approx 0.5391 - 1$$
$$\underline{\log 1.787 \approx 0.2521}$$
$$\log x \approx 0.2870 - 1$$
$$x \approx 0.1936.$$

Answer   0.1936.

. . . .

Remember that the mantissa of a logarithm must be non-negative. If we are given $\log x = -1.4923$, it does us no good to look up 0.4923 in the antilog table, because $-1.4923 \neq 0.4923 - 1$. We must write $\log x = 0.abcd - N$ where N is an integer.

For example, if $\log x = -1.4923$, we add and subtract 2:

$$
\begin{array}{rl}
0 = & 2.0000 - 2 \\
\log x = & -1.4923 \\
\hline
\log x = & 0.5077 - 2.
\end{array}
$$

Now we look up 0.5077 in the antilog table and find that $x \approx 3.219 \times 10^{-2}$.

In certain situations we can force the answer to come out in convenient form, say $\log x = 0.5077 - 2$ rather than $\log x = -1.4923$. When we must subtract a logarithm from a smaller one, we first add and subtract a suitable integer to the smaller one, then take the difference. This trick is illustrated in Example 6.3.

■ *Example 6.3*

Use 4-place tables to compute (2.400)/(3780).

SOLUTION

$$
\begin{array}{rl}
\log 2.400 \approx 0.3802 = & 4.3802 - 4 \\
\log 3780 \approx 3.5775 = & 3.5775 \\
\hline
\log x \approx & 0.8027 - 4 \\
x \approx & 6.349 \times 10^{-4}.
\end{array}
$$

Answer 6.349×10^{-4}.

. . . .

■ *Example 6.4*

Use 4-place table to compute $(2.400)^{37.80}$.

SOLUTION

$$
\begin{aligned}
\log x &= (37.80)(\log 2.400). \\
\log 2.400 &\approx 0.3802 \\
\log x &\approx 37.80 \times 0.3802.
\end{aligned}
$$

To multiply 37.80 by 0.3802, use logs again. In other words, find the log of $\log x$:

$$
\begin{array}{rl}
\log 37.80 \approx & 1.5775 \\
\log 0.3802 \approx & 0.5800 - 1 \\
\hline
\log \log x \approx & 1.1575 \\
\log x \approx & 14.37.
\end{array}
$$

The mantissa has 2 significant figures, so one can expect an answer only of the same accuracy.

Answer 2.3×10^{14}.

. . . .

■ *Example 6.5*

Use 4-place tables to compute $(0.0024)^{3.78}$.

SOLUTION

$$\log 0.0024 \approx 0.3802 - 3 = -2.6198,$$
$$\log x = (3.78)(\log 0.0024) \approx (3.78)(-2.6198) = -(3.78)(2.6198).$$

To continue, we compute the product $(3.78)(2.6198)$:

$$\log 3.78 \approx 0.5775$$
$$\log 2.6198 \approx \log 2.620 \approx 0.4183$$
$$\overline{\log[(3.78)(2.6198)] \approx 0.9958}$$
$$(3.78)(2.6198) \approx 9.904.$$

Hence

$$\log x \approx -9.904 = 0.096 - 10$$
$$x \qquad\qquad \approx 1.2 \times 10^{-10}.$$

Answer 1.2×10^{-10}.

. . . .

■ *Example 6.6*

Use 4-place tables to compute $\log_2 50$.

SOLUTION In general,

$$\log_b x = \frac{\log x}{\log b},$$

so if $y = \log_2 50$, then

$$y = \frac{\log 50}{\log 2} \approx \frac{1.6990}{0.3010}.$$

Hence

$$\log 1.6990 \approx 1.2303 - 1$$
$$\log 0.3010 \approx 0.4786 - 1$$
$$\overline{\log y \approx 0.7517}$$
$$y \approx 5.646.$$

Answer 5.646.

. . . .

■ *Example 6.7*

Use 4-place tables to compute $\sqrt{1 + \sqrt[3]{5}}$.

SOLUTION Because there is addition involved, we must do the computation in several steps. First we compute the quantity inside the square root:

$$\log 5 \approx 0.6990$$
$$\log \sqrt[3]{5} = \tfrac{1}{3}\log 5 \approx 0.2330$$
$$\sqrt[3]{5} \approx 1.7100$$
$$1 + \sqrt[3]{5} \approx 2.7100$$
$$\log(1 + \sqrt[3]{5}) \approx 0.4330.$$

Now we compute the square root:

$$\log \sqrt{1 + \sqrt[3]{5}} = \tfrac{1}{2}\log(1 + \sqrt[3]{5}) \approx 0.2165$$
$$\sqrt{1 + \sqrt[3]{5}} \approx 1.646.$$

Answer   1.646.

EXERCISES

Compute: (Use 4-place tables.)

1. 4.812×3.99
2. 10.4×0.2561
3. 3.891×710.2
4. 104.1×892.0
5. 0.0426×1.333
6. 78.45×0.002310
7. $20.09/17.62$
8. $2.718/3.142$
9. $0.7812/0.01204$
10. $14.27/(3.812 \times 10^9)$
11. $(0.489 \times 3.16)/(2.74)$
12. $41.80/(32.41 \times 7.822)$

13. $\dfrac{7.981}{(35.49 \times 1.827)^3}$
14. $\dfrac{4.694 \times 95.56}{0.003259 \times 104.7}$
15. $\dfrac{(52.10)(8.055)(127.6)}{(77.41)^2(15.62)}$

16. $\dfrac{(12.94)(0.137)(166.5)}{(29.84)(73.22)}$
17. $\sqrt{7}$
18. $\sqrt{33.05}$

19. $\sqrt{0.0063}$
20. $\sqrt{22930}$
21. $\sqrt[3]{102}$

22. $\sqrt[3]{788}$
23. $\dfrac{(57.1)\sqrt{7.294}}{\sqrt[3]{838.5}}$
24. $\sqrt[3]{\dfrac{15.08}{(3.441)(12.81)}}$

25. $5^{3.4}$
26. $12^{2.81}$
27. $10^{3.704}$
28. $10^{0.3355}$
29. $(20.9)^{1.64}$
30. $(157)^{-1.27}$
31. $(10.41)^{0.1212}$
32. $(0.2)^{-10}$
33. $(0.3120)^{0.0041}$
34. $(998)^{0.3677}$
35. $(9.394)^{-0.008214}$
36. 2^{1000}
37. $(2.4)^{1.3}(0.12)^{4.1}$
38. $(2.71)^{1.3}(11.1)^{-0.4}$
39. $(7.9)^{-1.42}(8.4)^{2.17}$
40. $(11)^{11.3}(11.3)^{11}$.

Which is larger?

41. 11^{12} or 12^{11}
42. 100^{101} or 101^{100}
43. 400^{401} or 401^{400}
44. 1000^{1001} or 1001^{1000}
45. $73 \cdot 74 \cdot 75 \cdot 76 \cdot 77 \cdot 78$ or 75^6
46. 2^4, $(1.99)^{4.01}$, or $(1.98)^{4.02}$.

Compute:

47. $\log_5 10$
48. $\log_3 27$
49. $\log_2 512$
50. $\log_{12} 24$
51. $\log_{100} 25$
52. $\log_{25} 100$.

53. In Section 1, it is stated that \$1 invested at 1% for 9000 years will earn a fortune. Show that 2200 years is enough to earn a billion!

54. How long will it take \$1 at 5% to be worth 10^{10}?

55. On p. 152 we used the crude estimate $(1.01)^{900} > 10$. Show that actually $(1.01)^{232} > 10$. You may use $\log 1.01 > 0.00432$.

56. (cont.) Show that $(1.01)^{2084} > 10^9$.

Compute x:

57. $x^3 = 1 + (4.012)^2$

58. $x^5 = 3 + \sqrt[3]{3}$

59. $x^2 = (1.032)^2 + (2.114)^2,\ x > 0$

60. $x = (\log 3)/(\log 5)$.

7. COMPUTATIONS WITH A SMALL CALCULATOR [optional]

The development of pocket calculators (electronic slide rules) makes an amazing amount of computing power available. With prices decreasing all the time, slide rules and other traditional computing aids will soon become obsolete. On pocket calculators, a problem like

$$\frac{(1.9924)^{0.21}(223.51)^{1.7}}{(38.965)(0.33125) + 10.021}$$

can be done in less than a minute, whereas by logs it takes about 10 minutes for a less accurate result.

We shall discuss calculations on a hypothetical model, the FP-02 (Fig. 7.1). This calculator has temporary memory registers (storage positions) that make it possible to perform sequences of operations without copying down intermediate answers for later use. For example, when you compute $ab + cd$, the calculator will remember the product ab for later addition to the result of multiplying c and d.

Elementary Operations

Before discussing the memory, let us see how elementary computations are done on the FP-02.

To start any computation, you enter the first number by pressing number keys. For instance to start with 4.302, you press in order 4 $\boxed{\cdot}$ 3 0 2 . Numbers larger than 10^{10} or smaller than 10^{-10} must be entered in scientific notation. For example, to start with 4.302×10^{19}, you press

$$4 \quad \boxed{\cdot} \quad 3 \quad 0 \quad 2 \quad \boxed{\text{EXP}} \quad 1 \quad 9.$$

The key $\boxed{\text{EXP}}$ controls the two places on the extreme right, which are reserved for exponents. The display will read

$$\boxed{4 . 3 0 2 \qquad\qquad 1 9}$$

The key $\boxed{\pm}$ changes the sign. To display -4.302, press 4 $\boxed{\cdot}$ 3 0 2 $\boxed{\pm}$. To display 4.302×10^{-19}, press

$$4 \quad \boxed{\cdot} \quad 3 \quad 0 \quad 2 \quad \boxed{\text{EXP}} \quad 1 \quad 9 \quad \boxed{\pm}.$$

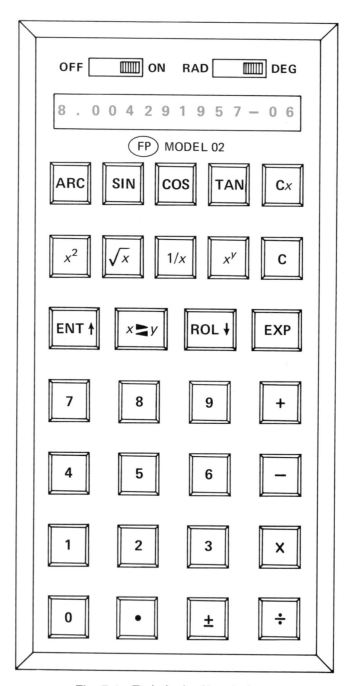

Fig. 7.1 Typical scientific calculator

We are ready for arithmetic. To add a and b, press in order

$$a \quad \boxed{\text{ENT} \uparrow} \quad b \quad \boxed{+} .$$

The sequence a $\boxed{\text{ENT} \uparrow}$ tells the calculator to remember a. The sequence b $\boxed{+}$ tells it to add b to the number just remembered. Similarly, to compute

$$
\left.\begin{aligned}
a - b \\[6pt]
a \times b \\[6pt]
a/b
\end{aligned}\right\} \quad \text{press} \quad
\left\{\begin{aligned}
&a \quad \boxed{\text{ENT} \uparrow} \quad b \quad \boxed{-} \\[6pt]
&a \quad \boxed{\text{ENT} \uparrow} \quad b \quad \boxed{\times} \\[6pt]
&a \quad \boxed{\text{ENT} \uparrow} \quad b \quad \boxed{\div} .
\end{aligned}\right.
$$

Remark: These sequences of steps may seem unnatural at first. Indeed, in calculators *without* memory, the sequence of steps for computing $a + b$ is usually something like $a \boxed{+} b \boxed{=}$. Yet for more complicated calculations, the sequence of steps we use enormously increases the capability of the instrument, as we shall soon see.

The calculator is built to handle a string of computations. After an operation, the machine automatically enters and remembers the result while you punch in a new number. After pressing any operating key, it is unnecessary to press $\boxed{\text{ENT} \uparrow}$ before entering a new number.

Examples:

$$(a + b - c)/d \quad : \quad a \quad \boxed{\text{ENT} \uparrow} \quad b \quad \boxed{+} \quad c \quad \boxed{-} \quad d \quad \boxed{\div} .$$

$$ab + cd \quad : \quad a \quad \boxed{\text{ENT} \uparrow} \quad b \quad \boxed{\times} \quad c \quad \boxed{\text{ENT} \uparrow} \quad d \quad \boxed{\times} \quad \boxed{+} .$$

$$(a + b)(c + d) \quad : \quad a \quad \boxed{\text{ENT} \uparrow} \quad b \quad \boxed{+} \quad c \quad \boxed{\text{ENT} \uparrow} \quad d \quad \boxed{+} \quad \boxed{\times} .$$

The keys $\boxed{x^2}$, $\boxed{\sqrt{x}}$, and $\boxed{1/x}$ perform simple functions almost instantly:

$$a^2 : \quad a \quad \boxed{x^2}, \quad \text{etc.}$$

The key $\boxed{x^y}$ is for computing arbitrary powers.

Examples:

$$a^b : \quad b \quad \boxed{\text{ENT} \uparrow} \quad a \quad \boxed{x^y} .$$

$$\sqrt[7]{a} = a^{1/7} : \quad 7 \quad \boxed{1/x} \quad a \quad \boxed{x^y} .$$

■ ***Example 7.1***

Set up computations for

(a) $\dfrac{3.508 + 1.742 - 2.993}{3897}$

(b) $(1.408 + 3.972)(254.3 - 197.7)$

(c) $(23^{17.4})(19^{15.2})$.

SOLUTION

(a) 3.508 [ENT ↑] 1.742 [+] 2.993 [−] 3897 [÷].

(b) 1.408 [ENT ↑] 3.972 [+] 254.3 [ENT ↑] 197.7 [−] [×].

(c) 17.4 [ENT ↑] 23 [xʸ] 15.2 [ENT ↑] 19 [xʸ] [×].

The numerical answers are

    (a) $5.791634591 \times 10^{-4}$ (b) 304.508   (c) $1.352443344 \times 10^{43}$.

Remark: There are always several ways to set up a computation. For instance,

    3.1 [ENT ↑] 5.7 [+]   and   5.7 [1/x] 3.1 [×]

are two ways of setting up $(3.1)/(5.7)$, and you can find others.

Memory Registers

The FP-02 has four memory registers, which we denote by X, Y, Z, T. These temporarily store numbers during a calculation. The number stored in X is displayed.

When you key in a number, it goes into register X. During the course of a computation, it may be shifted to another register or combined with another number. You may think of the memory registers as forming a *vertical stack,* arranged X, Y, Z, T from the bottom up.

Table 1 shows the effects of typical operation keys, and should be studied carefully. Note that pressing an operation key both performs the operation and changes the stack of memory registers.

Table 1   *Some FP-02 operations*

OLD STACK	NEW STACK		OLD STACK	NEW STACK		OLD STACK	NEW STACK		OLD STACK	NEW STACK
T	Z		T	T		T	T		T	T
Z	Y		Z	T		Z	Z		Z	T
Y	X		Y	Z		Y	Y		Y	Z
X	X		X	Y/X		X	$\sqrt{X}$		X	X^Y

(a) effect of [ENT ↑]   (b) effect of [÷]   (c) effect of [√x]   (d) effect of [xʸ]

■ *Example 7.2*

Set up a computation for

$$2840 + (1.76)(4.21)^{6.8}$$

and show the state of the stack at each step.

SOLUTION   One possibility is

2840 [**ENT ↑**] 1.76 [**ENT ↑**] 6.8 [**ENT ↑**] 4.21 [*x^y*] [×] [+]

$$(\approx 3.378788113 \times 10^4).$$

Table 2 shows the stack at each step (with four-figure round-off for simplicity). Steps 1–7 load the stack. The final three steps form

$$X^Y Z + T = (4.21)^{6.8}(1.76) + 2840 \approx 3.378 \times 10^4.$$

Remark: Remember not to overload the stack; it can hold only four numbers. For instance if you were to press [**ENT ↑**] after step 7, the result would be

X	Y	Z	T
4.21	4.21	6.8	1.76

so 2840 would be lost.

Table 2

Step	Entry	X = display		Y	Z	T
		TEMPORARY MEMORY STACK				
1	2840	2840		0	0	0
2	ENT ↑	2840		2840	0	0
3	1.76	1.76		2840	0	0
4	ENT ↑	1.76		1.76	2840	0
5	6.8	6.8		1.76	2840	0
6	ENT ↑	6.8		6.8	1.76	2840
7	4.21	4.21		6.8	1.76	2840
8	x^y	1.758	04	1.76	2840	2840
9	×	3.094	04	2840	2840	2840
10	+	3.378	04	2840	2840	2840

The key $\boxed{x \rightleftharpoons y}$ interchanges X and Y; the key $\boxed{\textbf{ROL} \downarrow}$ rolls the stack downward (Y $\longrightarrow$ X, Z $\longrightarrow$ Y, T $\longrightarrow$ Z, X $\longrightarrow$ T). Both are most useful, both for checking what is in the stack, and for set-ups. Finally, the key $\boxed{\textbf{C}}$ clears everything, so you can start a new problem, and the key $\boxed{\textbf{C x}}$ clears the X-register only (in case of a key-pressing error).

Back to the Tables

Are tables useless? After these last paragraphs, you might conclude that they are. However, the very fact that tables present a lot of information you can scan quickly makes them valuable.

◼ **Example 7.3**

Approximate a solution to the equation

$$10^x = 3 + x.$$

SOLUTION Since $10^0 < 3 + 0$ and $10^1 > 3 + 1$, we smell a solution between 0 and 1. Now we inspect the antilog (10^x) table at the back of this book. After a short search we see

$$10^{0.550} \approx 3.548 \quad \text{and} \quad 10^{0.551} \approx 3.556.$$

By interpolation,

$$10^{0.5501} \approx 3.549 \quad \text{and} \quad 10^{0.5502} \approx 3.550.$$

Hence 0.5502 is a reasonable approximation. (Actually, $x \approx 0.5502601816$.)

Answer $x \approx 0.5502$.

EXERCISES

Set up the problem for computation on the FP-02:

1. $1/4.321$

2. $35.74/15.21$

3. $(4.23 + 3)(4.23 + 5)$

4. $(4.23)^2 + 8(4.23) + 15$

5. $\dfrac{1}{(7.3 + 1)^2}$

6. $\dfrac{1}{(7.3)^2 + 2(7.3) + 1}$

7. $\dfrac{5.1}{\sqrt{3.7}}$

8. $\dfrac{5.1\sqrt{3.7}}{3.7}$

9. $\dfrac{1}{\sqrt{7.52} - \sqrt{4.21}}$

10. $\dfrac{\sqrt{7.52} + \sqrt{4.21}}{7.52 - 4.21}$

11. $6 \times 0.03721 + 19 \times 1.428 - 27 \times 3.142$

12. $\dfrac{(3.6814)(1.4281)^2 - (1.2231)^3}{(3.591)^2 + (4.003)^2}$

13. $\left[\dfrac{4(1.721)^2 + 7(3.998)^2 + 5(1.072)^2 + 2(1.911)^2}{4 + 7 + 5 + 2}\right]^{1/2}$

14. $[(0.429)^{1.08}(3.552)^{2.24}(2.774)^{1.73}]^{2/3}$

15. $\dfrac{(0.05)/12}{1 - [1 + (0.05)/12]^{-120}}$

16. $(7.011)^2\left[\dfrac{1.042}{(6.339)^2} + \dfrac{3.924}{(7.811)^2} + \dfrac{10.24}{(6.668)^2}\right].$

Given numbers a and b, find what is computed on the FP-02 by the sequence of steps:

17. b ⎡ENT ↑⎤ a ⎡x²⎤ ⎡×⎤ 3 ⎡÷⎤

18. a ⎡ENT ↑⎤ b ⎡ENT ↑⎤ ⎡ENT ↑⎤ 10 ⎡xʸ⎤ ⎡÷⎤ ⎡1/x⎤ ⎡×⎤

19. b ⎡ENT ↑⎤ ⎡ENT ↑⎤ a ⎡+⎤ ⎡x ⇌ y⎤ ⎡x²⎤ ⎡+⎤

20*. a ⎡ENT ↑⎤ ⎡ENT ↑⎤ b ⎡ENT ↑⎤ ⎡ROL ↓⎤ ⎡−⎤ ⎡ROL ↓⎤

⎡+⎤ ⎡x ⇌ y⎤ ⎡ROL ↓⎤ ⎡÷⎤.

Set up the computation on the FP-02:

21. $\sqrt[3]{a^2 b}$ **22.** $(a/b)^{5/7}.$

Use the tables in this book to estimate a solution:

23. $x = 10 \log x, \quad 1 < x < 2$ **24.** $x = 10 \log x, \quad x > 2$ **25.** $x + x^2 = \frac{1}{10}x^3$
26. $\cos x = x, \quad 0 < x < 1.00, \quad x$ in radians
27. $\tan x = \cos x, \quad 0 < x < 1.00, \quad x$ in radians
28. $\sin x + \tan x = 2, \quad x$ in radians.

8. APPLICATIONS

Exponential Equations

Equations that are linear or quadratic in an exponential a^x can be solved for x in terms of logarithms. Two examples will illustrate the idea.

■ *Example 8.1*

Solve for x:
$$4 \cdot 3^x - 9 = 11.$$

SOLUTION Add 9 to both sides, then divide by 4:
$$4 \cdot 3^x = 9 + 11 = 20,$$
$$3^x = 20/4 = 5.$$

Now take logs:
$$x \log 3 = \log 3^x = \log 5.$$

Now divide.

Answer $(\log 5)/(\log 3).$

. . . .

■ *Example 8.2*

Solve for x:

$$3^x - 4 - 5 \cdot 3^{-x} = 0.$$

SOLUTION   This is a quadratic equation in disguise. In fact, if we multiply both sides by 3^x, we obtain

$$(3^x)^2 - 4(3^x) - 5 = 0.$$

The quadratic factors:

$$(3^x - 5)(3^x + 1) = 0.$$

Hence either $3^x = 5$ or $3^x = -1$. But $3^x > 0$ for all x, hence $3^x = 5$ is the only possibility. Take logs to solve: $x \log 3 = \log 5$.

Answer $(\log 5)/(\log 3)$.

. . . . ▪

Remark: Another disguise for the same quadratic equation is

$$9^x - 4 \cdot 3^x - 5 = 0$$

because $9^x = (3^2)^x = (3^x)^2$.

Bacteria Growth

Under certain conditions the rate of growth of a colony of bacteria is proportional to the number of bacteria in the colony. It is shown in calculus courses that this implies the growth law

$$N(t) = N_0 2^{t/k},$$

where $N(t)$ is the number of bacteria in the colony at time t, where $N_0 = N(0)$ is the number present at time $t = 0$, and where k is the time it takes for the colony to double.

■ *Example 8.3*

There are 10^5 bacteria at the start of an experiment and 3×10^7 after 24 hours. Find the growth law.

SOLUTION   We have $N_0 = 10^5$ and $N(24) = 3 \times 10^7$, hence

$$3 \times 10^7 = 10^5 \cdot 2^{t/k} = 10^5 \cdot 2^{24/k},$$

$$2^{24/k} = 300.$$

Take logs:

$$\frac{24}{k} \log 2 = \log 300,$$

$$k = \frac{24 \log 2}{\log 300} \approx 2.92.$$

Answer $N(t) \approx 10^5 \cdot 2^{t/2.92}$.

. . . .

A radioactive element decays at a rate proportional to the amount present. Its **half-life** is the time in which a given quantity decays to one-half of its original mass. In calculus courses it is shown that this implies the decay law

$$M(t) = M_0 2^{-t/H},$$

where $M(t)$ is the mass at time t, where $M_0 = M(0)$ is the initial mass at time $t = 0$, and where H is the half-life.

▪ *Example 8.4*

A kilogram of carbon-14 decays in 12 years to 0.99853 kg. Find the half-life.

SOLUTION By the decay law,

$$0.99853 = 2^{-12/H}.$$

Take logs:

$$\log 0.99853 = -\frac{12}{H} \log 2,$$

$$H = -\frac{12 \log 2}{\log 0.99853} \approx 5654.$$

Answer 5654 years.

Solve for x:

1. $2^x = 10$
2. $3 \cdot 5^x = 5$
3. $4 \cdot 7^x + 1 = 5$
4. $3 \cdot 2^x - 5 = 98$
5. $6^x + 6 = 6$
6. $2 \cdot 5^x - 3 = -4$
7. $2^x - 2 + 2^{-x} = 0$
8. $4^x - 5 \cdot 2^x + 6 = 0$
9. $7^x - 6 \cdot 7^{-x} = 1.$
10. $4^x + 5 \cdot 2^x + 6 = 0$

11. A colony of bacteria has a population of 3×10^6 initially and 9×10^6 two hours later. How long does it take the colony to double?
12. (cont.) How long does it take the colony to multiply by 10?
13. Assume that the population grows at a rate proportional to the population itself. In 1950 the US population was 151×10^6 and in 1960 it was 178×10^6. Make a prediction for the year 2000.
14. (cont.) Estimate what the population was in 1930.
15. Thorium-X has a half-life of 3.64 days. How long will it take for a quantity to decay to $\frac{1}{10}$ of its mass?
16. A certain radioactive substance loses $\frac{1}{5}$ of its original mass in 3 days. What is its decay law?
17. Under certain conditions, the rate of decrease of atmospheric pressure as a function of altitude above sea level is proportional to the pressure. Suppose the barometer reads 30 in. at sea level and 25 in. at 4000 ft. Find the barometric pressure at 20,000 ft.

18. In a certain college it was found that the number of students dropping out each day was proportional to the number still enrolled. If 8000 started out and 15% dropped out after 4 weeks, find the number left after 12 weeks.

9. BUSINESS APPLICATIONS [optional]

As further applications of exponential functions, we shall discuss several basic problems involving the investment of money. This section might be considered as a very short course in "math of finance".

Compound Interest

I deposit an amount Q (principal) in a savings account for n time periods, where the interest rate is i per period. What is the worth A of my account?

Initially ($n = 0$) the worth is simply Q. After one period ($n = 1$) the worth is Q plus the interest on Q, that is, $Q + Qi = Q(1 + i)$. After two periods, the worth is $Q(1 + i)$ plus the interest on $Q(1 + i)$, that is,

$$Q(1 + i) + Q(1 + i)i = Q(1 + i)^2.$$

Likewise at $n = 3$ the worth is

$$Q(1 + i)^2 + Q(1 + i)^2 i = Q(1 + i)^3.$$

Similarly, after n periods the worth is

$$A = Q(1 + i)^n.$$

This is the basic formula of compound interest. In application, remember that interest rates are usually quoted annually, while the time period is usually a fraction of a year. Thus if there are p periods per year, the annual rate j must be changed to $i = j/p$ to apply the formula. Also remember that *percent* means *hundredths*.

■ *Example 9.1*

I deposit $6000 in a savings account at 5% annual interest, compounded quarterly. Find the amount of my account after 9 years.

SOLUTION   Here $i = \frac{1}{4}(0.05) = 0.0125$, $n = 9 \cdot 4 = 36$, and $Q = 6000$, so

$$A = Q(1 + i)^n = (6000)(1.0125)^{36}.$$

Answer $9383.66.

■ *Example 9.2*

Find the annual interest rate j at which money compounded monthly will double in 8 years.

SOLUTION   Here $A = 2Q$ and $n = 8 \cdot 12 = 96$. The monthly interest rate is $i = j/12$. Thus the set-up is

$$2Q = Q(1 + i)^{96},$$
$$(1 + i)^{96} = 2,$$
$$1 + i = 2^{1/96} \approx 1.007246,$$
$$i \approx 0.007246, \quad j = 12i \approx 0.08696.$$

Answer   8.70%.

. . . .   ▪

It is an established business principle that if a loan at interest rate i per period is repaid in a fractional time, then the formula $A = Q(1 + i)^n$ still applies, even though n is not an integer.

▪ *Example 9.3*

Company Z buys a treasury bill for $250,000 at 4% annual interest, compounded monthly. It redeems the bill after 18 months 10 days. Find the amount received.

SOLUTION   Here $A = Q(1 + i)^n$ with $Q = 250000$, $i = \frac{1}{12}(0.04) = 0.003333 \cdots$, $n = 18\frac{1}{3} = 18.333 \cdots$, so

$$A = (250000)(1.00333 \cdots)^{18.333\cdots} \approx 265727.25.$$

Answer $265,727.25.

. . . .   ▪

Annuity

"Annuity" originally meant a yearly payment of a fixed amount, but now it refers to any periodic payment of a fixed amount. The accumulated result of an annuity, the periodic payments plus the interest earned, is called a **sinking fund.**

Suppose I deposit at the *beginning* of each period the same amount P into a savings account paying interest i per period, and I do this for n periods. What will my account be worth immediately after the n-th payment?

Let A_k denote the amount of the account after k payments. Clearly $A_1 = P$. The amount A_k will consist of three parts: (1) the k-th payment P, (2) the amount A_{k-1} of the account after the $(k - 1)$-th payment, and (3) the interest $A_{k-1}i$ earned by A_{k-1} in the k-th period. Therefore

$$A_k = P + A_{k-1} + A_{k-1}i = P + A_{k-1}(1 + i).$$

Thus we have

$$A_1 = P$$
$$A_2 = P + A_1(1 + i) = P + P(1 + i) = P[1 + (1 + i)]$$
$$A_3 = P + A_2(1 + i) = P + P[1 + (1 + i)](1 + i) = P[1 + (1 + i) + (1 + i)^2].$$

Continuing, we have

$$A_n = P[1 + (1 + i) + (1 + i)^2 + \cdots + (1 + i)^{n-1}].$$

To sum the bracketed expression, we recall the polynomial identity (p. 39)

$$x^n - 1 = (x - 1)(x^{n-1} + x^{n-2} + \cdots + x^2 + x + 1).$$

We substitute $x = 1 + i$:

$$(1 + i)^n - 1 = i[(1 + i)^{n-1} + \cdots + (1 + i)^2 + (1 + i) + 1],$$

$$(1 + i)^{n-1} + \cdots + (1 + i)^2 + (1 + i) + 1 = \frac{(1 + i)^n - 1}{i}.$$

The final result is

$$\boxed{A_n = P\frac{(1 + i)^n - 1}{i}.}$$

This is the basic law of annuities. In tables, the quantity A_n is often called the "amount of an annuity" or the "amount of P per period".

To find the periodic deposit required to grow to a given amount A_n in n deposits, solve the formula for P:

$$P = A_n \frac{i}{(1 + i)^n - 1},$$

Now we refer to P as a "sinking fund", the periodic deposit that will grow to A_n in the future.

Example 9.4

ZYX Corp. floats a $28,000,000 bond issue, due in 12 years. It can obtain 4% interest, compounded quarterly, in a savings account. How much money must be deposited quarterly in a sinking fund, starting 3 months from now, to retire the bonds at maturity?

SOLUTION   Here $n = 4 \cdot 12 = 48$ (not 47 nor 49), $i = \frac{1}{4}(0.04) = 0.01$ and $A_{48} = 28 \times 10^6$. Hence the periodic deposit P is determined by the equation

$$28 \times 10^6 = P\frac{(1.01)^{48} - 1}{0.01},$$

from which

$$P = \frac{28 \times 10^4}{(1.01)^{48} - 1} \approx 457,347.40.$$

Answer $457,347.40.

Example 9.5

I deposit $100 monthly in a savings account for 10 years. Of course the interest rate changes several times over the years; however, immediately after my last deposit, the balance is \$14,960.27. Estimate the average annual interest rate, compounded monthly.

SOLUTION   Here we must solve the equation

$$14960.27 = 100 \, \frac{(1 + j/12)^{120} - 1}{j/12}$$

for j, the annual interest rate. This is different from our previous problems, because it is obviously hopeless to *solve* this equation of degree 120. Instead we test some values of j and interpolate.
    If we guess $j = 0.05$, then

$$100 \, \frac{(1 + j/12)^{120} - 1}{j/12} \approx 15528.23,$$

too large. If we try $j = 0.04$, we get

$$100 \, \frac{(1 + j/12)^{120} - 1}{j/12} \approx 14724.98,$$

too small. Clearly $0.04 < j < 0.05$. By linear interpolation,

$$j \approx 0.04 + (0.01) \, \frac{14960.27 - 14724.98}{15528.23 - 14724.98} \approx 0.0429.$$

Answer   4.29%

. . . .

Remark: We are using the kind of pocket calculator described in Section 7 for the computations in these examples. The computations can also be done by logarithms. In practice, however, it is often unnecessary to use logarithms since much information is available in interest tables. For instance, the CRC Standard Mathematical Tables has interest tables useful for the range $\frac{1}{4}\%$ to 8%, $n \le 120$. But the best book of tables we know of is *Financial Compound Interest and Annuity Tables,* Financial Publishing Co., Boston, 1955. It contains a wealth of information and useful tables up to $n = 360$.
    For instance, for the last example we find tabulated the function

$$s_n = \frac{(1 + i)^n - 1}{i}, \qquad i = j/12.$$

For $n = 120$ and $j = 4\frac{1}{4}\%$, the value is $s_n \approx 149.2072617712$; for $j = 4\frac{1}{2}\%$, it is $s_n \approx 151.1980736751$. (This, of course, is far more accuracy than we need for our example.) By interpolation, $j \approx 4.30\%$, which is more accurate than our answer, 4.29%.

Payment of Debts

    Suppose I borrow Q and propose to repay the loan by paying amount P at the end of each period, for n periods. The interest rate is i per period. What is P?
    There are several ways to solve this problem. Here is one based on our previous work and nice logical reasoning. Suppose the lender deposits each payment, as fast as he receives it, in a sinking fund (savings account at interest rate i). Then after

the n-th payment this money is worth

$$P\frac{(1 + i)^n - 1}{i}.$$

He reasons that this must be precisely the value of his original investment Q after n periods, that is, $Q(1 + i)^n$. Hence

$$P\frac{(1 + i)^n - 1}{i} = Q(1 + i)^n.$$

Solve for P:

$$\boxed{P = Q\frac{i}{1 - (1 + i)^{-n}}.}$$

This is the basic formula for the **periodic payment** P to pay off a loan Q. It is also referred to as the formula for "partial payment" or "annuity whose present value is Q".

To find the present worth of n future periodic payments of amount P at interest rate i, we solve the formula for Q:

$$Q = P\frac{1 - (1 + i)^{-n}}{i},$$

Now Q is called the "present value of an annuity", or "present worth of P per period".

■ *Example 9.6*

I want a 20-year mortgage on my house, and I am eligible for a 7% (compounded monthly) F.H.A. loan. I can afford monthly payments of $165. How much can I borrow?

SOLUTION   Here $n = 20 \cdot 12 = 240$ and $i = \frac{1}{12}(0.07) = 0.005833\cdots$, so

$$Q = 165\frac{1 - (1.0058333\cdots)^{-240}}{0.0058333\cdots}$$

$$\approx (165)(128.9825) \approx 21282.11.$$

Answer $21,282.11.

■ *Example 9.7*

I repay the above mortgage for 10 years, when I come into an inheritance and decide to pay off the outstanding debt. How much is it?

SOLUTION   Think of the periodic payments as a sinking fund, whose value after 10 years is

$$A = P\frac{(1+i)^n - 1}{i} = (165)\frac{(1.005833\cdots)^{120} - 1}{0.005833\cdots} \approx 28558.99.$$

After 10 years, the lender's investment of \$21,282.11 at 7% is worth

$$(21282.11)(1+i)^{120} \approx 42769.83.$$

The amount still owed is the difference

$$42769.83 - 28558.99 = 14210.84.$$

Answer   \$14,210.84.

Set up the following problems. Give a numerical answer if you have access to a calculator or appropriate tables:

1. Bank B offers 7% compounded quarterly while bank C offers $6\frac{3}{4}$% compounded daily. Which is better for the investor?
2. At 12% compounded monthly, money will triple in how long?
3. What annual interest rate, compounded monthly, is equivalent to 25% compounded annually?
4. What annual interest rate, compounded daily, is equivalent to 100% compounded annually.
5. How much must be salted away monthly at $\frac{1}{2}$% per month to save \$10,000 in 15 years?
6. How long will it take monthly deposits of \$50 to accumulate to \$12,500 if the annual interest rate is $7\frac{1}{2}$%, compounded monthly?
7. Estimate the average annual interest rate, compounded monthly, if \$10 deposits each month accumulate to \$2405.08 in 12 years.
8*. I deposit \$50 per month in a savings plan for 20 years. The annual interest rate changes from 5% to 6% after 8 years, and the compounding is monthly. Find the final amount.
9. I take a mortgage of \$28,000 for 18 years at $8\frac{1}{2}$%, compounded monthly. Find my monthly payment.
10. (cont.) Find my outstanding debt after 6 years. [Hint: See Example 9.7.]
11*. (cont.) How much total *interest* will I pay during the first year of the mortgage?
12*. (cont.) How much *interest* will I pay during the 12-th year of the mortgage?
13. I borrow \$3500 from a "friend" at 2% interest per month. I repay \$90 per month for 6 years. How much do I still owe?
14. I borrow \$2500 from another "friend" at 3% interest per month. I agree to pay him \$40 per month for 5 years and then repay the whole outstanding debt, which is how much?
15. Prove that the annual interest rate j, compounded p times per year, is equivalent to the annual rate i, compounded once, if

$$j = p[(1+i)^{1/p} - 1].$$

16. (cont.) I want to deposit amount A in a year, in p equal installments of A/p. Show that the amount of my account after a year is Ai/j.

17*. The Shark Loan Co. offers cheap loans up to $2500 for only 9% simple interest per year. I enter and request $2500 for one year. The lender deducts the interest, $(2500)(0.09) = \$225$ and hands me the remaining $2275, reminding me to repay $\frac{1}{12}(2500) = \$208.34$ per month for 12 months. (They always round up to the nearest cent in these deals.) Estimate the honest annual interest rate, compounded monthly.

18*. (cont.) The next customer wants $2000 for 2 years, so the lender deducts the $(2000)(0.18)$, etc. Now work out the true interest rate.

Test 1 [without optional material]

1. Without tables, show that $(1.5)^{20} > 1000$.

2. Graph roughly $y = 1/\sqrt{x^2 - 1}$.

3. Compute without tables:
   (a) $\log_3[(81) \sqrt[3]{3}]$ (b) $\log_3(\log_2 512)$.

4. Rationalize the numerator:

$$\frac{\sqrt{a} + \sqrt{b}}{\sqrt{a} - \sqrt{b}}.$$

5. Suppose $a > 1$, $b > 1$, $c > 1$. Simplify

$$b^{\log_b c}/\log_a a^c.$$

Test 2

1. Compute with 4-place tables

$$(\log 52)/(\log 13).$$

2. Solve for x:

$$5^x + (4)(5^{-x}) = 4.$$

3. Compute with 4-place tables:

$$\frac{(0.001964)(38.29)^4}{(0.0009222)^2}$$

4. Set up Problem 3 for the FP-02 pocket calculator.

5. Solve (a) or (b):
   (a) Find the monthly deposit, at 6% per year compounded monthly, that will accumulate $5000 in 5 years.
   (b) A certain radioactive substance decays to 25% of its mass in 3 days. Find the time required for it to decay to 1% of its original mass.

6

TRIGONOMETRIC FUNCTIONS

In this and the next two chapters we study a family of functions called trigo-nometric functions. These functions are indispensable in mathematics and its appli-cations, so it is essential to know them cold.

In this chapter we study the definitions, basic properties, and graphs of trigo-nometric functions. In Chapter 7 we study relations among these functions, and in Chapter 8 their applications to triangle solving and vectors. We shall see other applications in the chapter on complex numbers.

Before beginning our work, let us take a quick look at how this subject got started. Its roots are in the geometry of similar figures and the corresponding algebra of ratios of lengths. In Fig. 1.1a we have an angle θ. We draw a perpendicular to one of its sides, forming a right triangle (Fig. 1.1b).

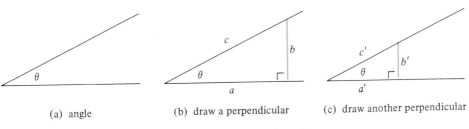

(a) angle   (b) draw a perpendicular   (c) draw another perpendicular

Fig. 1.1 The basic idea: $\dfrac{b}{a} = \dfrac{b'}{a'}$, etc.

Any ratio of two sides of this triangle is a number that *doesn't depend on the triangle at all*, only on θ. For instance the ratio b/a depends only on θ, that is, b/a is a *function* of θ. If we take another perpendicular (Fig. 1.1c), the corresponding ratio b'/a' is equal to b/a because the two triangles are *similar*.

Some useful facts from geometry will be found inside the back cover.

2. DISTANCES AND ANGLES

Given two points (x_1, y_1) and (x_2, y_2) in the coordinate plane, what is the distance between them? We show several cases in Fig. 2.1. In each case we introduce an auxiliary point (x_2, y_1) forming a right triangle. The legs have lengths $|x_2 - x_1|$ and $|y_2 - y_1|$, so by the Pythagorean theorem,

$$d^2 = |x_2 - x_1|^2 + |y_2 - y_1|^2 = (x_2 - x_1)^2 + (y_2 - y_1)^2.$$

> The distance between two points (x_1, y_1) and (x_2, y_2) is
> $$\sqrt{(x_2 - x_1)^2 + (y_2 - y_1)^2}.$$

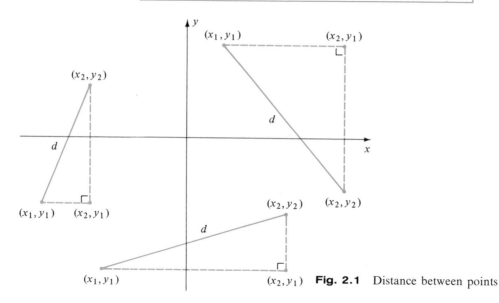

Fig. 2.1   Distance between points

If the points (x_1, y_1) and (x_2, y_2) lie on the same horizontal line or the same vertical line, the distance is $|x_1 - x_2|$ or $|y_1 - y_2|$; there is no need to introduce an auxiliary point. Nevertheless, the distance formula still yields the correct answer.

Examples:

(a) $(3, -1), (7, 2)$: $d^2 = (7 - 3)^2 + [2 - (-1)]^2 = 4^2 + 3^2 = 25$, $d = 5$.

(b) $(2, 4), (-3, 1)$:
$$d^2 = (-3 - 2)^2 + (1 - 4)^2 = (-5)^2 + (-3)^2 = 34, \quad d = \sqrt{34}.$$

(c) $(2, 7), (2, -5)$: $d^2 = (2 - 2)^2 + (-5 - 7)^2 = (-12)^2$, $d = 12$.

The locus of all points one unit from the origin is a circle of radius 1, called the **unit circle.** By the distance formula, a point (x, y) is on the unit circle if and only if $(x - 0)^2 + (y - 0)^2 = 1^2$,

$$x^2 + y^2 = 1.$$

This formula is the **equation** of the unit circle.

The unit circle consists of all points (x, y) in the plane that satisfy the condition

$$x^2 + y^2 = 1.$$

The equation is simply a restatement of the Pythagorean theorem for right triangles of hypotenuse 1. See Fig. 2.2.

It is easy to construct points on the unit circle. Let a and b be any real numbers, not both zero. Then $a^2 + b^2 > 0$ and

$$\left(\frac{a}{\sqrt{a^2 + b^2}}\right)^2 + \left(\frac{b}{\sqrt{a^2 + b^2}}\right)^2 = 1.$$

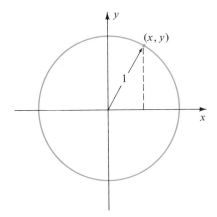

Fig. 2.2 Unit circle, $x^2 + y^2 = 1$

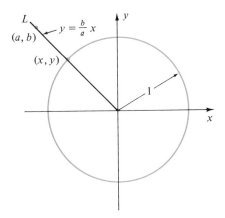

Fig. 2.3 Point on unit circle determined by (a, b)

Therefore

$$(x, y) = \left(\frac{a}{\sqrt{a^2 + b^2}}, \frac{b}{\sqrt{a^2 + b^2}}\right)$$

is a point on the unit circle. For example, if $a = -1$ and $b = 3$, the point is $(-1/\sqrt{10}, 3/\sqrt{10})$; if $a = 5$ and $b = 12$, the point is $(\frac{5}{13}, \frac{12}{13})$.

The construction has a simple geometric interpretation (Fig. 2.3). Two numbers a and b determine a point (a, b) in the plane. The ray L through (a, b) from the origin intersects the unit circle in the point

$$(x, y) = \left(\frac{a}{\sqrt{a^2 + b^2}}, \frac{b}{\sqrt{a^2 + b^2}} \right).$$

Why? Because (x, y) is on the unit circle, and it is also on L since x and y satisfy $y = bx/a$, the equation of the line L. (This argument needs to be modified slightly in the case $a = 0$. How?)

Radian Measure

The common units for measuring angles, inherited from the ancient world, are the degree ($°$), minute ($'$), and second ($''$). There are 60 seconds to a minute, 60 minutes to a degree, and 360 degrees to a whole angle. (We prefer decimal fractions of degrees, for instance 26.4°, 8.05°, etc.)

For scientific work, a particularly useful angle measure is the radian. One **radian** is the central angle of a circle subtended by an arc whose length equals the radius (Fig. 2.4). The circumference of the circle is $2\pi r$; therefore the whole angle is 2π radians. Thus the basic relation between radians and degrees:

$$2\pi \text{ radians} = 360°$$
$$\pi \text{ radians} = 180°.$$

Clearly, if π radians equals 180°, then 1 radian equals $180/\pi$ degrees, and if 180° equals π radians, then 1° equals $\pi/180$ radians.

$$1 \text{ radian} = \frac{180}{\pi} \text{ degrees} \approx 57.2958°,$$

$$1° = \frac{\pi}{180} \text{ radians} \approx 0.0174533 \text{ radians}.$$

It is customary to omit the unit "radian" in practice, e.g., we shall write $90° = \frac{1}{2}\pi$. Here are the radian measures of some common angles:

$$15° = \frac{\pi}{12}, \quad 30° = \frac{\pi}{6}, \quad 45° = \frac{\pi}{4}, \quad 60° = \frac{\pi}{3},$$

$$75° = \frac{5\pi}{12}, \quad 90° = \frac{\pi}{2}, \quad 105° = \frac{7\pi}{12}, \quad 120° = \frac{2\pi}{3},$$

$$135° = \frac{3\pi}{4}, \quad 150° = \frac{5\pi}{6}, \quad 180° = \pi, \quad 225° = \frac{5\pi}{4},$$

$$270° = \frac{3\pi}{2}, \quad 315° = \frac{7\pi}{4}, \quad 360° = 2\pi.$$

Fig. 2.4 The radian: arc = radius

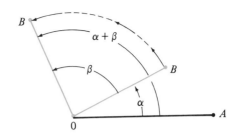

Fig. 2.5 Addition of angles

Ambiguity of Angles

Imagine a fixed bar $\overline{OA}$ and a movable bar $\overline{OB}$ free to pivot about the point O. To add angles α and β, pivot $\overline{OB}$ counterclockwise through the angle α, then pivot again through the angle β. The final position of $\overline{OB}$ indicates the sum $\alpha + \beta$. See Fig. 2.5.

Suppose $\alpha = 270°$ and $\beta = 89°$. Then $\alpha + \beta = 359°$. See Fig. 2.6a. Now suppose $\alpha = 270°$ but $\beta = 91°$ instead of $89°$. See Fig. 2.6b.

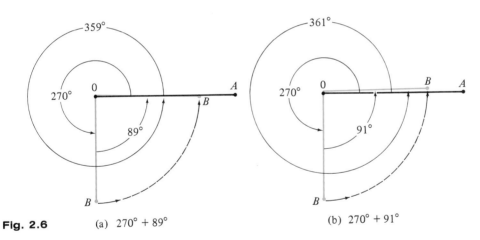

Fig. 2.6 (a) $270° + 89°$ (b) $270° + 91°$

Fig. 2.6b indicates that $270° + 91° = 1°$, not $361°$. Obviously, to do any reasonable kind of arithmetic with angles, we must consider $361°$ and $1°$ to be the same angle. What is more, if we add another $360°$, the bar $\overline{OB}$ goes around one full rotation and stops again at the $1°$ position. But according to arithmetic, the sum should be $361° + 360° = 721°$. Then we must identify $1°$ also with $721°$. Clearly each time we add $360°$ we obtain another angle which we must identify with $1°$. For the same reasons, we must identify all angles of the form

$$\alpha, \quad \alpha + 360°, \quad \alpha + 2 \cdot 360°, \quad \cdots, \quad \alpha + n \cdot 360°, \quad \cdots.$$

To subtract angles, pivot $\overline{OB}$ clockwise. For instance, to form $90° - 30°$, move $\overline{OB}$ counterclockwise until it forms a right angle with $\overline{OA}$, then move it *clockwise* $30°$. See Fig. 2.7.

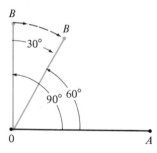

Fig. 2.7 $90° - 30°$

If we start with an angle α and swing $\overline{OB}$ clockwise n full turns, we come back to the same configuration, whereas by subtraction we arrive at $\alpha - n \cdot 360°$. Hence we must identify the angles

$$\alpha, \quad \alpha - 360°, \quad \alpha - 2 \cdot 360°, \cdots, \quad \alpha - n \cdot 360°, \cdots.$$

In terms of radian measure, we identify all the angles

$$\theta \pm 2\pi n \qquad (n = 0, 1, 2, 3 \cdots).$$

> Each real number measures a unique angle, but each angle corresponds to a whole family of real numbers, any two of which differ by an integer multiple of 2π.

EXERCISES

Compute the distance between the points:

1. $(3, 1), (2, 7)$ **2.** $(0, 4), (5, 1)$ **3.** $(6, 7), (6, 3)$
4. $(2, 8), (-1, 8)$ **5.** $(-1, 3), (6, -2)$ **6.** $(0, 0), (3, 4)$

7. Find all points on the unit circle one of whose coordinates is $\frac{1}{2}$.
8. Given a real number a with $-1 < a < 1$, show there are always two points on the unit circle with first coordinate a.

Convert to degrees:

9. $\dfrac{\pi}{3}$ **10.** $\dfrac{5\pi}{12}$ **11.** $\dfrac{6\pi}{5}$

12. $\dfrac{7\pi}{15}$ **13.** $\dfrac{\pi}{10}$ **14.** $\dfrac{4\pi}{9}$

15. $\dfrac{2\pi}{9}$ **16.** $\dfrac{17\pi}{12}$ **17.** $\dfrac{11\pi}{6}$

18. $\dfrac{3\pi}{8}$ **19.** $\dfrac{\pi}{90}$ **20.** $\dfrac{\pi}{45}$.

Convert to radians:

21. $135°$

22. $12°$

23. $6°$

24. $36°$

25. $67.5°$

26. $54°$

27. $9°$

28. $202.5°$

29. $357°$

30. $275°$.

Convert degrees to radians; use 3 significant figures:

31. $1.2°$

32. $10.3°$

33. $67°$

34. $31°$.

Convert radians to degrees; use 3 significant figures:

35. 0.3

36. 1.1

37. 1.05

38. 0.25.

Find an equivalent angle θ in the range $0 \le \theta < 2\pi$:

39. $\dfrac{103}{6}\pi$

40. $\dfrac{41}{19}\pi$

41. $-\dfrac{95}{8}\pi$

42. $-\dfrac{22}{3}\pi$

43. 63π

44. $\dfrac{977}{4}\pi$

45. $-17\dfrac{1}{4}\pi$

46. $-4\dfrac{1}{3}\pi$.

3. SINE AND COSINE

We study angles placed in a convenient standard position: centered at $(0,0)$ and measured from the positive x-axis. We measure positive angles counterclockwise, negative angles clockwise. Some examples are shown in Fig. 3.1.

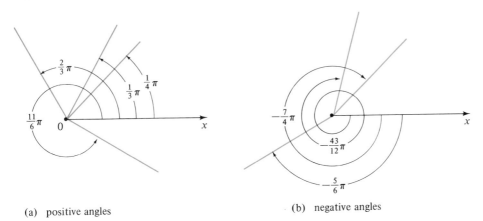

(a) positive angles

(b) negative angles

Fig. 3.1 Angles measured from positive x-axis

Let θ be any real number. The angle corresponding to θ determines a unique point on the unit circle (Fig. 3.2 on next page). Thus there is a correspondence

$$\theta \longrightarrow (x, y)$$

assigning to each real number θ a point (x, y) on the unit circle. We write

$$x = \cos \theta, \qquad y = \sin \theta$$

to indicate that x and y depend on θ; they are functions of θ. (Read "x equals cosine θ" and "y equals sine θ".)

Examples (Fig. 3.3):

$$0 \longrightarrow (1, 0) \qquad : \qquad \cos 0 = 1, \qquad \sin 0 = 0$$

$$\frac{\pi}{2} \longrightarrow (0, 1) \qquad : \qquad \cos \frac{\pi}{2} = 0, \qquad \sin \frac{\pi}{2} = 1$$

$$\frac{\pi}{4} \longrightarrow \left(\frac{\sqrt{2}}{2}, \frac{\sqrt{2}}{2} \right) \qquad : \qquad \cos \frac{\pi}{4} = \frac{\sqrt{2}}{2}, \qquad \sin \frac{\pi}{4} = \frac{\sqrt{2}}{2}$$

$$\frac{19\pi}{4} \longrightarrow \left(-\frac{\sqrt{2}}{2}, \frac{\sqrt{2}}{2} \right) : \qquad \cos \left(\frac{19\pi}{4} \right) = -\frac{\sqrt{2}}{2}, \qquad \sin \left(\frac{19\pi}{4} \right) = \frac{\sqrt{2}}{2}$$

$$-\frac{\pi}{3} \longrightarrow \left(\frac{1}{2}, -\frac{\sqrt{3}}{2} \right) \qquad : \qquad \cos \left(-\frac{\pi}{3} \right) = \frac{1}{2}, \qquad \sin \left(-\frac{\pi}{3} \right) = -\frac{\sqrt{3}}{2}$$

$$-\frac{65}{6}\pi \longrightarrow \left(-\frac{\sqrt{3}}{2}, -\frac{1}{2} \right) : \quad \cos \left(-\frac{65}{6}\pi \right) = -\frac{\sqrt{3}}{2}, \quad \sin \left(-\frac{65}{6}\pi \right) = -\frac{1}{2}.$$

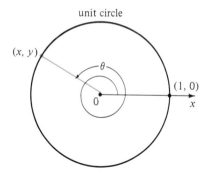

unit circle

(x, y)

θ

$(1, 0)$

x

0

$\theta = \frac{19}{4}\pi$:

$(-\frac{1}{2}\sqrt{2}, \frac{1}{2}\sqrt{2})$

$\theta = \frac{1}{2}\pi$: $(0, 1)$

$\theta = \frac{1}{4}\pi$: $(\frac{1}{2}\sqrt{2}, \frac{1}{2}\sqrt{2})$

$\theta = 0$: $(1, 0)$

$\theta = -\frac{1}{6}\pi$: $(\frac{1}{2}\sqrt{3}, -\frac{1}{2})$

$\theta = -\frac{1}{3}\pi$: $(\frac{1}{2}, -\frac{1}{2}\sqrt{3})$

Fig. 3.2 Each real θ determines an (x, y) on the unit circle.

Fig. 3.3 Examples of $\theta \longrightarrow (x, y)$

Certain sines and cosines are so frequently used, it pays to memorize them:

θ	0	$\frac{\pi}{6}$	$\frac{\pi}{4}$	$\frac{\pi}{3}$	$\frac{\pi}{2}$	π	$\frac{3\pi}{2}$
$\cos \theta$	1	$\frac{\sqrt{3}}{2}$	$\frac{\sqrt{2}}{2}$	$\frac{1}{2}$	0	-1	0
$\sin \theta$	0	$\frac{1}{2}$	$\frac{\sqrt{2}}{2}$	$\frac{\sqrt{3}}{2}$	1	0	-1

We have introduced two new functions, $\cos\theta$ and $\sin\theta$, each defined for all real values of θ. We now study the nature of these functions.

If θ is any real number, then θ and $\theta \pm 2\pi$ represent the same point (x, y) on the unit circle. Therefore:

$$\cos(\theta \pm 2\pi) = \cos\theta, \qquad \sin(\theta \pm 2\pi) = \sin\theta.$$

This property is expressed by saying that $\cos\theta$ and $\sin\theta$ are **periodic** functions with period 2π; they repeat themselves after $\pm 2\pi$ units. It follows that they repeat after $\pm 4\pi$, $\pm 6\pi$, $\pm 8\pi$, $\cdots$:

$$\cos(\theta + 2\pi n) = \cos\theta, \qquad \sin(\theta + 2\pi n) = \sin\theta, \qquad n = 0, \pm 1, \pm 2, \cdots.$$

For each θ, the point $(x, y) = (\cos\theta, \sin\theta)$ is on the unit circle $x^2 + y^2 = 1$, hence

$$\cos^2\theta + \sin^2\theta = 1.$$

[$\cos^2\theta$ and $\sin^2\theta$ is common notation for $(\cos\theta)^2$ and $(\sin\theta)^2$.] This relation restricts the values that $\cos\theta$ and $\sin\theta$ can have. Since squares are non-negative, both $\cos^2\theta \leq 1$ and $\sin^2\theta \leq 1$.

> For all real values of θ,
>
> $$-1 \leq \cos\theta \leq 1, \qquad -1 \leq \sin\theta \leq 1.$$

The numbers θ and $\theta + \pi$ define two points on the unit circle at opposite ends of a diameter, because π is the angle of a semicircle. Thus if (x, y) corresponds to θ, then $(-x, -y)$ corresponds to $\theta + \pi$. See Fig. 3.4. Therefore:

$$\cos(\theta + \pi) = -\cos\theta, \qquad \sin(\theta + \pi) = -\sin\theta.$$

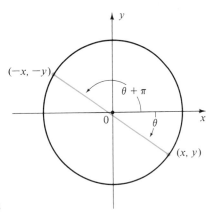

Fig. 3.4 θ and $\theta + \pi$ determine opposite points

Thus after π units, $\cos\theta$ and $\sin\theta$ repeat themselves, but with their signs reversed. This is consistent with their periodic property; after two increments of π, the sign reverses twice:

$$\cos(\theta + 2\pi) = \cos[(\theta + \pi) + \pi] = -\cos(\theta + \pi) = -(-\cos\theta) = \cos\theta.$$

The numbers θ and $-\theta$ define two points (x, y) and $(x, -y)$, reflections of each other in the x-axis (Fig. 3.5a). Therefore

$$\cos(-\theta) = \cos\theta, \qquad \sin(-\theta) = -\sin\theta.$$

Thus $\cos\theta$ is an even function and $\sin\theta$ is an odd function.

The numbers θ and $\frac{1}{2}\pi - \theta$ define two points (x, y) and (y, x), reflections of each other in the line $y = x$. See Fig. 3.5b. Therefore

$$\cos\left(\frac{\pi}{2} - \theta\right) = \sin\theta, \qquad \sin\left(\frac{\pi}{2} - \theta\right) = \cos\theta.$$

In other words, the sine and cosine of complementary angles are equal.

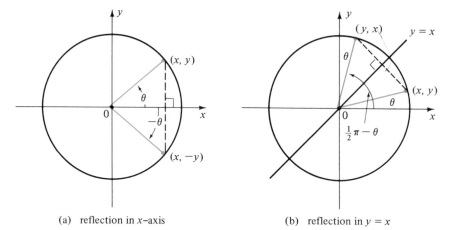

(a) reflection in x-axis   (b) reflection in $y = x$ **Fig. 3.5**

Analogous formulas for $\cos(\theta + \frac{1}{2}\pi)$ and $\sin(\theta + \frac{1}{2}\pi)$ follow from the preceding formulas and the even and odd properties of $\cos\theta$ and $\sin\theta$:

$$\cos\left(\theta + \frac{\pi}{2}\right) = \cos\left[\frac{\pi}{2} - (-\theta)\right] = \sin(-\theta) = -\sin\theta,$$

$$\sin\left(\theta + \frac{\pi}{2}\right) = \sin\left[\frac{\pi}{2} - (-\theta)\right] = \cos(-\theta) = \cos\theta.$$

Thus

$$\cos\left(\theta + \frac{\pi}{2}\right) = -\sin\theta, \qquad \sin\left(\theta + \frac{\pi}{2}\right) = \cos\theta.$$

See Fig. 3.6a.

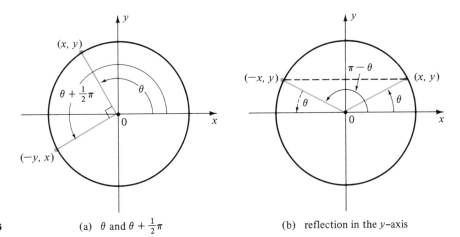

Fig. 3.6   (a) θ and $\theta + \frac{1}{2}\pi$ (b) reflection in the y-axis

The numbers θ and $\pi - \theta$ define two points (x, y) and $(-x, y)$, which are symmetric in the y-axis. See Fig. 3.6b. Therefore

$$\cos(\pi - \theta) = -\cos\theta, \qquad \sin(\pi - \theta) = \sin\theta.$$

Related to the second equation is an important fact about the sine. If c is a number, $-1 < c < 1$, then there are precisely two angles, θ and $\pi - \theta$, between 0 and 2π for which the sine has the value c. Indeed, the line $y = c$ intersects the unit circle in precisely two points, (x, c) and $(-x, c)$, corresponding to angles θ and $\pi - \theta$. Hence $\sin\theta = \sin(\pi - \theta) = c$. If $0 < c < 1$, these two angles are in the first and second quadrants (Fig. 3.7).

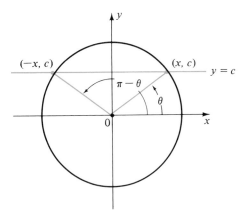

Fig. 3.7   Points where $\sin\theta = c$

Let us summarize the basic properties of cosine and sine.

$$\cos(\theta + 2\pi) = \cos \theta \qquad \sin(\theta + 2\pi) = \sin \theta$$
$$\cos(\theta + \pi) = -\cos \theta \qquad \sin(\theta + \pi) = -\sin \theta$$
$$\cos(-\theta) = \cos \theta \qquad \sin(-\theta) = -\sin \theta$$
$$\cos\left(\frac{\pi}{2} - \theta\right) = \sin \theta \qquad \sin\left(\frac{\pi}{2} - \theta\right) = \cos \theta$$
$$\cos\left(\frac{\pi}{2} + \theta\right) = -\sin \theta \qquad \sin\left(\frac{\pi}{2} + \theta\right) = \cos \theta$$
$$\cos(\pi - \theta) = -\cos \theta \qquad \sin(\pi - \theta) = \sin \theta.$$

■ *Example 3.1*

Given $P = \sin 65°$, express

$$\text{(a)} \quad \cos 65° \qquad \text{(b)} \quad \sin 25° \qquad \text{(c)} \quad \cos 155°$$

in terms of P.

SOLUTION   We have $\cos^2 65° + \sin^2 65° = 1$ and $0 < \cos 65° < 1$, so

$$\cos 65° = \sqrt{1 - \sin^2 65°} = \sqrt{1 - P^2}.$$

Next, $25° = 90° - 65°$, so

$$\sin 25° = \sin(90° - 65°) = \cos 65° = \sqrt{1 - P^2}.$$

Finally, $155° = 90° + 65°$, so

$$\cos 155° = \cos(90° + 65°) = -\sin 65° = -P.$$

Answer $\sqrt{1 - P^2}, \quad \sqrt{1 - P^2}, \quad -P.$

EXERCISES

Compute $\sin \theta$ and $\cos \theta$ for $\theta =$

1. $-\pi/2$
2. $-\pi$
3. $-\pi/4$
4. $-\pi/6$
5. $7\pi/3$
6. $-7\pi/3$
7. $17\pi/6$
8. $-17\pi/6$
9. $100\pi/3$
10. $-1000\pi/3$
11. -9π
12. $99\pi/4.$

Find all real numbers θ satisfying

13. $\cos \theta = \frac{1}{2}$
14. $\sin \theta = -\frac{1}{2}$
15. $\cos \theta = 0$
16. $\cos \theta = -1$
17. $\sin \theta = -\frac{1}{2}\sqrt{3}$
18. $\cos \theta = \frac{1}{2}\sqrt{3}$
19. $\sin \theta = \cos \theta$
20. $\sin \theta = \sin^2 \theta.$

21. Compute $\sin^2 10° + \cos^2 10°$.
22. Compute $\sin 10° + \sin 20° + \sin 30° + \cdots + \sin 360°$ without tables.
23. If $\alpha + \beta = \frac{1}{2}\pi$, show that $\sin \alpha = \cos \beta$.
24. If $\alpha + \beta = \frac{3}{2}\pi$, show that $\sin \alpha = -\cos \beta$.

Show that the function is periodic and find its least positive period:

25. $3 + 4 \cos \theta$
26. $2 \sin \theta - \cos \theta$
27. $\sin^2 \theta$
28. $\sin 2\theta$
29. $\cos 5\theta$
30. $\cos \pi\theta$
31. $|\cos \theta|$
32. $\sin \theta + \sin^2 \theta$.

Express in terms of $A = \cos 20°$:

33. $\sin 70°$
34. $\sin 20°$
35. $\cos 160°$
36. $\cos 110°$
37. $\sin 110°$
38. $\cos 200°$
39. $\sin 340°$
40. $\cos 250°$.

4. OTHER TRIGONOMETRIC FUNCTIONS

The Tangent Function

The cosine and the sine are the basic trigonometric functions; four other trigonometric functions are constructed from them. Consider next the **tangent** function $\tan \theta$ defined by the formula

$$\tan \theta = \frac{\sin \theta}{\cos \theta}.$$

The denominator must be different from 0, so $\tan \theta$ is defined for all real θ except $\theta = \pm\frac{1}{2}\pi, \pm\frac{3}{2}\pi, \pm\frac{5}{2}\pi, \cdots$.

For certain values of θ, the value of $\tan \theta$ is easily found. For example, we have $\tan 0 = \tan \pi = 0$. Also, as is clear from Fig. 4.1,

$$\tan \frac{\pi}{6} = \frac{1}{\sqrt{3}}, \qquad \tan \frac{\pi}{4} = 1, \qquad \tan \frac{\pi}{3} = \sqrt{3}.$$

(a) $\tan \frac{1}{6}\pi = \dfrac{\frac{1}{2}}{\frac{1}{2}\sqrt{3}} = \dfrac{1}{\sqrt{3}}$ (b) $\tan \frac{1}{4}\pi = \dfrac{\frac{1}{2}\sqrt{2}}{\frac{1}{2}\sqrt{2}} = 1$ (c) $\tan \frac{1}{3}\pi = \dfrac{\frac{1}{2}\sqrt{3}}{\frac{1}{2}} = \sqrt{3}$

Fig. 4.1

If θ is in the first or third quadrant, then $\sin\theta$ and $\cos\theta$ have the same sign, so $\tan\theta > 0$. If θ is in the second or fourth quadrant, then $\sin\theta$ and $\cos\theta$ have opposite signs, so $\tan\theta < 0$.

The tangent is an odd function:

$$\tan(-\theta) = \frac{\sin(-\theta)}{\cos(-\theta)} = \frac{-\sin\theta}{\cos\theta} = -\tan\theta.$$

The tangent inherits a periodic property from the sine and cosine:

$$\tan(\theta + \pi) = \frac{\sin(\theta + \pi)}{\cos(\theta + \pi)} = \frac{-\sin\theta}{-\cos\theta} = \tan\theta.$$

Thus $\tan\theta$ is periodic with period π.

We summarize the basic properties of the tangent:

$$\tan\theta = \frac{\sin\theta}{\cos\theta}, \qquad \theta \neq \pm\frac{\pi}{2}, \quad \pm\frac{3\pi}{2}, \quad \pm\frac{5\pi}{2}, \cdots,$$

$$\tan\theta = \frac{y}{x} \qquad \text{where} \quad \theta \longrightarrow (x, y),$$

$$\tan(-\theta) = -\tan\theta,$$

$$\tan(\theta + \pi) = \tan\theta.$$

The Cotangent, Secant, and Cosecant

There are three other common trigonometric functions: cotangent, secant, and cosecant. Here are their definitions and domains:

$$\cot\theta = \frac{1}{\tan\theta} = \frac{\cos\theta}{\sin\theta} \qquad \theta \neq 0, \quad \pm\pi, \quad \pm 2\pi, \quad \pm 3\pi, \cdots$$

$$\sec\theta = \frac{1}{\cos\theta} \qquad \theta \neq \frac{\pi}{2}, \quad \pm\frac{3\pi}{2}, \quad \pm\frac{5\pi}{2}, \cdots$$

$$\csc\theta = \frac{1}{\sin\theta} \qquad \theta \neq 0, \quad \pm\pi, \quad \pm 2\pi, \quad \pm 3\pi, \cdots.$$

This completes the list of the six trigonometric functions. Notice that tangent, cotangent, secant, and cosecant are all defined in terms of sine and cosine. They have symmetry and periodicity properties, stemming from similar properties of sine and cosine:

$$\cot(-\theta) = -\cot\theta \qquad \cot(\theta + \pi) = \cot\theta$$

$$\sec(-\theta) = \sec\theta \qquad \sec(\theta + 2\pi) = \sec\theta$$

$$\csc(-\theta) = -\csc\theta \qquad \csc(\theta + 2\pi) = \csc\theta.$$

Examples:

$$\cot \tfrac{1}{6}\pi = \frac{1}{\tan \tfrac{1}{6}\pi} = \frac{1}{1/\sqrt{3}} = \sqrt{3},$$

$$\sec \tfrac{3}{4}\pi = \frac{1}{\cos \tfrac{3}{4}\pi} = \frac{1}{-1/\sqrt{2}} = -\sqrt{2},$$

$$\csc 210° = \frac{1}{\sin 210°} = \frac{1}{\sin(180+30)°} = \frac{1}{-\sin 30°} = \frac{1}{-\tfrac{1}{2}} = -2.$$

The tangent and cotangent (secant and cosecant) of complementary angles are equal:

$$\tan\left(\frac{\pi}{2} - \theta\right) = \cot \theta \qquad \cot\left(\frac{\pi}{2} - \theta\right) = \tan \theta$$

$$\sec\left(\frac{\pi}{2} - \theta\right) = \csc \theta \qquad \csc\left(\frac{\pi}{2} - \theta\right) = \sec \theta.$$

The reason is

$$\tan\left(\frac{\pi}{2} - \theta\right) = \frac{\sin(\tfrac{1}{2}\pi - \theta)}{\cos(\tfrac{1}{2}\pi - \theta)} = \frac{\cos \theta}{\sin \theta} = \cot \theta,$$

and similarly for the cotangent, secant, and cosecant.

Trig Functions Defined by a Point in the Plane

Given any point (a, b) in the plane different from $(0, 0)$, the ray from $(0, 0)$ through (a, b) forms an angle θ with the positive x-axis (Fig. 4.2 on next page). We shall need the values of the six trigonometric functions for this θ. (Naturally, they depend on a and b.)

The ray intersects the unit circle at $(x, y) = (\cos \theta, \sin \theta)$. But in Section 2 we found that

$$x = \frac{a}{\sqrt{a^2 + b^2}}, \qquad y = \frac{b}{\sqrt{a^2 + b^2}}.$$

This gives us $\cos \theta$ and $\sin \theta$, from which we compute the other four trigonometric functions of θ.

Let $(a, b) \neq (0, 0)$. Suppose the ray from $(0, 0)$ through (a, b) makes an angle θ with the positive x-axis. Let $(a, b) \neq (0, 0)$ and set $r = \sqrt{a^2 + b^2}$. Then

$$\sin \theta = \frac{b}{r} \qquad\qquad \csc \theta = \frac{r}{b} \quad (b \neq 0)$$

$$\cos \theta = \frac{a}{r} \qquad\qquad \sec \theta = \frac{r}{a} \quad (a \neq 0)$$

$$\tan \theta = \frac{b}{a} \quad (a \neq 0) \qquad \cot \theta = \frac{a}{b} \quad (b \neq 0).$$

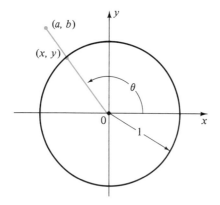

Fig. 4.2 The ray through (a, b) meets the unit circle at (x, y).

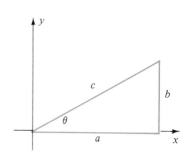

Fig. 4.3 Typical right triangle

Trig Functions and Right Triangles

The six trig functions are important in the geometry of right triangles. In a typical right triangle, we set up coordinate axes at the θ-vertex, as shown in Fig. 4.3. By the preceding results we may interpret the trig functions of θ as ratios of sides of the triangle:

$$
\sin \theta = \frac{b}{c} \qquad \csc \theta = \frac{c}{b}
$$

$$
\cos \theta = \frac{a}{c} \qquad \sec \theta = \frac{c}{a}
$$

$$
\tan \theta = \frac{b}{a} \qquad \cot \theta = \frac{a}{b}.
$$

In Chapter 8 we shall apply these relations to problems of measurement.

EXERCISES

1. Fill in the table

θ	0	$\frac{1}{6}\pi$	$\frac{1}{4}\pi$	$\frac{1}{3}\pi$	$\frac{1}{2}\pi$
$\tan \theta$					
$\cot \theta$					
$\sec \theta$					
$\csc \theta$					

2. (cont.) Do the same for the angles $-\frac{1}{6}\pi$, $-\frac{1}{4}\pi$, $-\frac{1}{3}\pi$, $-\frac{1}{2}\pi$, $-\pi$.

Prove

3. $\cot(\theta + \frac{1}{2}\pi) = -\tan \theta$

4. $\sec(\theta + \frac{1}{2}\pi) = -\csc \theta$

5. $1 + \tan^2 \theta = \sec^2 \theta$

6. $1 + \cot^2 \theta = \csc^2 \theta$

7. $\tan(\theta + \frac{1}{2}\pi) = -\cot\theta$
8. $\csc(\theta + \frac{1}{2}\pi) = \sec\theta$
9. $\cot(-\theta) = -\cot\theta$
10. $\cot(\theta + \pi n) = \cot\theta$
11. $\sec(-\theta) = \sec\theta$
12. $\sec(\theta + 2\pi n) = \sec\theta$
13. $\csc(-\theta) = -\csc\theta$
14. $\csc(\theta + 2\pi n) = \csc\theta.$

Find $\sin\theta$, $\cos\theta$, and $\tan\theta$ if θ is the angle between the positive x-axis and the ray from $(0,0)$ through (a, b):

15. $(a, b) = (3, 4)$
16. $(a, b) = (24, 7)$
17. $(a, b) = (0, 6)$
18. $(a, b) = (-3, 0)$
19. $(a, b) = (-5, 12)$
20. $(a, b) = (\frac{4}{5}, -\frac{3}{5})$
21. $(a, b) = (-2, -3)$
22. $(a, b) = (-5, 1).$

Show that the function is periodic and find its least positive period:

23. $\tan\theta + 2\cot\theta$
24. $3\sec\theta + 4\csc\theta$
25. $\sin\theta - \sec\theta$
26. $\cos\theta + 2\csc\theta$
27. $\sin\theta + \tan\theta$
28. $\cos^2\theta + \cot\theta$
29. $\tan^2\theta$
30. $\tan^2 3\theta$
31. $\csc 2\pi\theta$
32. $3\sec(\theta - \frac{1}{6}\pi).$

5. GRAPHS OF SINE AND COSINE

In this section we shall construct the graphs of the trigonometric functions, starting with $\sin\theta$ and $\cos\theta$. We can predict the general shapes of the sine and cosine graphs by considering a point P moving around the unit circle (Fig. 5.1a).

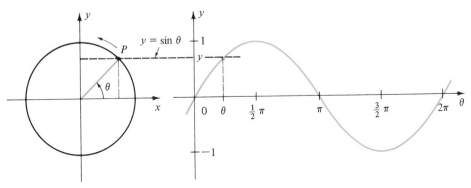

(a) projection on the y–axis
of a point moving uniformly
around the unit circle

(b) rough graph of $y = \sin\theta$
(different scales on axes)

Fig. 5.1

Suppose P moves counterclockwise with constant speed around the circle, starting from $(1, 0)$ at time 0. Suppose also that P makes a complete revolution every 2π seconds, so θ measures not only the central angle, but elapsed time.

The projection of P on the y-axis is $y = \sin\theta$. As P goes around and around the circle, y goes up and down the y-axis between -1 and $+1$. When θ starts from

0, the projection starts from 0 and moves up the y-axis. At first the projection moves quickly, but as θ continues to increase, y continues up the axis slower and slower, barely reaching its maximum height 1 as θ reaches $\frac{1}{2}\pi$. Then y begins to decrease faster and faster as θ approaches π. When θ passes π, we see that y becomes negative, reaching -1 when θ reaches $\frac{3}{2}\pi$ and returning to 0 as θ reaches 2π; then it starts all over again. This suggests the graph in Fig. 5.1b.

By considering the projection of the moving point on the x-axis, we arrive at a rough graph (Fig. 5.2) of $x = \cos\theta$. (The back and forth motion of either projection is called **simple harmonic motion.**)

Fig. 5.2   Rough graph of $x = \cos\theta$

Graph of Sin θ

We shall first construct the graph of $y = \sin\theta$. Since $\sin(\theta + 2\pi) = \sin\theta$, we need only graph $\sin\theta$ for the interval $-\pi \le \theta \le \pi$. We can then obtain the complete graph by shifting that piece left and right by integer multiples of 2π. Actually, since $\sin\theta$ is an odd function [$\sin(-\theta) = -\sin\theta$] we need plot the graph only for $0 \le \theta \le \pi$, then obtain the part for $-\pi \le \theta \le 0$ by reflecting in the origin.

We can get a pretty accurate graph by plotting the points $(\theta, \sin\theta)$ for $\theta = 0, 0.1\pi, 0.2\pi, \cdots, 1.0\pi$ with the use of a sine table.

$\frac{1}{\pi}\theta$	0.0	0.1	0.2	0.3	0.4	0.5	0.6	0.7	0.8	0.9	1.0
$\sin\theta$	0.00	0.31	0.59	0.81	0.95	1.00	0.95	0.81	0.59	0.31	0.00

[Note that $(1/\pi)\theta$ is in the first row, not θ.]

We plot this data and draw the graph (Fig. 5.3a), then extend it to $-\pi \le \theta \le \pi$ by symmetry (Fig. 5.3b).

Finally we shift left and right, continuing the graph indefinitely (Fig. 5.4).

The graph has several symmetries; not only is it symmetric in $(0, 0)$, but also in each point $(\pm\pi, 0)$, $(\pm 2\pi, 0)$, $(\pm 3\pi, 0)$, $\cdots$. For instance,

$$\sin(\pi - \theta) = -\sin(-\theta) = \sin\theta = -\sin(\pi + \theta);$$

hence there is symmetry in the point $(\pi, 0)$. But $\pm n\pi$ differs either from 0 or π by a multiple of 2π. Therefore the symmetry in the two points $(0, 0)$ and $(\pi, 0)$ carries over to all points $(\pm n\pi, 0)$.

There is symmetry also in each vertical line $\theta = \pm\frac{1}{2}\pi$, $\theta = \pm\frac{3}{2}\pi$, $\theta = \pm\frac{5}{2}\pi$, $\cdots$. For the line $\theta = \frac{1}{2}\pi$, this is because $\sin(\frac{1}{2}\pi - \theta) = \sin(\frac{1}{2}\pi + \theta)$, which follows from

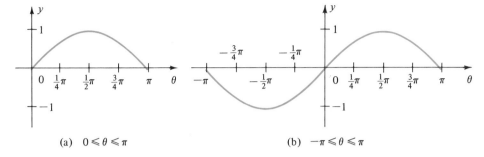

(a) $0 \leqslant \theta \leqslant \pi$ (b) $-\pi \leqslant \theta \leqslant \pi$

Fig. 5.3   Construction of the graph of $y = \sin \theta$

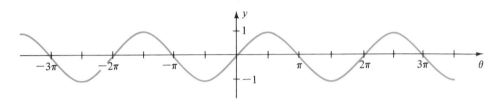

Fig. 5.4   Graph of $y = \sin \theta$

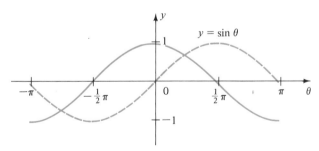

(a)   detailed graph: $-\pi \leqslant \theta \leqslant \pi$

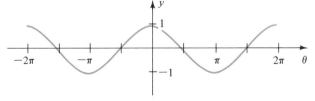

(b)   complete graph

Fig. 5.5   Graph of $y = \cos \theta$

the formula $\sin \theta = \sin(\pi - \theta)$ upon replacing θ by $\frac{1}{2}\pi - \theta$. For the line $\theta = -\frac{1}{2}\pi$, it is because of the symmetry in $\theta = \frac{1}{2}\pi$ and the oddness of $\sin \theta$. (Just look at Fig. 5.3b.) But each line $\theta = \pm\frac{3}{2}\pi$, $\theta = \pm\frac{5}{2}\pi$, $\cdots$ differs from either $\theta = -\frac{1}{2}\pi$ or $\theta = \frac{1}{2}\pi$ by a multiple of 2π. Therefore the symmetry in these two lines carries over to all of the others.

To summarize: the graph of $y = \sin x$ is "odd" with respect to each point where the curve crosses the x-axis, and "even" with respect to the vertical lines through its high and low points.

Graph of Cos θ

Once we have the graph of $y = \sin \theta$, we can obtain the graph of $y = \cos \theta$ free of charge. We simply use the relation

$$\cos \theta = \sin\left(\theta + \frac{\pi}{2}\right),$$

which shows that the graph of $y = \cos \theta$ is just the graph of $y = \sin \theta$ shifted $\frac{1}{2}\pi$ units to the left (Fig. 5.5 on preceding page).

Like the sine curve, the cosine curve has period 2π, is "odd" with respect to the points where it crosses the x-axis and is "even" with respect to the vertical lines through its high and low points.

Graph of $y = a \sin b\theta$

Given the graph of $y = \sin \theta$, what does the graph of $y = \sin 2\theta$ look like? We know $y = \sin \theta$ has period 2π; the graph makes one complete cycle on the interval $0 \le \theta \le 2\pi$. Now 2θ goes from 0 to 2π as θ goes from 0 to π. Therefore the graph of $y = \sin 2\theta$ makes one full cycle on the interval $0 \le \theta \le \pi$; its period is π. Thus the 2 in $\sin 2\theta$ "compresses" the curve $y = \sin \theta$ by a factor of 2. We might say that the 2 "hurries" the oscillations of $\sin \theta$ so that $y = \sin 2\theta$ makes two full cycles between 0 and 2π, while $y = \sin \theta$ makes one.

Similarly, the graph of $y = \sin 3\theta$ makes one cycle on the interval $0 \le \theta \le \frac{2}{3}\pi$ and three cycles on the interval $0 \le \theta \le 2\pi$. The graph of $y = \sin \frac{1}{2}\theta$ makes one cycle on the interval $0 \le \theta \le 4\pi$, a half-cycle on the interval $0 \le \theta \le 2\pi$. See Fig. 5.6.

In general, for each positive number b, the graph $y = \sin b\theta$ is a compressed (or stretched) version of the graph of $y = \sin \theta$. The period of $\sin b\theta$ is $2\pi/b$:

$$\sin b\left(\theta + \frac{2\pi}{b}\right) = \sin(b\theta + 2\pi) = \sin b\theta.$$

Note that everything we have said about $\sin b\theta$ holds equally for $\cos b\theta$.

The function $\sin b\theta$ oscillates between ± 1; the function $a \sin b\theta$ oscillates between $\pm a$ since multiplying by a stretches the graph vertically by a factor a. See Fig. 5.7 on p. 214.

Finally, we note that the graph of $y = a \sin(b\theta - c) + d$ is obtained from the graph of $y = a \sin b\theta$ by (1) shifting $|c|/b$ units, right if $c > 0$, left if $c < 0$, and (2) shifting $|d|$ units, up if $d > 0$, down if $d < 0$. See Fig. 5.8 on p. 214 for an example.

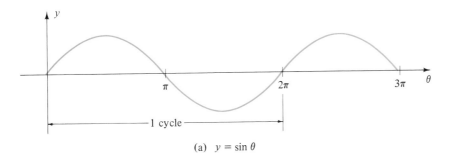

(a) $y = \sin \theta$

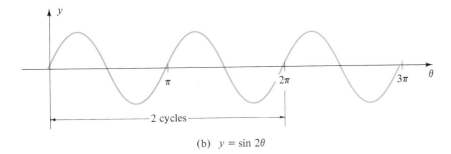

(b) $y = \sin 2\theta$

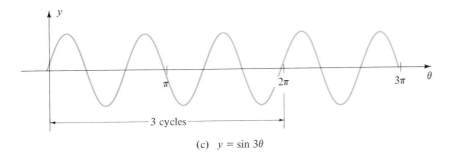

(c) $y = \sin 3\theta$

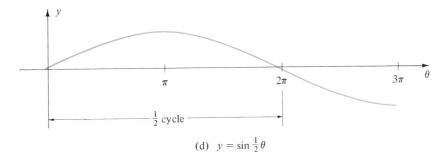

(d) $y = \sin \frac{1}{2} \theta$

Fig. 5.6

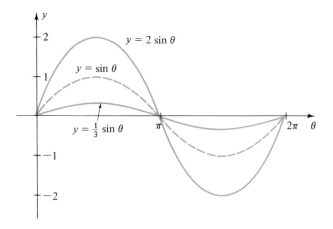

Fig. 5.7　Graphs of $y = 2 \sin \theta$ and $y = \frac{1}{3} \sin \theta$

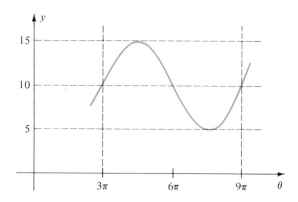

Fig. 5.8　Graph of $y = 5 \sin(\frac{1}{3}\theta - \pi) + 10$

EXERCISES

Graph the function (at least for $-\pi \le \theta \le \pi$):

1.　$2 \sin \theta$	**2.**　$3 \cos \theta$	**3.**　$-\frac{1}{2} \cos \theta$
4.　$-5 \sin \theta$	**5.**　$\sin 4\theta$	**6.**　$\cos 3\theta$
7.　$1 + \cos \theta$	**8.**　$2 - \sin \theta$	**9.**　$\frac{1}{2} - \cos \theta$
10.　$\frac{1}{2} + \sin \theta$	**11.**　$|\sin \theta|$	**12.**　$|\cos \theta|$
13.　$\cos 2\pi\theta$	**14.**　$\sin \pi\theta$	**15.**　$\sin^2 \theta$
16.　$\cos^2 \theta$	**17.**　$\sin(\theta + \frac{1}{6}\pi)$	**18.**　$\cos(\theta + \frac{1}{4}\pi)$
19.　$4 + 3 \cos \theta$	**20.**　$-6 + 2 \sin \theta$	**21.**　$\frac{1}{2} \sin(\theta + \frac{1}{3}\pi)$
22.　$2 \cos(\theta - \frac{1}{6}\pi)$	**23.**　$\sin(3\theta - \frac{1}{3}\pi)$	**24.**　$2 \cos 4(\theta - \frac{1}{6}\pi)$
25.　$1 + \frac{1}{2} \cos 2\pi\theta$	**26.**　$5 \sin 3\theta - 1.$	

A point P moves counterclockwise around the circle of radius 3 centered at $(0,0)$, making a complete revolution in 2π sec. Find formulas for the x and y coordinates and plot their graphs:

27. if P starts at $(3,0)$ when $t = 0$
28. if P starts at $(3,0)$ when $t = \frac{1}{3}\pi$
29. if P starts at $(0,3)$ when $t = 0$
30. if P starts at $(\frac{3}{2}\sqrt{3}, \frac{3}{2})$ when $t = 0$
31. if P starts at $(3,0)$ when $t = 0$, but makes a complete revolution in 1 sec
32. if P starts at $(0, -3)$ when $t = 0$, but makes one revolution in 10 sec.

33. Show graphically that the curves $y = \sin\theta$ and $y = \cos\theta$ are mirror-images of each other in the line $\theta = \frac{1}{4}\pi$.
34. (cont.) Explain the symmetry in Ex. 33 without using graphs.

6. GRAPHS OF THE OTHER FUNCTIONS

Graph of Tan θ

The function

$$\tan\theta = \frac{\sin\theta}{\cos\theta}$$

is defined for all values of θ except those for which $\cos\theta = 0$, that is, $\theta = \pm\frac{1}{2}\pi$, $\pm\frac{3}{2}\pi, \cdots$.

To study the graph of $\tan\theta$, we recall that

$$\tan(\theta + \pi) = \tan\theta.$$

Thus the graph repeats after π units, so it will be enough to study it in an interval of length π, say $-\frac{1}{2}\pi < \theta < \frac{1}{2}\pi$. Actually, because $\tan\theta$ is an odd function $[\tan(-\theta) = -\tan\theta]$ we need only construct the part of the graph where $0 \le \theta < \frac{1}{2}\pi$.

From a table of tangents we find the data:

$\frac{1}{\pi}\theta$	0.00	0.05	0.10	0.15	0.20	0.25	0.30	0.35	0.40	0.45	0.50
$\tan\theta$	0.00	0.16	0.32	0.51	0.73	1.00	1.38	1.97	3.08	6.31	∞

We sketch the graph (Fig. 6.1a) for $0 \le \theta < \frac{1}{2}\pi$. Then by symmetry we extend the graph (Fig. 6.1b) to the interval $-\frac{1}{2}\pi < \theta < \frac{1}{2}\pi$. Finally, we shift by multiples of π to obtain the complete graph (Fig. 6.1c). See next page.

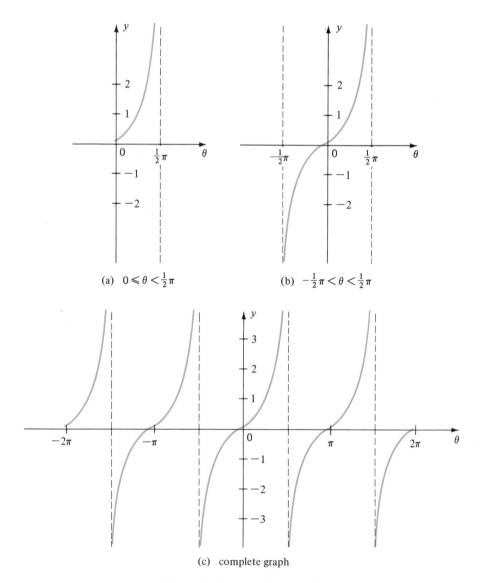

(a) $0 \leqslant \theta < \frac{1}{2}\pi$

(b) $-\frac{1}{2}\pi < \theta < \frac{1}{2}\pi$

(c) complete graph

Fig. 6.1 Graph of $y = \tan \theta$

Graphs of Cot θ, Sec θ, and Csc θ

We construct the graph of $y = \cot \theta$ by using the relation

$$\cot \theta = \tan \left(\frac{\pi}{2} - \theta \right).$$

We first draw $y = \tan \theta$, then reflect it in the y-axis; the result is $y = \tan(-\theta)$. Then we shift $\frac{1}{2}\pi$ to the left. The result is $y = \tan(\frac{1}{2}\pi - \theta) = \cot \theta$. See Fig. 6.2.

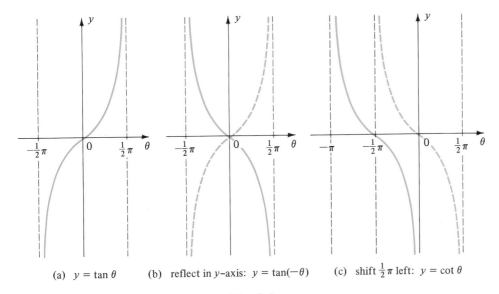

(a) $y = \tan \theta$ (b) reflect in y–axis: $y = \tan(-\theta)$ (c) shift $\frac{1}{2}\pi$ left: $y = \cot \theta$

Fig. 6.2

The complete graph is drawn in Fig. 6.3. Note the vertical asymptote at $\theta = n\pi$ for each integer n.

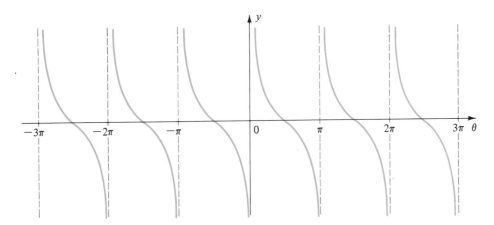

Fig. 6.3   Graph of $y = \cot \theta$

We graph $y = \sec \theta$ and $y = \csc \theta$ (Fig. 6.4, next page) by means of the relations

$$\sec \theta = \frac{1}{\cos \theta}, \qquad \csc \theta = \frac{1}{\sin \theta}.$$

Note the vertical asymptotes where the functions are undefined. Note also that $\csc(\theta + \frac{1}{2}\pi) = \sec \theta$, so the curve $y = \sec \theta$ is obtained by shifting the curve $y = \csc \theta$ left by $\frac{1}{2}\pi$.

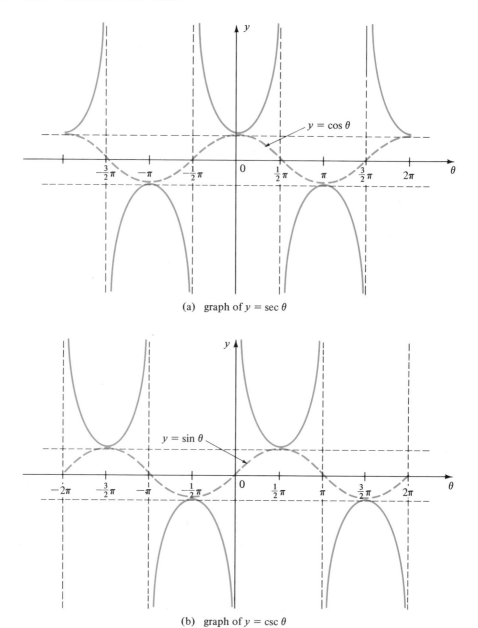

(a) graph of $y = \sec \theta$

(b) graph of $y = \csc \theta$

Fig. 6.4

Horizontal Tangents

Look closely at the cosine curve (Fig. 5.5) near the point (0, 1). A blow-up is shown in Fig. 6.5a, derived from the following data:

θ	.000	$\pm$.010	$\pm$.020	$\pm$.030	$\pm$.040	$\pm$.050
$\cos\theta$	1.000	1.000	1.000	1.000	0.999	0.999

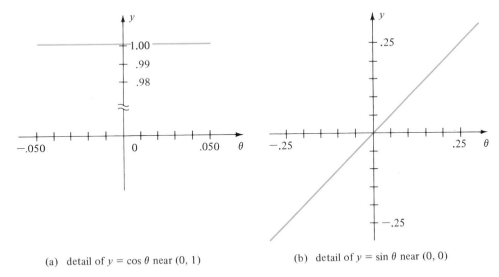

(a) detail of $y = \cos\theta$ near $(0, 1)$ (b) detail of $y = \sin\theta$ near $(0, 0)$

Fig. 6.5

The figure shows that the curve $y = \cos\theta$ is very flat at $(0, 1)$. We say that the graph has a **horizontal tangent** at $(0, 1)$. By symmetry, the graph also has horizontal tangents at each of its high and low points. The same is true of the curve $y = \sin\theta$, a shifted version of the curve $y = \cos\theta$.

Slopes

Look closely at the sine curve (Fig. 5.3b) near the point $(0, 0)$. A blow-up is shown in Fig. 6.5b, based on the data

θ	.000	$\pm$.050	$\pm$.100	$\pm$.150	$\pm$.200	$\pm$.250
$\sin\theta$	.000	$\pm$.050	$\pm$.100	$\pm$.149	$\pm$.199	$\pm$.247

For small values of θ, the table shows that $\sin\theta \approx \theta$; therefore the graph of $y = \sin\theta$ is very close to the graph of $y = \theta$, a line of slope 1. Because of its periodic behavior, the graph of $y = \sin\theta$ also looks like a line of slope 1 at each point where it crosses the x-axis while rising. By symmetry it looks like a line of slope -1 at each point where it crosses the axis while falling. Similar statements hold for $y = \cos\theta$.

To summarize: the curves $y = \sin \theta$ and $y = \cos \theta$ are very flat (have horizontal tangents) at each of their high and low points. They look like straight lines of slope -1 or $+1$ near each point where they cross the y-axis.

We conclude this discussion with an observation about the curve $y = \tan \theta$. Since the approximations $\sin \theta \approx \theta$ and $\cos \theta \approx 1$ are both very close for small values of θ, we expect

$$\tan \theta = \frac{\sin \theta}{\cos \theta} \approx \frac{\theta}{1} = \theta \qquad \text{for } \theta \text{ small.}$$

A tangent table confirms this suspicion.

θ	0.000	±0.050	±0.100	±0.150	±0.200	±0.250
$\tan \theta$	0.000	±0.050	±0.100	±0.151	±0.203	±0.255

We conclude that near $(0, 0)$ the curve $y = \tan \theta$ looks very much like the line $y = \theta$, in other words, Fig. 6.5b is also a good graph of $y = \tan \theta$ for small θ.

Periodic Functions

A function $f(\theta)$ is called **periodic** of period p if $p > 0$ and

$$f(\theta + p) = f(\theta).$$

The number p is the **period** of $f(\theta)$. Actually we should say "a period" rather than "the period". For instance $\sin(\theta + 2\pi) = \sin \theta$, but also $\sin(\theta + 4\pi) = \sin \theta$, so both 2π and 4π are periods of $\sin \theta$ as well as 6π, 8π, etc. However, from the graph of $\sin \theta$ it is clear that 2π is special among all periods of $\sin \theta$; it is the *smallest* period of the function. When we say "p is *the* period of $f(\theta)$" we usually mean that p is its smallest period. That is the sense in which we have used the term up to now.

Every trigonometric function we have discussed so far has been periodic, usually with period a simple rational multiple of π such as 2π, 3π, $\frac{1}{4}\pi$, etc.

Suppose you must graph a periodic function $y = f(\theta)$ of period p. Then graph it carefully on any segment of the θ-axis of length p. After that, or before that, it simply repeats itself. For instance, the function

$$y = \cos 2\theta + 2 \cos \theta$$

is periodic with period $p = 2\pi$. We graph it carefully for $0 \leq \theta \leq 2\pi$ by plotting points, and then repeat the graph for $2\pi \leq \theta \leq 4\pi, 4\pi \leq \theta \leq 6\pi, \cdots, -2\pi \leq \theta \leq 0,$ $-4\pi \leq \theta \leq -2\pi, \cdots$. See Fig. 6.6.

Fig. 6.6   Graph of $y = \cos 2\theta + 2 \cos \theta$. The period is 2π. Thus on each interval $2n\pi \le \theta \le (2n + 2)\pi$, the graph is merely a shift of the graph on $0 \le \theta \le 2\pi$.

EXERCISES

Graph the function

1. $5 \tan \theta$

2. $-\sec \theta$

3. $\sec(\theta + \frac{1}{6}\pi)$

4. $|\tan \theta|$

5. $1 + \cot \theta$

6. $1 - \csc \theta$

7. $\tan \pi\theta$

8. $\cot 2\theta$

9. $\sec^2 \theta$

10. $\cot(\theta - \frac{1}{4}\pi)$.

11*. Given that $\sin \theta \approx \theta$ for small values of θ, show that $\cos \theta \approx 1 - \frac{1}{2}\theta^2$ for small values of θ. [Hint: $\sin^2 \theta + \cos^2 \theta = 1$.] How does this confirm that $y = \cos \theta$ has a horizontal tangent at $\theta = 0$?

12. (cont.) Using cosine tables, check the accuracy of the approximation $\cos \theta \approx 1 - \frac{1}{2}\theta^2$ for $\theta = 0.05, 0.10, 0.15, 0.20$.

Without tables estimate the value:

13. $\dfrac{\sin 0.1}{0.1}$

14. $\dfrac{\tan 0.06}{0.06}$

15. $\dfrac{\sin 0.02}{0.01}$

16. $\dfrac{\sin 0.04}{\tan 0.06}$

17. $\cot 0.1$

18. $\csc(-0.2)$.

19. Show graphically that the equation $\tan \theta = \theta$ has infinitely many solutions, and exactly one solution in the range $-\frac{1}{2}\pi < \theta < \frac{1}{2}\pi$.

20. Show graphically that the equation $\sec \theta = \theta$ has infinitely many solutions but none in the range $-\frac{1}{2}\pi < \theta < \frac{1}{2}\pi$.

Graph for $-4\pi \le \theta \le 4\pi$:

21. $\sin 2\theta - \sin \theta$

22. $\sin \theta + \cos 2\theta$.

Test 1

1. Graph $y = 1 + \cos \theta$, $y = \tan(\theta - \frac{1}{4}\pi)$.

2. Compute $\sin 135°$, $\cos 240°$, $\tan \frac{1}{6}\pi$, $\sec 3\pi$.

3. Find *all* real numbers θ for which $\tan \theta = 1$.

4. Express in terms of $\sin \theta$ and $\cos \theta$:
$$\cos(\theta - \pi), \qquad \sin(\tfrac{3}{2}\pi + \theta), \qquad \tan \theta + \cot \theta.$$

5. Graph $y = \sin 3\theta$, $y = \cos \pi\theta$.

1. From the definitions of sine and cosine, explain the formulas:
   (a) $\sin^2 \theta + \cos^2 \theta = 1$ (b) $\sin(\theta + \pi) = -\sin \theta$.
2. Graph $y = 2 \sin(\theta + \frac{1}{6}\pi)$, $y = \sin 2\theta$.
3. Find *all* real numbers θ for which $\sin^2 \theta = 3 \cos^2 \theta$.
4. Compute $\sin 30°$, $\cos 120°$, $\tan(-\frac{2}{3}\pi)$, $\sec \frac{17}{4}\pi$.
5. A point moves with constant speed counterclockwise around the circle of radius 2 centered at $(0, 0)$. If it starts at $(\sqrt{2}, \sqrt{2})$ when $t = 0$ and makes one complete revolution in 2π sec, find formulas for its x- and y-coordinates.

7

IDENTITIES AND INVERSE FUNCTIONS

1. BASIC IDENTITIES

Because the six trigonometric functions are so closely related to each other, there is a multitude of identities interconnecting them; we shall discuss the most important. We begin with three very basic ones:

$$\sin^2 \theta + \cos^2 \theta = 1$$
$$1 + \tan^2 \theta = \sec^2 \theta$$
$$1 + \cot^2 \theta = \csc^2 \theta.$$

The first identity, already discussed in Chapter 6, Section 3, is an immediate consequence of the very definition of $\sin \theta$ and $\cos \theta$, whereby $(\cos \theta, \sin \theta)$ is a point on the unit circle.

To derive the second identity, divide both sides of the first by $\cos^2 \theta$:

$$\frac{\sin^2 \theta}{\cos^2 \theta} + \frac{\cos^2 \theta}{\cos^2 \theta} = \frac{1}{\cos^2 \theta}, \qquad \tan^2 \theta + 1 = \sec^2 \theta.$$

To derive the third, divide both sides of the first by $\sin^2 \theta$:

$$\frac{\sin^2 \theta}{\sin^2 \theta} + \frac{\cos^2 \theta}{\sin^2 \theta} = \frac{1}{\sin^2 \theta}, \qquad 1 + \cot^2 \theta = \csc^2 \theta.$$

Functions in Terms of Other Functions

Each trigonometric function can be expressed in terms of any one of the other five. For example:

$$\tan \theta = \sqrt{\sec^2 \theta - 1} \qquad \text{for} \quad \theta \text{ in quadrant 1 or 3,}$$

$$\csc \theta = -\sqrt{1 + \cot^2 \theta} \qquad \text{for} \quad \theta \text{ in quadrant 3 or 4.}$$

Not counting the signs of the square roots, there are $6 \times 5 = 30$ such formulas. It is not worth memorizing them since each desired formula can easily be derived. The simplest method consists of three steps:

(1) Label the sides of a right triangle by means of one given function alone.

(2) Read off the other 5 functions in terms of ratios of sides.

(3) Adjust the sign of each square root according to the quadrant.

Example 1.1

Express the trigonometric functions in terms of $\tan \theta$ for θ in the first quadrant and in the second quadrant.

SOLUTION   Label a right triangle with $\tan \theta$ on the side opposite θ and 1 on the side adjacent (Fig. 1.1).

Fig. 1.1

Now write the other 5 functions of θ as ratios of sides:

$$\sin \theta = \frac{\pm \tan \theta}{\sqrt{1 + \tan^2 \theta}} \qquad \cos \theta = \frac{\pm 1}{\sqrt{1 + \tan^2 \theta}}$$

$$\sec \theta = \pm \sqrt{1 + \tan^2 \theta} \qquad \csc \theta = \frac{\pm \sqrt{1 + \tan^2 \theta}}{\tan \theta}$$

$$\cot \theta = \frac{1}{\tan \theta}.$$

Find the correct sign in each case from the table:

	sin, csc	cos, sec	tan, cot
quad 1	+	+	+
quad 2	+	−	−

Answer   First quadrant:

$$\sin\theta = \frac{\tan\theta}{\sqrt{1+\tan^2\theta}} \qquad \cos\theta = \frac{1}{\sqrt{1+\tan^2\theta}}$$

$$\sec\theta = \sqrt{1+\tan^2\theta} \qquad \csc\theta = \frac{\sqrt{1+\tan^2\theta}}{\tan\theta}$$

$$\cot\theta = \frac{1}{\tan\theta};$$

Second quadrant:

$$\sin\theta = \frac{-\tan\theta}{\sqrt{1+\tan^2\theta}} \qquad \cos\theta = \frac{-1}{\sqrt{1+\tan^2\theta}}$$

$$\sec\theta = -\sqrt{1+\tan^2\theta} \qquad \csc\theta = -\frac{\sqrt{1+\tan^2\theta}}{\tan\theta}$$

$$\cot\theta = \frac{1}{\tan\theta}.$$

Proving Identities

Two trigonometric expressions may look quite different, yet be identical. That is because there are so many relations among the trigonometric functions. As a simple example, $\tan\theta$ is identical to

$$\frac{\sin\theta}{\cos\theta}, \qquad \frac{1}{\cot\theta}, \qquad \sin\theta\sec\theta, \qquad \cot\theta(\sec^2\theta - 1),$$

and many other expressions.

The typical problem is this. We are given two rational expressions A and B in the basic trigonometric functions $\sin\theta$, $\cot\theta$, etc. and we want to prove the identity $A = B$, that is, that $A = B$ for all values of θ for which the expressions are defined. A systematic procedure for proving such an identity is to express everything in A and B in terms of sines and cosines only, then to simplify. If the results are the same, after judicious use of the basic identity $\cos^2\theta + \sin^2\theta = 1$, then $A = B$ is proved.

Sometimes a preliminary simplification is useful. If we are asked to prove $A/C = B/D$, we might first clear fractions, then attempt to prove the *equivalent* identity $AD = BC$.

■ *Example 1.2*

Prove $\dfrac{\csc\theta + 1}{\cot\theta} = \dfrac{1}{\sec\theta - \tan\theta}.$

SOLUTION   The relation is equivalent to

$$(\csc\theta + 1)(\sec\theta - \tan\theta) = \cot\theta.$$

We shall prove this is an identity. The left-hand side (LHS), in terms of $\sin\theta$ and $\cos\theta$, is

$$\text{LHS} = \left(\frac{1}{\sin\theta} + 1\right)\left(\frac{1}{\cos\theta} - \frac{\sin\theta}{\cos\theta}\right) = \left(\frac{1 + \sin\theta}{\sin\theta}\right)\left(\frac{1 - \sin\theta}{\cos\theta}\right)$$

$$= \frac{1 - \sin^2\theta}{\sin\theta\cos\theta} = \frac{\cos^2\theta}{\sin\theta\cos\theta} = \frac{\cos\theta}{\sin\theta}.$$

The right-hand side (RHS) is

$$\text{RHS} = \cot\theta = \frac{\cos\theta}{\sin\theta}.$$

Hence (LHS) = (RHS), so we have proved the identity.

To prove $A = B$, you can just as well prove any equivalent relation such as $A + C = B + C$ or $(A + C)D = (B + C)D$, provided $D \neq 0$. In fact if you can find a chain of equivalent relations

$$A = B, \quad A_1 = B_1, \quad A_2 = B_2, \quad \cdots, \quad A_n = B_n,$$

and the last one is an identity, then so is the first, and the proof is complete.

A convenient shorthand is to abbreviate $\cos\theta$ by c and $\sin\theta$ by s. This not only saves a lot of writing, but it also decreases the chance of error.

■ *Example 1.3*

Prove $\quad \dfrac{1 - \cos\theta}{1 + \cos\theta} = (\csc\theta - \operatorname{ctn}\theta)^2.$

SOLUTION Express everything in terms of $c = \cos\theta$ and $s = \sin\theta$; then form a chain of equivalent relations:

$$\frac{1 - c}{1 + c} = \left(\frac{1}{s} - \frac{c}{s}\right)^2, \quad \frac{1 - c}{1 + c} = \frac{(1 - c)^2}{s^2},$$

$$s^2(1 - c) = (1 + c)(1 - c)^2 = (1 - c^2)(1 - c).$$

Since $s^2 = 1 - c^2$, the last statement is an identity, hence so is the first.

Converting to sines and cosines is not the only approach, nor necessarily the quickest in every case.

■ *Example 1.4*

Prove $\quad \dfrac{\tan\alpha - \tan\beta}{\cot\alpha - \cot\beta} = -\tan\alpha\tan\beta.$

SOLUTION Here it is easiest to express the cotangents in terms of tangents. The chain of relations is

$$\tan \alpha - \tan \beta = -\tan \alpha \tan \beta (\cot \alpha - \cot \beta),$$

$$\tan \alpha - \tan \beta = -\tan \alpha \tan \beta \left(\frac{1}{\tan \alpha} - \frac{1}{\tan \beta} \right),$$

$$\tan \alpha - \tan \beta = -\tan \beta + \tan \alpha.$$

EXERCISES

Simplify:

1. $\sin^4 \theta - \cos^4 \theta$

2. $(\sin \theta + \cos \theta)^2 - 2 \sin \theta \cos \theta$

3. $\dfrac{1 - \cos^2 \theta}{\sin \theta}$

4. $\dfrac{1 + \tan^2 \theta}{\tan^2 \theta}.$

Express the first function in terms of the second; adjust the sign so the formula is correct in the given quadrant:

5. $\sin \theta$ in terms of $\cot \theta$, quadrant 2

6. $\sin \theta$ in terms of $\cot \theta$, quadrant 4

7. $\csc \theta$ in terms of $\cot \theta$, quadrant 1

8. $\csc \theta$ in terms of $\cot \theta$, quadrant 3

9. $\sec \theta$ in terms of $\cot \theta$, quadrant 3

10. $\cos \theta$ in terms of $\cot \theta$, quadrant 2

11. $\tan \theta$ in terms of $\sin \theta$, quadrant 4

12. $\sec \theta$ in terms of $\sin \theta$, quadrant 1

13. $\csc \theta$ in terms of $\cos \theta$, quadrant 2

14. $\tan \theta$ in terms of $\cos \theta$, quadrant 3.

Prove the identity:

15. $\sin \theta \sec \theta = \tan \theta$

16. $\csc \theta = \sec \theta \cot \theta$

17. $\sec \theta - \cos \theta = \sin \theta \tan \theta$

18. $\csc \theta = \sin \theta + \cos \theta \cot \theta$

19. $\sec^2 \theta - \csc^2 \theta = \tan^2 \theta - \cot^2 \theta$

20. $\sin^2 \theta \cot^2 \theta + \cos^2 \theta \tan^2 \theta = 1$

21. $\sec^2 \theta + \csc^2 \theta = \sec^2 \theta \csc^2 \theta$

22. $\tan^2 \theta - \sin^2 \theta = \tan^2 \theta \sin^2 \theta$

23. $\sec^4 \theta - \tan^4 \theta = 1 + 2 \tan^2 \theta$

24. $\sin^4 \theta + \cos^4 \theta + 2 \sin^2 \theta \cos^2 \theta = 1$

25. $\cot^4 \theta + \cot^2 \theta = \csc^4 \theta - \csc^2 \theta$

26. $\cos^6 \theta + 3 \cos^2 \theta \sin^2 \theta + \sin^6 \theta = 1$

27. $(\cos \theta - \sin \theta)^2 + (\cos \theta + \sin \theta)^2 = 2$

28. $(x \cos \theta + y \sin \theta)^2 + (-x \sin \theta + y \cos \theta)^2 = x^2 + y^2$

29. $\dfrac{1 - \tan \theta}{1 + \tan \theta} = \dfrac{\cot \theta - 1}{\cot \theta + 1}$

30. $\dfrac{1}{1 - \sin \theta} - \dfrac{1}{1 + \sin \theta} = 2 \tan \theta \sec \theta$

31. $\dfrac{\tan \theta}{\sec \theta - \cos \theta} = \dfrac{\sec \theta}{\tan \theta}$

32. $\dfrac{\sin \theta}{\csc \theta + \cot \theta} = 1 - \cos \theta$

33. $\sec \theta + \csc \theta = (\tan \theta + \cot \theta)(\cos \theta + \sin \theta)$

34. $\dfrac{1 + \cos \theta}{\sin \theta} + \dfrac{\sin \theta}{1 + \cos \theta} = 2 \csc \theta$

35. $\dfrac{\cos \alpha - \sin \beta}{\cos \beta - \sin \alpha} = \dfrac{\cos \beta + \sin \alpha}{\cos \alpha + \sin \beta}$

36. $\dfrac{\tan \alpha + \tan \beta}{\cot \alpha + \cot \beta} = \dfrac{\tan \alpha \tan \beta - 1}{1 - \cot \alpha \cot \beta}.$

2. THE ADDITION LAWS

The **addition laws** for the sine and cosine are formulas that express $\sin(\alpha + \beta)$ and $\cos(\alpha + \beta)$ in terms of $\sin \alpha$, $\sin \beta$, $\cos \alpha$, and $\cos \beta$:

$$\sin(\alpha + \beta) = \sin \alpha \cos \beta + \cos \alpha \sin \beta$$
$$\cos(\alpha + \beta) = \cos \alpha \cos \beta - \sin \alpha \sin \beta.$$

These addition formulas are very basic in mathematics and should be memorized. Before we discuss a proof, let us note the equivalent forms, obtained by substituting $-\beta$ for β and using $\cos(-\beta) = \cos \beta$ and $\sin(-\beta) = -\sin \beta$:

$$\sin(\alpha - \beta) = \sin \alpha \cos \beta - \cos \alpha \sin \beta$$
$$\cos(\alpha - \beta) = \cos \alpha \cos \beta + \sin \alpha \sin \beta.$$

We shall prove the formula for $\cos(\alpha - \beta)$ by using a simple fact about circles: the length of a chord in a given circle depends only on the central angle subtended by the chord, not on the particular location of the chord.

Now take any chord of the unit circle (Fig. 2.1). The length L of the chord *depends only on* $\alpha - \beta$. By the distance formula,

$$L^2 = (\cos \alpha - \cos \beta)^2 + (\sin \alpha - \sin \beta)^2$$
$$= (\cos^2 \alpha + \sin^2 \alpha) + (\cos^2 \beta + \sin^2 \beta) - 2(\cos \alpha \cos \beta + \sin \alpha \sin \beta).$$
$$L^2 = 2 - 2(\cos \alpha \cos \beta + \sin \alpha \sin \beta).$$

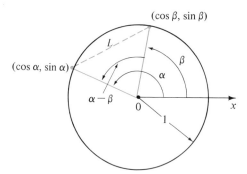

Fig. 2.1 L depends only on $\alpha - \beta$.

Because L^2, as well as L, *depends only on* the difference $\alpha - \beta$, it follows that *the quantity*

$$\cos \alpha \cos \beta + \sin \alpha \sin \beta$$

depends only on $\alpha - \beta$. In other words, if $a - b = \alpha - \beta$, then

$$\cos a \cos b + \sin a \sin b = \cos \alpha \cos \beta + \sin \alpha \sin \beta.$$

In particular, we may choose $a = \alpha - \beta$ and $b = 0$:

$$\cos(\alpha - \beta) \cos 0 + \sin(\alpha - \beta) \sin 0 = \cos \alpha \cos \beta + \sin \alpha \sin \beta,$$

that is,

$$\cos(\alpha - \beta) = \cos \alpha \cos \beta + \sin \alpha \sin \beta.$$

This is the desired formula. The formula for $\sin(\alpha - \beta)$ follows if α is replaced by $\frac{1}{2}\pi - \alpha$ and β is replaced by $-\beta$:

$$\sin(\alpha - \beta) = \cos[\tfrac{1}{2}\pi - (\alpha - \beta)] = \cos[(\tfrac{1}{2}\pi - \alpha) - (-\beta)]$$
$$= \cos(\tfrac{1}{2}\pi - \alpha) \cos(-\beta) + \sin(\tfrac{1}{2}\pi - \alpha) \sin(-\beta),$$

that is,

$$\sin(\alpha - \beta) = \sin \alpha \cos \beta - \cos \alpha \sin \beta.$$

Addition formulas for the other functions are readily derived from those for sine and cosine. The most important is the formula for tangent:

$$\tan(\alpha + \beta) = \frac{\tan \alpha + \tan \beta}{1 - \tan \alpha \tan \beta} \qquad (\tan \alpha \tan \beta \neq 1).$$

Proof.

$$\tan(\alpha + \beta) = \frac{\sin(\alpha + \beta)}{\cos(\alpha + \beta)} = \frac{\sin \alpha \cos \beta + \cos \alpha \sin \beta}{\cos \alpha \cos \beta - \sin \alpha \sin \beta}.$$

Divide numerator and denominator by $\cos \alpha \cos \beta$:

$$\tan(\alpha + \beta) = \frac{\dfrac{\sin \alpha}{\cos \alpha} + \dfrac{\sin \beta}{\cos \beta}}{1 - \dfrac{\sin \alpha \sin \beta}{\cos \alpha \cos \beta}} = \frac{\tan \alpha + \tan \beta}{1 - \tan \alpha \tan \beta}.$$

■ *Example 2.1*

Compute $\cos 75°$ without tables.

SOLUTION   Since $75 = 45 + 30$, we can use the addition laws to express $\cos 75°$ in terms of $\sin 45°$, $\cos 45°$, $\sin 30°$, and $\cos 30°$:

$$\cos 75° = \cos(45° + 30°) = \cos 45° \cos 30° - \sin 45° \sin 30°$$
$$= (\tfrac{1}{2}\sqrt{2})(\tfrac{1}{2}\sqrt{3}) - (\tfrac{1}{2}\sqrt{2})(\tfrac{1}{2})$$
$$= \tfrac{1}{4}\sqrt{6} - \tfrac{1}{4}\sqrt{2}.$$

Answer $\tfrac{1}{4}(\sqrt{6} - \sqrt{2})$.

· ·

Double-Angle Formulas

There are important special cases of the addition formulas that express $\sin 2\theta$, $\cos 2\theta$, and $\tan 2\theta$ in terms of $\sin \theta$, $\cos \theta$, and $\tan \theta$. In the formulas for $\sin(\alpha + \beta)$,

$\cos(\alpha + \beta)$, and $\tan(\alpha + \beta)$, set $\alpha = \beta = \theta$. The results are the **double-angle formulas:**

$$\sin 2\theta = 2 \sin \theta \cos \theta$$

$$\cos 2\theta = \cos^2 \theta - \sin^2 \theta$$

$$\tan 2\theta = \frac{2 \tan \theta}{1 - \tan^2 \theta}.$$

The second formula has two useful alternative forms. We replace $\cos^2 \theta$ by $1 - \sin^2 \theta$, then $\sin^2 \theta$ by $1 - \cos^2 \theta$:

$$\cos 2\theta = \cos^2 \theta - \sin^2 \theta = (1 - \sin^2 \theta) - \sin^2 \theta = 1 - 2 \sin^2 \theta,$$

$$\cos 2\theta = \cos^2 \theta - \sin^2 \theta = \cos^2 \theta - (1 - \cos^2 \theta) = 2 \cos^2 \theta - 1.$$

$$\cos 2\theta = 1 - 2 \sin^2 \theta = 2 \cos^2 \theta - 1.$$

Example 2.2

Express $\sin 3\theta$ in terms of $\sin \theta$ and $\cos \theta$.

SOLUTION   Write $3\theta = \theta + 2\theta$, and use the addition laws to express $\sin 3\theta$ in terms of $\sin \theta$, $\cos \theta$, $\sin 2\theta$, and $\cos 2\theta$. Then apply the double-angle formulas to $\sin 2\theta$ and $\cos 2\theta$:

$$\sin 3\theta = \sin(\theta + 2\theta) = \sin \theta \cos 2\theta + \cos \theta \sin 2\theta.$$

Hence, by the double-angle formulas,

$$\sin 3\theta = \sin \theta[\cos^2 \theta - \sin^2 \theta] + \cos \theta[2 \sin \theta \cos \theta]$$
$$= 3 \sin \theta \cos^2 \theta - \sin^3 \theta.$$

If you prefer the answer in terms of $\sin \theta$ alone, you can replace $\cos^2 \theta$ by $1 - \sin^2 \theta$.

Answer $\sin 3\theta = 3 \sin \theta \cos^2 \theta - \sin^3 \theta = 3 \sin \theta - 4 \sin^3 \theta.$

EXERCISES

Use the addition laws to compute

1. $\cos(\theta + 2\pi)$ and $\sin(\theta + 2\pi)$
3. $\cos(\theta + \frac{1}{2}\pi)$ and $\sin(\theta + \frac{1}{2}\pi)$

2. $\cos(\theta + \pi)$ and $\sin(\theta + \pi)$
4. $\cos(\frac{1}{2}\pi - \theta)$ and $\sin(\frac{1}{2}\pi - \theta)$.

Express:

5. $\cos(\theta + \frac{1}{4}\pi)$ and $\sin(\theta + \frac{1}{4}\pi)$ in terms of $\cos \theta$ and $\sin \theta$
6. $\cos(\theta + \frac{1}{6}\pi)$ and $\sin(\theta + \frac{1}{6}\pi)$ in terms of $\cos \theta$ and $\sin \theta$
7. $\cot(\alpha + \beta)$ in terms of $\cot \alpha$ and $\cot \beta$
8. $\tan(\alpha - \beta)$ in terms of $\tan \alpha$ and $\tan \beta$.

Compute without tables:

9. $\sin 75°$ **10.** $\tan 75°$ **11.** $\cos 15°$ **12.** $\sin 15°$.

Express:

13. $\cos 3\theta$ in terms $\cos \theta$

15. $\cot 3\theta$ in terms of $\cot \theta$

14. $\tan 3\theta$ in terms of $\tan \theta$

16. $\sec 3\theta$ in terms of $\sec \theta$.

Use double-angle formulas to prove

17. $\sin 4\theta = 4 \sin \theta \cos \theta - 8 \sin^3 \theta \cos \theta$

19. $\cot 2\theta + \csc 2\theta = \cot \theta$

21. $\tan \theta + \cot \theta = 2 \csc 2\theta$

18. $\cos 4\theta = 8 \cos^4 \theta - 8 \cos^2 \theta + 1$

20. $\csc 2\theta - \cot 2\theta = \tan \theta$

22. $\sin \theta \cos \theta \cos 2\theta \cos 4\theta = \frac{1}{8} \sin 8\theta$.

23. If $0 < \theta < \frac{1}{4}\pi$, which is larger, $\tan 2\theta$ or $2 \tan \theta$?

24*. If $\frac{1}{4}\pi < \theta < \frac{1}{3}\pi$, show that $\sin \theta < \sin 2\theta < \sqrt{2} \sin \theta$.

Prove:

25. $\cot 2\theta = \dfrac{\cot^2 \theta - 1}{2 \cot \theta}$

27. $\cot(\alpha + \beta) = \dfrac{\cot \alpha \cot \beta - 1}{\cot \alpha + \cot \beta}$

29. $\dfrac{1 + \cos 2\theta}{1 - \cos 2\theta} = \cot^2 \theta$

31. $\cos^3 \theta = \frac{1}{4}(3 \cos \theta + \cos 3\theta)$

26. $\sec 2\theta = \dfrac{\sec^2 \theta}{2 - \sec^2 \theta}$

28. $\tan 4\theta = \dfrac{4 \tan \theta - 4 \tan^3 \theta}{1 - 6 \tan^2 \theta + \tan^4 \theta}$

30. $\dfrac{\sin 2\theta + \sin \theta}{\cos 2\theta + \cos \theta + 1} = \tan \theta$

32. $\dfrac{\cot \theta - \tan \theta}{\cot \theta + \tan \theta} = \cos 2\theta$.

3. FURTHER IDENTITIES

Half-Angle Formulas

From the double-angle formulas, we can obtain half-angle formulas that express $\sin \frac{1}{2}\theta$, $\cos \frac{1}{2}\theta$, and $\tan \frac{1}{2}\theta$ in terms of $\sin \theta$ and $\cos \theta$. We start with the double-angle formulas

$$\cos 2\theta = 2 \cos^2 \theta - 1 = 1 - 2 \sin^2 \theta$$

and solve for $\sin \theta$ and $\cos \theta$:

$$\sin \theta = \pm\sqrt{\frac{1 - \cos 2\theta}{2}}, \qquad \cos \theta = \pm\sqrt{\frac{1 + \cos 2\theta}{2}},$$

where the sign depends on the quadrant. Replacing θ by $\frac{1}{2}\theta$, we obtain the **half-angle** formulas:

$$\sin \tfrac{1}{2}\theta = \pm\sqrt{\frac{1 - \cos \theta}{2}}, \qquad \cos \tfrac{1}{2}\theta = \pm\sqrt{\frac{1 + \cos \theta}{2}}.$$

We could obtain a formula for $\tan \frac{1}{2}\theta$ by dividing the two expressions above, but we avoid square roots by using a little trick:

$$\tan \tfrac{1}{2}\theta = \frac{\sin \frac{1}{2}\theta}{\cos \frac{1}{2}\theta} = \frac{\sin \frac{1}{2}\theta}{\cos \frac{1}{2}\theta} \cdot \frac{2 \cos \frac{1}{2}\theta}{2 \cos \frac{1}{2}\theta}.$$

We use double-angle formulas in both the numerator and denominator:

$$\tan \tfrac{1}{2}\theta = \frac{\sin \theta}{1 + \cos \theta}.$$

This half-angle formula for the tangent has a geometric interpretation. In Fig. 3.1a, the inscribed angle is $\tfrac{1}{2}\theta$, half the corresponding central angle. From the right triangle (Fig. 3.1b), $\tan \tfrac{1}{2}\theta = \sin \theta/(1 + \cos \theta)$.

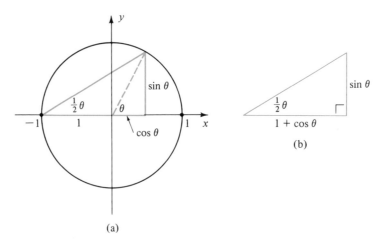

(a)

(b)

Fig. 3.1 $\tan \tfrac{1}{2}\theta = \dfrac{\text{opposite}}{\text{adjacent}} = \dfrac{\sin \theta}{1 + \cos \theta}$

Identities expressing $\sin \theta$ and $\cos \theta$ in terms of $\tan \tfrac{1}{2}\theta$ are useful in calculus:

$$\sin \theta = \frac{2 \tan \tfrac{1}{2}\theta}{1 + \tan^2 \tfrac{1}{2}\theta}$$

$$\cos \theta = \frac{1 - \tan^2 \tfrac{1}{2}\theta}{1 + \tan^2 \tfrac{1}{2}\theta}.$$

The proofs are easy:

$$\sin 2\theta = 2 \sin \theta \cos \theta = \frac{2 \sin \theta \cos \theta}{\sin^2 \theta + \cos^2 \theta} = \frac{2 \dfrac{\sin \theta}{\cos \theta}}{\dfrac{\sin^2 \theta}{\cos^2 \theta} + 1} = \frac{2 \tan \theta}{1 + \tan^2 \theta},$$

$$\cos 2\theta = \cos^2 \theta - \sin^2 \theta = \frac{\cos^2 \theta - \sin^2 \theta}{\cos^2 \theta + \sin^2 \theta} = \frac{1 - \dfrac{\sin^2 \theta}{\cos^2 \theta}}{1 + \dfrac{\sin^2 \theta}{\cos^2 \theta}} = \frac{1 - \tan^2 \theta}{1 + \tan^2 \theta}.$$

We give two useful applications of the addition formulas. The first is to express any function

$$f(\theta) = a \cos \theta + b \sin \theta$$

in terms of a single cosine or sine.

We have seen that for any two given constants a and b, the ray from $(0, 0)$ through (a, b) makes an angle θ_0 with the positive x-axis such that

$$\cos \theta_0 = \frac{a}{\sqrt{a^2 + b^2}}, \qquad \sin \theta_0 = \frac{b}{\sqrt{a^2 + b^2}}.$$

With this in mind we write

$$f(\theta) = a \cos \theta + b \sin \theta$$

$$= \sqrt{a^2 + b^2} \left(\frac{a}{\sqrt{a^2 + b^2}} \cos \theta + \frac{b}{\sqrt{a^2 + b^2}} \sin \theta \right)$$

$$= \sqrt{a^2 + b^2} \, (\cos \theta_0 \cos \theta + \sin \theta_0 \sin \theta) = \sqrt{a^2 + b^2} \, \cos(\theta - \theta_0).$$

If $a^2 + b^2 > 0$, then

$$a \cos \theta + b \sin \theta = \sqrt{a^2 + b^2} \, \cos(\theta - \theta_0),$$

where θ_0 is an angle satisfying

$$\cos \theta_0 = \frac{a}{\sqrt{a^2 + b^2}}, \qquad \sin \theta_0 = \frac{b}{\sqrt{a^2 + b^2}}.$$

■ **Example 3.1**

Express $-\cos \theta + \sqrt{3} \sin \theta$ as a cosine function.

SOLUTION   Here $a = -1$, $b = \sqrt{3}$, and $a^2 + b^2 = 4$, so we write

$$-\cos \theta + \sqrt{3} \sin \theta = 2 \left(-\frac{1}{2} \cos \theta + \frac{\sqrt{3}}{2} \sin \theta \right).$$

We must choose θ_0 to satisfy

$$\cos \theta_0 = -\frac{1}{2}, \qquad \sin \theta_0 = \frac{\sqrt{3}}{2}.$$

One possible choice is $\theta_0 = \frac{2}{3}\pi$.

Answer $2 \cos(\theta - \frac{2}{3}\pi)$.

Remark: Suppose instead we choose $\theta_0 = \frac{1}{6}\pi$. Then $\sin \theta_0 = \frac{1}{2}$ and $\cos \theta_0 = \frac{1}{2}\sqrt{3}$, so the given function is

$$2(-\sin \theta_0 \cos \theta + \cos \theta_0 \sin \theta) = 2 \sin(\theta - \theta_0).$$

In general, each function $a \cos \theta + b \sin \theta$ can be expressed as a cosine *or* as a sine.

■ *Example 3.2*

Prove that $|-\cos \theta + \sqrt{3} \sin \theta| \leq 2$ for all values of θ.

SOLUTION By the result of Example 3.1,

$$|-\cos \theta + \sqrt{3} \sin \theta| = |2 \cos(\theta - \tfrac{2}{3}\pi)| = 2|\cos(\theta - \tfrac{2}{3}\pi)| \leq 2 \cdot 1 = 2.$$

The second application of the addition formulas is to express products of sines and cosines as sums. First we add the formulas

$$\sin(\alpha + \beta) = \sin \alpha \cos \beta + \cos \alpha \sin \beta$$
$$\sin(\alpha - \beta) = \sin \alpha \cos \beta - \cos \alpha \sin \beta,$$

obtaining

$$\sin(\alpha + \beta) + \sin(\alpha - \beta) = 2 \sin \alpha \cos \beta.$$

Next we add and subtract the formulas

$$\cos(\alpha + \beta) = \cos \alpha \cos \beta - \sin \alpha \sin \beta$$
$$\cos(\alpha - \beta) = \cos \alpha \cos \beta + \sin \alpha \sin \beta,$$

obtaining

$$\cos(\alpha + \beta) + \cos(\alpha - \beta) = 2 \cos \alpha \cos \beta$$
$$\cos(\alpha - \beta) - \cos(\alpha + \beta) = 2 \sin \alpha \sin \beta.$$

We interpret the results as formulas for products of sines and cosines:

$$
\boxed{
\begin{aligned}
\sin \alpha \sin \beta &= \tfrac{1}{2}[\cos(\alpha - \beta) - \cos(\alpha + \beta)] \\
\cos \alpha \cos \beta &= \tfrac{1}{2}[\cos(\alpha + \beta) + \cos(\alpha - \beta)] \\
\sin \alpha \cos \beta &= \tfrac{1}{2}[\sin(\alpha + \beta) + \sin(\alpha - \beta)].
\end{aligned}
}
$$

Sometimes we go the other way and express the sum or difference of two sines or two cosines as a product. The trick is to write

$$\alpha = \tfrac{1}{2}(\alpha + \beta) + \tfrac{1}{2}(\alpha - \beta) = \gamma + \delta,$$
$$\beta = \tfrac{1}{2}(\alpha + \beta) - \tfrac{1}{2}(\alpha - \beta) = \gamma - \delta.$$

Then

$$\sin \alpha + \sin \beta = \sin(\gamma + \delta) + \sin(\gamma - \delta)$$
$$= 2 \sin \gamma \cos \delta$$
$$= 2 \sin \tfrac{1}{2}(\alpha + \beta) \cos \tfrac{1}{2}(\alpha - \beta).$$

In this way we derive four formulas:

$$\sin \alpha + \sin \beta = 2 \sin \tfrac{1}{2}(\alpha + \beta) \cos \tfrac{1}{2}(\alpha - \beta)$$
$$\sin \alpha - \sin \beta = 2 \cos \tfrac{1}{2}(\alpha + \beta) \sin \tfrac{1}{2}(\alpha - \beta)$$
$$\cos \alpha + \cos \beta = 2 \cos \tfrac{1}{2}(\alpha + \beta) \cos \tfrac{1}{2}(\alpha - \beta)$$
$$\cos \alpha - \cos \beta = -2 \sin \tfrac{1}{2}(\alpha + \beta) \sin \tfrac{1}{2}(\alpha - \beta).$$

The last two batches of formulas have important applications in situations where one vibration is imposed on another, such as in the modulation of radio signals, and in the phenomenon of beats in acoustics.

EXERCISES

Show that

1. $\sin 22.5° = \tfrac{1}{2} \sqrt{2 - \sqrt{2}}$ 2. $\tan 22.5° = \sqrt{2} - 1$ 3. $\tan 67.5° = \sqrt{2} + 1$.

4. Compute $\cos 15°$ two ways: by half-angle formulas and by addition laws. Conclude that

$$\sqrt{2 + \sqrt{3}} = \tfrac{1}{2}(\sqrt{6} + \sqrt{2}).$$

Prove:

5. $\tan \tfrac{1}{2}\theta = \dfrac{1 - \cos \theta}{\sin \theta}$ 6. $\cot \tfrac{1}{2}\theta = \dfrac{1 + \cos \theta}{\sin \theta}$.

Express each combination as a cosine. Use 2-place accuracy:

7. $4 \cos \theta + 3 \sin \theta$ 8. $3 \cos \theta - 4 \sin \theta$
9. $\cos \theta - 2 \sin \theta$ 10. $\cos \theta + 2 \sin \theta$
11. $2 \cos \theta + 3 \sin \theta$ 12. $-2 \cos \theta - 3 \sin \theta$
13. $3 \cos \theta - \sin \theta$ 14. $5 \sin \theta$.

15. Prove that $|\sin \theta + \cos \theta| \le \sqrt{2}$.
16. (cont.) Find all θ for which $\sin \theta + \cos \theta = \sqrt{2}$.
17. Find the maximum value of $\cos \theta - 3 \sin \theta$.
18. Find the minimum value of $4 \sin \theta + 3 \cos \theta$.

Express each function as a sum (difference) of sines or cosines:

19. $\sin \theta \sin 2\theta$ 20. $\cos 2\theta \cos \theta$
21. $\sin 3\theta \cos 4\theta$ 22. $\sin 101\theta \sin 100\theta$.

Express each function as a product of constants, sines, and cosines:

23. $\sin \theta + \sin 2\theta$ **24.** $\cos 2\theta - \cos \theta$

25. $\sin 5\theta - \sin 4\theta$ **26.** $\sin 5\theta + \sin 4\theta$

27. $\cos 6\theta - \cos 5\theta$ **28.** $\cos \theta + \cos 3\theta.$

Prove:

29. $\tan \alpha + \tan \beta = \sin(\alpha + \beta)/\cos \alpha \cos \beta$

30. $\cos^4 \theta = \frac{1}{8}(3 + 4 \cos 2\theta + \cos 4\theta)$ [Hint: Use half-angle formulas twice.]

31. $\sin^4 \theta = \frac{1}{8}(3 - 4 \cos 2\theta + \cos 4\theta)$

32. $\sin^2 \alpha - \sin^2 \beta = \sin(\alpha + \beta) \sin(\alpha - \beta)$

33. $\cos^2 \alpha - \sin^2 \beta = \cos(\alpha + \beta) \cos(\alpha - \beta)$

34. $1 + 2 \cos \theta + 2 \cos 2\theta + 2 \cos 3\theta = \dfrac{\sin \frac{7}{2}\theta}{\sin \frac{1}{2}\theta}$. [Hint: Start by multiplying the left side by $\sin \frac{1}{2}\theta$.]

35. Set $t = \tan \frac{1}{2}\theta$. Express $\sin 2\theta$ in terms of t. [Hint: Use half-angle formulas.]

36. Set $t = \tan \frac{1}{2}\theta$. Express $\cos 2\theta$ in terms of t. [Hint: Use half-angle formulas.]

Prove:

37. $\cot \alpha - \cot \beta = \dfrac{\sin(\beta - \alpha)}{\sin \alpha \sin \beta}$ **38.** $\dfrac{\tan \frac{1}{2}(\alpha + \beta)}{\tan \frac{1}{2}(\alpha - \beta)} = \dfrac{\sin \alpha + \sin \beta}{\sin \alpha - \sin \beta}$

39. $\dfrac{\sin \alpha - \sin \beta}{\cos \alpha + \cos \beta} = \tan \frac{1}{2}(\alpha - \beta)$ **40.** $\dfrac{\sin \alpha + \sin \beta}{\cos \alpha - \cos \beta} = \cot \frac{1}{2}(\beta - \alpha).$

4. INVERSE FUNCTIONS

Consider this problem: given $-1 \leq y \leq 1$, find θ for which $\sin \theta = y$. We know from experience that the problem has many solutions, a rather annoying situation. Previously we considered a similar problem: given $y > 0$, find x for which $10^x = y$. Here the situation is much nicer; for each such y there is exactly one solution.

Why are these problems so different? A glance at the graphs of $y = 10^x$ and $y = \sin \theta$ shows a major difference (Fig. 4.1). The graph of the exponential rises steadily, intersecting each line $y = c$ for $c > 0$ in exactly one point. The graph of $\sin \theta$ however, oscillates, intersecting each line $y = c$ for $-1 \leq c \leq 1$ in infinitely many points.

Generally a function $f(x)$ is most pleasant to work with when its graph is either increasing or decreasing. For then, to each y there is a unique x such that $y = f(x)$. That means x is a function of y, and we may write $x = g(y)$ as well as $y = f(x)$. Each point on the graph can be labeled either $(x, f(x))$ or $(g(y), y)$. See Fig. 4.2.

The functions $f(x)$ and $g(y)$ are closely related. For

$$f[g(y)] = f(x) = y, \qquad g[f(x)] = g(y) = x.$$

Each function undoes the effect of the other! Two functions $f(x)$, $g(y)$ that stand in this relationship to each other are called **inverse functions.**

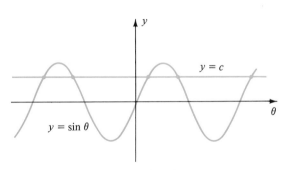

(a) solution of $10^x = c$ (b) solution of $\sin \theta = c$

Fig. 4.1

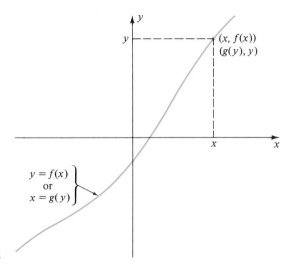

Fig. 4.2 Strictly increasing graph

Examples of inverse functions:

(1) $f(x) = 10^x$, $g(y) = \log y$:
$$f[g(y)] = 10^{\log y} = y, \qquad g[f(x)] = \log 10^x = x.$$

(2) $f(x) = \frac{9}{5}x + 32$, $g(y) = \frac{5}{9}(y - 32)$:
$$f[g(y)] = \frac{9}{5}[\frac{5}{9}(y - 32)] + 32 = y, \qquad g[f(x)] = \frac{5}{9}[(\frac{9}{5}x + 32) - 32] = x.$$

(3) $f(x) = x^3$, $g(y) = y^{1/3}$:
$$f[g(y)] = (y^{1/3})^3 = y, \qquad g[f(x)] = (x^3)^{1/3} = x.$$

Arc Sine

To construct a reasonable inverse function for $\sin \theta$, we must limit ourselves to a portion of the graph where $\sin \theta$ is either steadily increasing or steadily decreasing. The interval $-\frac{1}{2}\pi \leq \theta \leq \frac{1}{2}\pi$ is the most natural choice; there $\sin \theta$ increases from -1 to 1. See Fig. 4.3. Once this choice is made, then to each x in the range $-1 \leq x \leq 1$, there corresponds *exactly one* θ in the range $-\frac{1}{2}\pi \leq \theta \leq \frac{1}{2}\pi$ for which $\sin \theta = x$. Thus θ is a function of x, the inverse function of $\sin \theta$. We shall write

$$\theta = \text{arc sin } x$$

and read "θ equals the arc sine of x". (This function arc sine is frequently denoted $\sin^{-1}$.)

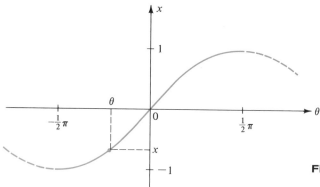

Fig. 4.3 $x = \sin \theta$
for $-\frac{1}{2}\pi \leq \theta \leq \frac{1}{2}\pi$:
strictly increasing

Examples:

$$\text{arc sin } 0 = 0, \qquad \text{arc sin } 1 = \tfrac{1}{2}\pi, \qquad \text{arc sin}(-1) = -\tfrac{1}{2}\pi, \qquad \text{arc sin } \tfrac{1}{2} = \tfrac{1}{6}\pi.$$

> There is a unique function
> $$\theta = \text{arc sin } x$$
> defined for $-1 \leq x \leq 1$ such that
> $$\sin \theta = x \qquad \text{and} \qquad \theta = \text{arc sin } x$$
> are equivalent, provided $-\frac{1}{2}\pi \leq \theta \leq \frac{1}{2}\pi$.

We emphasize that what we have done here is just common sense. We have simply agreed that an angle can be identified by its sine, *provided* we consider only angles between $-\frac{1}{2}\pi$ and $\frac{1}{2}\pi$. Thus arc sin $0 = 0$ (not π or 2π) and arc sin $\frac{1}{2} = \frac{1}{6}\pi$ (not $\frac{13}{6}\pi$ or $-\frac{11}{6}\pi$).

To graph the function $\theta = \text{arc sin } x$, we merely flip the graph of $x = \sin \theta$ for $-\frac{1}{2}\pi \leq \theta \leq \frac{1}{2}\pi$ over the line $x = \theta$. See Fig. 4.4.

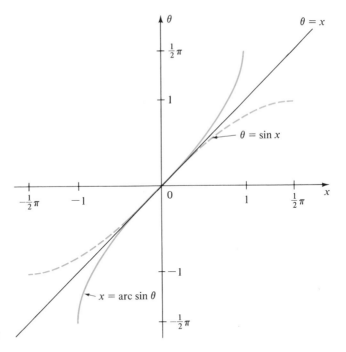

Fig. 4.4 $\theta = \text{arc sin } x$

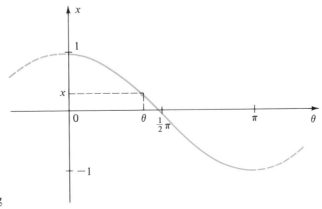

Fig. 4.5 $x = \cos \theta$
for $0 \leq \theta \leq \pi$:
strictly decreasing

Arc Cosine

To define an inverse for $\cos \theta$, it is natural to restrict θ to the interval $0 \leq \theta \leq \pi$ where $\cos \theta$ decreases steadily from 1 to -1. See Fig. 4.5.

For each x in the range $-1 \leq x \leq 1$, there corresponds exactly one θ in the range $0 \leq \theta \leq \pi$ for which $\cos \theta = x$. Thus θ is a function of x, the inverse function of $\cos \theta$. We shall write

$$\theta = \text{arc cos } x.$$

Examples:

$$\text{arc cos } 0 = \tfrac{1}{2}\pi, \quad \text{arc cos } 1 = 0, \quad \text{arc cos}(-\tfrac{1}{2}) = \tfrac{2}{3}\pi, \quad \text{arc cos}(-1) = \pi.$$

> There is a unique function
> $$\theta = \text{arc cos } x$$
> defined for $-1 \leq x \leq 1$ such that
> $$\cos \theta = x \quad \text{and} \quad \theta = \text{arc cos } x$$
> are equivalent, provided $0 \leq \theta \leq \pi$.

We easily obtain the graph of arc cos x from the cosine graph (Fig. 4.6.).

From Figs. 4.4 and 4.6 it is clear that the functions arc sin x and arc cos x are closely related. Indeed, shift the graph of arc cos x down by $\tfrac{1}{2}\pi$; the result is the negative of arc sin x. See Fig. 4.7. We conclude that arc cos $x - \tfrac{1}{2}\pi = -$ arc sin x.

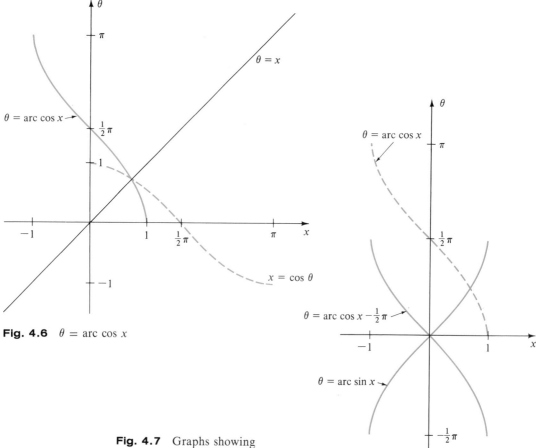

Fig. 4.6 $\theta = \text{arc cos } x$

Fig. 4.7   Graphs showing
arc sin $x = -(\text{arc cos } x - \tfrac{1}{2}\pi)$

Therefore:

$$\boxed{\arc \sin x + \arc \cos x = \frac{\pi}{2}.}$$

This identity can also be proved without graphs. For suppose $\theta = \arc \sin x$, so that $-\frac{1}{2}\pi \le \theta \le \frac{1}{2}\pi$. Then $0 \le \frac{1}{2}\pi - \theta \le \pi$.
But

$$\cos(\tfrac{1}{2}\pi - \theta) = \sin \theta = x,$$

hence $\frac{1}{2}\pi - \theta = \arc \cos x$. Therefore

$$\arc \sin x + \arc \cos x = \theta + (\tfrac{1}{2}\pi - \theta) = \tfrac{1}{2}\pi.$$

Fig. 4.8 Right triangle: $\alpha + \beta = \frac{1}{2}\pi$

The identity has a very simple geometric interpretation in the case $0 < x < 1$. Figure 4.8 shows a right triangle in which $\sin \alpha = \cos \beta = x$. That means $\alpha = \arc \sin x$ and $\beta = \arc \cos x$. But $\alpha + \beta = \frac{1}{2}\pi$.

■ *Example 4.1*

Find   (a) $\arc \sin(-\tfrac{1}{2})$ (b) $\arc \cos(\tfrac{1}{2}\sqrt{2})$.

SOLUTION   (a)   The only angle θ such that $-\frac{1}{2}\pi \le \theta \le \frac{1}{2}\pi$ and $\sin \theta = -\frac{1}{2}$ is $\theta = -\frac{1}{6}\pi$.
   (b)   The only angle θ such that $0 \le \theta \le \pi$ and $\cos \theta = \frac{1}{2}\sqrt{2}$ is $\theta = \frac{1}{4}\pi$.

Answer $-\frac{1}{6}\pi$, $\frac{1}{4}\pi$.

· ·

Arc Tan and Arc Cot

The graph of $x = \tan \theta$ (Fig. 6.1, p. 216) shows that $\tan \theta$ is strictly increasing for $-\frac{1}{2}\pi < \theta < \frac{1}{2}\pi$ and takes on every possible real value. Therefore an inverse function exists.

> There is a unique function
> $$\theta = \arc \tan x$$
> defined for $-\infty < x < \infty$ such that
> $$\tan \theta = x \quad \text{and} \quad \theta = \arc \tan x$$
> are equivalent, provided $-\frac{1}{2}\pi < \theta < \frac{1}{2}\pi$.

The graph of $\arc \tan x$ is shown in Fig. 4.9; the lines $\theta = \pi/2$ and $\theta = -\pi/2$ are asymptotes of the graph. See the next page.

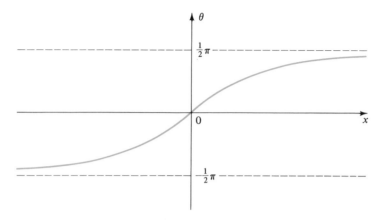

Fig. 4.9 $\theta = \text{arc tan } x$

The graph of $x = \cot \theta$ (Fig. 6.3, p. 217) shows that $\cot \theta$ is strictly decreasing for $0 < \theta < \pi$ and takes on every possible real value. Therefore an inverse function exists.

> There is a unique function
> $$\theta = \text{arc cot } x$$
> defined for $-\infty < x < \infty$ such that
> $$\cot \theta = x \quad \text{and} \quad \theta = \text{arc cot } x$$
> are equivalent, provided $0 < \theta < \pi$.

The graph of $\text{arc cot } x$ is shown in Fig. 4.10. Note the asymptotes $\theta = 0$ and $\theta = \frac{1}{2}\pi$. By comparing Figs. 4.9 and 4.10 we easily conclude that

$$\text{arc tan } x + \text{arc cot } x = \frac{\pi}{2}.$$

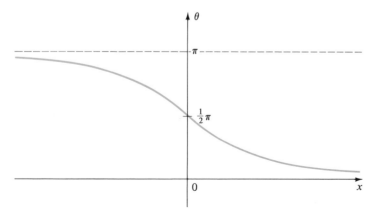

Fig. 4.10 $\theta = \text{arc cot } x$

Find arc sin x for the following values of x (use tables if necessary):

1. $\frac{1}{2}\sqrt{3}$ **2.** $\frac{1}{2}\sqrt{2}$ **3.** $-\frac{1}{2}\sqrt{2}$ **4.** $-\frac{1}{2}\sqrt{3}$
5. 0.8   **6.** -0.9 **7.** $\frac{1}{4}\pi$ **8.** 0.03.

Compute (use tables if necessary):

9. $\text{arc cos}(-\frac{1}{2}\sqrt{3})$ **10.** $\text{arc tan } 1$ **11.** $\text{arc cot } \sqrt{3}$
12. $\text{arc cos}(-\frac{1}{2})$ **13.** $\text{arc tan}(-2)$ **14.** $\text{arc cos}(.4)$
15. $\text{arc tan}(3.1)$ **16.** $\text{arc cot } 4.09$.

Prove:

17. $\text{arc sin}(-x) = -\text{arc sin } x$ **18.** $\text{arc cos}(-x) = \pi - \text{arc cos } x$

19. $\text{arc cot}(-x) = \pi - \text{arc cot } x$ **20.** $\text{arc sin } x = \text{arc tan } \dfrac{x}{\sqrt{1-x^2}}$

21. $2 \text{ arc cos } x = \text{arc cos}(2x^2 - 1)$ **22.** $2 \text{ arc tan } x = \text{arc tan}[2x/(1-x^2)]$
      for $0 \le x \le 1$
23. $\text{arc tan } \frac{1}{2} + \text{arc tan } \frac{1}{3} = \frac{1}{4}\pi$ **24.** $2 \text{ arc tan } \frac{1}{3} + \text{arc tan } \frac{1}{7} = \frac{1}{4}\pi$.

25. Graph $x = \sec\theta$ for $0 \le \theta \le \pi$, $\theta \ne \frac{1}{2}\pi$. Show that $\sec\theta$ restricted to this domain has an inverse function, arc sec x.
26. (cont.) Show that arc sec x is defined for $|x| \ge 1$, and sketch its graph.
27. Graph $x = \csc\theta$ for $-\frac{1}{2}\pi \le \theta \le \frac{1}{2}\pi$, $\theta \ne 0$. Show that $\csc\theta$ restricted to this domain has an inverse function, arc csc x.
28. (cont.) Show that arc csc x is defined for $|x| \ge 1$, and sketch its graph.
29. Explain why arc sin $x \approx x$ is a good approximation if x is small.
30. (cont.) Find a similar approximation for arc tan x.

Find without tables:

31. $\text{arc cos}(\cos \frac{1}{7}\pi)$ **32.** $\sin(\text{arc sin } 0.38)$
33. $\tan(\text{arc tan } 100)$ **34.** $\text{arc cot}(\cot(-\frac{1}{8}\pi))$.

Prove:

35. $\text{arc sin}(\cos\theta) = \frac{1}{2}\pi - \theta$ **36.** $\cos(\text{arc sin } x) = \sqrt{1-x^2}$
      $(0 \le \theta \le \pi)$
37. $\cot(\text{arc tan } x) = 1/x$ **38.** $\text{arc tan}(\cot\theta) = \frac{1}{2}\pi - \theta$
      $(0 < \theta < \pi)$

39. $\tan(\text{arc sin } x) = \dfrac{x}{\sqrt{1-x^2}}$ **40.** $\sin(\text{arc tan } x) = \dfrac{x}{\sqrt{1+x^2}}$.

An equation of the type $\sin\theta = \frac{1}{2}$, where you are asked to find θ, is the simplest example of a **trigonometric equation.** Other examples are

$$\sin 2\theta = \cos\theta, \qquad \sec^2\theta = 4\tan\theta - 2, \qquad \tan\theta = 2\cos\theta.$$

In each case you must find all values of θ, if any, that satisfy the given equation.

There is no fixed way to solve trigonometric equations. However, many can be handled by using trigonometric identities and techniques of algebra (factoring, common denominators, etc.).

If θ is the solution of a trigonometric equation that involves only integer multiples of θ, then so is $\theta \pm 2\pi n$. For simplicity we shall restrict our attention to θ in the range $0 \leq \theta < 2\pi$.

■ Example 5.1

Solve $\sin 2\theta = \cos \theta$.

SOLUTION   Use the double angle formula for $\sin 2\theta$:

$$2 \sin \theta \cos \theta = \cos \theta, \qquad (2 \sin \theta - 1)\cos \theta = 0.$$

The product is zero only if one of the factors is zero, that is, only if

$$2 \sin \theta - 1 = 0 \qquad \text{or} \qquad \cos \theta = 0.$$

From the first condition, $\theta = \frac{1}{6}\pi$ or $\frac{5}{6}\pi$; from the second, $\theta = \frac{1}{2}\pi$ or $\frac{3}{2}\pi$.

Answer $\theta = \frac{1}{6}\pi, \frac{5}{6}\pi, \frac{1}{2}\pi, \frac{3}{2}\pi.$

■ Example 5.2

Solve $\sec^2 \theta = 4 \tan \theta - 2$.

SOLUTION   Replace $\sec^2 \theta$ by $1 + \tan^2 \theta$. The result is a quadratic equation for $\tan \theta$:

$$1 + \tan^2 \theta = 4 \tan \theta - 2, \qquad \tan^2 \theta - 4 \tan \theta + 3 = 0.$$

The solutions of this quadratic are $\tan \theta = 1$ and $\tan \theta = 3$. The first solution yields $\theta = \frac{1}{4}\pi$ or $\frac{5}{4}\pi$. The second solution yields (from a table of tangents), $\theta \approx 1.249$ or $\theta \approx 1.249 + \pi$.

Answer $\theta = \frac{1}{4}\pi, \frac{5}{4}\pi; \theta \approx 1.249, 1.249 + \pi.$

■ Example 5.3

Find θ in the first quadrant satisfying $\tan \theta = 2 \cos \theta$.

SOLUTION   Write $\tan \theta = \sin \theta / \cos \theta$:

$$\frac{\sin \theta}{\cos \theta} = 2 \cos \theta, \qquad \sin \theta = 2 \cos^2 \theta.$$

Replace $\cos^2 \theta$ by $1 - \sin^2 \theta$; the result is a quadratic equation for $\sin \theta$:

$$2 \sin^2 \theta + \sin \theta - 2 = 0.$$

Solve for $\sin \theta$:

$$\sin \theta = \frac{-1 \pm \sqrt{17}}{4} \approx \frac{-1 \pm 4.123}{4} \approx -1.281, \, 0.781.$$

Since $\sin \theta \approx -1.281$ is impossible, the only feasible solution is $\sin \theta \approx 0.781$. From a table of sines, $\theta \approx 0.897$.

Answer $\theta \approx 0.897$.

. . . .

An Application to Geometry

It is often useful to express an angle in terms of inverse functions.

Example 5.4

Find θ in Fig. 5.1a in terms of a, b, c.

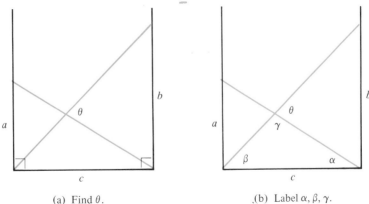

Fig. 5.1   (a) Find θ.   (b) Label α, β, γ.

SOLUTION We mark auxiliary angles α, β, γ in Fig. 5.1b. Then $\theta + \gamma = \pi$ (supplementary angles) and $\alpha + \beta + \gamma = \pi$ (sum of the angles of a triangle), so

$$\theta = \alpha + \beta.$$

Since α and β are angles in right triangles with two given sides, we see that

$$\tan \alpha = \frac{a}{c}, \qquad \tan \beta = \frac{b}{c}.$$

Hence

$$\alpha = \arctan \frac{a}{c}, \qquad \beta = \arctan \frac{b}{c}.$$

Answer $\theta = \arctan \dfrac{a}{c} + \arctan \dfrac{b}{c}.$

. . . .

Polar Coordinates

Until now, we have always used rectangular coordinates to identify points in the plane. Other coordinate systems exist as well. The most important of these is the system of **polar coordinates.**

In a rectangular coordinate system, two families of grid lines, x = constant and y = constant, fill the plane. Each point (a, b) is the intersection of two of these lines, $x = a$ and $y = b$, and receives the coordinates (a, b). See Fig. 5.2a.

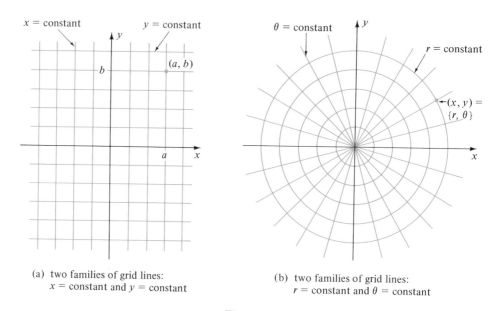

(a) two families of grid lines:
x = constant and y = constant

(b) two families of grid lines:
r = constant and θ = constant

Fig. 5.2

Polar coordinates work on a similar principle. The grid lines are (1) all circles centered at $(0, 0)$, and (2) all rays from $(0, 0)$. See Fig. 5.2b. Each point (x, y) different from $(0, 0)$ is the intersection of one circle and one ray. The circle is identified by a positive number r, its radius, and the ray is identified by a real number θ, its angle in radians from the positive x-axis. Thus (x, y) is assigned the **polar coordinates** $\{r, \theta\}$. Since θ is determined only up to a multiple of 2π, we agree that

$$\{r, \theta + 2\pi n\} = \{r, \theta\} \qquad (n \text{ any integer}).$$

The point $(0, 0)$ does not determine an angle θ. Nonetheless, it is customary to say that *any* pair $\{0, \theta\}$ represents $(0, 0)$.

The idea of polar coordinates is quite natural. You identify a point by telling how far it is from here, and in what direction. (This is the principle of the radar screen.)

Given the polar coordinates $\{r, \theta\}$ of a point, what are its rectangular coordinates? The point is r units from the origin in the direction θ. Hence $x = r\cos\theta$, $y = r\sin\theta$. See Fig. 5.3a. Conversely, given the rectangular coordinates (x, y), what are the polar coordinates? Figure 5.3b shows that $r = \sqrt{x^2 + y^2}$, and that θ is determined by $\cos \theta = x/r$ and $\sin \theta = y/r$.

Polar to rectangular	Rectangular to polar
$x = r \cos \theta$ $y = r \sin \theta$	$r = \sqrt{x^2 + y^2}$ $\cos \theta = \dfrac{x}{\sqrt{x^2 + y^2}} = \dfrac{x}{r}$ $\sin \theta = \dfrac{y}{\sqrt{x^2 + y^2}} = \dfrac{y}{r}$

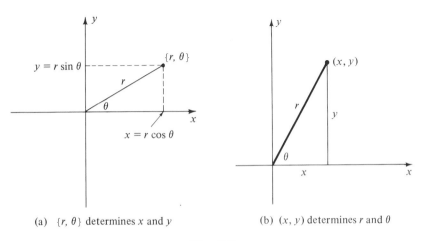

(a) $\{r, \theta\}$ determines x and y (b) (x, y) determines r and θ

Fig. 5.3

■ *Example 5.4*

(a) Convert $(2, -2\sqrt{3})$ to polar coordinates. (b) Convert $\{3, \frac{1}{6}\pi\}$ to rectangular coordinates.

SOLUTION (a) $r^2 = 4 + 12 = 16$, $r = 4$. Also $\cos \theta = \frac{2}{4} = \frac{1}{2}$ and $\sin \theta = \frac{1}{4}(-2\sqrt{3}) = -\frac{1}{2}\sqrt{3}$, so $\theta = \frac{5}{3}\pi$.
(b) $x = r \cos \theta = 3 \cos \frac{1}{6}\pi = \frac{3}{2}\sqrt{3}$, and $y = r \sin \theta = 3 \sin \frac{1}{6}\pi = \frac{3}{2}$.

Answer (a) $\{4, \frac{5}{3}\pi\}$ (b) $(\frac{3}{2}\sqrt{3}, \frac{3}{2})$.

Negative r

In applications it is convenient to allow points $\{r, \theta\}$ with $r < 0$. For example, consider a ray and a point $\{r, \theta\}$ on the ray (Fig. 5.4a). Suppose the point moves towards $(0, 0)$, through $(0, 0)$, and keeps on going! Then r decreases, becomes 0, but then what? So that θ won't jump abruptly to $\theta + \pi$, we agree that θ remains constant, but r becomes negative (Fig. 5.4b). This amounts to agreeing that

$$\{-r, \theta\} = \{r, \theta + \pi\}.$$

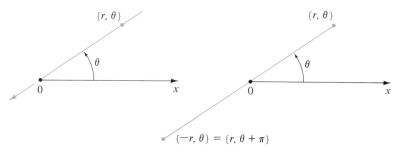

(a) Hold θ fixed and move   (b) Identify $\{-r, \theta\}$ and $\{r, \theta + \pi\}$.
$\{r, \theta\}$ through the origin.

Fig. 5.4

Equations of Graphs

Each curve $r = c$ is a circle with center $(0, 0)$, except for $c = 0$. Each curve $\theta = \theta_0$ is a line through the origin (Fig. 5.5).

What is the equation in polar coordinates for a line L not necessarily through the origin? Drop a perpendicular from $(0, 0)$ to L; let its foot have polar coordinates $\{p, \alpha\}$. See Fig. 5.6. If $\{r, \theta\}$ is any point on L, then from the right triangle, $\cos(\theta - \alpha) = p/r$. Therefore

> The equation in polar coordinates of the general straight line is
>
> $$r \cos(\theta - \alpha) = p,$$
>
> where p and α are constants.

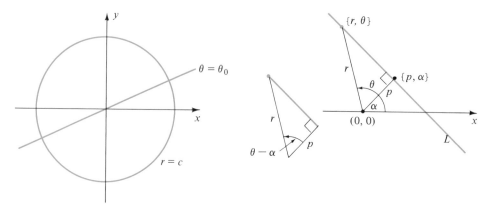

Fig. 5.5   Graphs of $r = c$ and $\theta = \theta_0$

Fig. 5.6

Solve: use 4-place tables if necessary:

1. $2 \cos^2 \theta + \sin \theta = 2$

2. $2 \cos^3 \theta = \cos \theta$

3. $\tan \theta = \cot \theta$

4. $\sin \theta + \cos \theta = 1$ [Hint: Square both sides]

5. $\tan \theta = \sin \theta$

6. $\tan \theta = \tan 2\theta$

7. $\tan \theta = \sin 2\theta$

8. $\cos \theta = \cos 2\theta$

9. $\sec^3 \theta - 2 \sec \theta = 0$

10. $\sin^6 \theta = \sin^2 \theta$

11. $2 \sin \theta \cos \theta = \cos 2\theta$

12. $\tan \theta + \cot \theta = \sec \theta \csc \theta$

13. $\sin 3\theta + \sin 5\theta = 0$

14. $\cos \theta - \cos 3\theta = 0$

15. $2 \sin^2 \theta = -3 \cos \theta$

16. $\tan^2 \theta + \tan \theta = 0$

17. $\tan^2 \theta \sec^2 \theta = 9 + \tan^2 \theta$

18. $\tan \theta = \cot^2 \theta$

19. $4 \cos 4\theta = 1, 0 < \theta < 90°$

20. $1 + \cos \theta = \sqrt{3} \sin \theta.$

Show the equation has no solution:

21. $\sin^2 \theta + \sin \theta = 3$

22. $\sin \theta - \cos \theta = \frac{3}{2}$. [Hint: See Example 3.2.]

23. Express s in terms of a and y (Fig. 5.7a).

24. Express θ in terms of x (Fig. 5.7b).

Fig. 5.7   (a) Ex. 23   (b) Ex. 24

Express in rectangular coordinates:

25. $\{1, \frac{1}{2}\pi\}$

26. $\{1, -\frac{1}{2}\pi\}$

27. $\{1, -\frac{1}{6}\pi\}$

28. $\{1, \frac{1}{3}\pi\}$

29. $\{2, -\frac{3}{4}\pi\}$

30. $\{2, \frac{5}{4}\pi\}.$

Express in polar coordinates:

31. $(1, 1)$

32. $(0, -1)$

33. $(-1, 1)$

34. $(-\frac{1}{2}, \frac{1}{2}\sqrt{3})$

35. $(\sqrt{3}, -1)$

36. $(\sqrt{2}, -\sqrt{2}).$

Find the equation in polar coordinates:

37. line through $(0, 0)$ and $\{3, \frac{1}{4}\pi\}$

38. circle, center $(0, 0)$, radius 5

39. line through $\{1, 0\}$ and $\{1, \frac{1}{2}\pi\}$

40. line $ax + by = c$. [Hint: Use the technique of Example 3.1.]

Test 1

1. Compute $\arcsin(-\frac{1}{2})$, $\arccos(-\frac{1}{2})$, $\arctan\sqrt{3}$.
2. Compute $\tan 15°$.
3. Show that $\dfrac{1 + \tan\theta}{1 - \tan\theta} = \tan\left(\theta + \dfrac{\pi}{4}\right)$.
4. Find *all* real numbers θ for which $\sin\theta = \sin 2\theta$.
5. Prove that

$$\cos\theta\cos 2\theta\cos 4\theta = \tfrac{1}{4}[\cos\theta + \cos 3\theta + \cos 5\theta + \cos 7\theta].$$

Test 2

1. If θ is in the second quadrant and $\sec\theta = -3$, compute $\sin\theta$ and $\tan\theta$.
2. Prove: $\cos^4\theta - \sin^4\theta = \cos 2\theta$.
3. Express $\sin 3\theta$ in terms of $\sin\theta$ and $\cos\theta$.
4. For what values of θ is $\tan\theta + \cot\theta = 2$?
5. (See figure). Express θ in terms of inverse trigonometric functions.

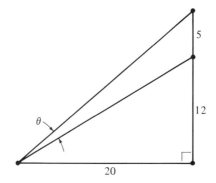

TRIGONOMETRY

The word "trigonometry" comes from Greek and means "the measurement of triangles (trigons)". A triangle is described by 3 angles and 3 sides. The general problem of trigonometry is: given some of these 6 quantities, measure (compute) the others. We discuss this problem for right triangles now and for oblique triangles in Section 3.

For angles θ between 0 and $\frac{1}{2}\pi$, the trigonometric functions can be interpreted as ratios of sides of a right triangle, one of whose angles is θ. A way of expressing these ratios is indicated in Fig. 1.1. With this information plus tables, we can solve

$$\sin \theta = \frac{opposite}{hypotenuse} \qquad \csc \theta = \frac{hypotenuse}{opposite}$$

$$\cos \theta = \frac{adjacent}{hypotenuse} \qquad \sec \theta = \frac{hypotenuse}{adjacent}$$

$$\tan \theta = \frac{opposite}{adjacent} \qquad \cot \theta = \frac{adjacent}{opposite}$$

Fig. 1.1

problems about right triangles: given two sides, or given one side and one acute angle, find the other sides and angles.

■ *Example 1.1*

A 16-ft ladder leans against a wall with its base 5 ft from the wall. What angle does the ladder make with the ground?

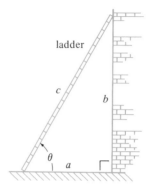

Fig. 1.2

SOLUTION   See Fig. 1.2. We are given a right triangle with $a = 5$ and $c = 16$ and asked to find θ. We have enough information to compute $\cos \theta$:

$$\cos \theta = \frac{\text{adjacent}}{\text{hypotenuse}} = \frac{a}{c} = \frac{5}{16} = 0.3125.$$

From a table of cosines we find $\theta \approx 71.79°$.

Answer   71.79°.

. . . .

■ *Example 1.2*

A 50-ft ramp is inclined at an angle of 6° with the ground. How high does it rise above the ground?

SOLUTION   In Fig. 1.2, we are given $c = 50$ and $\theta = 6°$, and asked to find b. We want to express b in terms of c and θ. Since

$$\sin \theta = \frac{\text{opposite}}{\text{hypotenuse}} = \frac{b}{c},$$

we have $b = c \sin \theta = 50 \sin 6°$. From a table of sines, $\sin \theta \approx 0.1045$, so $b \approx 50(0.1045)$.

Answer   5.225 ft.

. . . .

Example 1.3

When the angle of elevation of the sun is 62°, the campus flagpole casts a 35-ft shadow. Find the height of the pole to two decimal places.

SOLUTION This time we are given $\theta = 62°$ and $a = 35$ ft, and asked for b. Since $b/a = \tan \theta$,

$$b = a \tan \theta = 35 \tan 62° \approx 35(1.881) \approx 65.84.$$

Answer   65.84 ft.

Many problems require a good deal of multiplication and division, and are simplified by using logs. Hence in practice we often work with logs of trig functions rather than with the functions themselves. We save time by referring to tables of logs of trig functions.

Note: In our Table 6, the column headed L. Sin lists $10 + \log \sin \theta$; otherwise the entries would be negative. Thus you must subtract 10 from each entry. For instance we find 9.6356 at 25.6°; hence $\log \sin 25.6° \approx 9.6356 - 10 = 0.6356 - 1$. The same applies to the column headed L. Tan, and to that headed L. Cos except for the range $0 \leq \theta \leq 0.8°$.

Example 1.4

Given $b = 4.902$ and $\theta = 25.60°$ in Fig. 1.2, find a and c.

SOLUTION Since $b/c = \sin \theta$, we have $c = b/\sin \theta$ or $\log c = \log b - \log \sin \theta$:

$$
\begin{array}{lll}
\log b = \log 4.902 & \approx 0.6904 \\
\underline{\log \sin \theta = \log \sin 25.60° \approx 0.6356 - 1} \\
\quad \log c & \approx 1.0548 \\
\quad\quad c & \approx 11.34.
\end{array}
$$

Since $a/b = \cot \theta$, we have $a = b \cot \theta$ or $\log a = \log b + \log \cot \theta$:

$$
\begin{array}{lll}
\log b = \log 4.902 & \approx 0.6904 \\
\underline{\log \cot \theta = \log \cot 25.60° \approx 0.3196} \\
\quad \log a & \approx 1.0100 \\
\quad\quad a & \approx 10.23.
\end{array}
$$

Answer $a \approx 10.23, c \approx 11.34.$

Remark: After finding c, we could have computed a by the Pythagorean Theorem. But the arithmetic involved would have been tedious; logarithms were quicker.

Use 4-place tables.

Compute the angles α and β in Fig. 1.3 to one-place accuracy:

1. $a = 12, c = 13$
2. $a = 6, c = 13$
3. $a = 107.3, c = 175.5$
4. $a = 24.03, b = 58.81$
5. $b = 15, c = 22$
6. $b = 8, c = 12.5$
7. $b = 45.8, c = 66.75$
8. $b = 237.6, c = 289.7$
9. $a = 4, b = 5$
10. $a = 16, b = 9$
11. $a = 27.12, b = 9.315$
12. $a = 18.06, c = 143.1.$

Compute all sides of the right triangle:

13. $c = 30, \alpha = 14.5°$
14. $c = 30, \alpha = 71.4°$
15. $a = 15, \alpha = 25.6°$
16. $a = 9.5, \alpha = 40.8°$
17. $c = 235.5, \beta = 34.2°$
18. $c = 1293, \beta = 68.8°$
19. $b = 56.03, \beta = 43.9°$
20. $b = 526.2, \beta = 33.7°.$

Fig. 1.3

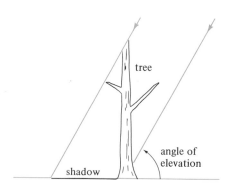

Fig. 1.4

21. When the angle of elevation of the sun is 60.5°, a tree casts a 44-ft shadow. What is the height of the tree? See Fig. 1.4.
22. (cont.) At what angle of elevation will the shadow be twice as long?
23. A 15-ft ladder leans against a wall making a 70° angle with the ground. At what height does its upper end touch the wall?
24. Two trees grow on opposite sides of a river (Fig. 1.5). On one side a surveyor lays off 100 ft so that the angle is 90°. He then finds angle β is 66.7°. What is the distance between the trees?
25. (cont.) If the distance between the trees is 43.65 ft, find β.
26. A chord of a circle subtends a central angle θ. Express the length of the chord in terms of θ.
27. (cont.) What is the perimeter of a regular octagon inscribed in a circle of radius 10?
28. (cont.) What is the perimeter of a regular polygon of 100 sides inscribed in the unit circle? Your answer should be near 2π. Why?
29. A road makes an angle of 4.5° with the horizontal. I start at sea level and drive exactly 5 miles. What is my elevation above sea level?
30. I observe the angle of elevation of a certain peak is 20°. I drive one mile closer along a level road and find that now the angle is 32°. What is the height of the peak?

Fig. 1.5

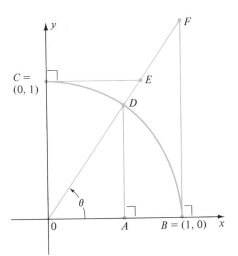

Fig. 1.6

31. A room measures $12 \times 15 \times 8$ ft. Compute the angle that the diagonal makes with each of the edges.

32. During the first few seconds of its flight, a rocket rises straight up. Five seconds after launch, an observer 1 mile away notes its angle of elevation is $7.5°$. Four seconds later the angle is $58.1°$. How far did the rocket rise during these four seconds?

33. A plane flying at an altitude of 500 ft passes directly overhead. Two seconds later its angle of elevation is $41°$. Compute its speed.

34. In Fig. 1.3, express ac/b^2 in terms of α.

Express each of the lengths in Fig. 1.6 in terms of θ:

35. OA **36.** AD **37.** BF
38. OF **39.** OE **40.** CE.

41. Express $x + y$ in terms of a, b, c, and θ. See Fig. 1.7.

42*. Express the height h of the tent in terms of a, b, and θ. See Fig. 1.8. The floor is rectangular and the peak centered.

Fig. 1.7

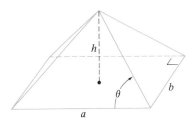

Fig. 1.8

43. Prove that the area of a right triangle is given by $A = \frac{1}{4}c^2 \sin 2\alpha$.

44. An isosceles triangle has base b, lateral side a, and apex angle θ. Show that its area is $A = \frac{1}{4}b \sqrt{4a^2 - b^2} = \frac{1}{4}b^2 \cot \frac{1}{2}\theta$. [Hint: Drop a perpendicular from the apex.]

45. (cont.) Find the apex angle of the isosceles triangle whose equal sides are 10 ft and area is 25 ft².

46. (cont.) An isosceles triangle has base 1 and apex angle 20°. For what apex angle would its area be twice as much?

47*. The 20-ft pipe will never turn the corner in the 6-ft hall (Fig. 1.9). At what angle θ does it get stuck?

48. The pyramid (Fig. 1.10) with square base a has height h. Find its edge length b and apex angle θ.

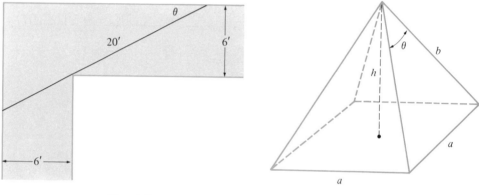

Fig. 1.9   **Fig. 1.10**

2. OBLIQUE TRIANGLES

In this section, we discuss general properties of triangles that will be needed for the numerical solution of oblique triangles in Section 3. Let us agree, once and for all, to label triangles as shown in Fig. 2.1.

Fig. 2.1 Six "parts" of a triangle

Law of Cosines

The first property of triangles is the **law of cosines:**

$$c^2 = a^2 + b^2 - 2ab \cos \gamma.$$

This is a generalization of the Pythagorean Theorem, for if γ is a right angle, then $\cos \gamma = 0$ and the formula says that $c^2 = a^2 + b^2$.

To prove the law of cosines, choose axes so that the angle γ is at the origin and the side a lies on the positive x-axis (Fig. 2.2).

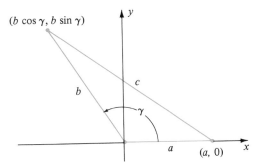

Fig. 2.2   Proof of the law of cosines

The upper vertex of the triangle is at $(b \cos \gamma, b \sin \gamma)$. By the distance formula,

$$c^2 = (b \cos \gamma - a)^2 + (b \sin \gamma)^2 = b^2(\cos^2 \gamma + \sin^2 \gamma) + a^2 - 2ab \cos \gamma;$$

$$c^2 = a^2 + b^2 - 2ab \cos \gamma.$$

The law of cosines is a basic fact in mathematics and has many applications other than solving triangles.

Law of Sines

The second property of triangles is the **law of sines:**

$$\boxed{\frac{a}{\sin \alpha} = \frac{b}{\sin \beta} = \frac{c}{\sin \gamma}.}$$

To prove the law, drop a perpendicular, of length h, on side a (extended if necessary) from the opposite vertex (Fig. 2.3). By the relations for right triangles,

$$\frac{h}{b} = \sin \gamma, \qquad \frac{h}{c} = \sin \beta \quad [=\sin(\pi - \beta)].$$

(a)

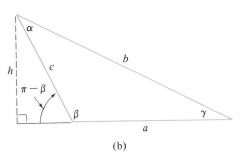

(b)

Fig. 2.3   Proof of the law of sines

Therefore

$$b \sin \gamma = h = c \sin \beta,$$

$$\frac{b}{\sin \beta} = \frac{c}{\sin \gamma}.$$

By the same argument,

$$\frac{a}{\sin \alpha} = \frac{b}{\sin \beta}.$$

■ *Example 2.1*

Let the vertex angle of an isosceles triangle be θ and the equal sides have length a. Prove that the base has length $b = 2a \sin \frac{1}{2}\theta$.

SOLUTION 1 (by law of cosines)

$$b^2 = a^2 + a^2 - 2a \cdot a \cdot \cos \theta = 2a^2 - 2a^2 \cos \theta = 2a^2(1 - \cos \theta).$$

By the half-angle formulas, $1 - \cos \theta = 2 \sin^2 \frac{1}{2}\theta$, hence

$$b^2 = 4a^2 \sin^2 \tfrac{1}{2}\theta, \qquad b = 2a \sin \tfrac{1}{2}\theta.$$

SOLUTION 2 (by law of sines) Call the base angle α. Then $2\alpha + \theta = \pi$. See Fig. 2.4a. Therefore $\alpha = \frac{1}{2}\pi - \frac{1}{2}\theta$ so $\sin \alpha = \sin(\frac{1}{2}\pi - \frac{1}{2}\theta) = \cos \frac{1}{2}\theta$.

By the law of sines,

$$\frac{b}{\sin \theta} = \frac{a}{\sin \alpha} = \frac{a}{\cos \frac{1}{2}\theta}.$$

Therefore

$$b = a \frac{\sin \theta}{\cos \frac{1}{2}\theta} = a \frac{2 \sin \frac{1}{2}\theta \cos \frac{1}{2}\theta}{\cos \frac{1}{2}\theta} = 2a \sin \tfrac{1}{2}\theta.$$

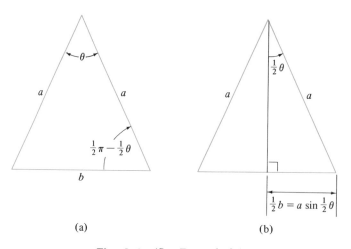

(a)   (b)

Fig. 2.4 (See Example 2.1.)

SOLUTION 3 (by right triangles)   Drop the perpendicular from the vertex to the base (Fig. 2.4b). From the right triangle,

$$\tfrac{1}{2}b = a \sin \tfrac{1}{2}\theta, \qquad b = 2a \sin \tfrac{1}{2}\theta.$$

. . . .

Law of Tangents

The third property of triangles is the **law of tangents.** Although not quite as important as the laws of sines and cosines, it is useful in certain computations.

$$\boxed{\frac{\tan \tfrac{1}{2}(\beta - \gamma)}{\tan \tfrac{1}{2}(\beta + \gamma)} = \frac{b - c}{b + c}.}$$

This law is a consequence of the law of sines and the identities on p. 235 for sums and differences of sines:

$$\frac{b - c}{b + c} = \frac{b - c}{b + c} \cdot \frac{\sin \gamma}{\sin \gamma} = \frac{b \sin \gamma - c \sin \gamma}{b \sin \gamma + c \sin \gamma}$$

$$= \frac{c \sin \beta - c \sin \gamma}{c \sin \beta + c \sin \gamma} = \frac{\sin \beta - \sin \gamma}{\sin \beta + \sin \gamma}$$

$$= \frac{2 \cos \tfrac{1}{2}(\beta + \gamma) \sin \tfrac{1}{2}(\beta - \gamma)}{2 \sin \tfrac{1}{2}(\beta + \gamma) \cos \tfrac{1}{2}(\beta - \gamma)}$$

$$= \frac{\sin \tfrac{1}{2}(\beta - \gamma)}{\cos \tfrac{1}{2}(\beta - \gamma)} \bigg/ \frac{\sin \tfrac{1}{2}(\beta + \gamma)}{\cos \tfrac{1}{2}(\beta + \gamma)} = \frac{\tan \tfrac{1}{2}(\beta - \gamma)}{\tan \tfrac{1}{2}(\beta + \gamma)}.$$

Area

There is simple formula for the area of a triangle in terms of two sides and the included angle:

$$\boxed{A = \tfrac{1}{2}ab \sin \gamma.}$$

Its proof is based on the area formula for right triangles. Drop a perpendicular on b from the opposite vertex (Fig. 2.5 on next page). Its length h (the altitude of the triangle) is

$$h = a \sin \gamma = a \sin(\pi - \gamma).$$

Therefore

$$A = \tfrac{1}{2}bh = \tfrac{1}{2}ba \sin \gamma.$$

The formulas $A = \tfrac{1}{2}bc \sin \alpha$ and $A = \tfrac{1}{2}ca \sin \beta$ follow by changing the notation.

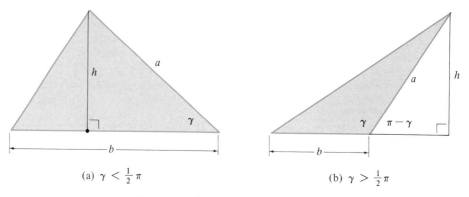

(a) $\gamma < \frac{1}{2}\pi$　　　　　　　(b) $\gamma > \frac{1}{2}\pi$

Fig. 2.5 Proof of the area formula

3. NUMERICAL SOLUTION

Use of the Law of Cosines

Here are two problems concerning triangles.

(1) Given the three sides of a triangle, find the three angles.
(2) Given two sides and the angle between them, find the third side and the other two angles.

Both problems can be solved by the law of cosines.

■ *Example 3.1*

The sides of a triangle are $a = 7$, $b = 6$, and $c = 2$. Use the law of cosines to find the angles. Check your answer.

SOLUTION By the law of cosines,

$$c^2 = a^2 + b^2 - 2ab \cos \gamma.$$

Therefore

$$\cos \gamma = \frac{a^2 + b^2 - c^2}{2ab} = \frac{49 + 36 - 4}{84} = \frac{81}{84} \approx 0.9643, \qquad \gamma \approx 15.36°.$$

Similarly,

$$\cos \beta = \frac{c^2 + a^2 - b^2}{2ca} = \frac{17}{28} \approx 0.6071, \qquad \beta \approx 52.62°.$$

Finally,

$$\cos \alpha = \frac{b^2 + c^2 - a^2}{2bc} = \frac{-9}{24} = -0.375.$$

Note that $\cos \alpha$ is negative, indicating an angle greater than 90°. Since the table lists only positive cosines, we use the relation $\cos(180 - \alpha)° = -\cos \alpha°$:

$$\cos(180 - \alpha)° = 0.375, \qquad 180 - \alpha \approx 67.98°, \qquad \alpha \approx 112.02°.$$

Check $\quad \alpha + \beta + \gamma \approx 112.02 + 52.62 + 15.36 = 180°.$

Answer $\quad \alpha \approx 112.02°, \quad \beta \approx 52.62°, \quad \gamma \approx 15.36°.$

Example 3.1 illustrates the solution of problem (1) above. In the case of problem (2), solve for the third side by the law of cosines and then you are back to problem (1).

Use of the Law of Sines

Here are two more problems about triangles.

(3) Given two angles and one side, find the other sides and angle.
(4) Given an angle, the side opposite, and another side, find the other angles and side.

If α and β are given, then $\gamma = \pi - \alpha - \beta$ is automatically known. Hence we can state the problems this way:

(3) Given α, β, γ, a, find b, c.
(4) Given c, γ, b, find α, a, β.

Both problems are solved by the law of sines.

Solution of problem 3. The solution is direct and straightforward. Given α, β, γ, a, we have

$$\frac{b}{\sin \beta} = \frac{c}{\sin \gamma} = \frac{a}{\sin \alpha};$$

hence,

$$b = \frac{a \sin \beta}{\sin \alpha}, \qquad c = \frac{a \sin \gamma}{\sin \alpha}.$$

Thus b and c can be expressed in terms of the data.

Solution of problem 4. The solution is complicated because we must study a number of cases. To simplify the discussion, we consider only triangles for which $\gamma < \frac{1}{2}\pi$.

Suppose c, γ, b, are given. By the law of sines,

$$\frac{b}{\sin \beta} = \frac{c}{\sin \gamma}, \qquad \text{so} \quad \sin \beta = \frac{b}{c} \sin \gamma.$$

There are three main cases:

(i) $\dfrac{b}{c} \sin \gamma > 1$.

There is no solution because $\sin \beta > 1$ is impossible.

(ii) $\dfrac{b}{c} \sin \gamma = 1$.

There is one solution, $\beta = \frac{1}{2}\pi$; the triangle is a right triangle.

(iii) $\dfrac{b}{c} \sin \gamma < 1$.

There is an acute angle β and an obtuse angle $\pi - \beta$ for which $\sin \beta = (b/c) \sin \gamma$. If $c < b$, then there are two solutions (Fig. 3.1c). However, if $c > b$, then the angle $\pi - \beta$ comes out on the wrong side of the angle at γ; there is only one solution (Fig. 3.1d). Note that if $c = b$, the triangle is isosceles and $\beta = \gamma$.

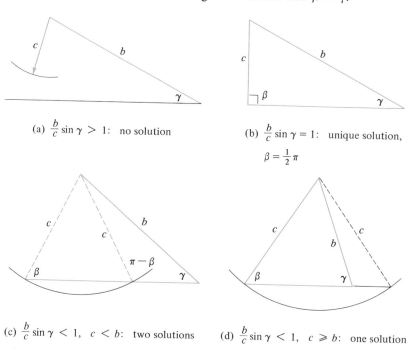

(a) $\dfrac{b}{c} \sin \gamma > 1$: no solution

(b) $\dfrac{b}{c} \sin \gamma = 1$: unique solution, $\beta = \frac{1}{2}\pi$

(c) $\dfrac{b}{c} \sin \gamma < 1$, $c < b$: two solutions

(d) $\dfrac{b}{c} \sin \gamma < 1$, $c \geqslant b$: one solution

Fig. 3.1   Given: c, γ, b

A little geometry accounts for the various cases. Think of the angle γ and the side b as fixed, with the side c free to vary (Fig. 3.1a). In (a), the arm c is too short to reach the base; there is no triangle. In (b), the arm just reaches the base; a right triangle is formed. In (c), the arm is long enough to reach the base at two points; two triangles are formed. In (d) the arm is too long; two triangles are formed, but the one on the right has γ as an *exterior* angle, so it is not a solution of the problem.

■ *Example 3.2*

Given $b = 4.592$, $c = 3.819$, and $\gamma = 32.72°$, find β, α, and a.

SOLUTION   Compute $\sin \beta$:

$$\sin \beta = \frac{b \sin \gamma}{c}, \quad \log \sin \beta = \log b + \log \sin \gamma - \log c:$$

$$\log b = \log 4.592 \approx 0.6620$$
$$\underline{\log \sin \gamma = \log \sin 32.72° \approx 0.7328 - 1}$$
$$\text{sum} = 1.3948 - 1$$
$$\underline{\log c = \log 3.819 \approx 0.5819}$$
$$\log \sin \beta \approx 0.8129 - 1.$$

Therefore $\sin \beta < 1$, so case (iii) applies; since $c < b$, there are two solutions. From the log sin table, one solution is $\beta \approx 40.54°$. The other solution is

$$180° - \beta \approx 180 - 40.54 = 139.46°.$$

To finish the job, we find a from the relation

$$\frac{a}{\sin \alpha} = \frac{c}{\sin \gamma}, \quad a = \frac{c \sin \alpha}{\sin \gamma}.$$

There are two values of α corresponding to the two values of β; we perform two computations:

$\beta \approx 40.54°$	$\beta \approx 180 - 40.54°$
$\alpha = 180 - \beta - \gamma \approx 180 - 73.26°$	$\alpha = 180 - \beta - \gamma \approx 7.82°$
$\log c \approx 0.5819$	$\log c \approx 0.5819$
$\log \sin \alpha \approx \log \sin 73.26° \approx 0.9812 - 1$	$\log \sin \alpha \approx 0.1337 - 1$
$\text{sum} = 1.5631 - 1$	$\text{sum} = 1.7156 - 2$
$\log \sin \gamma \approx 0.7328 - 1$	$\log \sin \gamma \approx 0.7328 - 1$
$\log a \approx 0.8303$	$\log a \approx 0.9828 - 1$
$a \approx 6.766$	$a \approx 0.9612$

Answer $\beta \approx 40.54°$, $\alpha \approx 106.74°$, $a \approx 6.766$; or
$\beta \approx 139.46°$, $\alpha \approx 7.82°$, $a \approx 0.9612$.

■ ■

Use of the Law of Tangents

The law of tangents provides an alternative method for solving a triangle given two sides and their included angle. It is better suited to computations with logarithms than is the law of cosines.

Suppose b, c, and α are given. Then we know $\beta + \gamma$ because $\alpha + \beta + \gamma = \pi$, that is,

$$\beta + \gamma = \pi - \alpha.$$

By the law of tangents,

$$\tan \tfrac{1}{2}(\beta - \gamma) = \frac{b - c}{b + c} \tan \tfrac{1}{2}(\beta + \gamma),$$

so we can compute $\tfrac{1}{2}(\beta - \gamma)$. Then we compute β and γ by

$$\beta = \tfrac{1}{2}(\beta + \gamma) + \tfrac{1}{2}(\beta - \gamma),$$

$$\gamma = \tfrac{1}{2}(\beta + \gamma) - \tfrac{1}{2}(\beta - \gamma).$$

Finally we find a by the law of sines.

■ *Example 3.3*

Given $b = 4.592$, $c = 3.819$, and $\alpha = 106.74°$, find β, γ, and a.

SOLUTION
$$b - c = 4.592 - 3.819 = 0.773,$$
$$b + c = 4.592 + 3.819 = 8.411,$$
$$\beta + \gamma = 180 - 106.74 = 73.26°,$$
$$\tfrac{1}{2}(\beta + \gamma) = \tfrac{1}{2}(73.26) = 36.63°.$$

Now we use the law of tangents.

$$\log(b - c) \approx 1.8882 - 2$$
$$\underline{\log(b + c) \approx 0.9248}$$
$$\log[(b - c)/(b + c)] \approx 0.9634 - 2$$
$$\underline{\log \tan \tfrac{1}{2}(\beta + \gamma) \approx 0.8713 - 1}$$
$$\log \tan \tfrac{1}{2}(\beta - \gamma) \approx 0.8347 - 2$$
$$\tfrac{1}{2}(\beta - \gamma) \approx 3.91°.$$

We have sufficient information to find β and γ:

$$\beta = \tfrac{1}{2}(\beta + \gamma) + \tfrac{1}{2}(\beta - \gamma) \approx 36.63 + 3.91 = 40.54°,$$

$$\gamma = \tfrac{1}{2}(\beta + \gamma) - \tfrac{1}{2}(\beta - \gamma) \approx 36.63 - 3.91 = 32.72°.$$

We find $a \approx 6.766$ by the law of sines. (See Example 3.2.)

Answer $\beta \approx 40.54°$, $\gamma \approx 32.72°$, $a \approx 6.766$.

Look once again at the FP-02 in Fig. 7.1, p. 178, and note the keys [SIN], [COS], [TAN], and [ARC]. These do just what they say. For example, if we push 24.85 [SIN] we see in the display $0.4202 \approx \sin 24.85$. If we *then* press [ARC] [SIN], we get arc sin(0.4202) so the display gives us back 24.85. Clearly, it is possible to use this computational power for rapid and direct solutions of triangles.

■ *Example 3.4*

Given $b = 4.592$, $c = 3.819$, and $\gamma = 32.73°$, set up the computation of β.

SOLUTION   We use the law of sines

$$\sin \beta = \frac{b \sin \gamma}{c}.$$

A possible sequence of steps is

> 4.592 [ENT ↑] 32.72 [SIN] [×] 3.819 [÷] [ARC] [SIN].

The result on the display is 40.54°. However, because $c < b$, there are *two* solutions. The other is $180.00 - 40.54 = 139.46°$. A calculator does your *computing,* not your *thinking*!

> *Answer* $\beta \approx 40.54°$, 139.46°

EXERCISES

Use 4-place tables.

Given three parts (angles or sides) of a triangle find the remaining three parts. Use the laws of cosines and sines.

1. $\alpha = 20°, \beta = 50°, a = 6$
2. $\alpha = 30°, \gamma = 50°, c = 10$
3. $\alpha = 26.5°, \beta = 40°, c = 15$
4. $\beta = 33.6°, \gamma = 48.1°, b = 6.73$
5. $\alpha = 110°, \beta = 42°, a = 21.47$
6. $\alpha = 39.2°, \beta = 106.5°, a = 10.54$
7. $a = 10, b = 12, \gamma = 36°$
8. $a = 3.6, c = 12.5, \beta = 46.9°$
9. $b = 15.81, c = 4.53, \gamma = 35.6°$
10. $a = 2.07, b = 10.63, \alpha = 28.4°$
11. $a = 137.7, b = 192.5, \alpha = 17.6°$
12. $b = 6.042, c = 7.881, \gamma = 62.3°$
13. $a = 5, b = 6, c = 10$
14. $a = 4.1, b = 9.5, c = 10.6$
15. $a = 12, b = 7, \gamma = 21°$
16. $a = 100, b = 45, \gamma = 66.1°$
17. $b = 20, c = 37, \alpha = 147.2°$
18. $a = 16, b = 25, \gamma = 129.2°$.

19. The leaning tower of Pisa when constructed had height 179 ft. It now leans at an angle of 5.45° from the vertical. Compute the length of its shadow when the angle of elevation of the sun is 56° and the sun is behind the tower (away from the tilt).

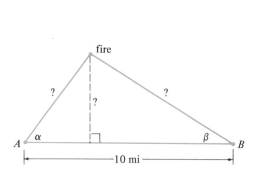

Fig. 3.2   Fig. 3.3

20. Two forest rangers 10 miles apart at points A and B observe a fire (Fig. 3.2). The ranger at A finds $\alpha = 52.6°$. The ranger at B finds $\beta = 31.5°$. How far is the fire from each ranger? from the line through A and B?

21. Two points A and B are on opposite sides of a river (Fig. 3.3). A third point C is 400 ft from A. If $\alpha = 77.3°$ and $\gamma = 41.8°$, find the distance from A to B.

22. The diagonals of a parallelogram are 10 ft and 15 ft, and they intersect at an angle of 28.7°. Compute the sides of the parallelogram.

23. A point A is 10 miles due north of here. Another point B is 12.36 miles from here on a line 17.4° east of north. What is the distance from A to B?

24. A plane travels 50 miles due east after take-off, then adjusts its course 10° north and flies 100 miles. How far is it from the point of departure?

25. We are given b, c, and γ in a triangle. Suppose $\gamma > \frac{1}{2}\pi$. When is there a solution(s)? Analyze the cases.

26. (cont.) Suppose $\gamma = 110°$, $b = 5$, and $c = 6$. Find the other side and angles.

27. The centerfielder stands 340 ft from home plate in straight centerfield. How far is he from third base? (The distance between bases is 90 ft.)

28. Two boats, one traveling 19 ft/sec, the other 23 ft/sec, leave a point on straight paths 65° from each other. How far apart are they after 2.5 minutes?

Use the law of tangents to solve for the remaining three parts:

29. $b = 28.62, c = 22.34, \alpha = 52.43°$ **30.** $b = 15.91, c = 13.77, \alpha = 11.12°$

31. $b = 7.482, c = 9.316, \alpha = 25.58°$ **32.** $b = 104.9, c = 209.3, \alpha = 60.00°$

33. $a = 21.66, b = 31.63, \gamma = 124.15°$ **34.** $a = 109.5, c = 35.58, \beta = 140.23°$

35. $a = 2.172, c = 4.008, \beta = 31.12°$ **36.** $a = 1.047, c = 8.223, \beta = 162.89°$.

37. Solve Ex. 29 by the law of cosines.

38. (cont.) Now explain why the law of tangents is better for logarithmic computation.

Set up the problem for solution on the FP-02:

39. Ex. 1   **40.** Ex. 3   **41.** Ex. 7   **42.** Ex. 8

43. Ex. 9   **44.** Ex. 10   **45.** Ex. 14   **46.** Ex. 29.

                                                (Use law of tangents.)

4. APPLICATIONS TO GEOMETRY

We have proved four basic triangle laws, the law of cosines, the law of sines, the law of tangents, and the area formula. Now we examine some further relations between the parts of a triangle (Fig. 4.1).

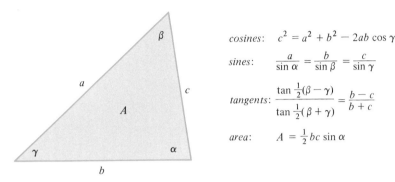

cosines: $\quad c^2 = a^2 + b^2 - 2ab\cos\gamma$

sines: $\quad \dfrac{a}{\sin\alpha} = \dfrac{b}{\sin\beta} = \dfrac{c}{\sin\gamma}$

tangents: $\quad \dfrac{\tan\frac{1}{2}(\beta-\gamma)}{\tan\frac{1}{2}(\beta+\gamma)} = \dfrac{b-c}{b+c}$

area: $\quad A = \frac{1}{2}bc\sin\alpha$

Fig. 4.1 Oblique triangle

Formulas of Newton and Mollweide

Newton's Formula	$\dfrac{b-c}{a} = \dfrac{\sin\frac{1}{2}(\beta-\gamma)}{\cos\frac{1}{2}\alpha}$
Mollweide's Formula	$\dfrac{b+c}{a} = \dfrac{\cos\frac{1}{2}(\beta-\gamma)}{\sin\frac{1}{2}\alpha}.$

These formulas involve all six sides and angles of the triangle, so they are not useful for solving a triangle. However they are useful for checking your solution by logarithms. Their proofs are similar to the proof in Section 2 for the law of tangents, and are left as exercises.

Circumradius

First we must recall a fact from plane geometry (Fig. 4.2a on next page). An angle inscribed in a circle is equal to half of the corresponding central angle.

Given a triangle (Fig. 4.2b), let C be its circumscribed circle (circumcircle). The problem is to express the radius R of this circle in terms of the sides and angles of the triangle. To do this, look at the isosceles triangle (Fig. 4.2c) with sides a, R, R. Its apex angle is 2α, twice the corresponding inscribed angle. If θ is its base angle, then $2\alpha + \theta + \theta = \pi$, so $\theta = \frac{1}{2}\pi - \alpha$. Now drop a perpendicular from the apex. Since it bisects the base,

$$\tfrac{1}{2}a = R\cos\theta = R\cos(\tfrac{1}{2}\pi - \alpha) = R\sin\alpha.$$

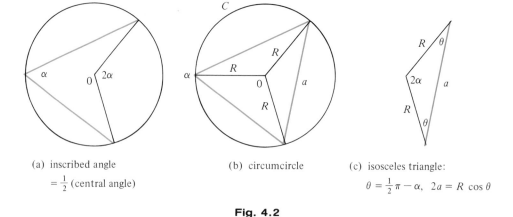

(a) inscribed angle
$= \frac{1}{2}$ (central angle)

(b) circumcircle

(c) isosceles triangle:
$\theta = \frac{1}{2}\pi - \alpha, \ 2a = R \cos \theta$

Fig. 4.2

This gives us $R = a/(2 \sin \alpha)$. Similar conclusions hold for β and b or γ and c.

$$R = \frac{a}{2 \sin \alpha} = \frac{b}{2 \sin \beta} = \frac{c}{2 \sin \gamma}.$$

Thus we have found another derivation of the law of sines. We have also shown that the common value of the ratios $a/\sin \alpha$, etc, is the diameter of the circumscribed circle.

This result can be applied to express the area A of a triangle in terms of its sides a, b, c and its circumradius R:

$$A = \tfrac{1}{2}ab \sin \gamma = \tfrac{1}{2}ab \left(\frac{c}{2R}\right),$$

$$A = \frac{abc}{4R}.$$

Heron's Formula

There is a formula that expresses the area A in terms of the sides a, b, and c. To prove it, we eliminate γ from $A = \tfrac{1}{2}ab \sin \gamma$. We obtain $\cos \gamma$ from the law of cosines, then use $\cos^2 \gamma + \sin^2 \gamma = 1$:

$$c^2 = a^2 + b^2 - 2ab \cos \gamma,$$

hence

$$(a^2 + b^2 - c^2)^2 = (2ab \cos \gamma)^2$$
$$= 4a^2b^2 \cos^2 \gamma$$
$$= 4a^2b^2(1 - \sin^2 \gamma).$$

But
$$4a^2b^2 \sin^2 \gamma = 16(\tfrac{1}{2}ab \sin \gamma)^2 = 16A^2,$$
therefore
$$(a^2 + b^2 - c^2)^2 = 4a^2b^2 - 16A^2.$$

It follows that
$$
\begin{aligned}
16A^2 &= 4a^2b^2 - (a^2 + b^2 - c^2)^2 \\
&= [2ab + (a^2 + b^2 - c^2)][2ab - (a^2 + b^2 - c^2)] \\
&= [(a + b)^2 - c^2][c^2 - (a - b)^2] \\
&= [(a + b) + c][(a + b) - c][c + (a - b)][c - (a - b)].
\end{aligned}
$$

It is convenient at this point to introduce the symbol s for the semi-perimeter of the triangle:
$$s = \tfrac{1}{2}(a + b + c).$$

Then the formula for $16A^2$ can be written
$$16A^2 = (2s)(2s - 2c)(2s - 2b)(2s - 2a).$$

We divide both sides by 16 to obtain the final result:

> **Heron's Formula** $\quad A = \sqrt{s(s - a)(s - b)(s - c)}.$

Inradius

Given a triangle (Fig. 4.3a), let C denote its inscribed circle (incircle), and let r be the radius of C. The problem is to express r in terms of a, b, and c.

Join the three vertices of the triangle to the incenter 0; this divides the triangle into 3 parts. The area of one of these (Fig. 4.3b) is $\tfrac{1}{2}ar$. Hence
$$A = \tfrac{1}{2}ar + \tfrac{1}{2}br + \tfrac{1}{2}cr = sr.$$

(a) incircle

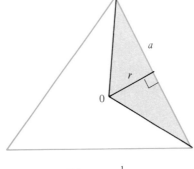

(b) area $= \tfrac{1}{2}ar$

Fig. 4.3

If we use Heron's formula for A, and solve for r, we arrive at

$$r = \sqrt{\frac{(s-a)(s-b)(s-c)}{s}}.$$

Half-Angle Formulas

The next, and final, formulas give the angles in terms of the sides. Of course, the law of cosines does just this, but it is not as well suited to calculations with logarithms as are these formulas.

Half-Angle Formulas

$$\cos\tfrac{1}{2}\alpha = \sqrt{\frac{s(s-a)}{bc}} \qquad \sin\tfrac{1}{2}\alpha = \sqrt{\frac{(s-b)(s-c)}{bc}}$$

$$\tan\tfrac{1}{2}\alpha = \frac{r}{s-a}.$$

We shall prove the third one, leaving proofs of the others as exercises.

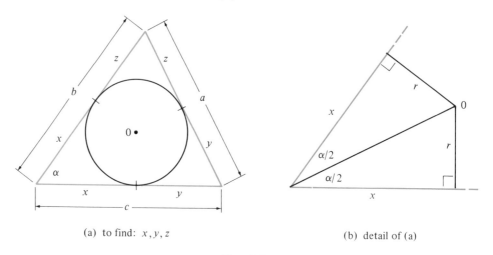

(a) to find: x, y, z (b) detail of (a)

Fig. 4.4

The incircle divides each side of the triangle into two parts (Fig. 4.4a). The parts adjacent to angle α are equal, by congruent triangles (Fig. 4.4b). Clearly

$$2(x + y + z) = \text{perimeter} = 2s,$$

$$x + y + z = s.$$

The line joining 0 to the vertex of α bisects the angle α, again by congruent triangles. From Fig. 4.4b, $\tan\tfrac{1}{2}\alpha = r/x$. But

$$x = s - (y + z) = s - a$$

since $y + z = a$ from Fig. 4.4a. Therefore

$$\tan \tfrac{1}{2}\alpha = \frac{r}{s - a}.$$

◼ *Example 4.1*

Use half-angle formulas to find the angles of a triangle, given the sides $a = 7$, $b = 6$, $c = 2$.

SOLUTION We compute

$$s = \tfrac{1}{2}(7 + 6 + 2) = 7.5$$

$$s - a = 0.5, \qquad s - b = 1.5, \qquad s - c = 5.5.$$

Hence

$$r = \sqrt{\frac{(s - a)(s - b)(s - c)}{s}} = \sqrt{\frac{(0.5)(1.5)(5.5)}{7.5}} \approx 0.7416.$$

Now we use the half-angle formula for the tangent:

$$\tfrac{1}{2}\alpha = \text{arc} \tan \frac{r}{s - a} \approx \text{arc} \tan(1.4832) \approx 56.01, \qquad \alpha \approx 112.02°,$$

$$\tfrac{1}{2}\beta = \text{arc} \tan \frac{r}{s - b} \approx \text{arc} \tan(0.4944) \approx 26.31, \qquad \beta \approx 52.62°,$$

$$\tfrac{1}{2}\gamma = \text{arc} \tan \frac{r}{s - c} \approx \text{arc} \tan(0.1348) \approx 7.68, \qquad \gamma \approx 15.36°.$$

Check $\alpha + \beta + \gamma \approx 112.02 + 52.62 + 15.36 = 180.00.$

Answer $\alpha \approx 112.02°, \quad \beta \approx 52.62°, \quad \gamma \approx 15.36°.$

EXERCISES

1. Prove Newton's formula.
2. Prove Mollweide's formula.
3. Prove $\sin \alpha + \sin \beta + \sin \gamma = s/R$.
4. If $\gamma = \tfrac{1}{3}\pi$, show that $(a - b)^2 = c^2 - ab$.
5. Prove that $A = a^2 \sin \beta \sin \gamma/(2 \sin \alpha)$.
6. Let α be a base angle and a the lateral side of an isosceles triangle. Prove that $A = \tfrac{1}{2}a^2 \sin 2\alpha$.
7*. Prove the half-angle formula for the cosine. [Hint: In Fig. 4.4b, we have $\cos \tfrac{1}{2}\alpha = (s - a)/h$, where h is the hypotenuse. Express h in terms of a, b, c, s.]
8*. Prove the half-angle formula for the sine.
9. Express R in terms of a, b, and c.
10. Prove that

$$\csc \alpha = \frac{bc}{2\sqrt{s(s - a)(s - b)(s - c)}}.$$

Use half-angle formulas to find α, β, and γ:

11. $a = 1.382, b = 3.199, c = 2.773$
13. $a = 150.0, b = 197.2, c = 138.0$

12. $a = 92.71, b = 84.62, c = 75.06$
14. $a = 7640, b = 9731, c = 4343.$

5. VECTORS [optional]

In this section we shall introduce vectors and discuss addition of vectors and multiplication of vectors by numbers. Then, in the next section, we shall introduce the inner product of vectors, an important idea which provides an interesting application of trigonometry.

The idea of vectors grew out of the study of forces in physics. A force applied at a point was represented by an arrow; it pointed in the direction of the force and its length represented the magnitude of the force. The concept of a quantity having direction and magnitude turned out to be a powerful tool not only in physics and engineering, but also in non-physical fields such as statistics, numerical analysis, and mathematical economics.

We choose once and for all a point **0** in the plane and call it the origin. A **vector** is a directed line segment from **0** to a point of the plane (Fig. 5.1a). The vector that goes from **0** to **0** (a degenerate line segment) we call the *vector* **0**.

The vector from **0** to a point **z** is completely determined by its terminal point. For this reason, we shall often identify a vector and its terminal point as one and the same thing (Fig. 5.1b).

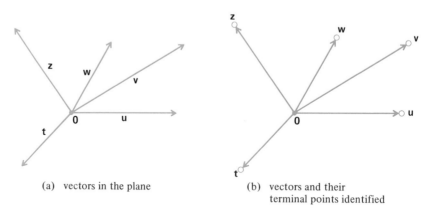

(a)   vectors in the plane

(b)   vectors and their terminal points identified

Fig. 5.1

We have defined vectors without reference to coordinate systems. This is as it should be, for in real life a force is a force and a velocity is a velocity, regardless of how we choose to set up coordinate axes. Nevertheless, to compute with vectors we shall need coordinates. Given a cartesian coordinate system in the plane, we assign to each vector **v** the coordinates of its terminal point, and write

$$\mathbf{v} = (x, y).$$

Some examples are shown in Fig. 5.2.

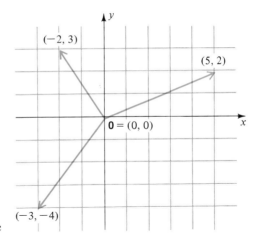

Fig. 5.2 Vectors in the coordinate plane

In the context of vector algebra, the word **scalar** means real number. We combine a scalar a and a vector $\mathbf{v}$ to form a new vector $a\mathbf{v}$ called the scalar multiple of a and $\mathbf{v}$.

If a is a scalar and $\mathbf{v}$ is a vector, then the **scalar multiple** $a\mathbf{v}$ is the vector defined by

$$a\mathbf{v} = \begin{cases} \mathbf{0} & \text{if } a = 0, \\ \text{a vector in the same direction as } \mathbf{v} \text{ with length } a \text{ times} \\ \text{the length of } \mathbf{v} & \text{if } a > 0. \\ \text{a vector in the opposite direction from } \mathbf{v} \text{ with length } |a| \\ \text{times the length of } \mathbf{v} & \text{if } a < 0. \end{cases}$$

Examples of multiplication by a scalar are shown in Fig. 5.3.

Fig. 5.3 Scalar multiples of vectors

If $\mathbf{v}$ is expressed in coordinates, it is easy to express $a\mathbf{v}$ in coordinates. The rule seems very natural:

$$a(x, y) = (ax, ay).$$

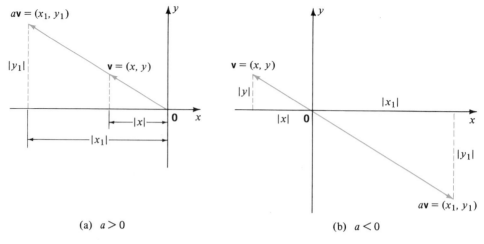

(a) $a > 0$ $\qquad\qquad\qquad$ (b) $a < 0$

Fig. 5.4 Proof by similar triangles that $a(x, y) = (ax, ay)$

To see why the rule holds, look at Fig. 5.4. If the new vector $a\mathbf{v}$ has coordinates (x_1, y_1), we must show that $x_1 = ax$ and $y_1 = ay$.

Project $\mathbf{v}$ and $a\mathbf{v}$ on the x-axis. By similar triangles,

$$\frac{|x_1|}{|x|} = \frac{\text{length of } a\mathbf{v}}{\text{length of } \mathbf{v}} = |a|.$$

If $a > 0$, the terminal points of $\mathbf{v}$ and $a\mathbf{v}$ lie in the same quadrant, so $|x_1|/|x| = x_1/x$. Hence the absolute values can be dropped:

$$\frac{x_1}{x} = a, \qquad x_1 = ax.$$

If $a < 0$, the terminal points of $\mathbf{v}$ and $a\mathbf{v}$ lie in opposite quadrants, so $|x_1|/|x| = -x_1/x$. Hence

$$-\frac{x_1}{x} = |a| = -a, \qquad x_1 = ax.$$

Thus $x_1 = ax$ in all cases. A similar argument shows that $y_1 = ay$.

Examples of scalar multiplication:

$$1(x, y) = (x, y) \qquad\qquad -3(5, -7) = (-15, 21)$$
$$3(0, 0) = (0, 0) = \mathbf{0} \qquad\qquad 0(5, -1) = \mathbf{0}$$
$$-2(1, 0) = (-2, 0) \qquad\qquad 2(3, -2) = (6, -4).$$

Two identities follow from the formula $a(x, y) = (ax, ay)$:

$$\boxed{(a + b)\mathbf{v} = a\mathbf{v} + b\mathbf{v} \qquad (ab)\mathbf{v} = a(b\mathbf{v}).}$$

One can interpret scalar multiplication physically in several ways. For example, a velocity vector **v** has direction and magnitude (speed). If the direction remains the same, but the speed triples, then the new velocity vector is 3**v**.

We now define the sum **v** + **w** of two vectors **v** and **w**. Draw a directed line segment parallel to **w**, starting from the terminal point of **v**. See Fig. 5.5a. Then the vector from **0** to the end point of this parallel segment is the vector **v** + **w**. See Fig. 5.5b.

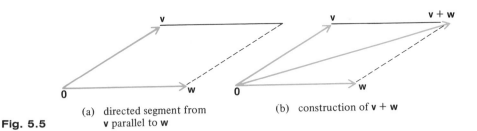

Fig. 5.5   (a) directed segment from   (b) construction of **v** + **w**
              v parallel to **w**

By closing the parallelogram, we see that the roles of **v** and **w** can be interchanged in the definition (Fig. 5.6). The result is

$$\mathbf{v} + \mathbf{w} = \mathbf{w} + \mathbf{v}.$$

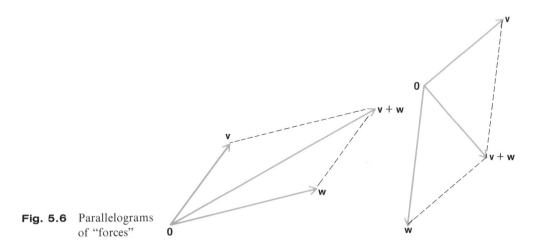

Fig. 5.6 Parallelograms
            of "forces"

The simplest interpretation of addition of vectors is in terms of force vectors. If two forces **v** and **w** are applied at the same point, the resultant force is **v** + **w**; that is the law of "parallelogram of forces".

Given **v** and **w** expressed in a coordinate system, how do we compute **v** + **w**?

The answer is again quite natural:

$$(x_1, y_1) + (x_2, y_2) = (x_1 + x_2, y_1 + y_2).$$

We say vectors are added "componentwise" or "coordinatewise".

To verify the formula, let $\mathbf{v} = (x_1, y_1)$, $\mathbf{w} = (x_2, y_2)$, and $\mathbf{v} + \mathbf{w} = (x_3, y_3)$. We must show that $x_3 = x_1 + x_2$ and $y_3 = y_1 + y_2$. See Fig. 5.7.

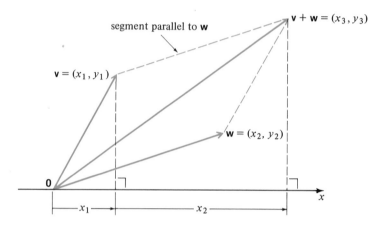

Fig. 5.7   (Sum of projections) = (projection of sum)

The projection on the x-axis of $\mathbf{v}$ is x_1. The projection on the x-axis of the directed segment parallel to $\mathbf{w}$ is a directed segment of length $|x_2|$. No matter what the sign of x_2 is, the terminal point of this segment is $x_1 + x_2$. But this point is the projection of $\mathbf{v} + \mathbf{w}$, that is, x_3. Hence $x_3 = x_1 + x_2$. Similarly $y_3 = y_1 + y_2$.

We denote by $-\mathbf{v}$ the vector $(-1)\mathbf{v}$, opposite to $\mathbf{v}$. Thus $-(3, -4) = (-3, 4)$.

The basic rules of vector algebra follow easily from the coordinate formulas for scalar multiplication and addition:

$$
\begin{aligned}
1 \cdot \mathbf{v} &= \mathbf{v} \\
\mathbf{v} + (-\mathbf{v}) &= \mathbf{0} \\
\mathbf{v} + \mathbf{w} &= \mathbf{w} + \mathbf{v} \quad &\text{(commutative law)} \\
\mathbf{u} + (\mathbf{v} + \mathbf{w}) &= (\mathbf{u} + \mathbf{v}) + \mathbf{w} \quad &\text{(associative law)} \\
a(\mathbf{v} + \mathbf{w}) &= a\mathbf{v} + a\mathbf{w} \\
(a + b)\mathbf{v} &= a\mathbf{v} + b\mathbf{v} \quad &\text{(distributive laws)} \\
a(b\mathbf{v}) &= (ab)\mathbf{v}.
\end{aligned}
$$

We define the **difference** by

$$\mathbf{v} - \mathbf{w} = \mathbf{v} + (-\mathbf{w}).$$

In coordinates,

$$(x_1, y_1) - (x_2, y_2) = (x_1, y_1) + (-x_2, -y_2) = (x_1 - x_2, y_1 - y_2).$$

Examples of vector addition and subtraction:

$$(2, 7) + (1, 3) = (3, 10) \qquad (2, 7) - (1, 3) = (1, 4)$$
$$(2, 7) + (1, -3) = (3, 4) \qquad (2, 7) - (1, -3) = (1, 10)$$
$$(2, 7) + (-5, 0) = (-3, 7) \qquad (2, 7) - (-5, 0) = (7, 7).$$

Note the geometric interpretation of the difference of vectors. The directed line segment from **w** to **v** has the same length and direction as the vector **v** − **w**. See Fig. 5.8.

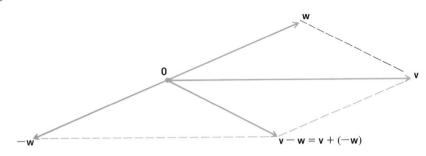

Fig. 5.8 Difference of vectors

EXERCISES

Compute:

1. $3(0, 1)$
4. $7(3, -2)$
7. $(3, -1) + (0, 1)$
10. $(2, -2) + (3, 3)$
13. $3(1, 1) + (7, 1)$

2. $-5(2, 0)$
5. $4(\frac{1}{2}, 3)$
8. $(4, 2) + (3, 1)$
11. $(4, -1) + (-3, 2)$
14. $(2, -2) + 8(1, 2)$
16. $3(2, 0) + 2(4, 5).$

3. $2(1, -1)$
6. $-\frac{1}{3}(4, -2)$
9. $(-1, 1) + (1, -1)$
12. $(7, 11) + (11, 7)$
15. $-2(1, 3) + 4(5, 4)$

Plot the vectors and compute their sum graphically by the "parallelogram of forces". Check your answer:

17. $(4, 1) + (3, 2)$
18. $(-3, 2) + (4, 4)$
20. $(2, -3) + (-1, -3).$
19. $(1, 5) + (-5, -1)$

21. Find a vector parallel to and having the same length as the directed segment from $(-2, 4)$ to $(-1, -6)$.

22. Find a point **x** such that the directed segment from $(2, 3)$ to **x** is parallel to the directed segment from $(1, 1)$ to $(-4, 0)$ and of the same length.

Find a and b; check your answer graphically:

23. $(-4, 5) = a(2, 0) + b(0, 1)$
24. $(3, 3) = a(1, 0) - b(1, 1).$

6. LENGTH AND INNER PRODUCT [optional]

Length

We start with some basic properties of the length of a vector. We denote the length of $\mathbf{v}$ by $\|\mathbf{v}\|$.

If $\mathbf{v} = (x, y)$, then

(1) $\qquad \|\mathbf{v}\|^2 = \|(x, y)\|^2 = x^2 + y^2,$

(2) $\qquad \|\mathbf{0}\| = 0, \qquad \|\mathbf{v}\| > 0 \quad$ if $\mathbf{v} \neq \mathbf{0},$

(3) $\qquad \|a\mathbf{v}\| = |a| \cdot \|\mathbf{v}\|,$

(4) $\quad \|\mathbf{v} + \mathbf{w}\| \leq \|\mathbf{v}\| + \|\mathbf{w}\| \qquad$ (triangle inequality).

The first property comes directly from the distance formula, and the second follows from the first. The third is also proved by direct calculation:

$$\|a(x, y)\|^2 = \|(ax, ay)\|^2 = (ax)^2 + (ay)^2 = a^2(x^2 + y^2) = a^2\|(x, y)\|^2,$$

therefore

$$\|a\mathbf{v}\|^2 = a^2\|\mathbf{v}\|^2, \qquad \|a\mathbf{v}\| = |a| \cdot |\mathbf{v}|.$$

A useful fact: if $\mathbf{v}$ is any non-zero vector, then $\mathbf{v}/\|\mathbf{v}\|$ is a unit vector (length 1) in the same direction. Just apply (3) with $a = 1/\|\mathbf{v}\|$. [We used this device previously; in Chapter 6, Section 2, p. 195, we stretched (or shrank) a vector (a, b) to the unit circle by dividing its coordinates by $\sqrt{a^2 + b^2}$. This amounted to making the vector $\mathbf{v} = (a, b)$ into a unit vector by dividing it by $\|\mathbf{v}\| = \sqrt{a^2 + b^2}$.]

The triangle inequality (4) expresses a basic geometric fact: the length of each side of a triangle cannot exceed the sum of the lengths of the other two sides (Fig. 6.1).

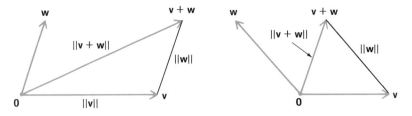

Fig. 6.1   Geometric proof that $\|\mathbf{v} + \mathbf{w}\| \leq \|\mathbf{v}\| + \|\mathbf{w}\|$

Remark: The inequality between real numbers

$$|a + b| \leq |a| + |b|,$$

mentioned in Chapter 2, Section 8, is called "the triangle inequality" because of its close resemblance to the triangle inequality for vectors.

Inner Product

We now define the third basic operation on vectors, the **inner product** (also called **dot product**). The inner product combines two vectors to produce a scalar. Suppose **v** and **w** are non-zero vectors, and $\cos \theta$ is the cosine of the angle between them. We define

$$\mathbf{v} \cdot \mathbf{w} = \|\mathbf{v}\| \cdot \|\mathbf{w}\| \cdot \cos \theta.$$

There are four possibilities for "the angle between **v** and **w**". If one is θ, then the others are $-\theta$, $2\pi - \theta$, and $\theta - 2\pi$. But

$$\cos \theta = \cos(-\theta) = \cos(2\pi - \theta) = \cos(\theta - 2\pi),$$

so there is only one possibility for "the cosine of the angle between **v** and **w**". See Fig. 6.2. If either $\mathbf{v} = \mathbf{0}$ or $\mathbf{w} = \mathbf{0}$, the angle between **v** and **w** is not defined. In

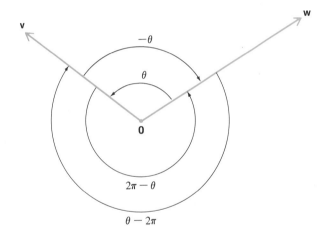

Fig. 6.2 The four angles have equal cosines

this case we define

$$\mathbf{0} \cdot \mathbf{w} = \mathbf{v} \cdot \mathbf{0} = 0.$$

Notice that the inner product of two vectors is a *real number,* not a vector.

Three special cases that arise all the time follow directly from the definition of inner product:

(1) $\mathbf{v} \cdot \mathbf{v} = \|\mathbf{v}\|^2$.

(2) Vectors **v** and **w** are perpendicular if and only if $\mathbf{v} \cdot \mathbf{w} = 0$.

(3) If **v** and **w** are unit vectors ($\|\mathbf{v}\| = \|\mathbf{w}\| = 1$), then $\mathbf{v} \cdot \mathbf{w} = \cos \theta$.

Some examples of dot products are shown in Fig. 6.3.

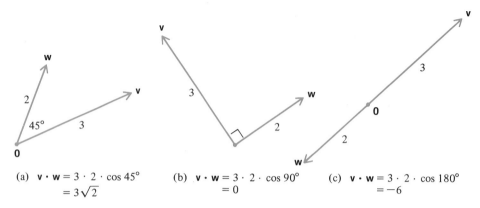

(a) $\mathbf{v} \cdot \mathbf{w} = 3 \cdot 2 \cdot \cos 45°$
$= 3\sqrt{2}$

(b) $\mathbf{v} \cdot \mathbf{w} = 3 \cdot 2 \cdot \cos 90°$
$= 0$

(c) $\mathbf{v} \cdot \mathbf{w} = 3 \cdot 2 \cdot \cos 180°$
$= -6$

Fig. 6.3   Examples of inner products

There is another important geometric interpretation of the inner product. Suppose $\mathbf{v}$ and $\mathbf{w}$ are non-zero vectors and θ is the angle between them (we can always take θ so that $0 \leq \theta \leq \pi$). See Fig. 6.4. The quantity $\|\mathbf{v}\| \cdot \cos \theta$ is the length of the projection of $\mathbf{v}$ on $\mathbf{w}$ if $0 \leq \theta \leq \frac{1}{2}\pi$ and the negative of the length if $\frac{1}{2}\pi \leq \theta \leq \pi$. We call $\|\mathbf{v}\| \cdot \cos \theta$ the **signed projection** of $\mathbf{v}$ on $\mathbf{w}$. Its sign is $+$ when $\mathbf{v}$ projects directly onto $\mathbf{w}$, and $-$ when $\mathbf{v}$ projects onto $\mathbf{w}$ extended backwards. Since $\mathbf{v} \cdot \mathbf{w} = (\|\mathbf{v}\| \cos \theta) \cdot \|\mathbf{w}\| = (\|\mathbf{w}\| \cos \theta) \cdot \|\mathbf{v}\|$,

$$\mathbf{v} \cdot \mathbf{w} = (\text{signed projection of } \mathbf{v} \text{ on } \mathbf{w}) \cdot \|\mathbf{w}\|$$
$$= (\text{signed projection of } \mathbf{w} \text{ on } \mathbf{v}) \cdot \|\mathbf{v}\|.$$

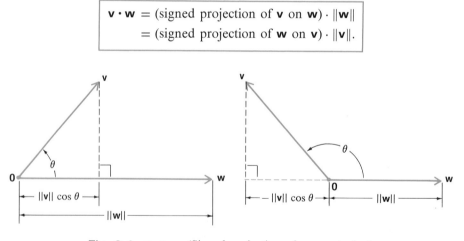

Fig. 6.4 $\mathbf{v} \cdot \mathbf{w} = (\text{Signed projection of } \mathbf{v} \text{ on } \mathbf{w}) \cdot \|\mathbf{w}\|$

Computations with Inner Products

If we are given the lengths of two vectors and the angle between them, we can compute their inner product directly from the definition, using trig tables. For instance, if vectors of lengths 8 and 11 meet at angle 26°, their inner product is

$$8 \cdot 11 \cdot \cos 26° \approx 88(0.8988) \approx 79.09.$$

If we are given the coordinates of two vectors, then we compute their inner product by the following formula:

$$(x_1, y_1) \cdot (x_2, y_2) = x_1 x_2 + y_1 y_2.$$

To see where this formula comes from, consider the triangle shown in Fig. 6.5.

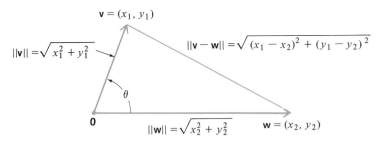

Fig. 6.5   Computation of $\|\mathbf{v} - \mathbf{w}\|$ by the law of cosines

By the law of cosines,

$$\|\mathbf{v} - \mathbf{w}\|^2 = \|\mathbf{v}\|^2 + \|\mathbf{w}\|^2 - 2\|\mathbf{v}\| \cdot \|\mathbf{w}\| \cdot \cos \theta = \|\mathbf{v}\|^2 + \|\mathbf{w}\|^2 - 2\mathbf{v} \cdot \mathbf{w}.$$

Hence

$$
\begin{aligned}
\mathbf{v} \cdot \mathbf{w} &= \tfrac{1}{2}(\|\mathbf{v}\|^2 + \|\mathbf{w}\|^2 - \|\mathbf{v} - \mathbf{w}\|^2) \\
&= \tfrac{1}{2}[\|(x_1, y_1)\|^2 + \|(x_2, y_2)\|^2 - \|(x_1 - x_2, y_1 - y_2)\|^2] \\
&= \tfrac{1}{2}[(x_1{}^2 + y_1{}^2) + (x_2{}^2 + y_2{}^2) - (x_1 - x_2)^2 - (y_1 - y_2)^2] \\
&= \tfrac{1}{2}[x_1{}^2 + y_1{}^2 + x_2{}^2 + y_2{}^2 - (x_1{}^2 - 2x_1 x_2 + x_2{}^2) - (y_1{}^2 - 2y_1 y_2 + y_2{}^2)] \\
&= x_1 x_2 + y_1 y_2.
\end{aligned}
$$

This formula for inner products provides an explicit expression for the cosine of the angle between two vectors.

If $\mathbf{v} = (x_1, y_1)$ and $\mathbf{w} = (x_2, y_2)$ are non-zero vectors, and θ is the angle between them, then

$$\cos \theta = \frac{\mathbf{v} \cdot \mathbf{w}}{\|\mathbf{v}\| \cdot \|\mathbf{w}\|} = \frac{x_1 x_2 + y_1 y_2}{\sqrt{x_1{}^2 + y_1{}^2}\sqrt{x_2{}^2 + y_2{}^2}}.$$

■ *Example 6.1*

Find the angle between the vectors $(4, 3)$ and $(1, 5)$.

SOLUTION   Set $\mathbf{v} = (4, 3)$ and $\mathbf{w} = (1, 5)$. Then

$$\|\mathbf{v}\|^2 = 4^2 + 3^2 = 25, \qquad \|\mathbf{w}\|^2 = 1^2 + 5^2 = 26,$$

$$\mathbf{v} \cdot \mathbf{w} = 4 \cdot 1 + 3 \cdot 5 = 19,$$

$$\cos \theta = \frac{\mathbf{v} \cdot \mathbf{w}}{\|\mathbf{v}\| \cdot \|\mathbf{w}\|} = \frac{19}{5\sqrt{26}}.$$

Answer $\theta = \arccos 19/(5\sqrt{26})$.

. . . .

The main properties of the inner product look just like the familiar rules of algebra.

(1)	$\mathbf{v} \cdot \mathbf{w} = \mathbf{w} \cdot \mathbf{v}$
(2)	$(a\mathbf{v}) \cdot \mathbf{w} = \mathbf{v} \cdot (a\mathbf{w}) = a(\mathbf{v} \cdot \mathbf{w})$
(3)	$(\mathbf{u} + \mathbf{v}) \cdot \mathbf{w} = \mathbf{u} \cdot \mathbf{w} + \mathbf{v} \cdot \mathbf{w}$
(4)	$\mathbf{u} \cdot (\mathbf{v} + \mathbf{w}) = \mathbf{u} \cdot \mathbf{v} + \mathbf{u} \cdot \mathbf{w}.$

Property (1) simply says that $x_1 x_2 + y_1 y_2 = x_2 x_1 + y_2 y_1$. Property (2) is nearly as easy; it says $(ax_1)x_2 + (ay_1)y_2 = x_1(ax_2) + y_1(ay_2) = a(x_1 x_2 + y_1 y_2)$.

In view of (1), relations (3) and (4) are equivalent to each other, so it suffices to prove (3). We have

$$\begin{aligned}
(\mathbf{u} + \mathbf{v}) \cdot \mathbf{w} &= [(x_1, y_1) + (x_2, y_2)] \cdot (x_3, y_3) \\
&= (x_1 + x_2, y_1 + y_2) \cdot (x_3, y_3) = (x_1 + x_2)x_3 + (y_1 + y_2)y_3 \\
&= (x_1 x_3 + y_1 y_3) + (x_2 x_3 + y_2 y_3) = (x_1, y_1) \cdot (x_3, y_3) + (x_2, y_2) \cdot (x_3, y_3) \\
&= \mathbf{u} \cdot \mathbf{w} + \mathbf{v} \cdot \mathbf{w}.
\end{aligned}$$

Perpendicular Vectors

If $\mathbf{v} = (a, b)$ is a non-zero vector, then $(-b, a)$ is perpendicular to $\mathbf{v}$ and has the same length as $\mathbf{v}$:

$$(a, b) \cdot (-b, a) = -ab + ba = 0,$$

$$\|(-b, a)\| = \sqrt{(-b)^2 + a^2} = \sqrt{a^2 + b^2} = \|(a, b)\|.$$

Therefore $(-b, a)$ is one of the two vectors shown in Fig. 6.6a.

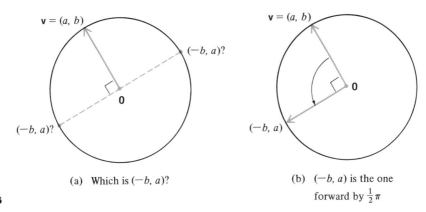

Fig. 6.6

(a) Which is $(-b, a)$?

(b) $(-b, a)$ is the one forward by $\frac{1}{2}\pi$

A direct check of the 4 quadrants shows that $(-b, a)$ is *always* $\frac{1}{2}\pi$ forward of **v**. See Fig. 6.6b.

> The result of rotating a vector $\mathbf{v} = (a, b)$ counterclockwise $\frac{1}{2}\pi$ radians is $(-b, a)$.

Compute:

1. $(2, 0) \cdot (3, 1)$

2. $(-1, -1) \cdot (1, 1)$

3. $(4, 1) \cdot (-1, 4)$

4. $(2, 2) \cdot (-1, 2)$

5. $(0, 3) \cdot (1, 1)$

6. $(2, 3) \cdot (-3, -1)$.

Compute in two ways:

7. $(2, 3) \cdot [(1, 4) + (2, 2)]$

8. $(-1, 2) \cdot [(1, 3) + (1, 4)]$

9. Prove $\|\mathbf{v} + \mathbf{w}\|^2 = \|\mathbf{v}\|^2 + 2\mathbf{v} \cdot \mathbf{w} + \|\mathbf{w}\|^2$

10. Prove $\|\mathbf{v} + \mathbf{w}\|^2 + \|\mathbf{v} - \mathbf{w}\|^2 = 2\|\mathbf{v}\|^2 + 2\|\mathbf{w}\|^2$

11. Prove $\|\mathbf{v} + \mathbf{w}\|^2 - \|\mathbf{v} - \mathbf{w}\|^2 = 4\mathbf{v} \cdot \mathbf{w}$

12. Prove $|\mathbf{v} \cdot \mathbf{w}| \leq \|\mathbf{v}\| \cdot \|\mathbf{w}\|$.

Compute the smallest positive angle between the vectors:

13. $(3, 4), (-4, 3)$

14. $(1, -2), (-2, 1)$

15. $(1, 3), (3, 1)$

16. $(1, -3), (3, 6)$

17. $(1, 1), (-1, 7)$

18. $(3, 4), (-5, -2)$.

19. Prove that the diagonals of a rhombus are perpendicular.

20. Prove Schwarz's Inequality: If a_1, a_2, b_1, b_2 are any real numbers, then $(a_1 a_2 + b_1 b_2)^2 \leq (a_1^2 + a_2^2)(b_1^2 + b_2^2)$. [Hint: Interpret in terms of lengths and inner products.]

1. A ladder leans against a wall making a 70° angle with the ground. If its foot is 5 ft from the wall, how long is the ladder?

2. Compute the angles of a triangle whose sides are 5, 6, and 8.

3. Given $\alpha = 31°$, $\beta = 51°$, and $b = 12.3$, compute a and c.

4. Given $a = 4$, $b = 5$, and $\beta = 23.6°$, compute c.

1. What is the height of a flagpole whose shadow measures 37 ft when the angle of elevation of the sun is 54.5°?

2. Given $\beta = 37°$, $\gamma = 66°$, and $c = 100$, compute a and b.

3. Given $a = 12$, $b = 15$, and $\gamma = 74.2°$, compute α and c.

4. Given $a = 622.7$, $b = 833.1$, and $\beta = 32.4°$, compute c.

TOPICS IN ALGEBRA

In arithmetic, we often reduce an "improper" fraction (rational number) to the sum of an integer and a "proper" fraction. For example,

$$\tfrac{6}{5} = 1 + \tfrac{1}{5}, \qquad \tfrac{19}{7} = 2 + \tfrac{5}{7}, \qquad \tfrac{31}{4} = 7 + \tfrac{3}{4}, \qquad \text{etc.}$$

There is an analogous reduction for a rational function if the degree of its numerator is greater than or equal to the degree of its denominator. For example,

$$\frac{x^2 - 3x + 1}{x} = x - 3 + \frac{1}{x},$$

$$\frac{x^3 + 1}{x^2 + 1} = x + \frac{-x + 1}{x^2 + 1},$$

$$\frac{x^4}{x^2 - 1} = x^2 + 1 + \frac{1}{x^2 - 1}.$$

In each case, the top-heavy rational function on the left is reduced to a polynomial plus a bottom-heavy rational function with the same denominator as the given rational function.

How do we carry out such a reduction? Consider how we do it for an improper fraction, say $\tfrac{19}{7}$. We divide 19 by 7; the quotient is 2 and the remainder is 5:

$$
\begin{array}{r}
2 \\
7 \overline{)\, 19} \\
14 \\
\hline
5
\end{array}
$$

Thus $\frac{19}{7} = 2 + \frac{5}{7}$. Note that the remainder $\frac{5}{7}$ has the same denominator as the given number.

The same idea applies to a rational function $f(x)/g(x)$ where $\deg f \geq \deg g$. Subtract from $f(x)$ a suitable expression $ax^k g(x)$ so the difference has degree less than $\deg f$. Do the same thing to the difference and keep doing it until the degree is less than $\deg g$; then stop.

Example:

$$f(x) = x^3 + 1, \quad g(x) = x^2 + 1.$$

$$f(x) - xg(x) = -x + 1, \quad \text{stop!}$$

$$f(x) = xg(x) + (-x + 1),$$

$$\frac{f(x)}{g(x)} = x + \frac{-x+1}{g(x)}.$$

Example:

$$f(x) = 3x^4 - 1, \quad g(x) = x^2 + 2.$$

$$f(x) - 3x^2 g(x) = -6x^2 - 1,$$

$$-6x^2 - 1 = -6g(x) + 11, \quad \text{stop!}$$

$$f(x) = 3x^2 g(x) + (-6x^2 - 1) = 3x^2 g(x) - 6g(x) + 11,$$

$$\frac{f(x)}{g(x)} = 3x^2 - 6 + \frac{11}{g(x)}.$$

We could write these examples in the usual long division form, showing all the work:

$$
\begin{array}{r}
x \phantom{{}+1} \\
x^2 + 1 \overline{) x^3 \phantom{{}+xx} + 1} \\
x^3 + x \phantom{{}+1} \\
\hline
-x + 1
\end{array}
\qquad
\begin{array}{r}
3x^2 - 6 \phantom{{}+1} \\
x^2 + 2 \overline{) 3x^4 \phantom{{}+6xxxx} - 1} \\
3x^4 + 6x^2 \phantom{{}-1} \\
\hline
-6x^2 - 1 \\
-6x^2 - 12 \\
\hline
11
\end{array}
$$

The above are examples of the following basic property of polynomials, called the **division algorithm**:

Let $f(x)$ and $g(x)$ be polynomials, with $g(x) \neq 0$. Then there are unique polynomials $q(x)$ and $r(x)$ such that

(1) $f(x) = g(x)q(x) + r(x),$

(2) $r(x) = 0$ or $\deg r(x) < \deg g(x).$

We call $q(x)$ the **quotient** and $r(x)$ the **remainder** resulting from the division of $f(x)$ by $g(x)$. (A complete proof of the division algorithm requires mathematical induction and is left as an exercise, p. 369.)

Examples:

1. $f(x) = x^3 + 1, \quad g(x) = x^2 + 1,$

$x^3 + 1 = (x^2 + 1)(x) + (-x + 1), \quad q(x) = x, \quad r(x) = -x + 1.$

2. $f(x) = 3x^4 - 1, \quad g(x) = x^2 + 2,$

$3x^4 - 1 = (x^2 + 2)(3x^2 - 6) + 11, \quad q(x) = 3x^2 - 6, \quad r(x) = 11.$

Shortcuts

You don't need long division to find that

$$\frac{x^2 - 3x + 1}{x} = x - 3 + \frac{1}{x};$$

all you need do is write

$$\frac{x^2 - 3x + 1}{x} = \frac{x^2}{x} - \frac{3x}{x} + \frac{1}{x} = x - 3 + \frac{1}{x},$$

which you can do in your head anyway.

You don't need long division to compute the quotient

$$\frac{x + 3}{x + 1}.$$

If the numerator were $x + 1$, there would be no problem. So make it $x + 1$, then compensate by adding 2. In other words write $x + 3 = (x + 1) + 2$:

$$\frac{x + 3}{x + 1} = \frac{(x + 1) + 2}{x + 1} = \frac{x + 1}{x + 1} + \frac{2}{x + 1} = 1 + \frac{2}{x + 1}.$$

Here is another example where the same trick applies: $x^2/(x^2 + x + 1)$. If the numerator were $x^2 + x + 1$, all would be fine. Write

$$\frac{x^2}{x^2 + x + 1} = \frac{x^2 + x + 1 - (x + 1)}{x^2 + x + 1} = 1 - \frac{x + 1}{x^2 + x + 1}.$$

Further examples:

$$\frac{x^3}{x^2 + 2} = \frac{x^3 + 2x - 2x}{x^2 + 2} = \frac{x^3 + 2x}{x^2 + 2} - \frac{2x}{x^2 + 2} = x - \frac{2x}{x^2 + 2}.$$

$$\frac{x^4}{x^2 - 1} = \frac{x^4 - 1 + 1}{x^2 - 1} = \frac{x^4 - 1}{x^2 - 1} + \frac{1}{x^2 - 1} = x^2 + 1 + \frac{1}{x^2 - 1}.$$

$$\frac{x^5}{x^2 + 1} = \frac{(x^5 + x^3) - (x^3 + x) + x}{x^2 + 1} = x^3 - x + \frac{x}{x^2 + 1}.$$

Synthetic Division

Here is a full-fledged long division problem in complete detail:

$$
\begin{array}{r}
3x^3 + x^2 + 4x + 11 \\
x^2 - 2x - 3\,\overline{\big)\,3x^5 - 5x^4 - 7x^3 \qquad\qquad - 25x - 15} \\
3x^5 - 6x^4 - 9x^3 \\
\hline
x^4 + 2x^3 \\
x^4 - 2x^3 - 3x^2 \\
\hline
4x^3 + 3x^2 - 25x \\
4x^3 - 8x^2 - 12x \\
\hline
11x^2 - 13x - 15 \\
11x^2 - 22x - 33 \\
\hline
9x + 18.
\end{array}
$$

Surely we can find some way to shorten this calculation, just to write less. The powers of x are dispensable, provided we keep their coefficients separated by spaces. The result is

$$
\begin{array}{r}
3 \quad 1 \quad 4 \quad 11 \\
1 \quad -2 \quad -3\,\overline{\big)\,3 \quad -5 \quad -7 \quad 0 \quad -25 \quad -15} \\
3 \quad -6 \quad -9 \\
\hline
1 \quad 2 \quad 0 \\
1 \quad -2 \quad -3 \\
\hline
4 \quad 3 \quad -25 \\
4 \quad -8 \quad -12 \\
\hline
11 \quad -13 \quad -15 \\
11 \quad -22 \quad -33 \\
\hline
9 \quad 18.
\end{array}
$$

This is certainly easier to write and easier to read. We can simplify slightly more by not writing 0, -25, and -15 the second time, but that is hardly worth it.

However, a considerable economy is possible if the divisor has the form $x - a$. For instance, suppose we divide the same polynomial by $x - 2$:

$$
\begin{array}{r}
3 \quad 1 \quad -5 \quad -10 \quad -45 \\
1 \quad -2\,\overline{\big)\,3 \quad -5 \quad -7 \quad 0 \quad -25 \quad -15} \\
3 \quad -6 \\
\hline
1 \quad -7 \\
1 \quad -2 \\
\hline
-5 \quad 0 \\
-5 \quad 10 \\
\hline
-10 \quad -25 \\
-10 \quad 20 \\
\hline
-45 \quad -15 \\
-45 \quad 90 \\
\hline
-105.
\end{array}
$$

The result $1 = (-5) - (-6)$ of the first subtraction is immediately copied above the line. Surely we don't have to write it three times, nor even twice. The same goes for the other subtractions. If we delete all repetitions and condense the notation somewhat, the result is

$$
\begin{array}{rrrrrrr}
 & & 3 & 1 & -5 & -10 & -45 \\
\hline
1 & -2 \,| \, 3 & -5 & -7 & 0 & -25 & -15 \\
 & & -6 & -2 & 10 & 20 & 90 \\
\cline{5-7}
 & & & & & & -105.
\end{array}
$$

We could also write the remainder -105 above the line instead of below, but that might create confusion. The final result is

$$3x^5 - 5x^4 - 7x^3 - 25x - 15 = (x - 2)(3x^4 + x^3 - 5x^2 - 10x - 45) - 105.$$

The shortcut process for doing the calculation is called **synthetic division.** Here is another example:

$$
\begin{array}{rrrrr}
 & & 1 & -5 & 7 \\
\hline
1 & 2\,|\,1 & -3 & -3 & 9 \\
 & & 2 & -10 & 14 \\
\cline{4-5}
 & & & & -5.
\end{array}
$$

This computation yields the identity

$$x^3 - 3x^2 - 3x + 9 = (x + 2)(x^2 - 5x + 7) - 5.$$

 Warning: Synthetic division applies *only* when the divisor is a linear polynomial of the form $x - a$.

The Remainder and Factor Theorems

Suppose a polynomial $f(x)$ is divided by the linear polynomial $x - a$. Then

$$f(x) = (x - a)q(x) + r(x),$$

where $r(x)$ is zero or a polynomial of degree zero, i.e., a constant. Thus

$$f(x) = (x - a)q(x) + c.$$

What is the constant c? Just set $x = a$ in the formula;

$$f(a) = 0 + c,$$

hence $c = f(a)$. Thus the value of a polynomial $f(x)$ for $x = a$ is the remainder when $f(x)$ is divided by $x - a$. This fact is called the Remainder Theorem.

Remainder Theorem If $f(x)$ is a polynomial and a is a real number, then $f(a)$ is the remainder when $f(x)$ is divided by $x - a$. Thus

$$f(x) = (x - a)q(x) + f(a),$$

where $q(x)$ is a polynomial.

■ *Example 1.1*

Find the remainder when $x^{17} + 3x^9 - 2$ is divided by $x + 1$.

SOLUTION In this example

$$f(x) = x^{17} + 3x^9 - 2 \quad \text{and} \quad a = -1,$$

so the remainder is $f(-1)$. But

$$f(-1) = (-1)^{17} + 3(-1)^9 - 2 = -1 - 3 - 2 = -6.$$

Answer -6.

. . . .

We deduce from the Remainder Theorem an important test for whether a is a root of the equation $f(x) = 0$. From the formula

$$f(x) = (x - a)q(x) + f(a)$$

we see that $f(a) = 0$ if and only if $f(x) = (x - a)q(x)$, that is, if and only if $x - a$ is a factor of $f(x)$.

> **Factor Theorem** If $f(x)$ is a polynomial, then a real number a is a root of the equation $f(x) = 0$ if and only if $x - a$ is a factor of $f(x)$.

It is particularly easy to apply the Factor Theorem by means of synthetic division.

■ *Example 1.2*

Show that -3 is a solution of $x^3 - 2x^2 - 13x + 6 = 0$.

SOLUTION Divide the cubic by $x + 3$:

$$
\begin{array}{r}
1 \quad -5 \quad 2 \\
\hline
1 \ \ 3|1 \quad -2 \quad -13 \quad 6 \\
 3 \quad -15 \quad 6 \\
\hline
 0.
\end{array}
$$

The remainder is 0, so $x + 3$ is a factor of $x^3 - 2x^2 - 13x + 6$ and -3 is a root of the equation.

EXERCISES

Divide $f(x)$ by $g(x)$; give quotient and remainder:

1. $f(x) = x + 3, g(x) = x - 1$
2. $f(x) = 2x + 1, g(x) = x + 2$
3. $f(x) = x + 1, g(x) = 2x - 1$
4. $f(x) = x, g(x) = 3x + 2$
5. $f(x) = x^2 - 2x + 2, g(x) = x - 3$
6. $f(x) = x^2 + x + 1, g(x) = x + 2$
7. $f(x) = 2x^2, g(x) = x^2 + 1$
8. $f(x) = 3x^2 + 2x + 5, g(x) = x^2 - x + 1$

9. $f(x) = 2x^3 + 2x + 1, g(x) = x - 1$
10. $f(x) = x^3 - 7x, g(x) = x + 2$
11. $f(x) = x^3, g(x) = x^2 + x$
12. $f(x) = x^3 + 2x, g(x) = x^2 + x + 1$
13. $f(x) = x^3 + x^2 + x + 1, g(x) = x^2 - 1$
14. $f(x) = 5x^3 - x^2 + 3x - 4, g(x) = x^2 + 2$
15. $f(x) = x^3, g(x) = 2x^3 - 4x + 1$
16. $f(x) = x^3, g(x) = x^4$
17. $f(x) = x^4 + 1, g(x) = x^2 + x + 1$
18. $f(x) = x^4 - 3x^3 + x^2 + 7x + 5, g(x) = x^2 + 1$
19. $f(x) = x^5, g(x) = x^2 + 1$
20. $f(x) = x^5 + x^4 - 7x + 2, g(x) = x^2 - 3x - 3.$

Find the remainder when $f(x)$ is divided by $g(x)$:

21. $f(x) = x^{10} + x^8, g(x) = x - 1$
22. $f(x) = x^{12} + x^{10} - x^5, g(x) = x$
23. $f(x) = x^{20} + 4x^{10} + 1, g(x) = x + 1$
24. $f(x) = x^{2n} + x^n + 1, g(x) = x - 1$
25. $f(x) = x^{100}, g(x) = x^2 - 1$
26. $f(x) = x^{50} + x^{25}, g(x) = x^2 + 1$. Hint: $x^{25} = x[(x^2 + 1) - 1]^{12}.$

Find $f(a)$ by synthetic division:

27. $f(x) = x^3 - x - 1, a = 2$
28. $f(x) = 4x^3 - 2x^2 + 3, a = -2$
29. $f(x) = x^4 + x^3 - 2x^2 + x + 1, a = 3$
30. $f(x) = 2x^4 + 3x^3 + 4x^2 + 5x + 6, a = -2$
31. $f(x) = x^3 - x^2 + 3x + 4, a = 5$
32. $f(x) = x^5 + 5x + 1, a = -4.$

Use synthetic division to decide whether -4 is a solution of the equation:

33. $2x^3 + 14x^2 - 17x - 24 = 0$
34. $x^4 - 4x^3 - 31x^2 + 5x + 4 = 0$
35. $3x^5 + 11x^4 - 4x^3 + 2x^2 + 10x + 8 = 0$
36. $x^6 + 4x^5 + 2x^4 + 9x^3 + 4x^2 + x + 52.$

37. Show that $x^{61} - x^{12} - x^4 + 1$ is divisible by $x - 1$.
38. Show that $x^6 - 4x^4 + 7x^3 - 10x^2 - x - 14$ is divisible by $x - 2$.

2. ZEROS OF POLYNOMIALS

A **zero** of a function $f(x)$ is a real number r such that $f(r) = 0$. As we know, the number r is also called a **root** of the equation $f(x) = 0$.

Graphically, a zero of any function $f(x)$ is simply a value of x where the graph $y = f(x)$ meets the x-axis (Fig. 2.1 on next page). The function shown in Fig. 2.1 has three zeros, r_1, r_2, and r_3. A function need not have any real zeros however. For example, $f(x) = x^2 + 1$ is positive for every real x, hence $f(x)$ has no real zeros. Its graph never crosses the x-axis (Fig. 2.2 on next page).

In this section we shall study the zeros of polynomial functions. Knowledge of the zeros usually provides considerable knowledge of the function.

We begin with the linear function $f(x) = ax + b$ where $a \neq 0$. We expect $f(x)$ to have exactly one real zero, since the graph is a non-horizontal straight line (Fig. 2.3 on next page). Let us verify our suspicion algebraically. If r is a zero of $f(x) = ax + b$, then

$$f(r) = ar + b = 0,$$
$$r = -b/a.$$

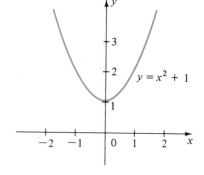

Fig. 2.1   Zeros of $f(x)$ **Fig. 2.2**   A function without real zeros

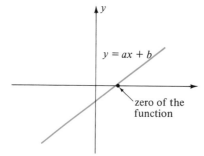

Fig. 2.3   A linear function has one zero.

Thus the only possible zero is $-b/a$. But this indeed is a zero since

$$f(-b/a) = a(-b/a) + b = -b + b = 0.$$

We note one other fact:

$$f(x) = ax + b = a\left(x + \frac{b}{a}\right) = a(x - r).$$

Each linear function $f(x) = ax + b$ has exactly one real zero: $r = -b/a$. We can write

$$f(x) = a(x - r).$$

Quadratic Polynomials

We next investigate the zeros of a quadratic polynomial $f(x) = ax^2 + bx + c$. We know what the graph $y = f(x)$ looks like from Chapter 3, Section 6. The question is whether, and how often, the graph meets the x-axis. There are three cases (Fig. 2.4).

The geometry indicates either two zeros, one zero, or no zeros. Our previous study (Chapter 2, Section 3) by algebra confirms this. Indeed, if D denotes the discriminant,

$$D = b^2 - 4ac,$$

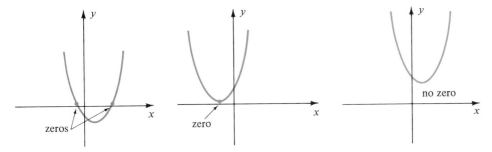

Fig. 2.4   Zeros of $f(x) = ax^2 + bx + c$ for $a > 0$

then the three cases are $D > 0$ (two real zeros), $D = 0$ (one real zero), and $D < 0$ (no real zeros).

Let us look in detail at the case $D > 0$ of two real zeros. By the quadratic formula, the zeros are

$$r_1 = \frac{-b + \sqrt{D}}{2a} \quad \text{and} \quad r_2 = \frac{-b - \sqrt{D}}{2a}.$$

There are simple formulas for the sum and the product of the two zeros:

$$r_1 + r_2 = \frac{-b}{a}, \quad r_1 r_2 = \frac{b^2 - D}{4a^2} = \frac{b^2 - b^2 + 4ac}{4a^2} = \frac{c}{a}.$$

We exploit these facts by observing that

$$f(x) = ax^2 + bx + c = a\left(x^2 + \frac{b}{a}x + \frac{c}{a}\right).$$

Therefore

$$f(x) = a[x^2 - (r_1 + r_2)x + r_1 r_2] = a(x - r_1)(x - r_2).$$

This factorization of $f(x)$ is consistent with the Factor Theorem, which says that $f(x)$ is divisible by $x - r_1$ and $x - r_2$.

If $D = 0$, the same reasoning shows that $f(x) = a(x - r)^2$, where $r = -b/2a$.

This completes our study of the zeros of quadratic polynomials. We summarize the situation:

$$f(x) = ax^2 + bx + c, \quad a \neq 0, \quad D = b^2 - 4ac$$

$D < 0$ f has no real zeros.

$D = 0$ f has exactly one real zero, $r = -b/2a$, and $f(x) = a(x - r)^2$.

$D > 0$ f has exactly two real zeros,

$$r_1 = \frac{-b + \sqrt{D}}{2a} \quad \text{and} \quad r_2 = \frac{-b - \sqrt{D}}{2a},$$

and $f(x) = a(x - r_1)(x - r_2)$.

Higher Degree Polynomials

We have shown by direct calculation that if r is a zero of a linear or quadratic polynomial $f(x)$, then $x - r$ is a factor. By the Factor Theorem, the same is true for polynomials of any degree. Let us restate this fact in the present context.

> If r is a zero of $f(x) = a_n x^n + a_{n-1} x^{n-1} + \cdots + a_1 x + a_0$, then $x - r$ is a factor of $f(x)$. That is, there is a polynomial $g(x)$ of degree $n - 1$ such that
> $$f(x) = (x - r)g(x).$$

The result has practical application. Suppose you need all zeros of $f(x)$ and you have found one of them, r. Then write $f(x) = (x - r)g(x)$. Finding further zeros of $f(x)$ is exactly the same as finding zeros of $g(x)$, a polynomial of *lower degree*. Clearly it is easier to handle $g(x)$ than $f(x)$.

This line of reasoning yields an important corollary concerning the number of zeros a polynomial can have.

> A polynomial function of degree n has at most n distinct zeros.

For suppose $f(x) = a_n x^n + \cdots + a_0$. If r_1 is a zero of $f(x)$, then $f(x) = (x - r_1)g(x)$, where $g(x)$ has degree $n - 1$. If r_2 is another zero of $f(x)$, then
$$f(r_2) = (r_2 - r_1)g(r_2) = 0.$$
Since $r_2 - r_1 \neq 0$, we conclude that $g(r_2) = 0$; in other words, r_2 is a zero of $g(x)$. It follows that $g(x) = (x - r_2)h(x)$, where $h(x)$ has degree $n - 2$. Thus
$$f(x) = (x - r_1)(x - r_2)h(x).$$
If r_3 is yet another zero of f, then
$$f(r_3) = (r_3 - r_1)(r_3 - r_2)h(r_3) = 0.$$
Since $r_3 \neq r_1$ and $r_3 \neq r_2$, we have $(r_3 - r_1)(r_3 - r_2) \neq 0$, and consequently $h(r_3) = 0$. Hence, $h(x) = (x - r_3)k(x)$, where $k(x)$ has degree $n - 3$. Thus
$$f(x) = (x - r_1)(x - r_2)(x - r_3)k(x).$$

Keep going in this way. Each new zero r_i yields a new factor $(x - r_i)$ and lowers the degree of the quotient. Since the degree can be lowered at most n times, there are at most n zeros.

Remark: A straight line is determined by two points. Hence a polynomial $f(x)$ of degree 1 (whose graph is a straight line) is determined by its values for two distinct values of x. Now we can assert much more: a polynomial $f(x)$ of degree n is determined by its values at $n + 1$ distinct values of x.

> Suppose $f(x)$ and $g(x)$ are polynomials, each of degree at most n. If $f(x) = g(x)$ for $n + 1$ distinct values of x, then $f(x)$ and $g(x)$ are identical.

For the proof, just look at the difference $d(x) = f(x) - g(x)$. Then $d(x) = 0$ for $n + 1$ distinct values of x. But $d(x)$ is either the zero polynomial or a non-zero polynomial of degree at most n. The latter case is impossible since a polynomial of degree n can have at most n zeros. Therefore $d(x)$ is the zero polynomial (all coefficients zero), i.e., $f(x)$ and $g(x)$ are identical.

Multiplicity

Consider the polynomial

$$f(x) = x^2(x + 1)^3(x - 2)(x^2 + 1).$$

The factor $x^2 + 1$ has no real zeros, hence the *distinct* real zeros of $f(x)$ are 0, -1, 2. The factor $x + 1$, corresponding to the zero -1, occurs in $f(x)$ exactly three times. The factor $x = x - 0$, corresponding to the zero 0, occurs exactly twice; the factor $x - 2$, corresponding to the zero 2, occurs exactly once.

In general, let $r_1, r_2, \cdots, r_k$ be the distinct real zeros (if any) of a polynomial $f(x)$. Then

$$f(x) = (x - r_1)^{m_1}(x - r_2)^{m_2} \cdots (x - r_k)^{m_k} g(x),$$

where the factor $g(x)$ is a polynomial with no real zeros, possibly just a constant. In case $f(x)$ has no real zeros, then $f(x) = g(x)$. Otherwise each m_i is a positive integer called the **multiplicity** of the zero r_i. If $m_i = 1$, then r_i is called a **simple zero**. If $m_i > 1$, then r_i is called a **multiple zero**. For example if $f(x) = x^2(x + 1)^3(x - 2)(x^2 + 1)$, then 2 is a simple zero, 0 is a zero of multiplicity 2 (a double zero), and -1 is a zero of multiplicity 3 (a triple zero).

Existence of Zeros

We have proved that a polynomial function of degree n has *at most* n real zeros, that is, the maximum possible number of zeros is n. But this does not assert that any real zeros exist; there may be none at all. For example, $f(x) = x^2 + 1$ and $f(x) = (x^2 + 1)(x^2 + 4)$ are positive for all x, hence they have no real zeros. The cubic $f(x) = x(x^2 + 1)$ has exactly one real zero, $r = 0$, less than the maximum possible number of real roots of a cubic, three.

All we can say for sure about the existence of zeros is that a polynomial of *odd* degree has at least one real zero.

> Each polynomial function of odd degree has at least one real zero.

This fact follows from behavior of a polynomial as $x \longrightarrow \infty$ and as $x \longrightarrow -\infty$. When $|x|$ is large, $f(x) = a_n x^n + \cdots + a_0$ behaves very much like its leading term $a_n x^n$. (See the remark below.) If n is odd, then $a_n x^n \longrightarrow \infty$ as $x \longrightarrow \infty$ and $a_n x^n \longrightarrow -\infty$ as $x \longrightarrow -\infty$ (or vice versa, depending on the sign of a_n). Hence the same is true of $f(x)$. That means $f(x)$ somehow changes from negative to positive (or vice versa) as x increases from very large negative to very large positive. It can do this only by taking on the value 0 at least once (Fig. 2.5 on next page).

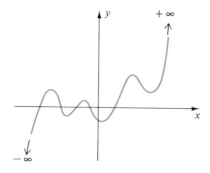

Fig. 2.5 Polynomial of odd degree

Remark: To study behavior of the general polynomial as $x \longrightarrow \infty$ or $x \longrightarrow -\infty$, we write

$$f(x) = a_n x^n + a_{n-1} x^{n-1} + \cdots + a_1 x + a_0 = a_n x^n \left(1 + \frac{a_{n-1}}{a_n x} + \cdots + \frac{a_1}{a_n x^{n-1}} + \frac{a_0}{a_n x^n} \right).$$

For $|x|$ very large, the quantity in parentheses is very close to 1, so $f(x)$ is about the size of the term $a_n x^n$. Hence for $|x|$ very large, the graph of $y = f(x)$ is pretty much like the graph of $y = a_n x^n$; it zooms up (or down) as $x \longrightarrow \pm\infty$.

Location of Zeros

We mention a few facts that are useful in pinning down the real zeros of polynomials.

> Let $f(x)$ be a polynomial and suppose x_0 and x_1 are numbers for which $f(x_0)$ and $f(x_1)$ have opposite signs. Then there is a zero of $f(x)$ between x_0 and x_1.

A complete proof of this statement requires the theory of continuous functions, but the idea is simple. The smooth graph of $y = f(x)$ somehow gets from $(x_0, f(x_0))$ to $(x_1, f(x_1))$. One of these points is above the x-axis and one is below, so the graph must cross the x-axis as x goes from x_0 to x_1; the point of crossing is a zero of $f(x)$.

■ *Example 2.1*

Given $f(x) = x^3 + 2x^2 - 7x + 1$. Show that $f(x)$ has a zero between 0 and 1. Locate two other zeros between consecutive integers.

SOLUTION By direct computation, $f(0) = 1$ and $f(1) = -3$. Since the signs are opposite, $f(x)$ has a zero between 0 and 1.

When x is large, $f(x)$ is surely positive (look at $x = 10$ for example). But $f(1) < 0$; hence $f(x)$ has a zero r with $r > 1$. When x is large negative, -10 for example, $f(x)$ is negative. But $f(0) > 0$; hence there is a zero r with $r < 0$.

To locate these zeros, compute $f(x)$ for $x = 2, 3, 4, \cdots$ and for $x =$

$-1, -2, -3, \cdots$ until a change of sign indicates a zero.

x	-4	-3	-2	-1	0	1	2
$f(x)$	-3	13	15	9	1	-3	3

Answer $-4 < r_1 < -3,\;\; 0 < r_2 < 1,\;\; 1 < r_3 < 2.$

Sometimes it is useful to know whether a polynomial has any zeros that are rational numbers. The following criterion is proved in courses on modern algebra.

> Let $f(x) = a_0 x^n + a_1 x^{n-1} + \cdots + a_n$ have *integer* coefficients with $a_0 \neq 0$. Then each *rational* zero of $f(x)$ has the form $r = p/q$, where p and q are integers, p divides a_n and q divides a_0.

In particular, if $a_0 = 1$, then the integer q divides 1. Hence $q = \pm 1$ and p/q is an *integer*.

> Let $f(x) = x^n + a_1 x^{n-1} + \cdots + a_n$ have *integer* coefficients and leading coefficient one. Then each *rational* zero of $f(x)$ is an *integer* which divides a_n.

■ *Example 2.2*

Find all rational zeros of

$$f(x) = 3x^3 + 7x^2 + x - 4.$$

SOLUTION   If $r = p/q$ is a rational zero, then p must divide -4 and q must divide 3. The possibilities are: $p = \pm 1, \pm 2, \pm 4$ and $q = \pm 1, \pm 3$. Hence the only possible rational zeros are $r = \pm 1, \pm 2, \pm 4, \pm\frac{1}{3}, \pm\frac{2}{3}, \pm\frac{4}{3}$.

For each of these 12 values we must check whether $f(r) = 0$. Actually it is easy to eliminate a number of cases. For instance in computing $f(1)$, we see not only that $f(1) > 0$ but that $f(x) > 0$ for all $x > 1$. Hence, we can rule out the values $\frac{4}{3}, 2, 4$. By similar reasoning we can eliminate -2 and -4.

Now we must test the remaining possibilities. We find $f(-\frac{4}{3}) = 0$, but $f(x) \neq 0$ in all other cases.

If we are lucky enough to try $-\frac{4}{3}$ *first*, then we can reduce the computations considerably. For then we know $f(x)$ is divisible by $x + \frac{4}{3}$. By division we find

$$f(x) = 3(x + \tfrac{4}{3})(x^2 + x - 1).$$

The remaining zeros of $f(x)$ are zeros of the quadratic factor $x^2 + x - 1$. These are $\frac{1}{2}(-1 \pm \sqrt{5})$, not rational.

Answer $f(x)$ has one rational zero, $-\frac{4}{3}$.

1. Find the most general linear function with $x = 3$ as a zero.
2. Find the most general quadratic function with $x = 2$ as a zero.
3. Find a quadratic function $f(x)$ with zeros -1 and 3, and $f(0) = 6$.
4. Find the most general quadratic function whose graph is tangent to the x-axis at $(2, 0)$.
5. Find the most general quadratic function, the sum of whose zeros is 5.
6. Find the most general quadratic function, the product of whose zeros is 4.
7. Prove that the graph of a linear function intersects each horizontal line exactly once.
8. Prove that the graph of a quadratic function intersects each horizontal line at most twice.
9. Let f be a linear function and g a quadratic function. Prove that their graphs intersect in at most two points.
10. Let f and g be quadratics. Prove that their graphs either coincide, or have at most two points in common.
11. Prove that if a and c have opposite signs, then $ax^2 + bx + c = 0$ has two real roots.
12. Prove that if $a > 0$, $b > 0$, and $c > 0$, then $ax^2 + bx + c$ has no positive zero.
13. Prove that if $a > 0$, $b < 0$, and $c > 0$, then $ax^2 + bx + c$ has no negative zero.
14. For what values of b does $x^2 + bx + b^2$ have a real zero?

Find the real zeros and their multiplicities:

15. $(2x + 1)(x^2 + 1)$ 16. $(x - 3)^2(x^2 + 2)$ 17. $(x^2 - 4)^2(x^2 + 4)$
18. $(x^2 - 5x + 6)(x^2 + 1)^2$ 19. $(x^2 + x - 2)(x^2 - 1)^2$ 20. $x^3(x^2 - 4)^2(x^2 - 3)$.

21. Prove that the graph of a cubic function intersects each horizontal line at most three times.
22. Prove that the graph of a polynomial of degree 10 intersects each non-vertical line at most 10 times.
23. Let f be linear and g cubic. Prove that their graphs have at most three points of intersection.
24. Let f be quadratic and g cubic. Prove that their graphs have at most three points in common.
25. Find a polynomial $f(x)$ of degree 6 such that $f(x) \geq 0$ for all x and $f(1) = f(2) = f(3) = 0$.
26. Does there exist a real number that exceeds its cube by one?
27. Does there exist a real number that exceeds its 5-th power by 1000?
28. Is there a real number whose 9-th power is 4 less than its square?
29. Find a cubic polynomial $f(x)$ for which $f(0) = f(5) = f(8) = 0$ and $f(10) = 17$.
30*. If $f(x)$ is a polynomial having only even powers of x and if $f(r) = 0$, prove that $f(x)$ is divisible by $x^2 - r^2$.

Find all rational zeros:

31. $x^3 + x^2 + 2x + 2$ 32. $x^3 + 3x^2 + 2x + 6$
33. $x^4 - x^3 - x^2 - x - 2$ 34. $x^6 + 2x^4 + x^2 + 2$
35. $3x^3 - 2x^2 + 3x - 2$ 36. $2x^3 + 3x^2 + 4x + 6$
37. $2x^4 + 3x^3 + 8x^2 + 3x - 4$ 38. $3x^4 + 5x^3 + x^2 + 5x - 2$.

Given that $f(x)$ has exactly one real zero, locate it between two integers:

39. $f(x) = 12x^3 - 28x^2 - 7x - 10$ 40. $f(x) = x^3 + 4x^2 + 10x + 15$.

41. Show that $f(x) = x^5 + x^2 - 7$ has a real zero between 1 and 2 but no zeros greater than 2.
42. If $f(x)$ is a polynomial with positive leading coefficient and if $f(0) < 0$, show that $f(x)$ has a real zero.
43. If $f(x)$ is a polynomial whose coefficients are all positive, show that $f(x)$ has no positive zeros.
44. If $f(x)$ is a polynomial whose coefficients alternate in sign, show that $f(x)$ has no negative zeros.
45*. Let $f(x) = x^n + a_1 x^{n-1} + \cdots + a_n$ have zeros $r_1, r_2, \cdots, r_n$. Prove that $r_1 + r_2 + \cdots + r_n = -a_1$.
46*. (cont.) Prove that $r_1 r_2 \cdots r_n = (-1)^n a_n$.

3. PARTIAL FRACTIONS

In many problems, it is convenient to decompose a rational function into a sum of simpler functions. We shall describe how this is done in some simple cases.

By long division (if necessary) we can reduce any rational function to the sum of a polynomial and a bottom-heavy rational function. For instance,

$$\frac{x^3 - 2x}{x^2 + x + 1} = x - 1 + \frac{-2x + 1}{x^2 + x + 1}.$$

Therefore, we can concentrate on bottom-heavy rational functions.

Consider the special case in which the denominator is a quadratic polynomial with distinct real roots. The typical bottom-heavy function of this type is

$$\frac{ax + b}{(x - r)(x - s)}, \qquad r \neq s.$$

Such a function can always be split into two simpler functions, as we shall see.

Examples:

$$\frac{1}{(x + 1)(x - 1)} = \frac{-\frac{1}{2}}{x + 1} + \frac{\frac{1}{2}}{x - 1}.$$

$$\frac{x}{(x + 1)(x + 2)} = \frac{-1}{x + 1} + \frac{2}{x + 2}.$$

$$\frac{2x - 3}{x(x + 1)} = \frac{-3}{x} + \frac{5}{x + 1}.$$

In general,

$$\frac{ax + b}{(x - r)(x - s)} = \frac{A}{x - r} + \frac{B}{x - s}$$

for suitable constants A and B. This expression is called the **partial fraction decomposition** of the given rational function.

To find A and B, clear of fractions:

$$ax + b = A(x - s) + B(x - r).$$

If the linear functions $ax + b$ and $A(x - s) + B(x - r)$ are equal, they are equal for all values of x. Substitute $x = r$ and $x = s$:

$$ar + b = A(r - s) + 0, \qquad as + b = 0 + B(s - r).$$

It follows that

$$A = \frac{ar + b}{r - s}, \qquad B = \frac{as + b}{s - r}.$$

■ **Example 3.1**

Decompose into partial fractions $\dfrac{3x + 1}{(x + 1)(x - 2)}$.

SOLUTION   Write

$$\frac{3x + 1}{(x + 1)(x - 2)} = \frac{A}{x + 1} + \frac{B}{x - 2},$$

$$3x + 1 = A(x - 2) + B(x + 1).$$

Set $x = 2$, then $x = -1$:

$$7 = 3B, \qquad -2 = -3A.$$

Hence

$$A = \tfrac{2}{3}, \qquad B = \tfrac{7}{3}.$$

Answer $\dfrac{\tfrac{2}{3}}{x + 1} + \dfrac{\tfrac{7}{3}}{x - 2}.$

Check:

$$\frac{\tfrac{2}{3}}{x + 1} + \frac{\tfrac{7}{3}}{x - 2} = \frac{\tfrac{2}{3}(x - 2) + \tfrac{7}{3}(x + 1)}{(x + 1)(x - 2)} = \frac{3x + 1}{(x + 1)(x - 2)}.$$

■ **Example 3.2**

Decompose into partial fractions $\dfrac{-x^2}{(x + 2)(x + 3)}$.

SOLUTION   The fraction is top-heavy, so first divide, then factor the denominator $x^2 + 5x + 6 = (x + 2)(x + 3)$:

$$\frac{-x^2}{x^2 + 5x + 6} = \frac{-x^2 - 5x - 6 + 5x + 6}{x^2 + 5x + 6} = -1 + \frac{5x + 6}{x^2 + 5x + 6}$$

$$= -1 + \frac{5x + 6}{(x + 2)(x + 3)}.$$

Now proceed:

$$\frac{5x + 6}{(x + 2)(x + 3)} = \frac{A}{x + 2} + \frac{B}{x + 3},$$

$$5x + 6 = A(x + 3) + B(x + 2).$$

Set $x = -3$, then $x = -2$:

$$-9 = -B, \qquad -4 = A.$$

Hence $A = -4$ and $B = 9$.

$$\textit{Answer} \quad -1 + \frac{-4}{x+2} + \frac{9}{x+3}.$$

. . . .

Denominator in Factored Form

We can handle denominators of any degree by a similar procedure, *provided* the denominator has the factored form $(x - r_1) \cdots (x - r_n)$, where $r_1, r_2, \cdots, r_n$ are distinct real numbers.

> If $r_1, r_2, \cdots, r_n$ are distinct real numbers and $\deg f(x) \leq n - 1$, then there are constants $A_1, \cdots, A_n$ such that
>
> $$\frac{f(x)}{(x - r_1) \cdots (x - r_n)} = \frac{A_1}{x - r_1} + \cdots + \frac{A_n}{x - r_n}.$$

■ *Example 3.3*

Decompose into partial fractions $\dfrac{1}{(x-1)(x-2)(x-3)}$.

SOLUTION Write

$$\frac{1}{(x-1)(x-2)(x-3)} = \frac{A}{x-1} + \frac{B}{x-2} + \frac{C}{x-3}.$$

Clear of fractions, i.e., multiply by $(x-1)(x-2)(x-3)$:

$$1 = A(x-2)(x-3) + B(x-1)(x-3) + C(x-1)(x-2).$$

Substitute the special values $x = 1$, $x = 2$, $x = 3$:

$$1 = 2A, \quad 1 = -B, \quad 1 = 2C.$$

Hence $A = \frac{1}{2}$, $B = -1$, $C = \frac{1}{2}$.

$$\textit{Answer} \quad \frac{\frac{1}{2}}{x-1} + \frac{-1}{x-2} + \frac{\frac{1}{2}}{x-3}.$$

Check:

To avoid fractions, let us check that twice the answer equals twice the given rational function:

$$\frac{1}{x-1} - \frac{2}{x-2} + \frac{1}{x-3} = \frac{(x-2)(x-3) - 2(x-1)(x-3) + (x-1)(x-2)}{(x-1)(x-2)(x-3)}$$

$$= \frac{(x^2 - 5x + 6) - 2(x^2 - 4x + 3) + (x^2 - 3x + 2)}{(x-1)(x-2)(x-3)}$$

$$= \frac{2}{(x-1)(x-2)(x-3)}.$$

. . . .

Other Denominators

It is somewhat complicated to formulate the most general partial fraction decomposition, so we shall only consider several special cases.

Case 1: $\dfrac{f(x)}{(x-a)^n}$, $\deg f(x) \leq n-1$.

The partial fraction decomposition of this rational function is

$$\frac{f(x)}{(x-a)^n} = \frac{A_0}{(x-a)^n} + \frac{A_1}{(x-a)^{n-1}} + \cdots + \frac{A_{n-1}}{x-a},$$

where $A_0, A_1, \cdots, A_{n-1}$ are constants. To find the constants, clear of fractions:

$$f(x) = A_0 + A_1(x-a) + \cdots + A_{n-1}(x-a)^{n-1}.$$

Thus if $f(x)$ is expressed as a polynomial in powers of $x - a$ rather than x, then its coefficients are the numerators in the partial fraction decomposition. But that is easy to do: in the polynomial $f(x)$, just replace x everywhere by $a + (x - a)$. Then expand and collect powers of $x - a$.

■ *Example 3.4*

Decompose into partial fractions $\dfrac{3 + 4x - x^2}{(x-1)^3}$.

SOLUTION To write the numerator as a polynomial in powers of $x - 1$, just replace x everywhere by $1 + (x - 1)$:

$$3 + 4x - x^2 = 3 + 4[1 + (x-1)] - [1 + (x-1)]^2$$
$$= 3 + 4 + 4(x-1) - 1 - 2(x-1) - (x-1)^2$$
$$= 6 + 2(x-1) - (x-1)^2.$$

Divide by $(x-1)^3$ for the answer.

Answer $\dfrac{6}{(x-1)^3} + \dfrac{2}{(x-1)^2} - \dfrac{1}{x-1}$.

- -

Case 2: $\dfrac{f(x)}{(x-a)(x^2+1)}$, $\deg f(x) \leq 2$.

The partial fraction decomposition in this case is

$$\frac{f(x)}{(x-a)(x^2+1)} = \frac{A}{x-a} + \frac{Bx+C}{x^2+1},$$

where A, B, and C are constants.

To find A, B, and C, we first clear of fractions:

$$f(x) = A(x^2+1) + (Bx+C)(x-a).$$

Now set $x = a$:

$$f(a) = A(a^2 + 1), \qquad A = \frac{f(a)}{a^2 + 1}.$$

With this value of A, consider the quadratic polynomial

$$f(x) - A(x^2 + 1).$$

It has a for a zero since

$$f(a) - A(a^2 + 1) = f(a) - f(a) = 0.$$

Therefore it is divisible by $x - a$, and the quotient is a linear polynomial:

$$f(x) - A(x^2 + 1) = (Bx + C)(x - a).$$

This determines B and C.

■ *Example 3.5*

Decompose into partial fractions $\dfrac{x^2 - 5x - 2}{(x - 3)(x^2 + 1)}$.

SOLUTION   Here $f(x) = x^2 - 5x - 2$, so

$$A = \frac{f(3)}{3^2 + 1} = \frac{9 - 15 - 2}{10} = \frac{-8}{10} = -\frac{4}{5}.$$

The polynomial

$$f(x) - A(x^2 + 1) = x^2 - 5x - 2 + \tfrac{4}{5}(x^2 + 1) = \tfrac{1}{5}(9x^2 - 25x - 6)$$

must be divisible by $x - 3$. Indeed,

$$\tfrac{1}{5}(9x^2 - 25x - 6) = \tfrac{1}{5}(x - 3)(9x + 2).$$

Thus

$$f(x) = -\tfrac{4}{5}(x^2 + 1) + \tfrac{1}{5}(x - 3)(9x + 2).$$

Divide by $(x - 3)(x^2 + 1)$ to get the answer.

$$\textit{Answer} \quad \frac{-\tfrac{4}{5}}{x - 3} + \frac{\tfrac{1}{5}(9x + 2)}{x^2 + 1}.$$

· ·

Case 3: $\dfrac{f(x)}{(x^2 + 1)^2}$, $\deg f(x) \le 3$.

The partial fraction decomposition is

$$\frac{f(x)}{(x^2 + 1)^2} = \frac{Ax + B}{x^2 + 1} + \frac{Cx + D}{(x^2 + 1)^2},$$

where A, B, C, and D are constants.

This case is similar to Case 1. First we clear of fractions:

$$f(x) = (Ax + B)(x^2 + 1) + (Cx + D).$$

To express $f(x)$ in this form, divide by $x^2 + 1$. The remainder is linear, $Cx + D$. But the quotient is also linear, $Ax + B$, since $\deg f(x) \leq 3$.

■ *Example 3.6*

Decompose into partial fractions $\dfrac{x^3 - 2x^2 - 3x + 1}{(x^2 + 1)^2}$.

SOLUTION   Divide $x^3 - 2x^2 - 3x + 1$ by $x^2 + 1$:

$$x^3 - 2x^2 - 3x + 1 = (x^2 + 1)(x - 2) - 4x + 3.$$

Now divide both sides of this relation by $(x^2 + 1)^2$ for the answer.

$$Answer \quad \frac{x - 2}{x^2 + 1} + \frac{-4x + 3}{(x^2 + 1)^2}.$$

EXERCISES

Decompose into partial fractions:

1. $\dfrac{2}{x(x + 4)}$

2. $\dfrac{x}{x^2 - 1}$

3. $\dfrac{-x}{(x + 1)(x + 2)}$

4. $\dfrac{3}{x(x - 3)}$

5. $\dfrac{x + 1}{x(x - 1)}$

6. $\dfrac{x - 1}{x(x + 1)}$

7. $\dfrac{2x + 1}{(x - 1)(x + 2)}$

8. $\dfrac{x + 3}{x(x - 3)}$

9. $\dfrac{2x - 3}{(x + 2)(x + 3)}$

10. $\dfrac{3x - 3}{(x - 2)(x - 5)}$

11. $\dfrac{1}{(2x + 1)(2x - 1)}$

12. $\dfrac{x}{(3x - 2)(3x + 2)}$

13. $\dfrac{x^2}{x^2 - 4}$

14. $\dfrac{x^3}{(x - 1)(x - 2)}$

15. $\dfrac{x^2}{x(x - 1)(x - 2)}$

16. $\dfrac{x}{(x + 1)(x + 2)(x + 3)}$

17. $\dfrac{x^2}{(x - 1)(x - 2)(x - 3)}$

18. $\dfrac{x^2 - 1}{x(x^2 - 4)}$

19. $\dfrac{x + 1}{(x - 2)(x - 3)(x - 4)}$

20. $\dfrac{x - 1}{(x + 2)(x + 3)(x + 4)}$

21. $\dfrac{2x^2 + 7x - 4}{x^3 - 4x}$

22. $\dfrac{x^2 - x + 10}{(x - 1)(2x - 1)(3x - 1)}$

23. $\dfrac{x}{(x + 1)^2}$

24. $\dfrac{x^2}{(x - 2)^3}$

25. $\dfrac{x^2 + 1}{(x - 1)^4}$

26. $\dfrac{x^3 - 2x + 1}{(x + 2)^4}$

27. $\dfrac{x^2}{(x + 3)(x^2 + 1)}$

28. $\dfrac{x^2 - 1}{(x - 2)(x^2 + 1)}$

29. $\dfrac{x - 2}{(x + 2)(x^2 + 4)}$

30. $\dfrac{x}{(x - 1)(x^2 + 5)}$

31. $\dfrac{x^3}{(x^2 + 1)^2}$

32. $\dfrac{-x^3 + 4}{(x^2 + 1)^2}$

33. $\dfrac{x^2}{(x^2 + 2)^2}$

34. $\dfrac{x^3 - x^2}{(x^2 + 3)^2}$.

4. SYSTEMS OF LINEAR EQUATIONS

In Chapter 2, Section 2, we discussed systems of two linear equations in two unknowns. You should review that material before continuing. Here we shall discuss the solution of linear systems of three equations in three unknowns.

We want to solve simultaneously three linear equations

$$\begin{cases} a_1 x + b_1 y + c_1 z = d_1 \\ a_2 x + b_2 y + c_2 z = d_2 \\ a_3 x + b_3 y + c_3 z = d_3 \end{cases}$$

for x, y, z, where $a_1, \cdots, d_3$ are real numbers, the **coefficients** of the system. We shall usually write the solution as a triple (x, y, z).

The general theory of systems of linear equations belongs to a subject called linear algebra. We shall be content to discuss a practical method of solution, the method of **elimination.**

■ *Example 4.1*

Find the solution (x, y, z) of the system

$$\begin{cases} 2x - y + z = 4 \\ 3y + 2z = -1 \\ -z = 3. \end{cases}$$

SOLUTION By the third equation, $z = -3$. Substitute this into the first two equations. The result is a new system of two equations for x and y:

$$\begin{cases} 2x - y = 4 - (-3) = 7 \\ 3y = -1 - 2(-3) = 5. \end{cases}$$

By the second equation, $y = \frac{5}{3}$. Substitute this into the first equation; the result is a single equation for x:

$$2x = 7 + \tfrac{5}{3} = \tfrac{26}{3}.$$

Its solution is $x = \frac{13}{3}$.

Answer $(\frac{13}{3}, \frac{5}{3}, -3)$.

. . . .

This example was very easy because we could solve for the unknowns one at a time. To solve a more general system, we reduce it to a system of similar type by eliminating the unknowns one by one.

■ *Example 4.2*

Find the solution of the system

$$\begin{cases} 2x - y + z = 4 \\ 2x + 2y + 3z = 3 \\ 6x - 9y - 2z = 17. \end{cases}$$

SOLUTION   Eliminate x from the second and third equations. Subtract the first equation from the second, and subtract 3 times the first equation from the third; the result is an equivalent system of three equations (the first the same as before):

$$\begin{cases} 2x - y + z = 4 \\ 3y + 2z = -1 \\ -6y - 5z = 5. \end{cases}$$

Now eliminate y from the third equation. Add twice the second equation to the third, but keep the first two equations:

$$\begin{cases} 2x - y + z = 4 \\ 3y + 2z = -1 \\ -z = 3. \end{cases}$$

This is the system in Example 3.1, which we can solve by elimination.

Answer $(\frac{13}{3}, \frac{5}{3}, -3)$.

. . . .

The method of elimination works for a very simple reason: adding a constant multiple of one equation to another equation does not affect the solution of the system. In other words, suppose

$$a_1 x + b_1 y + c_1 z = d_1 \qquad \text{and} \qquad a_2 x + b_2 y + c_2 z = d_2$$

are two linear equations. If k is any number, then the two systems

(1)
$$\begin{cases} a_1 x + b_1 y + c_1 z = d_1 \\ a_2 x + b_2 y + c_2 z = d_2 \end{cases}$$

and

(2)
$$\begin{cases} a_1 x + b_1 y + c_1 z = d_1 \\ (a_2 + ka_1)x + (b_2 + kb_1)y + (c_2 + kc_1)z = d_2 + kd_1 \end{cases}$$

have *exactly* the same solutions.

Proof: If (x, y, z) satisfies (1), then

$$(a_2 + ka_1)x + (b_2 + kb_1)y + (c_2 + kc_1)z$$
$$= (a_2 x + b_2 y + c_2 z) + k(a_1 x + b_1 y + c_1 z)$$
$$= d_2 + kd_1,$$

so (x, y, z) is a solution of (2). Conversely, if (x, y, z) satisfies (2), then

$$a_2 x + b_2 y + c_2 z$$
$$= (a_2 + ka_1)x + (b_2 + kb_1)y + (c_2 + kc_1)z$$
$$\quad - k(a_1 x + b_1 y + c_1 z)$$
$$= (d_2 + kd_1) - kd_1 = d_2,$$

so (x, y, z) is a solution of (1).

Practical hint: When you apply the method of elimination, you do not *have* to eliminate first x and then y. Eliminate any two of the unknowns in an order that makes the computation easiest.

▨ *Example 4.3*

Solve the system

$$\begin{cases} 3x + y - 2z = 4 \\ -5x \quad\quad + 2z = 5 \\ -7x - y + 3z = -2. \end{cases}$$

SOLUTION   Since y is missing from the second equation, add the first to the third; then y is eliminated from two equations:

$$\begin{cases} 3x + y - 2z = 4 \\ -5x \quad\quad + 2z = 5 \\ -4x \quad\quad + z = 2. \end{cases}$$

Now add -2 times the third to the second; this eliminates z:

$$\begin{cases} 3x + y - 2z = 4 \\ 3x \quad\quad\quad = 1 \\ -4x \quad\quad + z = 2. \end{cases}$$

By the second equation, $x = \frac{1}{3}$. By the third equation, $z = 2 + 4x = 2 + 4(\frac{1}{3}) = \frac{10}{3}$. Finally,

$$y = 4 - 3x + 2z = 4 - 1 + \frac{20}{3} = \frac{29}{3}.$$

Answer $(\frac{1}{3}, \frac{29}{3}, \frac{10}{3})$.

The Matrix of a System

A **matrix** is a rectangle of numbers. Associated with the system in Example 4.3 is the matrix

$$\begin{bmatrix} 3 & 1 & -2 & \vdots & 4 \\ -5 & 0 & 2 & \vdots & 5 \\ -7 & -1 & 3 & \vdots & -2 \end{bmatrix}.$$

It is a 3 × 4 matrix (3 rows and 4 columns). The dotted line separates the coefficients of the unknowns from the constants.

Using the matrix, we can describe the system without writing the variables x, y, z and the equal signs. We can carry out elimination in this shorthand, passing through a series of matrices of equivalent systems until we come to the desired from.

■ *Example 4.4*

Solve the system

$$\begin{cases} 2x + 3y - z = -2 \\ -3x - y + z = -2 \\ x + y - 3z = -8 \end{cases}$$

by elimination, using matrix notation.

SOLUTION The matrix of the system is

$$\begin{bmatrix} 2 & 3 & -1 & \vdots & -2 \\ -3 & -1 & 1 & \vdots & -2 \\ 1 & 1 & -3 & \vdots & -8 \end{bmatrix}.$$

Add the first row to the second and -3 times the first row to the third:

$$\begin{bmatrix} 2 & 3 & -1 & \vdots & -2 \\ -1 & 2 & 0 & \vdots & -4 \\ -5 & -8 & 0 & \vdots & -2 \end{bmatrix}.$$

This eliminates z from the second and third equation. Now eliminate x from the third equation by adding -5 times the second row to the third row:

$$\begin{bmatrix} 2 & 3 & -1 & \vdots & -2 \\ -1 & 2 & 0 & \vdots & -4 \\ 0 & -18 & 0 & \vdots & 18 \end{bmatrix}.$$

The third line of this matrix represents the equation $-18y = 18$. Thus $y = -1$. From the second line,

$$-x - 2 = -4, \qquad x = 2,$$

and finally from the first

$$4 - 3 - z = -2, \qquad z = 3.$$

Answer $(2, -1, 3)$.

■ ■

Curve Fitting

The following example indicates a useful application of linear systems. This type of problem arises frequently in statistics and analysis of data.

■ *Example 4.5*

Find a quadratic polynomial whose graph passes through

$$(1, -2), \qquad (3, 1), \qquad \text{and} \qquad (4, -1).$$

SOLUTION   Let the quadratic polynomial be $y = Ax^2 + Bx + C$, where the coefficients A, B, C must be determined. Three conditions are required:

$$\begin{cases} A + B + C = -2 \\ 9A + 3B + C = 1 \\ 16A + 4B + C = -1. \end{cases}$$

This is a linear system for (A, B, C). We solve it by elimination, using matrix notation. First we subtract the second row from the third and the first row from the second; then we subtract twice the third from the second:

$$\begin{bmatrix} 1 & 1 & 1 & \vdots & -2 \\ 9 & 3 & 1 & \vdots & 1 \\ 16 & 4 & 1 & \vdots & -1 \end{bmatrix}, \quad \begin{bmatrix} 1 & 1 & 1 & \vdots & -2 \\ 8 & 2 & 0 & \vdots & 3 \\ 7 & 1 & 0 & \vdots & -2 \end{bmatrix}, \quad \begin{bmatrix} 1 & 1 & 1 & \vdots & -2 \\ -6 & 0 & 0 & \vdots & 7 \\ 7 & 1 & 0 & \vdots & -2 \end{bmatrix}.$$

Hence $-6A = 7$, $A = -7/6$,

$$7A + B = -2, \qquad -49/6 + B = -2, \qquad B = 37/6,$$

$$A + B + C = -2, \qquad -7/6 + 37/6 + C = -2, \qquad C = -7.$$

Answer $-\frac{7}{6}x^2 + \frac{37}{6}x - 7$.

■ ■

Certain systems of equations do not have precisely one solution. There may be either no solutions at all, or more than one solution. In the latter case it turns out that there are infinitely many. Both cases can be handled by elimination.

Inconsistent Systems

Sometimes the elimination process leads to an equation

$$0x + 0y + 0z = d,$$

where $d \neq 0$. Obviously no choice of x, y, and z will satisfy this equation. Then the system simply has no solution, and it is called an **inconsistent** system. An example with two unknowns will suffice to illustrate this. Look at the system

$$\begin{cases} x + 3y = -1 \\ 2x + 6y = 3. \end{cases}$$

Add -2 times the first equation to the second. The result is the system

$$\begin{cases} x + 3y = -1 \\ 0 = 5. \end{cases}$$

There is no solution.

What happened? Why is this system different from any other system? Notice that the left-hand side of the second equation is twice the left-hand side of the first, but the right-hand side is not twice the right-hand side of the first. This is a contradiction; if $x + 3y = -1$, then $2(x + 3y)$ should be -2, not 3.

Underdetermined Systems

Some systems have more than one solution. This happens when one of the equations is a consequence of the other two, so that really there are only two (or fewer) equations. Such systems are called **underdetermined.**

Example 1:

$$\begin{cases} x + y + z = 1 \\ x - 2y + 2z = 4 \\ 2x - y + 3z = 5. \end{cases}$$

Eliminate x from the second and third equations:

$$\begin{cases} x + y + z = 1 \\ -3y + z = 3 \\ -3y + z = 3. \end{cases}$$

The last two equations both say the same thing. Therefore the system is equivalent to the system

$$\begin{cases} x + y + z = 1 \\ -3y + z = 3. \end{cases}$$

The system has not one solution, but a whole family of solutions. One of the unknowns can be assigned an arbitrary value and then the resulting system of two equations in the other two variables can be solved.

For instance, set $y = t$, arbitrary. Then

$$-3t + z = 3, \qquad z = 3t + 3,$$

and

$$x + t + (3t + 3) = 1, \qquad x = -4t - 2,$$

so the solutions are $(-4t - 2, t, 3t + 3)$, where t can be any real number.

Example 2:

$$\begin{cases} x + y + z = 1 \\ 2x + 2y + 2z = 2 \\ 3x + 3y + 3z = 3. \end{cases}$$

This system is obviously equivalent to the single equation $x + y + z = 1$. Therefore two of the unknowns can be assigned arbitrary values and the resulting single linear equation solved for the third. For instance, set $y = s$ and $z = t$, arbitrary. Then

$$x + s + t = 1, \qquad x = 1 - s - t,$$

so the solutions are $(1 - s - t, s, t)$, where s and t can be any two real numbers.

EXERCISES

Solve:

1. $\begin{cases} x + y + z = 0 \\ 2y - 3z = -1 \\ 3y + 5z = 2 \end{cases}$
2. $\begin{cases} 2x - y - z = 1 \\ 2y - 3z = -1 \\ -3y + 5z = 2 \end{cases}$
3. $\begin{cases} x + y - z = 0 \\ x - y + z = 0 \\ -x + y + z = 0 \end{cases}$

4. $\begin{cases} 2x + y + 3z = 1 \\ -x + 4y + 2z = 0 \\ 3x + y + z = -1 \end{cases}$
5. $\begin{cases} 2x - y - 3z = 1 \\ -x - 4y - 2z = 1 \\ 3x - y - z = 1 \end{cases}$
6. $\begin{cases} 4x + 2y - z = 0 \\ x + 3y + 2z = 0 \\ x + y + 3z = 4. \end{cases}$

7. $\begin{cases} x + 5y + 5z = 1 \\ 2x + 5y + 3z = 3 \\ x - 4y - 2z = 2 \end{cases}$
8. $\begin{cases} x - y - z = 2 \\ -x + y - z = -2 \\ -x - y + z = 1 \end{cases}$
9. $\begin{cases} x + y = -4 \\ y + z = 3 \\ z + x = 2 \end{cases}$

10. $\begin{cases} x + 2y = 3 \\ y + 2z = 4 \\ z + 2x = 5. \end{cases}$

Find a quadratic function whose graph passes through the three given points:

11. $(-1, 1)$, $(0, 0)$, $(1, 2)$
12. $(-2, -5)$, $(0, 3)$, $(2, 0)$
13. $(-1, 1)$, $(1, 1)$, $(3, 5)$
14. $(-3, 2)$, $(1, 1)$, $(2, -1)$.

Show that the system is inconsistent:

15. $\begin{cases} x + y + 2z = 1 \\ 3x + 5y + 7z = 2 \\ -x - y - 2z = 0 \end{cases}$
16. $\begin{cases} x + y + 2z = 1 \\ -x + 2y + z = 3 \\ y + z = 1. \end{cases}$

Find all solutions of the underdetermined system:

17. $\begin{cases} 2x - y + z = 1 \\ 3x + y + z = 0 \\ 7x - y + 3z = 2 \end{cases}$ **18.** $\begin{cases} 11x + 10y + 9z = 5 \\ x + 2y + 3z = 1 \\ 3x + 2y + z = 1. \end{cases}$

Solve by elimination:

19. $\begin{cases} x + y + z + w = 10 \\ x + y + 2z = 11 \\ x - 3y + w = -14 \\ y + 3z - w = 7 \end{cases}$ **20.** $\begin{cases} 2x + y + z = 3 \\ y + z + w = 5 \\ 4x + z + w = 0 \\ 3y - 2z - w = 6. \end{cases}$

5. DETERMINANTS

With each square matrix we shall associate a *number* called the **determinant** of the matrix. For a 2×2 matrix, the determinant is especially simple. Given

$$\begin{bmatrix} a & b \\ c & d \end{bmatrix},$$

we define its determinant by

$$\det \begin{bmatrix} a & b \\ c & d \end{bmatrix} = \begin{vmatrix} a & b \\ c & d \end{vmatrix} = ad - bc.$$

Several important properties of determinants follow directly from the definition:

(1) $\begin{vmatrix} 0 & 0 \\ c & d \end{vmatrix} = \begin{vmatrix} a & b \\ 0 & 0 \end{vmatrix} = 0$ (2) $\begin{vmatrix} c & d \\ a & b \end{vmatrix} = - \begin{vmatrix} a & b \\ c & d \end{vmatrix}$ (3) $\begin{vmatrix} a & b \\ a & b \end{vmatrix} = 0$

(4) $\begin{vmatrix} a_1 + a_2 & b_1 + b_2 \\ c & d \end{vmatrix} = \begin{vmatrix} a_1 & b_1 \\ c & d \end{vmatrix} + \begin{vmatrix} a_2 & b_2 \\ c & d \end{vmatrix},$

$\begin{vmatrix} a & b \\ c_1 + c_2 & d_1 + d_2 \end{vmatrix} = \begin{vmatrix} a & b \\ c_1 & d_1 \end{vmatrix} + \begin{vmatrix} a & b \\ c_2 & d_2 \end{vmatrix}$

(5) $\begin{vmatrix} ka & kb \\ c & d \end{vmatrix} = k \begin{vmatrix} a & b \\ c & d \end{vmatrix},$ $\begin{vmatrix} a & b \\ kc & kd \end{vmatrix} = k \begin{vmatrix} a & b \\ c & d \end{vmatrix}$ (6) $\begin{vmatrix} 1 & 0 \\ 0 & 1 \end{vmatrix} = 1.$

Each property is a statement about the rows of the matrix. For instance, (2) says that the determinant of a matrix changes sign when the rows of the matrix are interchanged; it amounts to the simple equation

$$bc - ad = -(ad - bc).$$

Examples:

$$\begin{vmatrix} 300 & 200 \\ 4 & -1 \end{vmatrix} = 100 \begin{vmatrix} 3 & 2 \\ 4 & -1 \end{vmatrix} = 100(-3-8) = -1100.$$

$$\begin{vmatrix} 3 & 4 \\ 4 & 5 \end{vmatrix} = \begin{vmatrix} 3 & 4 \\ 3+1 & 4+1 \end{vmatrix} = \begin{vmatrix} 3 & 4 \\ 3 & 4 \end{vmatrix} + \begin{vmatrix} 3 & 4 \\ 1 & 1 \end{vmatrix} = 0 + \begin{vmatrix} 3 & 4 \\ 1 & 1 \end{vmatrix} = -1.$$

To each of the properties (1)–(5) of rows, there corresponds a property of columns. For instance, to (5), there corresponds

$$(5') \quad \begin{vmatrix} ka & b \\ kc & d \end{vmatrix} = k \begin{vmatrix} a & b \\ c & d \end{vmatrix}, \quad \begin{vmatrix} a & kb \\ c & kd \end{vmatrix} = k \begin{vmatrix} a & b \\ c & d \end{vmatrix}.$$

3 × 3 Determinants

The **determinant** of a 3×3 matrix is a number, defined by

$$\begin{vmatrix} a_1 & b_1 & c_1 \\ a_2 & b_2 & c_2 \\ a_3 & b_3 & c_3 \end{vmatrix} = a_1 b_2 c_3 + a_2 b_3 c_1 + a_3 b_1 c_2 - a_1 b_3 c_2 - a_2 b_1 c_3 - a_3 b_2 c_1.$$

There are six products; each has one factor from each column and one factor from each row. To remember the formula, follow the scheme in Fig. 5.1.

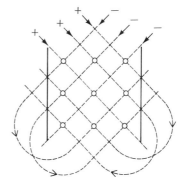

Fig. 5.1   Memory aid for 3×3 determinants

The 2×2 determinant formula has 2 terms; the 3×3 determinant formula has 6 terms. The 4×4 and 5×5 determinant formulas, have 24 and 120 terms and follow a much more complicated scheme than that indicated in Fig. 5.1.

A 3×3 determinant can be evaluated in several ways by grouping terms suitably. For example,

$$\begin{vmatrix} a_1 & b_1 & c_1 \\ a_2 & b_2 & c_2 \\ a_3 & b_3 & c_3 \end{vmatrix} = a_1(b_2 c_3 - b_3 c_2) - b_1(a_2 c_3 - a_3 c_2) + c_1(a_2 b_3 - a_3 b_2).$$

Each quantity in parentheses is the determinant of a 2×2 matrix, so we can write

$$\begin{vmatrix} a_1 & b_1 & c_1 \\ a_2 & b_2 & c_2 \\ a_3 & b_3 & c_3 \end{vmatrix} = a_1 \begin{vmatrix} b_2 & c_2 \\ b_3 & c_3 \end{vmatrix} - b_1 \begin{vmatrix} a_2 & c_2 \\ a_3 & c_3 \end{vmatrix} + c_1 \begin{vmatrix} a_2 & b_2 \\ a_3 & b_3 \end{vmatrix}.$$

The coefficients of the 2×2 determinants are the terms of the *first* row of the matrix, with suitable signs. There are corresponding formulas based on the second and third rows. Similarly, there are three formulas involving columns. For instance,

$$\begin{vmatrix} a_1 & b_1 & c_1 \\ a_2 & b_2 & c_2 \\ a_3 & b_3 & c_3 \end{vmatrix} = a_1 \begin{vmatrix} b_2 & c_2 \\ b_3 & c_3 \end{vmatrix} - a_2 \begin{vmatrix} b_1 & c_1 \\ b_3 & c_3 \end{vmatrix} + a_3 \begin{vmatrix} b_1 & c_1 \\ b_2 & c_2 \end{vmatrix}.$$

All of these formulas can be described in a systematic way. First write the matrix in index notation,

$$\begin{bmatrix} a_{11} & a_{12} & a_{13} \\ a_{21} & a_{22} & a_{23} \\ a_{31} & a_{32} & a_{33} \end{bmatrix}.$$

Thus a_{32} denotes the element in the third row and second column, etc. Now blot out the i-th row and j-th column. The result is a 2×2 matrix, whose determinant m_{ij} is called the **minor** of a_{ij}. For instance,

$$m_{32} = \begin{vmatrix} a_{11} & a_{12} & a_{13} \\ a_{21} & a_{22} & a_{23} \\ a_{31} & a_{32} & a_{33} \end{vmatrix} = \begin{vmatrix} a_{11} & a_{13} \\ a_{21} & a_{23} \end{vmatrix}.$$

The six expansion formulas mentioned above are:

Expansion by Minors

$$\begin{vmatrix} a_{11} & a_{12} & a_{13} \\ a_{21} & a_{22} & a_{23} \\ a_{31} & a_{32} & a_{33} \end{vmatrix} = \begin{cases} a_{11}m_{11} - a_{12}m_{12} + a_{13}m_{13} & \text{(first row)} \\ -a_{21}m_{21} + a_{22}m_{22} - a_{23}m_{23} & \text{(second row)} \\ a_{31}m_{31} - a_{32}m_{32} + a_{33}m_{33} & \text{(third row)} \\[6pt] a_{11}m_{11} - a_{21}m_{21} + a_{31}m_{31} & \text{(first column)} \\ -a_{12}m_{12} + a_{22}m_{22} - a_{32}m_{32} & \text{(second column)} \\ a_{13}m_{13} - a_{23}m_{23} + a_{33}m_{33} & \text{(third column).} \end{cases}$$

The signs in these formulas follow this pattern:

$$\begin{bmatrix} + & - & + \\ - & + & - \\ + & - & + \end{bmatrix}.$$

■ *Example 5.1*

Evaluate $D = \begin{vmatrix} 2 & -3 & 1 \\ 2 & 4 & -5 \\ 3 & 0 & 2 \end{vmatrix}$

(a) by the first row   (b) by the second column.

SOLUTION

(a) $D = 2\begin{vmatrix} 4 & -5 \\ 0 & 2 \end{vmatrix} - (-3)\begin{vmatrix} 2 & -5 \\ 3 & 2 \end{vmatrix} + 1\begin{vmatrix} 2 & 4 \\ 3 & 0 \end{vmatrix} = 2(8) + 3(19) + (-12) = 61.$

(b) $D = -(-3)\begin{vmatrix} 2 & -5 \\ 3 & 2 \end{vmatrix} + 4\begin{vmatrix} 2 & 1 \\ 3 & 2 \end{vmatrix} - 0\begin{vmatrix} 2 & 1 \\ 2 & -5 \end{vmatrix} = 3(19) + 4(1) - 0 = 61.$

Answer 61.

■ ■

Note: The presence of a zero in the second column reduced the amount of computation. In general it pays to expand a determinant by a column that contains one or more zeros. For example,

$$\begin{vmatrix} 3 & 26 & -177 \\ 0 & 89 & 6 \\ 0 & 14 & 1 \end{vmatrix} = 3\begin{vmatrix} 89 & 6 \\ 14 & 1 \end{vmatrix} + 0 + 0 = 3(5) = 15,$$

(expansion by the first column).

Properties of Determinants

Determinants of 3×3 matrices have properties similar to those of 2×2 determinants.

(1) If a row (column) has all zeros, then the determinant equals 0.

(2) If two rows (columns) are interchanged, the determinant changes sign.

(3) If two rows (columns) are equal term-by-term, the determinant equals 0.

(4) If a row (column) is the sum term-by-term of two other rows (columns) then the determinant is the sum of the two determinants corresponding to the summand rows (columns).

(5) If all terms of one row (column) are multiplied by a factor k, then the determinant is multiplied by k.

(6) $\begin{vmatrix} 1 & 0 & 0 \\ 0 & 1 & 0 \\ 0 & 0 & 1 \end{vmatrix} = 1.$

Each of these properties follows directly from the definition

$$\begin{vmatrix} a_1 & b_1 & c_1 \\ a_2 & b_2 & c_2 \\ a_3 & b_3 & c_3 \end{vmatrix} = a_1 b_2 c_3 + a_2 b_3 c_1 + a_3 b_1 c_2 - a_1 b_3 c_2 - a_2 b_1 c_3 - a_3 b_2 c_1.$$

For example, if a row contains all zeros, say $a_1 = b_1 = c_1 = 0$, then each product is 0, hence the determinant is 0. Similarly, if one column contains all zeros, say $b_1 = b_2 = b_3 = 0$, then the determinant is 0. This proves property (1).

If all terms of one row are multiplied by a factor k, say a_3, b_3, c_3 becomes ka_3, kb_3, kc_3, then each of the six products is multiplied by k, so the determinant itself is multiplied by k. This proves (5). We omit the verification of the other properties.

Note what (4) says. For instance

$$\begin{vmatrix} a_1 & b_1 & c_1 + d_1 \\ a_2 & b_2 & c_2 + d_2 \\ a_3 & b_3 & c_3 + d_3 \end{vmatrix} = \begin{vmatrix} a_1 & b_1 & c_1 \\ a_2 & b_2 & c_2 \\ a_3 & b_3 & c_3 \end{vmatrix} + \begin{vmatrix} a_1 & b_1 & d_1 \\ a_2 & b_2 & d_2 \\ a_3 & b_3 & d_3 \end{vmatrix}.$$

The first two columns stay fixed; the determinant splits according to its third column.

In practice, property (5) is most often used to factor out a common factor in any single row or column. A consequence of (4) and (5) is the following property, useful in practical computations.

> (7) If a multiple of one row (column) is added to another row (column) the determinant is unchanged.

For example, let us show that adding k times its second column to its third column does not change the value of a determinant. By property (4),

$$\begin{vmatrix} a_1 & b_1 & c_1 + kb_1 \\ a_2 & b_2 & c_2 + kb_2 \\ a_3 & b_3 & c_3 + kb_3 \end{vmatrix} = \begin{vmatrix} a_1 & b_1 & c_1 \\ a_2 & b_2 & c_2 \\ a_3 & b_3 & c_3 \end{vmatrix} + \begin{vmatrix} a_1 & b_1 & kb_1 \\ a_2 & b_2 & kb_2 \\ a_3 & b_3 & kb_3 \end{vmatrix}.$$

We need only show that the second determinant on the right is 0. But it is: we just factor out k from the third column leaving a determinant with two identical columns, hence 0.

To simplify the computation of a determinant, we often use property (7) to introduce zeros in strategic places.

■ *Example 5.2*

Evaluate

$$D = \begin{vmatrix} 4 & 3 & -3 \\ 2 & -2 & 3 \\ -6 & 2 & -5 \end{vmatrix}.$$

SOLUTION Add -2 times the second row to the first, add 3 times the second row to the third, then expand by minors of the first column:

$$D = \begin{vmatrix} 0 & 7 & -9 \\ 2 & -2 & 3 \\ -6 & 2 & -5 \end{vmatrix} = \begin{vmatrix} 0 & 7 & -9 \\ 2 & -2 & 3 \\ 0 & -4 & 4 \end{vmatrix}$$

$$= -2 \begin{vmatrix} 7 & -9 \\ -4 & 4 \end{vmatrix} = -(2)(4) \begin{vmatrix} 7 & -9 \\ -1 & 1 \end{vmatrix} = -8(7 - 9) = 16.$$

Answer 16.

. . . .

We mention one further property of determinants:

---

(8)   If the rows and columns of a matrix are interchanged with each other, the determinant is unchanged:

$$\begin{vmatrix} a_1 & a_2 & a_3 \\ b_1 & b_2 & b_3 \\ c_1 & c_2 & c_3 \end{vmatrix} = \begin{vmatrix} a_1 & b_1 & c_1 \\ a_2 & b_2 & c_2 \\ a_3 & b_3 & c_3 \end{vmatrix}.$$

Cramer's Rule

One application of determinants is to the solution of systems of linear equations. Using determinants, we can write formulas for the solution of the system

(*)
$$\begin{cases} a_1 x + b_1 y + c_1 z = d_1 \\ a_2 x + b_2 y + c_2 z = d_2 \\ a_3 x + b_3 y + c_3 z = d_3. \end{cases}$$

The formulas can be proved by a lot of tedious computation, or systematically in a course on linear algebra.

Cramer's Rule

The system (*) has a unique solution if and only if

$$D = \begin{vmatrix} a_1 & b_1 & c_1 \\ a_2 & b_2 & c_2 \\ a_3 & b_3 & c_3 \end{vmatrix} \neq 0.$$

If $D \neq 0$, then the solution (x, y, z) is given by

$$x = \frac{1}{D} \begin{vmatrix} d_1 & b_1 & c_1 \\ d_2 & b_2 & c_2 \\ d_3 & b_3 & c_3 \end{vmatrix}, \qquad y = \frac{1}{D} \begin{vmatrix} a_1 & d_1 & c_1 \\ a_2 & d_2 & c_2 \\ a_3 & d_3 & c_3 \end{vmatrix}, \qquad z = \frac{1}{D} \begin{vmatrix} a_1 & b_1 & d_1 \\ a_2 & b_2 & d_2 \\ a_3 & b_3 & d_3 \end{vmatrix}.$$

Note carefully what Cramer's Rule says. We start with the matrix

$$\begin{bmatrix} a_1 & b_1 & c_1 \\ a_2 & b_2 & c_2 \\ a_3 & b_3 & c_3 \end{bmatrix}$$

of coefficients of the left side of (*). Its determinant is D, and Cramer's Rule covers only the case $D \neq 0$. To find x we replace the x-column of the matrix by the column of constants in (*), take the determinant and divide by D. Ditto for y and z. With obvious modifications, Cramer's Rule applies to 2×2 systems also.

■ *Example 5.3*

Solve by Cramer's Rule

$$\begin{cases} 3x & - z = 15 \\ 2x + y - 2z = 0 \\ 4x - 3y + z = 5. \end{cases}$$

SOLUTION

$$D = \begin{vmatrix} 3 & 0 & -1 \\ 2 & 1 & -2 \\ 4 & -3 & 1 \end{vmatrix} = -5 \neq 0, \qquad x = \frac{1}{-5} \begin{vmatrix} 15 & 0 & -1 \\ 0 & 1 & -2 \\ 5 & -3 & 1 \end{vmatrix} = \frac{-70}{-5} = 14,$$

$$y = \frac{1}{-5} \begin{vmatrix} 3 & 15 & -1 \\ 2 & 0 & -2 \\ 4 & 5 & 1 \end{vmatrix} = \frac{-130}{-5} = 26, \qquad z = \frac{1}{-5} \begin{vmatrix} 3 & 0 & 15 \\ 2 & 1 & 0 \\ 4 & -3 & 5 \end{vmatrix} = \frac{-135}{-5} = 27.$$

Answer $(14, 26, 27)$.

Evaluate:

1. $\begin{vmatrix} 2 & 1 \\ -1 & 3 \end{vmatrix}$ **2.** $\begin{vmatrix} 1 & 3 \\ 1 & 1 \end{vmatrix}$ **3.** $\begin{vmatrix} 4 & -3 \\ -3 & -2 \end{vmatrix}$ **4.** $\begin{vmatrix} 2 & 3 \\ 3 & 5 \end{vmatrix}$

5. $\begin{vmatrix} a & 1 \\ 0 & a \end{vmatrix}$ **6.** $\begin{vmatrix} 1 & b \\ 0 & 1 \end{vmatrix}$ **7.** $\begin{vmatrix} 1 & 0 \\ b & 1 \end{vmatrix}$ **8.** $\begin{vmatrix} a & 0 \\ 1 & a \end{vmatrix}.$

Evaluate $\begin{vmatrix} -3 & 0 & 3 \\ 4 & -2 & 1 \\ 5 & 2 & -1 \end{vmatrix}$:

9. by the defining formula   **10.** by minors of the second column
11. by minors of the third column   **12.** by minors of the first row
13. by minors of the second row   **14.** by minors of the first column.

Evaluate in any way:

15. $\begin{vmatrix} 1 & 2 & 3 \\ 3 & 2 & 1 \\ 2 & 1 & 3 \end{vmatrix}$
16. $\begin{vmatrix} 4 & 5 & 6 \\ 5 & 6 & 4 \\ 6 & 4 & 5 \end{vmatrix}$
17. $\begin{vmatrix} -1 & 1 & 2 \\ 3 & -1 & 3 \\ 4 & 0 & 2 \end{vmatrix}$
18. $\begin{vmatrix} 1 & 2 & 3 \\ 1 & 4 & 9 \\ 1 & 8 & 27 \end{vmatrix}$

19. $\begin{vmatrix} 3 & 6 & -1 \\ 2 & 1 & -2 \\ 4 & -3 & 1 \end{vmatrix}$
20. $\begin{vmatrix} 15 & 0 & -1 \\ 0 & 1 & -2 \\ 5 & -3 & 1 \end{vmatrix}$
21. $\begin{vmatrix} 3 & 15 & -1 \\ 2 & 0 & -2 \\ 4 & 5 & 1 \end{vmatrix}$
22. $\begin{vmatrix} 3 & 0 & 15 \\ 2 & 1 & 0 \\ 4 & -3 & 5 \end{vmatrix}$

23. $\begin{vmatrix} 1 & 1 & 1 \\ 1 & -1 & 2 \\ 1 & 1 & 4 \end{vmatrix}$
24. $\begin{vmatrix} 2 & 0 & 0 \\ -1 & 3 & 0 \\ 1 & 2 & -1 \end{vmatrix}$
25. $\begin{vmatrix} 2 & 3 & 6 \\ -1 & 2 & -9 \\ 1 & 3 & -2 \end{vmatrix}$
26. $\begin{vmatrix} -1 & 2 & -9 \\ 1 & 3 & -2 \\ 2 & 2 & 6 \end{vmatrix}$

27. $\begin{vmatrix} 4 & -3 & 1 \\ 2 & 1 & -2 \\ 3 & 0 & -1 \end{vmatrix}$
28. $\begin{vmatrix} 1 & 1 & 3 \\ 2 & 4 & 7 \\ 2 & 3 & 9 \end{vmatrix}$
29. $\begin{vmatrix} x & 1 & 0 \\ -1 & x & 1 \\ 0 & -1 & x \end{vmatrix}$
30. $\begin{vmatrix} x & y & y \\ y & x & y \\ y & y & x \end{vmatrix}$

31. $\begin{vmatrix} 2x & x+y & x+z \\ x+y & 2y & y+z \\ x+z & y+z & 2z \end{vmatrix}$
32. $\begin{vmatrix} y+z & x & x \\ y & x+z & y \\ z & z & x+y \end{vmatrix}.$

Solve by Cramer's Rule:

33. $\begin{cases} x + 2y = 1 \\ 3y = 2 \end{cases}$
34. $\begin{cases} 2x = 3 \\ -x + y = 0 \end{cases}$
35. $\begin{cases} x + 2y = 1 \\ x + 3y = 2 \end{cases}$

36. $\begin{cases} x + y = a \\ x - y = b \end{cases}$
37. $\begin{cases} 2x - 3y = -1 \\ 3x + 5y = 2 \end{cases}$
38. $\begin{cases} 2x - 3y = -1 \\ -3x + 5y = 2. \end{cases}$

39. $\begin{cases} x + y - z = 3 \\ x - y + z = 2 \\ -x + y + z = 1 \end{cases}$
40. $\begin{cases} 2y - 3z = 0 \\ x + y + z = 1 \\ 3y + 5z = 0 \end{cases}$

41. $\begin{cases} 2x + y + 2z = 1 \\ -x + 4y + 2z = 0 \\ 3x + y + z = -1 \end{cases}$
42. $\begin{cases} 2x - y - 3z = 1 \\ -x + 4y - 2z = 1 \\ 3x - y - z = 1 \end{cases}$

43. $\begin{cases} 2x - y + z = 4 \\ 3y + 2z = -1 \\ - z = 3 \end{cases}$
44. $\begin{cases} x + z = -1 \\ 2x + 3y + z = 5 \\ 4x + 4y + 2z = -3 \end{cases}$

45. $\begin{cases} 3x - y - z = 1 \\ -x - 4y - 2z = 1 \\ 2x - y - 3z = 1 \end{cases}$
46. $\begin{cases} 2x + y + 3z = 1 \\ 3x + y + z = -1 \\ -x + 4y + 2z = 0. \end{cases}$

Test 1

1. Use synthetic division to compute $f(-3)$, where
$$f(x) = x^4 + 2x^3 - 5x^2 + 1.$$

2. Decompose into partial fractions: $\dfrac{x^3}{x^2 - 4x + 3}$.

3. Solve by elimination:

$$\begin{cases} x + y - 3z = 1 \\ 2x - y - z = 0 \\ -x + 4y + z = 2. \end{cases}$$

4. Solve the same system by Cramer's Rule.

Test 2

1. Divide $2x^4 - x^3 + 1$ by $x^2 - x - 2$; give quotient and remainder.

2. Decompose into partial fractions: $\dfrac{x^2 + 1}{x^3 - 4x}$.

3. Find a polynomial of degree 4 that has zeros at 1, 2, 3, and 4, and such that $f(5) = 2$.
4. Solve by any method:

$$\begin{cases} x + y - 3z = 1 \\ 2x - y - z = 0 \\ -x + 4y + z = 0. \end{cases}$$

5. Evaluate by minors of the third column:

$$\begin{vmatrix} 1 & -1 & 3 \\ 1 & 2 & -2 \\ -1 & 1 & 2 \end{vmatrix}.$$

10

COMPLEX NUMBERS

It is a shortcoming of the real number system that not every polynomial equation has a real root. For example, the equation

$$x^2 + 1 = 0$$

has no real root because squares of real numbers are non-negative; a solution would be a real number x for which $x^2 = -1$. Furthermore the general quadratic equation

(1) $$ax^2 + bx + c = 0,$$

where a, b, c are real numbers and $a \neq 0$, has no real solution if $D = b^2 - 4ac < 0$.

Imagine that there exists a number system containing the real number system, in which the equation $x^2 + 1 = 0$ has a root, $\sqrt{-1}$. Then the quadratic equation (1) can be solved, even when the discriminant D is negative. If $D < 0$, the roots are

$$\frac{-b \pm \sqrt{D}}{2a} = \frac{-b \pm \sqrt{-D}\sqrt{-1}}{2a},$$

provided the rules of ordinary arithmetic are valid in the extended number system.

Let us go about constructing a suitable number system. We shall enlarge the real numbers to a system that contains $\sqrt{-1}$ and that satisfies the rules of arithmetic.

We start with the set of real numbers and a symbol i which will play the role of $\sqrt{-1}$. The new number system, called the **complex number system,** consists of all formal expressions

$$a + bi,$$

where a and b are real numbers. We must now say how to operate with these formal symbols.

321

Since i is a new sort of object, it is natural to say two complex numbers $a + bi$ and $c + di$ are equal if and only if $a = c$ and $b = d$.

If ordinary rules of arithmetic are to hold, we must have

$$(a + bi) + (c + di) = (a + c) + (b + d)i.$$

This we take as the *definition* of addition in the new system.

Since we want to have $i = \sqrt{-1}$, we agree to replace i^2 by -1. Thus it seems natural to compute the product of $a + bi$ and $c + di$ as follows:

$$
\begin{aligned}
(a + bi)(c + di) &= ac + a(di) + (bi)c + (bi)(di) \\
&= ac + (ad)i + (bc)i + (bd)i^2 \\
&= [ac + (bd)i^2] + [ad + bc]i \\
&= (ac - bd) + (ad + bc)i.
\end{aligned}
$$

This we take as the *definition* of multiplication in the new system.

---

1. Two complex numbers $a + bi$ and $c + di$ are **equal** if and only if $a = c$ and $b = d$.

2. The **sum** of two complex numbers is defined by

    $$(a + bi) + (c + di) = (a + c) + (b + d)i.$$

3. The **product** of two complex numbers is defined by

    $$(a + bi)(c + di) = (ac - bd) + (ad + bc)i.$$

These three definitions represent an attempt to breathe life into the set of formal symbols $a + bi$. We shall show that they succeed: with the above definitions of addition and multiplication, the complex numbers satisfy all the rules of arithmetic valid for the real numbers.

The crucial definition is (3). From it follows $i^2 = -1$:

$$i^2 = (0 + 1i)(0 + 1i) = (0 \cdot 0 - 1 \cdot 1) + (0 \cdot 1 + 1 \cdot 0)i = -1.$$

We shall *identify* the complex number $a + 0i$ with the real number a. This is perfectly reasonable since complex numbers $a + 0i$ and $b + 0i$ add and multiply just as do the real numbers a and b:

$$(a + 0i) + (b + 0i) = (a + b) + (0 + 0)i = (a + b) + 0i,$$

$$(a + 0i)(b + 0i) = (ab - 0 \cdot 0) + (a \cdot 0 + 0 \cdot b)i = ab + 0i.$$

Thus, the complex number system contains a subsystem that we can identify with the real number system. In other words, the complex number system is an *extension* of the real number system, and its arithmetic is consistent with that of the real number system.

Notation: Sometimes it is convenient to write $a + ib$ instead of $a + bi$. This is particularly useful with expressions such as $\cos \theta + i \sin \theta$ and $-1 + i \sqrt{3}$. In electrical engineering i is used for current and the symbol j is used for the complex number we are calling i.

Let us now set down the rules of arithmetic in our new system. In the statements of the rules, α, β, and γ are any complex numbers.

Commutative laws:
$$\alpha + \beta = \beta + \alpha$$
$$\alpha\beta = \beta\alpha$$

Associative laws:
$$\alpha + (\beta + \gamma) = (\alpha + \beta) + \gamma$$
$$\alpha(\beta\gamma) = (\alpha\beta)\gamma.$$

Identity laws:
$$\alpha + 0 = \alpha$$
$$\alpha \cdot 1 = \alpha.$$

Distributive law: $\alpha(\beta + \gamma) = \alpha\beta + \alpha\gamma.$

Additive inverse: If $\alpha = a + bi$, set $-\alpha = (-a) + (-b)i$. Then
$$\alpha + (-\alpha) = 0.$$

Multiplicative inverse: If $\alpha = a + bi \neq 0$, set
$$\alpha^{-1} = \left(\frac{a}{a^2 + b^2}\right) + \left(\frac{-b}{a^2 + b^2}\right)i = \frac{a - bi}{a^2 + b^2}.$$

Then $\alpha\alpha^{-1} = 1.$

The last law says that each non-zero complex number has a reciprocal. It looks complicated, but it follows easily from the previous rules:

$$\alpha\alpha^{-1} = (a + bi)\left(\frac{a - bi}{a^2 + b^2}\right) = \frac{(a + bi)(a - bi)}{a^2 + b^2} = \frac{a^2 + b^2}{a^2 + b^2} = 1.$$

The other laws follow easily from properties of the real numbers. They are left as exercises.

Just as for real numbers, division by a complex number β is defined as multiplication by β^{-1}.

$$\frac{\alpha}{\beta} = \alpha\beta^{-1}, \qquad \beta \neq 0.$$

Just as for real numbers $(\beta^{-1})^2 = (\beta^2)^{-1}$ since

$$(\beta^2)(\beta^{-1})^2 = (\beta\beta)(\beta^{-1}\beta^{-1}) = (\beta\beta^{-1})(\beta\beta^{-1}) = 1.$$

Thus we can define β^{-2}, and similarly β^{-3}, β^{-4}, $\cdots$ so that the usual rules of exponents hold for integer powers (positive and negative) of complex numbers.

For a numerical example of these rules, let us take $\alpha = 1 - i$, $\beta = 2 + 3i$, and $\gamma = -2 + i$. Then

$$\alpha\beta = (1 - i)(2 + 3i) = 5 + i, \qquad \beta\gamma = (2 + 3i)(-2 + i) = -7 - 4i,$$
$$(\alpha\beta)\gamma = (5 + i)(-2 + i) = -11 + 3i,$$
$$\alpha(\beta\gamma) = (1 - i)(-7 - 4i) = -11 + 3i.$$

Hence $(\alpha\beta)\gamma = \alpha(\beta\gamma)$, as predicted by the associative law for multiplication. Next,

$$\beta^{-1} = (2 + 3i)^{-1} = \frac{2}{4 + 9} + \frac{-3}{4 + 9}i = \frac{2}{13} - \frac{3}{13}i.$$

[*Check:* $\beta\beta^{-1} = (2 + 3i)(\frac{2}{13} - \frac{3}{13}i) = \frac{13}{13} + \frac{0}{13}i = 1.$]

$$\frac{\alpha}{\beta} = \alpha\beta^{-1} = (1 - i)(2 + 3i)^{-1} = (1 - i)(\tfrac{2}{13} - \tfrac{3}{13}i) = -\tfrac{1}{13} - \tfrac{5}{13}i.$$

$$\beta^{-2} = (\beta^{-1})^2 = (\tfrac{2}{13} - \tfrac{3}{13}i)(\tfrac{2}{13} - \tfrac{3}{13}i) = -\tfrac{5}{169} - \tfrac{12}{169}i.$$

Complex Conjugates and Absolute Values

For each complex number $\alpha = a + bi$, we define the **complex conjugate** $\bar{\alpha}$ by

$$\bar{\alpha} = a - bi = a + (-b)i.$$

Notice that the conjugate of $a - bi$ is $a + bi$, that is, $\bar{\bar{\alpha}} = \alpha$.

Examples:

$$\overline{4} = 4, \qquad \overline{i} = -i, \qquad \overline{-3i} = 3i, \qquad \overline{4 - 3i} = 4 + 3i.$$

The operation of taking the complex conjugate satisfies two basic algebraic rules:

$$\boxed{\overline{\alpha + \beta} = \bar{\alpha} + \bar{\beta}, \qquad \overline{\alpha\beta} = \bar{\alpha}\bar{\beta}.}$$

To prove the rules, set $\alpha = a + bi$ and $\beta = c + di$. Then

$$\overline{\alpha + \beta} = \overline{(a + c) + (b + d)i} = (a + c) - (b + d)i = (a - bi) + (c - di) = \bar{\alpha} + \bar{\beta},$$

so the first rule holds. Next,

$$\overline{\alpha\beta} = \overline{(ac - bd) + (ad + bc)i} = (ac - bd) - (ad + bc)i,$$

and

$$\bar{\alpha}\bar{\beta} = (a - bi)(c - di) = (ac - bd) - (ad + bc)i,$$

so $\overline{\alpha\beta} = \bar{\alpha}\bar{\beta}$, and the second rule holds.

The rules extend to any number of summands or factors; for instance,

$$\overline{\alpha + \beta + \gamma} = \bar{\alpha} + \bar{\beta} + \bar{\gamma}, \qquad \overline{\alpha\beta\gamma} = \bar{\alpha}\bar{\beta}\bar{\gamma}, \qquad \text{etc.}$$

We define the **absolute value** or **modulus** of α by

$$|\alpha| = \sqrt{a^2 + b^2}.$$

Clearly $|\alpha| > 0$ unless $\alpha = 0$, in which case $|\alpha| = 0$.

Examples:

$$|3| = 3, \qquad |-5i| = 5, \qquad |-12 + 5i| = \sqrt{(-12)^2 + 5^2} = \sqrt{169} = 13.$$

The absolute value operation satisfies several basic rules:

$$|\bar{\alpha}| = |\alpha|, \qquad |\alpha|^2 = \alpha\bar{\alpha}, \qquad |\alpha\beta| = |\alpha|\,|\beta|, \qquad |\alpha/\beta| = |\alpha|/|\beta|.$$

The proofs are left as exercises.

Conjugates and absolute values help simplify division of complex numbers. Since $\alpha\bar{\alpha} = |\alpha|^2$, we can divide both sides by $|\alpha|^2$ and obtain

$$\alpha^{-1} = \frac{\bar{\alpha}}{|\alpha|^2} = \frac{a - bi}{a^2 + b^2}, \qquad \alpha \neq 0.$$

(This agrees with the formula for the multiplicative inverse given earlier.) It follows that

$$\frac{\alpha}{\beta} = \alpha\beta^{-1} = \frac{\alpha\bar{\beta}}{|\beta|^2} = \frac{\alpha\bar{\beta}}{\beta\bar{\beta}}.$$

Thus, to evaluate α/β, multiply numerator and denominator by $\bar{\beta}$. For example,

$$\frac{2 + 3i}{1 + 2i} = \frac{2 + 3i}{1 + 2i}\frac{1 - 2i}{1 - 2i} = \frac{8 - i}{5} = \frac{8}{5} - \frac{1}{5}i.$$

Remark: The introduction of complex numbers was one of the greatest advances ever made in mathematics. It took hundreds of years to develop the idea once the need was felt, so don't expect to learn it perfectly in a few minutes. You should read and reread this section until you are confident you understand it.

EXERCISES

Express in the form $a + bi$:

1. $(3 + 2i) + (6 - i)$

2. $(1 - i) + (4 + 3i)$

3. $2(1 + 4i) - 3(2 + i)$

4. $(-2 + 3i) + 5(1 - i)$

5. $i(2 - 3i)$

6. $i(8i + 5)$

7. $(1 + i)(3 - 4i)$

8. $(1 - i)(2 + 7i)$

9. $(5 + 4i)(3 + 2i)$

10. $(1 - 6i)(3 + i)$

11. $(1 + i)^2$

12. $(2 - i)(2 + i)$

13. $\dfrac{1}{3 + 4i}$

14. $\dfrac{1}{6 - i}$

15. $\dfrac{3 + i}{2 - i}$

16. $\dfrac{2 + 7i}{1 - 3i}$

17. $\dfrac{1}{(2 + 3i)(5 - 4i)}$

18. $\dfrac{1}{(1 - i)(8 + 3i)}$

19. $\dfrac{1 - 3i}{(2 + i)(2 + 5i)}$

20. $\dfrac{(1 + i)(1 + 2i)}{(3 - 2i)(4 - 3i)}$

21. $\dfrac{i}{2 + i} + \dfrac{3 + i}{4 + i}$

22. $\dfrac{1}{4 - 3i} + \dfrac{5 + 3i}{2 - i}$.

Compute $\overline{\alpha\beta}$ and $\bar{\alpha}\bar{\beta}$:

23. $\alpha = 1 - i,\ \beta = 3i$

24. $\alpha = 1 + i,\ \beta = 1 - i$

25. $\alpha = 2 - i,\ \beta = 3 + 2i$

26. $\alpha = -1 - 2i,\ \beta = 3 - i$

27. $\alpha = (1 + i)^2,\ \beta = 1 + 2i$

28. $\alpha = (1 - i)^2,\ \beta = 1 - 2i$.

Compute:

29. $|3 + 4i|$

30. $|3 - 4i|$

31. $|(1 + i)(2 - i)|$

32. $|(2 - 3i)(6 + i)|$

33. $|i(2 + i)| + |1 + 4i|$

34. $|6(3 - 2i)| + |2 - i|$

35. $\left|\dfrac{1}{2 - i}\right|$

36. $\left|\dfrac{2 + 5i}{3 + i}\right|$

37. $\left|\dfrac{1 + i}{1 - i}\right|$

38. $\left|\dfrac{3 + 4i}{4 - 3i}\right|$.

Prove:

39. $\alpha + \beta = \beta + \alpha$

40. $\alpha\beta = \beta\alpha$

41. $\alpha(\beta\gamma) = (\alpha\beta)\gamma$

42. $\alpha + (\beta + \gamma) = (\alpha + \beta) + \gamma$

43. $\alpha(\beta + \gamma) = \alpha\beta + \alpha\gamma$

44. $(\alpha + \beta)\gamma = \alpha\gamma + \beta\gamma$

45. $|\bar{\alpha}| = |\alpha|$

46. $|\alpha|^2 = \alpha\bar{\alpha}$

47. $|\alpha\beta| = |\alpha| \, |\beta|$

48. $|\alpha/\beta| = |\alpha|/|\beta|$.

49. Simplify: $i + i^2 + i^3 + \cdots + i^{11}$.

50. Compute $(1 + i)^2$ and use the result to compute $(1 + i)^{12}$.

2. THE COMPLEX PLANE

Each complex number $\alpha = a + bi$ is completely determined by the ordered pair (a, b) of real numbers. We call a the **real part** of α and b the **imaginary part** of α, and we write

$$a = \text{Re}(\alpha) \quad \text{and} \quad b = \text{Im}(\alpha).$$

If $\alpha = a + bi$, then $\bar{\alpha} = a - bi$. Therefore

$$\alpha + \bar{\alpha} = (a + bi) + (a - bi) = 2a, \quad \alpha - \bar{\alpha} = (a + bi) - (a - bi) = 2bi.$$

From these equations follow the useful relations

$$\text{Re}(\alpha) = \frac{\alpha + \bar{\alpha}}{2}, \quad \text{Im}(\alpha) = \frac{\alpha - \bar{\alpha}}{2i}.$$

The correspondence

$$\alpha = a + bi \longleftrightarrow (a, b),$$

between complex numbers and ordered pairs of real numbers, strongly suggests a geometric representation for complex numbers. Indeed, we identify the *complex number $\alpha = a + bi$* with the *point (a, b)* in the plane (Fig. 2.1).

This gives an entirely new interpretation to the cartesian (coordinate) plane: as the set of all complex numbers. When looked at this way, the plane is called the **complex plane** (also **Gaussian plane** after K. F. Gauss).

The definition of addition of complex numbers,

$$(a + bi) + (c + di) = (a + c) + (b + d)i,$$

has a useful geometric interpretation. To add $\beta = c + di$ to α, we must increase the x-coordinate of α by c and the y-coordinate of α by d. (See Fig. 2.2a.) Therefore

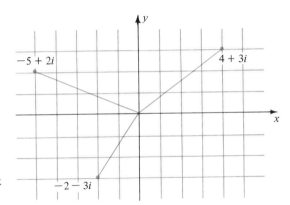

Fig. 2.1 Points in the complex plane

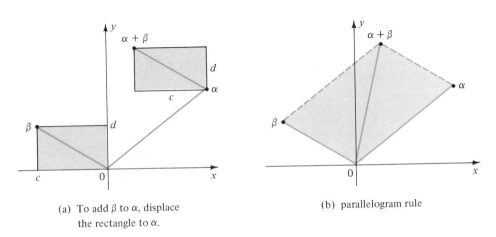

(a) To add β to α, displace the rectangle to α.

(b) parallelogram rule

Fig. 2.2

we take the little rectangle built on β and move it lock, stock, and barrel, so that 0 is moved to α and β moved to $\alpha + \beta$. Thus *the segment from α to $\alpha + \beta$ is the parallel displacement of the segment from 0 to β.*

This means that the three segments: from 0 to β, from 0 to α, and from α to $\alpha + \beta$ make up three sides of a parallelogram. We complete the parallelogram (Fig. 2.2b), and we conclude that complex numbers can be added *graphically* by (what is called) the **parallelogram rule.**

Remark: The parallelogram rule of addition is the same as the rule for adding forces, or vectors in general. Compare Section 8.5, p. 275.

Absolute values and complex conjugates also have geometric interpretations. The absolute value $|\alpha|$ is the distance of $\alpha = a + bi$ from 0, because $|\alpha|^2 = a^2 + b^2$. The complex conjugate $\bar{\alpha} = a - bi$ is the *reflection* (mirror image) of α in the x-axis (Fig. 2.3 on next page).

Thus $\alpha + \beta$, $|\alpha|$, and $\bar{\alpha}$ have simple geometric interpretations. The product $\alpha\beta$ also has a geometric interpretation, but that is more easily seen in polar coordinates.

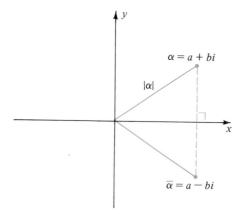

Fig. 2.3 Graphic interpretation of absolute value and conjugate

Polar Form

Let $\alpha = a + bi$ be a non-zero complex number. The point (a, b) has polar coordinates $\{r, \theta\}$, where $r > 0$,

$$a = r \cos \theta \qquad \text{and} \qquad b = r \sin \theta.$$

Therefore $\alpha = a + bi = (r \cos \theta) + (r \sin \theta)i$, that is,

$$\boxed{\alpha = r(\cos \theta + i \sin \theta).}$$

The representation of α in the form $r(\cos \theta + i \sin \theta)$ is called the **polar form** of the complex number α. See Fig. 2.4a. (This contrasts with the **rectangular form** $\alpha = a + bi$.)

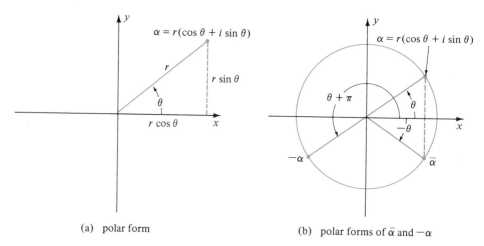

(a) polar form

(b) polar forms of $\bar{\alpha}$ and $-\alpha$

Fig. 2.4

When a complex number α is expressed in the polar form $\alpha = r(\cos\theta + i\sin\theta)$, the positive number r is the absolute value of α. Just check it:

$$|\alpha|^2 = \alpha\bar{\alpha} = [r(\cos\theta + i\sin\theta)][r(\cos\theta - i\sin\theta)] = r^2(\cos^2\theta + \sin^2\theta) = r^2;$$

hence $|\alpha| = r$.

The angle θ in $\alpha = r(\cos\theta + i\sin\theta)$ is called the **argument** of α.

We note the polar forms for $\bar{\alpha}$ and $-\alpha$. Suppose $\alpha = r(\cos\theta + i\sin\theta)$. Then $\arg\bar{\alpha} = -\theta$, and $\arg(-\alpha) = \theta + \pi$; clearly $|\bar{\alpha}| = |-\alpha| = |\alpha| = r$. See Fig. 2.4b.

> If $\alpha = r(\cos\theta + i\sin\theta)$, then
>
> $\bar{\alpha} = r[\cos(-\theta) + i\sin(-\theta)]$, $-\alpha = r[\cos(\theta + \pi) + i\sin(\theta + \pi)]$.

Multiplication and Division

Suppose $\alpha_1 = r_1(\cos\theta_1 + i\sin\theta_1)$ and $\alpha_2 = r_2(\cos\theta_2 + i\sin\theta_2)$ are two non-zero complex numbers. We compute their product:

$$\alpha_1\alpha_2 = (r_1 r_2)(\cos\theta_1 + i\sin\theta_1)(\cos\theta_2 + i\sin\theta_2)$$
$$= (r_1 r_2)[(\cos\theta_1\cos\theta_2 - \sin\theta_1\sin\theta_2) + i(\cos\theta_1\sin\theta_2 + \sin\theta_1\cos\theta_2)]$$
$$= (r_1 r_2)[\cos(\theta_1 + \theta_2) + i\sin(\theta_1 + \theta_2)].$$

With the help of the addition formulas, the product comes out in polar form!

> Let $\alpha_1 = r_1(\cos\theta_1 + i\sin\theta_1)$ and $\alpha_2 = r_2(\cos\theta_2 + i\sin\theta_2)$. Then
>
> $$\alpha_1\alpha_2 = r_1 r_2[\cos(\theta_1 + \theta_2) + i\sin(\theta_1 + \theta_2)].$$
>
> Therefore
> $$|\alpha_1\alpha_2| = |\alpha_1| \cdot |\alpha_2|,   arg\,\alpha_1\alpha_2 = arg\,\alpha_1 + \arg\alpha_2.$$
>
> To multiply two complex numbers, multiply their moduli and add their arguments.

There is a geometric interpretation of these formulas: to multiply a complex number by α, multiply its modulus by $|\alpha|$ and add $\arg\alpha$ to its argument (rotate by the angle $\arg\alpha$).

For example, since $i = 1(\cos\frac{1}{2}\pi + i\sin\frac{1}{2}\pi)$, multiplication by i rotates a complex number counterclockwise by $\frac{1}{2}\pi$. Since $-1 + i = \sqrt{2}\,(\cos\frac{3}{4}\pi + i\sin\frac{3}{4}\pi)$, multiplication by $-1 + i$ stretches a complex number by a factor $\sqrt{2}$ and rotates it counterclockwise by $\frac{3}{4}\pi$. See Fig. 2.5 on next page.

From the rule for multiplication of complex numbers in polar form follows a companion rule for division. First we note the form of the reciprocal of a complex number:

$$\frac{1}{r(\cos\theta + i\sin\theta)} = \frac{1}{r(\cos\theta + i\sin\theta)}\frac{\cos\theta - i\sin\theta}{\cos\theta - i\sin\theta}$$

$$= \frac{\cos\theta - i\sin\theta}{r(\cos^2\theta + \sin^2\theta)} = \frac{\cos(-\theta) + i\sin(-\theta)}{r}.$$

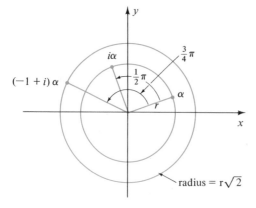

radius $= r\sqrt{2}$ **Fig. 2.5** Multiplication by i and by $-1 + i$

Hence

$$[r(\cos\theta + i\sin\theta)]^{-1} = r^{-1}[\cos(-\theta) + i\sin(-\theta)].$$

Now we can express the formula $\alpha/\beta = \alpha\beta^{-1}$ in polar form:

$$\frac{r_1(\cos\theta_1 + i\sin\theta_1)}{r_2(\cos\theta_2 + i\sin\theta_2)} = \left\{ r_1(\cos\theta_1 + i\sin\theta_1) \right\} \left\{ \frac{1}{r_2}[\cos(-\theta_2) + i\sin(-\theta_2)] \right\}.$$

Using the rule for multiplication, we find:

$$\frac{r_1(\cos\theta_1 + i\sin\theta_1)}{r_2(\cos\theta_2 + i\sin\theta_2)} = \frac{r_1}{r_2}[\cos(\theta_1 - \theta_2) + i\sin(\theta_1 - \theta_2)].$$

Therefore

$$\left| \frac{\alpha_1}{\alpha_2} \right| = \frac{|\alpha_1|}{|\alpha_2|}, \qquad \arg\frac{\alpha_1}{\alpha_2} = \arg\alpha_1 - \arg\alpha_2.$$

To divide two non-zero complex numbers, divide their moduli and subtract their arguments.

To use the neat formulas for multiplication and division, we must know how to convert complex numbers into polar form. Given $a + bi$, we can write

$$a + bi = r(\cos\theta + i\sin\theta),$$

provided

$$r = \sqrt{a^2 + b^2} \qquad \text{and} \qquad \cos\theta = \frac{a}{r}, \qquad \sin\theta = \frac{b}{r}.$$

From these formulas we can find r and θ.

■ *Example 2.1*

Convert $\alpha = 1 + i\sqrt{3}$ and $\beta = 1 + i$ into polar form and compute $\alpha\beta$ and α/β.

SOLUTION

$$|1 + i\sqrt{3}| = \sqrt{1 + 3} = 2, \qquad |1 + i| = \sqrt{1 + 1} = \sqrt{2}.$$

Hence

$$1 + i\sqrt{3} = 2(\cos\theta_1 + i\sin\theta_1), \qquad 1 + i = \sqrt{2}(\cos\theta_2 + i\sin\theta_2).$$

We find θ_1 and θ_2, either by plotting α and β in the complex plane or from the relations

$$\cos\theta_1 = \frac{1}{2}, \qquad \sin\theta_1 = \frac{\sqrt{3}}{2}, \qquad \cos\theta_2 = \sin\theta_2 = \frac{1}{\sqrt{2}} = \frac{\sqrt{2}}{2}.$$

Hence $\theta_1 = \frac{1}{3}\pi$ and $\theta_2 = \frac{1}{4}\pi$:

$$\alpha = 2\left(\cos\frac{\pi}{3} + i\sin\frac{\pi}{3}\right), \qquad \beta = \sqrt{2}\left(\cos\frac{\pi}{4} + i\sin\frac{\pi}{4}\right).$$

By the rules for multiplication and division,

$$\alpha\beta = 2\sqrt{2}\left(\cos\frac{7\pi}{12} + i\sin\frac{7\pi}{12}\right), \qquad \frac{\alpha}{\beta} = \sqrt{2}\left(\cos\frac{\pi}{12} + i\sin\frac{\pi}{12}\right).$$

Answer $\quad \alpha = 2(\cos\frac{1}{3}\pi + i\sin\frac{1}{3}\pi), \ \beta = \sqrt{2}(\cos\frac{1}{4}\pi + i\sin\frac{1}{4}\pi);$

$$\alpha\beta = 2\sqrt{2}(\cos\tfrac{7}{12}\pi + i\sin\tfrac{7}{12}\pi), \ \alpha/\beta = \sqrt{2}(\cos\tfrac{1}{12}\pi + i\sin\tfrac{1}{12}\pi).$$

If we had worked the example in rectangular form, we would have found

$$\frac{\alpha}{\beta} = \frac{1 + i\sqrt{3}}{1 + i} = \frac{1 + i\sqrt{3}}{1 + i}\frac{1 - i}{1 - i} = \frac{1 + \sqrt{3}}{2} + \frac{\sqrt{3} - 1}{2}i.$$

Comparing this with the answer above, we must have

$$\sqrt{2}\cos\frac{\pi}{12} = \frac{\sqrt{3} + 1}{2}, \qquad \sqrt{2}\sin\frac{\pi}{12} = \frac{\sqrt{3} - 1}{2},$$

that is,

$$\cos 15° = \frac{\sqrt{6} + \sqrt{2}}{4}, \qquad \sin 15° = \frac{\sqrt{6} - \sqrt{2}}{4},$$

which are formulas we have seen before, or which can be derived from $\sin 30°$ and $\cos 30°$ by the half-angle formulas.

Squares and Square Roots

If $\alpha = r(\cos\theta + i\sin\theta) \neq 0$, then the multiplication rule implies

$$\alpha^2 = r^2(\cos 2\theta + i\sin 2\theta).$$

Therefore, to square a complex number, you square its modulus and double its argument.

It stands to reason that to find a complex square root of α, you go backwards, i.e., you take the square root of the modulus and half of the argument. But there are *two* geometric angles whose doubles are θ (plus a multiple of 2π), namely $\frac{1}{2}\theta$ and $\frac{1}{2}\theta + \pi$.

Each non-zero complex number has two complex square roots:

$$\sqrt{r(\cos\theta + i\sin\theta)} = \begin{cases} \sqrt{r}\,(\cos\frac{1}{2}\theta + i\sin\frac{1}{2}\theta) \\ \sqrt{r}\,[\cos(\frac{1}{2}\theta + \pi) + i\sin(\frac{1}{2}\theta + \pi)]. \end{cases}$$

Note that the second square root is the negative of the first.

■ *Example 2.2*

Find the square roots of $9i$.

SOLUTION In polar form,

$$9i = 9(\cos\tfrac{1}{2}\pi + i\sin\tfrac{1}{2}\pi).$$

According to the preceding formulas, one square root is

$$3(\cos\tfrac{1}{4}\pi + i\sin\tfrac{1}{4}\pi) = 3(\tfrac{1}{2}\sqrt{2} + \tfrac{1}{2}i\sqrt{2})$$

and the other square root is the negative of this number.

Answer $\pm\frac{3}{2}\sqrt{2}(1 + i)$.

Check $[\frac{3}{2}\sqrt{2}(1 + i)]^2 = (\frac{3}{2}\sqrt{2})^2(1 + i)^2 = \frac{9}{2}(1 + 2i + i^2)$
$$= \tfrac{9}{2}(1 + 2i - 1) = 9i.$$

EXERCISES

Express in polar form:

1. $-3i$
2. -5
3. $1 - i$
4. $3 + 3i$
5. $-\sqrt{3} - i$
6. $-2 + 2i\sqrt{3}$
7. $3\sqrt{3} + 3i$
8. $4 - 4i.$

Express the factors and the products (quotients) in polar form:

9. $(1 + i)(1 - i)$
10. $(1 - i)(\sqrt{3} + i)$
11. $(1 + i\sqrt{3})^2$
12. $(2 + 2i)(4\sqrt{3} - 4i)$
13. $\dfrac{\sqrt{3} - i}{1 + i}$
14. $\dfrac{-\sqrt{3} + i}{3 + 3i}$
15. $\dfrac{1 - i}{2 + 2i\sqrt{3}}$
16. $\dfrac{3 - 3i}{\sqrt{3} - i}.$

Find Re α, Im α, and arg α:

17. $\alpha = \sqrt{3} + 3i$ **18.** $\alpha = 1 - i$ **19.** $\alpha = \dfrac{\sqrt{3} - i}{1 + i}$

20. $\alpha = \dfrac{4 - 4i}{2\sqrt{3} - 2i}$.

Compute both $|\alpha| \cdot |\beta|$ and $|\alpha\beta|$ and compare:

21. $\alpha = 1 - i, \ \beta = 3i$ **22.** $\alpha = 1 + i, \ \beta = -1 + i$

23. $\alpha = 2 - i, \ \beta = 3 + 2i$ **24.** $\alpha = -1 + 2i, \ \beta = 3 - i$

25. $\alpha = (1 + i)^2, \ \beta = 1 + 2i$ **26.** $\alpha = (1 - i)^2, \ \beta = 1 - 2i$.

Compute $|\alpha|/|\beta|$ and $|\alpha/\beta|$:

27. $\alpha = 1 + i, \ \beta = 2i$ **28.** $\alpha = -i, \ \beta = 2 - i$

29. $\alpha = 4, \ \beta = 3 - i$ **30.** $\alpha = 1 + 2i, \ \beta = 3i$

31. $\alpha = 3 - i, \ \beta = 2 + 5i$ **32.** $\alpha = 2 - 5i, \ \beta = -3 - i$.

Solve:

33. $z^2 = -i$ **34.** $z^2 = -9$

35. $z^2 = 1 + i$ **36.** $z^2 = 2 - 2i$

37. $z^2 = \sqrt{3} - i$ **38.** $z^2 = 1 + i\sqrt{3}$

39. $z^2 = (1 + i)^3$ **40.** $z^2 = (1 - i)^3$.

41*. Use complex numbers to prove the identity
$$(x^2 + y^2)(u^2 + v^2) = (xu - yv)^2 + (xv + yu)^2.$$

42. (cont.) From $13 = 2^2 + 3^2$ and $37 = 1^2 + 6^2$, express 481 as a sum of two perfect squares.

43. Justify geometrically the formula $\text{Re}(\alpha) = \frac{1}{2}(\alpha + \bar{a})$.

44. (cont.) Do the same for $\text{Im}(\alpha) = (\alpha - \bar{a})/2i$.

3. ZEROS OF POLYNOMIALS

Let $ax^2 + bx + c$ be any quadratic polynomial with real coefficients and $a \neq 0$. The quadratic formula
$$r = \frac{-b \pm \sqrt{b^2 - 4ac}}{2a}$$

gives the zeros of the polynomial. If $b^2 - 4ac \geq 0$, the zeros are real; if $b^2 - 4ac < 0$, they are complex. For example, consider $2x^2 - x + 1$. Then the discriminant is $b^2 - 4ac = 1 - 8 = -7$, so the zeros are
$$\frac{1 + i\sqrt{7}}{4} \quad \text{and} \quad \frac{1 - i\sqrt{7}}{4}.$$

To check, set $r = \frac{1}{4}(1 \pm i\sqrt{7})$. Then
$$4r - 1 = \pm i\sqrt{7}, \quad (4r - 1)^2 = -7, \quad 16r^2 - 8r + 1 = -7,$$
$$16r^2 - 8r + 8 = 0, \quad 2r^2 - r + 1 = 0.$$

At this point we can assert at least this much: we have succeeded in enlarging the real number system to the complex number system, and in this new system each *quadratic* polynomial with *real* coefficients has zeros.

The emphasis on "quadratic" and "real" suggests two very natural questions.

(1)   Does each quadratic polynomial with *complex* coefficients have complex zeros?

(2)   Does each real polynomial (real coefficients)

$$a_n x^n + a_{n-1} x^{n-1} + \cdots + a_1 x + a_0$$

of any degree have complex zeros?

If the answer to either question were no, we would be faced with enlarging the complex number system (perhaps over and over again) to handle more and more complicated polynomials. Fortunately the answer to both questions is yes. The complex number system is big enough; it is the "right" system for polynomials.

The answer to question (1) is easy because of the quadratic formula. Suppose α, β, γ are complex. Then the equation

$$\alpha z^2 + \beta z + \gamma = 0 \qquad (\alpha \neq 0)$$

has solutions

$$z = \frac{-\beta \pm \sqrt{\beta^2 - 4\alpha\gamma}}{2\alpha}.$$

This formula is meaningful since, as we have seen, complex numbers have complex square roots.

The proof that question (2) has a positive answer is very deep. It was given by K. F. Gauss and constitutes one of the most remarkable achievements in mathematics. Here is a precise statement of the result, in a generality that goes beyond questions (1) and (2).

> Let
>
> $$f(z) = \alpha_n z^n + \alpha_{n-1} z^{n-1} + \cdots + \alpha_1 z + \alpha_0 \qquad (\alpha_n \neq 0)$$
>
> be any polynomial with complex coefficients and degree $n \geq 1$. Then there is a complex number β such that $f(\beta) = 0$.

This result is called the **Fundamental Theorem of Algebra.** Its proof requires advanced theory. Note that the theorem only guarantees the existence of a zero; it does not say a word about how to find one. However there are numerical methods for approximating zeros to any degree of accuracy, some well suited for computers.

The same reasoning we used in Chapter 9 for real polynomials (p. 290) establishes the following **Factor Theorem.**

> Let $f(z)$ be a polynomial of degree $n \geq 1$ with complex coefficients. Suppose β is a complex zero of $f(z)$. Then $f(z) = (z - \beta)g(z)$, where $g(z)$ is a polynomial of degree $n - 1$.

In other words, if β is a zero of $f(z)$, then $z - \beta$ is a linear factor of $f(z)$. The quotient $g(z)$ is a complex polynomial of degree $n - 1$. If $n - 1 \geq 1$, then $g(z)$ has a complex zero γ, hence a factor $z - \gamma$, by the Fundamental Theorem. Therefore

$$f(z) = (z - \beta)(z - \gamma)h(z),$$

where $h(z)$ is a complex polynomial of degree $n - 2$. By the same argument, if $n - 2 \geq 1$, then $h(z)$ has a linear factor $z - \delta$, etc. The final result is:

Let $f(z) = \alpha_n z^n + \cdots + \alpha_0$ be a complex polynomial of degree $n \geq 1$. Then there are complex numbers $\beta_1, \cdots, \beta_n$ such that

$$f(z) = \alpha_n(z - \beta_1)(z - \beta_2) \cdots (z - \beta_n).$$

In other words, each complex polynomial is a product of linear factors.

Note: The numbers $\beta_1, \cdots, \beta_n$ are the **zeros** of $f(z)$, and there can be repetitions among them. We could write the complete factorization of $f(z)$ in the form

$$f(z) = \alpha_n(z - \beta_1)^{m_1} \cdots (z - \beta_k)^{m_k},$$

where $\beta_1, \cdots, \beta_k$ are distinct from each other. The exponents $m_1, \cdots, m_k$ are positive integers with $m_1 + m_2 + \cdots + m_k = \deg f(z)$. We call m_j the **multiplicity** of the zero β_j. If $m_j = 1$, we call β_j a **simple zero** of $f(z)$.

Complex Zeros of Real Polynomials

Each polynomial with complex coefficients has a complex zero. If the coefficients happen to be real, we can say even more:

Let $f(z) = a_n z^n + \cdots + a_0$ be a polynomial with *real coefficients*. If β is a complex zero of $f(z)$ *that is not real*, then $\bar{\beta}$ also is a zero of $f(z)$.

In simple words, complex zeros of real polynomials occur in conjugate pairs. The proof uses the rules $\overline{\alpha + \beta} = \bar{\alpha} + \bar{\beta}$, $\overline{\alpha\beta} = \bar{\alpha}\bar{\beta}$, and $\overline{\alpha^n} = (\bar{\alpha})^n$. Suppose $f(\beta) = 0$. Then

$$a_n \beta^n + a_{n-1} \beta^{n-1} + \cdots + a_0 = 0.$$

Take conjugates on both sides. Since each a_j is *real*, $\bar{a}_j = a_j$:

$$a_n \bar{\beta}^n + a_{n-1} \bar{\beta}^{n-1} + \cdots + a_0 = \bar{0} = 0.$$

This says $f(\bar{\beta}) = 0$. Done!

An immediate consequence is the factorization

$$f(z) = (z - \beta)(z - \bar{\beta})h(z),$$

where $\deg h(z) = n - 2$. Now observe that

$$(z - \beta)(z - \bar{\beta}) = z^2 - (\beta + \bar{\beta})z + \beta\bar{\beta}$$
$$= z^2 - 2[\mathrm{Re}(\beta)]z + |\beta|^2.$$

This is a *real* quadratic polynomial $g(z)$. Therefore $f(z) = g(z)h(z)$, where $h(z) = f(z)/g(z)$ is also a real polynomial since division of real polynomials can only lead to real polynomials. The same reduction can now be repeated on $h(z)$, etc.

Now let us fit all the pieces together. Start with any non-constant real polynomial $f(x)$. It has some (maybe no) real zeros and some (maybe no) conjugate pairs of non-real zeros. Each real zero yields a real linear factor $x - r$; each pair of non-real conjugate zeros yields a real quadratic factor.

Let $f(x)$ be a real polynomial of degree $n \geq 1$. Then

$$f(x) = a_n(x - r_1) \cdots (x - r_k)g_1(x) \cdots g_s(x),$$

where $r_1, \cdots, r_k$ are real and

$$g_j(x) = x^2 + b_j x + c_j$$

with b_j and c_j real and $b_j^2 - 4c_j < 0$. Here $k \geq 0$, $s \geq 0$, and $k + 2s = n$.

The quadratic factors $g_1(x), \cdots, g_s(x)$ cannot be split into real linear factors; we call such factors **irreducible.** Now we can restate the result above as follows:

Each real polynomial is the product of real irreducible linear and quadratic factors.

■ *Example 3.1*

Express $f(x) = x^5 - 3x^4 + 2x^3 - 6x^2 + x - 3$ as the product of real irreducible factors.

SOLUTION By trial and error we find $f(3) = 0$. Hence $x - 3$ is a factor, and by division

$$f(x) = (x - 3)(x^4 + 2x^2 + 1).$$

But $x^4 + 2x^2 + 1 = (x^2 + 1)^2$, and $x^2 + 1$ is irreducible.

Answer $(x - 3)(x^2 + 1)^2$.

Remark: The irreducible factor $x^2 + 1$ corresponds to the complex zeros $\pm i$ since $x^2 + 1 = (x - i)(x + i)$. The factorization of $f(x)$ into complex linear factors is

$$f(x) = (x - 3)(x - i)^2(x + i)^2.$$

EXERCISES

Solve:

1. $z^2 - 2z + 5 = 0$ 2. $z^2 + 4z + 13 = 0$ 3. $z^2 + z + 6 = 0$
4. $2z^2 - 3z + 10 = 0$ 5. $z^4 + 5z^2 + 4 = 0$ 6. $z^4 + 4z^2 + 29 = 0$
7. $z^3 - 1 = 0$ 8. $z^3 + 8 = 0$ 9. $z^4 - 1 = 0$
10. $z^3 + z - 2 = 0$.

11. Factor $x^3 + 1$ into real irreducible factors.

12. Factor $x^4 - 1$ into real irreducible factors.

13. Show from the factorization

$$z^5 + 1 = (z + 1)(z^4 - z^3 + z^2 - z + 1)$$

that -1 is a simple zero of $z^5 + 1$.

14. Show from the factorization

$$z^n - 1 = (z - 1)(z^{n-1} + z^{n-2} + \cdots + z + 1)$$

that $+1$ is a simple zero of $z^n - 1$.

15. Prove that a real polynomial of odd degree has a real zero.

16. If a polynomial $f(z)$ has zeros $\pm i$, prove it is divisible by $z^2 + 1$.

Write down the most general real polynomial satisfying the given conditions:

17. degree 4 and zeros ± 2, $1 \pm i$

18. degree 5 and a real zero of multiplicity 4

19. degree 4 and no real zeros

20. degree 6 and zeros $\pm i$, $\pm 2i$, and 0 with multiplicity 2.

4. DE MOIVRE'S THEOREM AND ROOTS OF UNITY

De Moivre's Theorem

De Moivre's Theorem is an important formula for the integer powers of a complex number.

> **De Moivre's Theorem**
>
> $$[r(\cos \theta + i \sin \theta)]^n = r^n(\cos n\theta + i \sin n\theta).$$

When $n = 0$, the formula is true; the left side is 1 by definition ($\alpha^0 = 1$ whenever $\alpha \neq 0$), and the right side is 1 because $r^0(\cos 0 + i \sin 0) = 1 \cdot 1 = 1$.

When $n \geq 1$, the formula results from repeated application of the rule for multiplying complex numbers in polar form. For instance when $n = 3$, pull out the factor r^3 and then

$$(\cos \theta + i \sin \theta)^3 = (\cos \theta + i \sin \theta)^2(\cos \theta + i \sin \theta)$$
$$= (\cos 2\theta + i \sin 2\theta)(\cos \theta + i \sin \theta)$$
$$= \cos 3\theta + i \sin 3\theta,$$

so the De Moivre formula holds.

When $n < 0$, the formula follows from the positive case and the expression of the reciprocal in polar form. Set $n = -m$ with $m > 0$. Then

$$[r(\cos \theta + i \sin \theta)]^n = \{[r(\cos \theta + i \sin \theta)]^{-1}\}^m$$
$$= \{r^{-1}[\cos(-\theta) + i \sin(-\theta)]\}^m$$
$$= (r^{-1})^m[\cos m(-\theta) + i \sin m(-\theta)]$$
$$= r^{-m}[\cos(-m\theta) + i \sin(-m\theta)]$$
$$= r^n(\cos n\theta + i \sin n\theta).$$

■ *Example 4.1*

Find $(1 + i)^{15}$ in rectangular form.

SOLUTION In polar form,

$$1 + i = \sqrt{2}\left(\cos\frac{\pi}{4} + i\sin\frac{\pi}{4}\right).$$

hence,

$$(1 + i)^{15} = (\sqrt{2})^{15}\left[\cos\left(15 \cdot \frac{\pi}{4}\right) + i\sin\left(15 \cdot \frac{\pi}{4}\right)\right].$$

Now $(\sqrt{2})^{15} = 2^{15/2} = 2^7\sqrt{2} = 128\sqrt{2}$. Also $15(\frac{1}{4}\pi) = \frac{15}{4}\pi = 4\pi - \frac{1}{4}\pi$. Therefore, by De Moivre's Theorem,

$$(1 + i)^{15} = 128\sqrt{2}\left[\cos\left(-\frac{\pi}{4}\right) + i\sin\left(-\frac{\pi}{4}\right)\right]$$

$$= 128\sqrt{2}\left(\frac{\sqrt{2}}{2} - \frac{\sqrt{2}}{2}i\right) = 128 - 128i.$$

Answer $128 - 128i$.

· ·

Roots of Unity

We now discuss the solution of the special polynomial equation

$$z^n = 1.$$

According to the general theory, the polynomial $z^n - 1$ has n complex zeros, which we call the n-th **roots of unity.**

If n is even, then ± 1 are roots of $z^n - 1$; if n is odd, then $+1$ is a root, but -1 is not. In either case, these are the only real roots. Thus, most of the roots of unity are non-real. How do we find them?

Let's try to solve $z^n = 1$. If z is a solution, then $|z|^n = |z^n| = 1$, hence $|z| = 1$. Therefore we can write

$$z = \cos\theta + i\sin\theta.$$

Now substitute into the equation $z^n = 1$ and use De Moivre's Theorem:

$$\cos n\theta + i\sin n\theta = 1 = \cos 0 + i\sin 0.$$

For this equation to hold, $n\theta$ must be 0 or a multiple of 2π, that is,

$$n\theta = 2\pi k \qquad (k \text{ an integer}).$$

Therefore θ must take one of the values

$$\theta = \frac{2\pi k}{n} \qquad (k = 0, \pm 1, \pm 2, \pm 3, \cdots).$$

While these look like a lot of values of θ, they represent only n distinct angles. After n consecutive integers, the same values appear increased by 2π:

$$\frac{2\pi(k+n)}{n} = \frac{2\pi k}{n} + 2\pi.$$

Thus we can restrict θ to the values

$$\theta = \frac{2\pi k}{n} \qquad (k = 0, 1, \cdots, n-1).$$

Other choices of k will produce a value of $2\pi k/n$ that differs from one of the above numbers by a multiple of 2π.

The n-th roots of unity are the n complex numbers

$$z_k = \cos\frac{2\pi k}{n} + i\sin\frac{2\pi k}{n} \qquad (k = 0, 1, \cdots, n-1).$$

They form the vertices of a regular n-gon inscribed in the unit circle (Fig. 4.1).

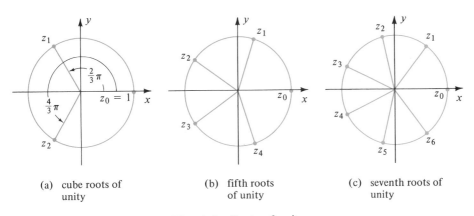

(a) cube roots of unity

(b) fifth roots of unity

(c) seventh roots of unity

Fig. 4.1   Roots of unity

Remark: It follows from De Moivre's Theorem that $z_k = z_1{}^k$ for $k = 0, 1, \cdots, n-1$. Therefore we have the nice factorization

$$z^n - 1 = (z-1)(z - z_1)(z - z_1{}^2) \cdots (z - z_1{}^{n-1}),$$

where $z_1 = \cos(2\pi/n) + i\sin(2\pi/n)$.

■ *Example 4.2*

Compute the cube roots of unity and express each in the form $a + bi$.

SOLUTION   Use the formula with $n = 3$ and $k = 0, 1, 2$:

$$k = 0: \qquad z_0 = \cos 0 + i \sin 0 = 1$$

$$k = 1: \qquad z_1 = \cos \frac{2\pi}{3} + i \sin \frac{2\pi}{3} = -\tfrac{1}{2} + \tfrac{1}{2} i \sqrt{3}$$

$$k = 2: \qquad z_2 = \cos \frac{4\pi}{3} + i \sin \frac{4\pi}{3} = -\tfrac{1}{2} - \tfrac{1}{2} i \sqrt{3}.$$

Answer $1, \ -\tfrac{1}{2} + \tfrac{1}{2} i \sqrt{3}, \ -\tfrac{1}{2} - \tfrac{1}{2} i \sqrt{3}.$

. . . .

Remark: Note that there is a conjugate pair of non-real roots. We could have predicted that since the polynomial $z^3 - 1$ has real coefficients. Also note it is not obvious that the cube of $-\tfrac{1}{2} \pm \tfrac{1}{2} i \sqrt{3}$ is 1; you should check it.

General *n*-th Roots

The solution of the equation

$$z^n = \alpha \qquad (\alpha \neq 0)$$

is similar to that of the special case $z^n = 1$. Write α in polar form:

$$\alpha = r_0 (\cos \theta_0 + i \sin \theta_0) \qquad (r_0 > 0),$$

and write the unknown z as

$$z = r(\cos \theta + i \sin \theta) \qquad (r > 0).$$

Then $z^n = r^n (\cos n\theta + i \sin n\theta)$ by De Moivre's Theorem, so we must have

$$r^n (\cos n\theta + i \sin n\theta) = r_0 (\cos \theta_0 + i \sin \theta_0).$$

This equation requires

$$r^n = r_0, \qquad n\theta = \theta_0 + 2\pi k, \qquad k \text{ an integer.}$$

Therefore we choose

$$r = r_0^{1/n}, \qquad \theta = \frac{1}{n} \theta_0 + \frac{2\pi k}{n} \qquad (k = 0, 1, \cdots, n - 1).$$

(As with roots of unity, there is no use taking other values of k; they yield no further angles.)

The n roots of the equation

$$z^n = r_0 (\cos \theta_0 + i \sin \theta_0) \qquad (r_0 > 0)$$

are

$$z_k = r_0^{1/n} \left[\cos \left(\frac{1}{n} \theta_0 + \frac{2\pi k}{n} \right) + i \sin \left(\frac{1}{n} \theta_0 + \frac{2\pi k}{n} \right) \right] \qquad (k = 0, 1, \cdots, n - 1).$$

They form the vertices of a regular n-gon centered at 0.

■ *Example 4.3*

Solve the equation $z^3 = 2i$.

SOLUTION   Write $z = r(\cos\theta + i\sin\theta)$ and $2i = 2(\cos\frac{1}{2}\pi + i\sin\frac{1}{2}\pi)$. Then $z^3 = 2i$ means

$$r^3(\cos 3\theta + i\sin 3\theta) = 2(\cos\tfrac{1}{2}\pi + i\sin\tfrac{1}{2}\pi).$$

Therefore $r = \sqrt[3]{2}$ and $3\theta = \frac{1}{2}\pi + 2\pi k$ for $k = 0, 1, 2$. Hence, $\theta = \frac{1}{6}\pi, \frac{1}{6}\pi + \frac{2}{3}\pi,$ $\frac{1}{6}\pi + \frac{4}{3}\pi$. The roots are

$$\begin{cases} z_0 = \sqrt[3]{2}\left(\cos\dfrac{\pi}{6} + i\sin\dfrac{\pi}{6}\right) = \sqrt[3]{2}\left(\dfrac{\sqrt{3}}{2} + \dfrac{1}{2}i\right) \\[2mm] z_1 = \sqrt[3]{2}\left(\cos\dfrac{5\pi}{6} + i\sin\dfrac{5\pi}{6}\right) = \sqrt[3]{2}\left(-\dfrac{\sqrt{3}}{2} + \dfrac{1}{2}i\right) \\[2mm] z_2 = \sqrt[3]{2}\left(\cos\dfrac{3\pi}{2} + i\sin\dfrac{3\pi}{2}\right) = -i\,\sqrt[3]{2}. \end{cases}$$

Answer $\frac{1}{2}\sqrt[3]{2}(\pm\sqrt{3} + i),\quad -i\sqrt[3]{2}$. See Fig 4.2.

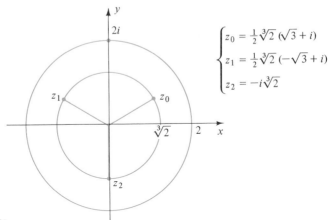

Fig. 4.2   Roots of $z^3 = 2i$

Application

Here is an amusing problem whose solution is very simple with complex numbers. If at 3:00 P.M., we interchange the hands of a clock, we obtain a configuration that never occurs on a normal clock. See Fig. 4.3 on next page. Any position of the hands that actually occurs we call a **true time.**

Problem.   When is it possible to interchange the hands of a clock and obtain a true time?

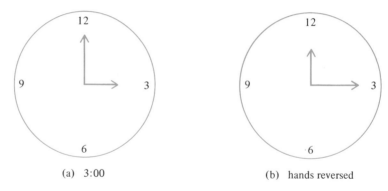

(a) 3:00   (b) hands reversed

Fig. 4.3

Obviously when the two hands overlap, twelve times in twelve hours. But are there any other solutions?

Let us represent a clock by the unit circle $|z| = 1$ in the complex plane, with the 12 at the point $z = 1$. For convenience we let our clock run counterclockwise; this doesn't affect the answer to the problem. See Fig. 4.4a. We indicate the position of the minute hand by a complex number z on the circle and the position of the hour hand by a complex number w. See Fig. 4.4b.

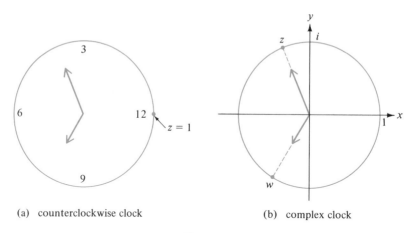

(a) counterclockwise clock   (b) complex clock

Fig. 4.4

At noon, $z = w = 1$. As the clock runs, z moves around the circle 12 times as fast as w. In terms of complex numbers that means $z = w^{12}$. Thus a true time is represented by an ordered pair of complex numbers (z, w), where $|z| = |w| = 1$ and $z = w^{12}$.

Suppose we interchange the hands at the true time (z, w). We obtain a position (w, z) which is a true time only if $w = z^{12}$. Therefore the problem is reduced to this: find pairs (z, w) such that $z = w^{12}$ and also $w = z^{12}$. For such a pair,

$$w = z^{12} = (w^{12})^{12}, \qquad w = w^{144}.$$

Because $w \neq 0$, we must have

$$w^{143} = 1.$$

Therefore, the desired values of w are the 143-rd roots of unity. When the hour hand points at one of these 143 equally spaced points, the reversed position of the hands is a true time.

Answer   143 times in 12 hours, at equal intervals (once every $5\frac{5}{143}$ minutes starting at 12 o'clock).

Compute:

1. $(1 - i)^{11}$
2. $(1 - i)^{27}$
3. $(\sqrt{3} + i)^{6}$
4. $(1 + i\sqrt{3})^{-17}$
5. $[\frac{1}{2}(-1 + i\sqrt{3})]^{13}$
6. $(\cos \frac{1}{7}\pi + i \sin \frac{1}{7}\pi)^{50}$.

Solve completely:

7. $z^4 = 1$
8. $z^5 = 1$
9. $z^6 = 1$
10. $z^8 = 1$
11. $z^3 = -i$
12. $z^3 = 8$
13. $z^4 = i$
14. $z^4 = -1$
15. $z^4 = -1 + i$
16. $z^3 = -1 + i$.

17. Find two distinct complex numbers, each one the square of the other.
18. Solve the clock problem for a 24 hour clock.
19. Let α be an n-th root of unity and $\alpha \neq 1$. Show that

$$\alpha^{n-1} + \alpha^{n-2} + \cdots + \alpha + 1 = 0.$$

20. (cont.) Solve $z^4 + z^3 + z^2 + z + 1 = 0$.
21. Let $\alpha = \cos \frac{2}{5}\pi + i \sin \frac{2}{5}\pi$. Set $\beta = \alpha + \alpha^{-1}$. Prove that $\beta^2 + \beta - 1 = 0$.
22. (cont.) Prove that $\beta = \frac{1}{2}(-1 + \sqrt{5})$.
23. (cont.) Prove that $\alpha^2 - \beta\alpha + 1 = 0$.
24. (cont.) Prove that $\alpha = \frac{1}{2}(\beta + i\sqrt{4 - \beta^2})$. Conclude that

$$\cos 72° = \frac{-1 + \sqrt{5}}{4}, \qquad \sin 72° = \frac{\sqrt{10 + 2\sqrt{5}}}{4}.$$

A number α is a **primitive** n-th root of unity if $\alpha^n = 1$, but $\alpha^m \neq 1$ for $0 < m < n$. Find the primitive n-th roots of unity for

25. $n = 4$
26. $n = 5$
27. $n = 6$
28. $n = 7$
29. $n = 8$
30. $n = 9$
31. $n = 10$
32. $n = 12$.

33*. Show that each primitive 6-th root of unity satisfies $z^2 - z + 1 = 0$.
34*. Show that each primitive 8-th root of unity satisfies $z^4 + 1 = 0$.
35*. Show that each primitive 9-th root of unity satisfies $z^6 + z^3 + 1 = 0$.
36*. Show that each primitive 12-th root of unity satisfies $z^4 - z^2 + 1 = 0$.
37. Let α be a 7-th root of unity and $\alpha \neq 1$. Show that α is a primitive 7-th root of unity.

38. Prove $\quad 1 + \cos\dfrac{2\pi}{n} + \cos\dfrac{4\pi}{n} + \cdots + \cos\dfrac{2(n-1)\pi}{n} = 0,$

$$\sin\dfrac{2\pi}{n} + \sin\dfrac{4\pi}{n} + \cdots + \sin\dfrac{2(n-1)\pi}{n} = 0.$$

[Hint: Use Ex. 19.]

39. Verify that $\alpha = 2 - i$ is a root of the equation $z^4 = -7 - 24i$. Show that all roots are $\alpha,\ i\alpha,\ -\alpha,\ -i\alpha$, that is, α multiplied by each of the 4-th roots of unity.

40. (cont.) If α is a root of the equation $z^n = \beta$, show that all roots are $z_0\alpha, z_1\alpha, \cdots, z_{n-1}\alpha$, where $z_0, z_1, \cdots, z_{n-1}$ are the n-th roots of unity.

41. If $1, z_1, z_2, \cdots, z_{n-1}$ are the n-th roots of unity, show that

$$(z - z_1)(z - z_2) \cdots (z - z_{n-1}) = z^{n-1} + z^{n-2} + \cdots + z + 1.$$

42*. (cont.) Let $P_0, P_1, \cdots, P_{n-1}$ be the vertices of a regular n-gon inscribed in a circle of radius 1. Compute the product of the lengths $\overline{P_0 P_1}, \overline{P_0 P_2}, \cdots, \overline{P_0 P_{n-1}}$. [Hint: Represent the points by complex numbers lying on the circle $|z| = 1$. Then use Ex. 41.]

Test 1

1. Express in the form $a + bi$:

(a) $\dfrac{2 - 5i}{1 + 3i}$ (b) $\overline{(1 + i)(3 - 2i)}$ (c) $\sqrt{-16}$.

2. If α, β, γ are the vertices of an equilateral triangle in the complex plane, prove that $\gamma - a = \omega(\beta - a)$, where ω is a 6-th root of unity.

3. Find the square roots of $1 - i\sqrt{3}$.

4. Find all 6-th roots of i.

5. Is the statement true or false? Why?

(a) $\mathrm{Re}(a) + \mathrm{Re}(\beta) = \mathrm{Re}(a + \beta)$

(b) $\mathrm{Re}(a)\,\mathrm{Re}(\beta) = \mathrm{Re}(a\beta)$

(c) If $|a| = 1$, then $a^{-1} = \bar{\alpha}$.

Test 2

1. Express in the form $a + bi$:

(a) $\dfrac{(4 - i)(2 + i)}{3 + 4i}$ (b) $(1 + i)^2\overline{(3 + 2i)}$.

2. The point a moves counterclockwise around the circle $|z| = 2$ in the complex plane. Describe the corresponding motion of a^{-1}.

3. Compute $(\sqrt{3} + i)^{14}$.

4. Plot in the complex plane: the number $a = 1 + i$ and all solutions of the equation $z^3 = a$.

5. Let d_1, d_2, d_3, d_4 be the distances from a point z in the complex plane to the points $2, -2, 2i, -2i$. Show that $d_1 \cdot d_2 \cdot d_3 \cdot d_4 = |z^4 - 16|$.

DISCRETE ALGEBRA

A **sequence** is a collection of numbers in a fixed order. For example

$$2, 4, 6, 8, 10, 12 \quad \text{and} \quad 1, -\tfrac{1}{2}, \tfrac{1}{4}, -\tfrac{1}{8}, \cdots, \tfrac{1}{64}$$

are sequences. A sequence of n numbers is often written

$$a_1, a_2, a_3, \cdots, a_n.$$

Although this is the most usual numbering of the terms, the indices may have a range of values other than the sequence $1, 2, 3, \cdots, n$. For example

$$b_0, b_1, b_2, \cdots, b_{n-1}; \quad c_7, c_8, \cdots, c_{n+6}; \quad d_{r+1}, d_{r+2}, \cdots, d_{r+n}$$

are sequences of n terms. Note that the sequence

$$a_k, a_{k+1}, a_{k+2}, \cdots, a_m$$

has $m - k + 1$ terms, *not* $m - k$. To check, simply write the subscripts as

$$k + 0, \quad k + 1, \quad k + 2, \cdots, k + (m - k).$$

Added to k are the $m - k + 1$ numbers $0, 1, 2, \cdots, m - k$, *including* 0.

In this section we study three useful types of sequences.

Arithmetic Progressions

An **arithmetic progression** of n terms is a sequence of n equally spaced numbers. Such a sequence of numbers must have the form

$$a_1 = a, \quad a_2 = a + d, \quad a_3 = a + 2d, \quad \cdots, \quad a_n = a + (n - 1)d.$$

Consecutive terms differ by the same number d, called the **common difference** of the arithmetic progression.

Examples:

$$2, 7, 12, 17, 22, 27 \qquad a = 2, \qquad d = 5, \qquad n = 6,$$

$$-1, -\tfrac{1}{2}, 0, \tfrac{1}{2} \qquad a = -1, \quad d = \tfrac{1}{2}, \qquad n = 4,$$

$$50, 48, 46, 44, \cdots, 2 \qquad a = 50, \quad d = -2, \quad n = 25.$$

In many applications, we need the sum of (the terms of) an arithmetic progression. Let us find a formula for the sum. We observe in an arithmetic progression $a_1, \cdots, a_n$ that

$$a_2 + a_{n-1} = (a_1 + d) + (a_n - d) = a_1 + a_n.$$

Similarly, $a_3 + a_{n-2} = a_2 + a_{n-1} = a_1 + a_n$, etc.

To exploit this information, we sum the sequence twice, once forwards, once backwards; then we add the two sums:

$$
\begin{aligned}
S &= \quad a \qquad + \quad (a + d) \quad + \quad (a + 2d) \quad + \cdots + \ [a + (n-1)d] \\
S &= [a + (n-1)d] \ + \ [a + (n-2)d] \ + \ [a + (n-3)d] \ + \cdots + \quad a \\
\hline
2S &= [2a + (n-1)d] + [2a + (n-1)d] + [2a + (n-1)d] + \cdots + [2a + (n-1)d].
\end{aligned}
$$

Each of the n columns has the same sum. Therefore

$$2S = n[2a + (n-1)d], \qquad S = \tfrac{1}{2}n[2a + (n-1)d].$$

We can always replace $[2a + (n-1)d]$ by $a_1 + a_n$ when convenient.

Sum of Arithmetic Progression

$$S = a + (a + d) + (a + 2d) + \cdots + [a + (n-1)d]$$

$$= \frac{n}{2}[2a + (n-1)d] = \frac{n}{2}(a_1 + a_n).$$

The second expression for S says that the sum of an arithmetic progression is its number of terms times the average of its first and last terms.

◼ *Example 1.1*

Compute the sum of the arithmetic progression:

(a)   all odd integers from 1 to 99,

(b) $-2.0 - 0.5 + 1.0 + 2.5 + \cdots + 31.0$.

SOLUTION (a)   The progression obviously has $\tfrac{1}{2}(100) = 50$ terms. By the formula for the sum, $S = \tfrac{1}{2}(50)(1 + 99) = 2500$.

(b)   In this case $a_1 = -2.0$, $d = 1.5$, and $a_n = 31.0$, so we can find n from

$$31.0 = -2.0 + (n - 1)(1.5),$$

$$(1.5)(n - 1) = 33, \quad n - 1 = 33/1.5 = 22, \quad n = 23.$$

Therefore $S = \frac{1}{2}(23)(-2.0 + 31) = \frac{1}{2}(23)(29) = 333.5$.

Answer   (a)   2500   (b)   333.5.

. . . .

■ *Example 1.2*

Find a formula for

$$(n + 1) + (n + 2) + \cdots + 3n.$$

SOLUTION   This arithmetic progression consists of the first $3n$ positive integers with the first n positive integers deleted, so it has $2n$ terms. By the formula for the sum,

$$S = \frac{2n}{2}[(n + 1) + 3n] = n(4n + 1).$$

Answer $n(4n + 1)$.

. . . .

Geometric Progressions

In an arithmetic progression, we *add* a fixed number to each term to find the next term. Now we discuss sequences where we *multiply* by a fixed number. A **geometric progression** of n terms is a sequence of numbers

$$a_1 = a, \quad a_2 = ar, \quad a_3 = ar^2, \quad \cdots, \quad a_n = ar^{n-1}.$$

We call r the **common ratio** of the geometric progression.

Examples:

$$2, 6, 18, 54, 162, 486 \qquad a = 2, \qquad r = 3, \qquad n = 6$$

$$13, \frac{13}{2}, \frac{13}{4}, \frac{13}{8} \qquad a = 13, \qquad r = \tfrac{1}{2}, \qquad n = 4$$

$$1.05, (1.05)^2, (1.05)^3, \cdots, (1.05)^{60} \qquad a = 1.05, \quad r = 1.05, \quad n = 60.$$

Let us find a formula for the sum of a geometric progression. Let

$$S = a + ar + ar^2 + \cdots + ar^{n-1} = a(1 + r + r^2 + \cdots + r^{n-1}).$$

If $r = 1$, then all n terms equal a, hence $S = na$. If $r \neq 1$, then

$$(1 - r)S = a(1 - r)(1 + r + r^2 + \cdots + r^{n-1}) = a(1 - r^n)$$

by one of the rules for factoring polynomials [Rule (6), p. 39]. Divide by $1 - r$:

Sum of Geometric Progression

$$S = a + ar + \cdots + ar^{n-1} = \begin{cases} na, & r = 1 \\ a\dfrac{1 - r^n}{1 - r} = a\dfrac{r^n - 1}{r - 1}, & r \neq 1. \end{cases}$$

Remark: The form $a(1 - r^n)/(1 - r)$ is better when $r < 1$ and the form $a(r^n - 1)/(r - 1)$ when $r > 1$, because the numerators and denominators are positive.

■ *Example 1.3*

Compute the sum of the geometric progression:

(a) $2 + 6 + 18 + 54 + 162 + 486$

(b) $3 - \dfrac{3}{2} + \dfrac{3}{4} - \dfrac{3}{8} + \dfrac{3}{16} - \dfrac{3}{32} + \dfrac{3}{64} - \dfrac{3}{128} + \dfrac{3}{256}.$

SOLUTION   (a)   In this case $a = 2$, $r = 3$, and $n = 6$, so

$$S = a\frac{r^n - 1}{r - 1} = 2\frac{3^6 - 1}{3 - 1} = 2\frac{729 - 1}{2} = 728.$$

(b)   In this case $a = 3, r = -\frac{1}{2}$, and $n = 9$, so

$$S = 3\frac{1 - (-\frac{1}{2})^9}{1 - (-\frac{1}{2})} = 3\frac{1 + 1/512}{3/2} = 2\left(1 + \frac{1}{512}\right) = \frac{513}{256}.$$

Answer   (a)   728   (b)   513/256.

■ *Example 1.4*

On the first day of each year, for 10 consecutive years, I invest $1000 in an account that bears 5% simple yearly interest. What is the value of the account at the end of 10 years?

SOLUTION   The last deposit earns interest once and grows to $1000(1.05)$. The next to last deposit earns interest twice and grows to $1000(1.05)^2$. Adding the final values of all 10 deposits, we obtain

$$1000(1.05) + 1000(1.05)^2 + 1000(1.05)^3 + \cdots + 1000(1.05)^{10}$$
$$= 1000(1.05)[1 + (1.05) + (1.05)^2 + \cdots + (1.05)^9]$$
$$= 1050\frac{(1.05)^{10} - 1}{1.05 - 1} \approx 13206.79.$$

Answer $13,206.79.

We define an important sequence of numbers:

$$0! = 1, \quad 1! = 1, \quad 2! = 1 \cdot 2 = 2, \quad 3! = 1 \cdot 2 \cdot 3 = 6,$$

$$n! = 1 \cdot 2 \cdot 3 \cdots n = (n - 1)! \, n.$$

The number $n!$ is read n **factorial.** These numbers increase rapidly, as is shown in a short table:

j	0	1	2	3	4	5	6	7	8	9	10
$j!$	1	1	2	6	24	120	720	5040	40320	362880	3628800

Also $20! \approx 2.4329 \times 10^{18}$ and $100! \approx 9.3326 \times 10^{157}$.

Any product of consecutive integers can be expressed in terms of factorials. For example,

$$5 \cdot 6 \cdot 7 \cdot 8 \cdot 9 = \frac{1 \cdot 2 \cdot 3 \cdot 4 \cdot 5 \cdot 6 \cdot 7 \cdot 8 \cdot 9}{1 \cdot 2 \cdot 3 \cdot 4} = \frac{9!}{4!}.$$

In general, if $k < n$, then

$$(k + 1)(k + 2)(k + 3) \cdots n = \frac{n!}{k!}.$$

EXERCISES

Find the 12-th term and the n-th term:

1. $4, 11, 18, 25, 32, \cdots$
2. $8, 5, 2, -1, -4, \cdots$
3. $\frac{1}{8}, \frac{1}{4}, \frac{1}{2}, 1, 2, \cdots$
4. $ab^2c^5, ab^3c^7, ab^4c^9, ab^5c^{11}, \cdots$.

Compute the sum:

5. the first n positive integers
6. the first n even positive integers
7. $11 + 12 + 13 + \cdots + 35$
8. $8 + 11 + 14 + \cdots + 53$
9. $-15 - 13 - 11 + \cdots + 5$
10. $8.7 + 13.7 + 18.7 + \cdots + 108.7$
11. $x + (x + 2) + (x + 4) + \cdots + (x + 18)$
12. $(x + 3y) + (x + 5y) + (x + 7y) + \cdots + (x + 21y)$
13. $(2a + 3b) + (2a + 6b) + (2a + 9b) + \cdots + (2a + 3nb)$
14. $(2n + 1) + (2n + 2) + (2n + 3) + \cdots + 5n$
15. $3 + 6 + 12 + \cdots + 3 \cdot 2^9$
16. $3 - 6 + 12 - + \cdots + 3(-2)^9$
17. $8 + 4 + 2 + 1 + \cdots + \frac{1}{16}$
18. $1 + \frac{1}{3} + \frac{1}{3^2} + \cdots + \frac{1}{3^5}$
19. $2^4 + 2^5 + 2^6 + \cdots + 2^{10}$
20. $1 - \frac{2}{3} + \left(\frac{2}{3}\right)^2 - \left(\frac{2}{3}\right)^3 + \left(\frac{2}{3}\right)^4$
21. $\frac{1}{x} + \frac{1}{x^2} + \frac{1}{x^3} + \cdots + \frac{1}{x^6}$
22. $x^{1/2} + x + x^{3/2} + \cdots + x^4$
23. $ab^2c^5 + ab^3c^7 + ab^4c^9 + \cdots + ab^8c^{17}$
24. $1 - y^2 + y^4 - + \cdots + y^{20}$.

25. Find all sequences a_1, a_2, a_3 that are both arithmetic and geometric progressions.
26. Find all arithmetic progressions a_1, a_2, a_3, such that $a_1{}^2, a_2{}^2, a_3{}^2$, is also an arithmetic progression.

27. Suppose $a_1, a_2, \cdots, a_n$ and $b_1, b_2, \cdots, b_n$ are arithmetic progressions. Prove that
$$a_1 + 3b_1, a_2 + 3b_2, \cdots, a_n + 3b_n$$
is also an arithmetic progression.

28. Suppose $a_1, a_2, \cdots, a_n$ and $b_1, b_2, \cdots, b_n$ are geometric progressions. Prove that
$$a_1 b_1, a_2 b_2, \cdots, a_n b_n$$
is also a geometric progression.

29. Show that each term of the sequence
$$1, 2, 4, 8, \cdots, 2^n$$
equals one plus the sum of all preceding terms.

30. Show that each term of the sequence
$$1, 2, 6, 18, \cdots, 2 \cdot 3^n$$
equals double the sum of all preceding terms.

31. A ball rebounds to half the height from which it is dropped. If it is dropped from 10 ft, how far does it travel from the moment it is dropped until the moment of its eighth bounce?

32. A rajah once demanded of his subjects one grain of rice on the first square of his chessboard, two on the second, four on the third, and so on, doubling each time until the 64-th square. How much total rice did he demand?

Express by means of factorials:

33. $18 \cdot 17 \cdot 16 \cdot 15 \cdot 14$ **34.** $7 \cdot 8 \cdot 9 \cdot 10 \cdot 11 \cdot 12 \cdot 13$

35. $2 \cdot 4 \cdot 6 \cdot 8 \cdot 10 \cdot 12$ **36.** $\dfrac{52 \cdot 51 \cdot 50 \cdot 49 \cdot 48}{120}$.

Compute:

37. $\dfrac{9!}{7!}$ **38.** $\dfrac{12!}{8! \, 4!}$ **39.** $\dfrac{10!}{(5!)^2}$

40. $\dfrac{14!}{8! \, 4! \, 2!}$ **41.** $\dfrac{15!}{11! \, 4!}$ **42.** $\dfrac{25!}{3! \, 22!}$.

2. PERMUTATIONS AND COMBINATIONS

In this section, we discuss certain techniques of counting that are useful in many applications of mathematics.

Let us start with a basic principle. Suppose you must make a sequence of n independent decisions. For the first decision you have k_1 choices, for the second k_2 choices, etc. The total number of possible *sequences of decisions* is the product $k_1 \cdot k_2 \cdot k_3 \cdots k_n$.

For example, suppose there are three roads, r_1, r_2, r_3, from A to B, two roads, s_1, s_2, from B to C, and two roads, t_1, t_2, from C to D. In how many ways can you choose a route from A to B to C to D? The answer is $3 \cdot 2 \cdot 2 = 12$. This is clear if you list the routes:

$$
\begin{array}{cccc}
r_1 s_1 t_1 & r_1 s_1 t_2 & r_1 s_2 t_1 & r_1 s_2 t_2 \\
r_2 s_1 t_1 & r_2 s_1 t_2 & r_2 s_2 t_1 & r_2 s_2 t_2 \\
r_3 s_1 t_1 & r_3 s_1 t_2 & r_3 s_2 t_1 & r_3 s_2 t_2
\end{array}
$$

Here is an important application of the basic principle. Given n distinct objects, in how many ways can we order them, that is, assign a first, a second, etc? We have n choices for the first object. After that, we have $n - 1$ choices for the second, then $n - 2$ choices for the third, etc. Therefore, by the basic principle, the total number of orderings is

$$n(n - 1)(n - 2) \cdots 3 \cdot 2 \cdot 1 = n!.$$

Each ordering of a set of objects is called a **permutation.**

> There are $n!$ permutations of n objects.

Examples:

There are $4! = 24$ ways to arrange 4 pictures on a wall from left to right.

There are $9! = 362880$ batting orders for the 9 players on a baseball team.

There are $52! \approx 8.066 \times 10^{67}$ arrangements of a deck of 52 cards.

Next, suppose we again have n objects and ask how many ways we can select k of them in order. (Last time we took all n of them.) Each such choice is called a **permutation of n things taken k at a time.** We proceed as before, except we stop after k selections. We have n choices for the first, $n - 1$ for the second, and $n - (k - 1) = n - k + 1$ choices for the k-th. Therefore, the total number of selections possible is

$$P_{n,k} = n(n - 1)(n - 2) \cdots (n - k + 1).$$

(If the last factor is a little annoying, just remember that there are exactly k consecutive factors starting with n and going down.) The number $P_{n,k}$ can be expressed in terms of factorials:

> The number of permutations of n objects taken k at a time is
> $$P_{n,k} = n(n - 1)(n - 2) \cdots (n - k + 1) = \frac{n!}{(n - k)!}.$$

Examples:

The number of choices for the first three batters from among 9 baseball players is

$$P_{9,3} = \frac{9!}{(9 - 3)!} = \frac{9!}{6!} = 9 \cdot 8 \cdot 7 = 504.$$

The number of ways the first 5 cards can be dealt from a deck of 52 cards is

$$P_{52,5} = \frac{52!}{(52 - 5)!} = \frac{52!}{47!} = 52 \cdot 51 \cdot 50 \cdot 49 \cdot 48 = 311,875,200.$$

Combinations

Sometimes we need the number of ways we can choose k objects from a set of n objects without regard to order. Each such choice is a **combination of n things taken k at a time.** The number of such choices is denoted by $C_{n,k}$ or more usually by

$$\binom{n}{k},$$

and is called a **binomial coefficient.**

There is a simple relation between combinations and permutations. Each *permutation* of n objects taken k at a time consists of a choice of k objects (a combination) followed by an ordering of these k objects. There are $\binom{n}{k}$ ways to choose k objects, then $k!$ ways to permute (number) them. Therefore by the general principle,

$$P_{n,k} = \binom{n}{k} \cdot k!, \qquad \text{hence} \qquad \binom{n}{k} = \frac{P_{n,k}}{k!}.$$

From our earlier expression for $P_{n,k}$ we obtain the following useful formula:

The number of combinations of n things taken k at a time is

$$\binom{n}{k} = \frac{n!}{k!(n-k)!}.$$

■ *Example 2.1*

(a) In how many ways can an officer choose 5 out of his 12-man squad for a patrol?

(b) In a race with 15 horses, how many outcomes are possible for the first three places (win, place, and show)?

SOLUTION (a) What counts is which five soldiers are chosen, not in what order they are chosen. So we want *combinations* of 12 objects taken 5 at a time:

$$\binom{12}{5} = \frac{12!}{5!7!} = \frac{12 \cdot 11 \cdot 10 \cdot 9 \cdot 8}{1 \cdot 2 \cdot 3 \cdot 4 \cdot 5} = 792.$$

(b) Here the order definitely counts. We want the number of *permutations* of 15 taken 3 at a time:

$$P_{15,3} = \frac{15!}{12!} = 15 \cdot 14 \cdot 13 = 2730.$$

Answer (a) 792 (b) 2730.

The combination symbol has a number of basic properties:

$$\binom{n}{0} = \binom{n}{n} = 1 \qquad \binom{n}{1} = \binom{n}{n-1} = n \qquad \binom{n}{k} = \binom{n}{n-k}$$

$$\binom{n+1}{k+1} = \binom{n}{k+1} + \binom{n}{k}, \qquad 0 \le k \le n-1.$$

These can be proved directly from the formula expressing $\binom{n}{k}$ in terms of factorials. They can also be proved by direct counting. For example, if you remove k objects from a box of n objects, then $n - k$ objects remain behind, so to each choice of k objects corresponds a complementary choice of $n - k$ objects. Hence

$$\binom{n}{k} = \binom{n}{n-k}.$$

Again, choosing $k + 1$ objects from a set $\{t_1, t_2, \cdots, t_n, t_{n+1}\}$ amounts either to choosing $k + 1$ objects from $\{t_1, t_2, \cdots, t_n\}$ or choosing the element t_{n+1} together with k objects from $\{t_1, t_2, \cdots, t_n\}$. Therefore

$$\binom{n+1}{k+1} = \binom{n}{k+1} + \binom{n}{k}.$$

EXERCISES

1. If 8 horses run in a race, how many different orders of finishing are possible?
2. (cont.) How many possibilities are there for the first three places?
3. How many 10 digit numbers are there composed of all ten digits $0, 1, \cdots, 9$, if 0 is allowed as the first digit; if zero is not allowed as the first digit?
4. How many 10 digit numbers are there composed of 6 ones and 4 fives?
5. How many 5 digit numbers are there composed only of 6's and 7's?
6. From a committee of 9 people, how many subcomittees of 3 people are possible?
7. From a committee of 9 people, in how many ways can one choose a subcommittee of 4 people and a subcommittee of 3 people, with no overlap?
8. The cabinet of a coalition government consists of 7 Radicals, 6 Loyalists, and 5 Progressives. In how many ways can the prime minister select a committee of 3 Radicals, 2 Loyalists, and 2 Progressives?
9. In how many ways can a class of 8 boys and 8 girls be seated in a room with 16 chairs if the boys must take the odd numbered seats?
10. (cont.) Suppose there are 20 seats; then how many ways are there?
11. From a deck of 52 cards, how many 5-card poker hands are possible?
12. (cont.) How many of these are flushes (all cards of the same suit)?
13. (cont.) How many are full houses (three of a kind plus a pair of another kind)?
14. Prove that $\binom{2n}{n}$ is the largest of the binomial coefficients $\binom{2n}{j}$, where $0 \le j \le 2n$.
15. The grid lines on a piece of paper are one unit apart. Consider all paths that run along the grid lines from $(0, 0)$ to (n, n) moving either upwards or to the right. Prove that the number of such paths is $\binom{2n}{n}$.

16*. (cont.) Each path in Ex. 15 passes through exactly one of the points $(k, n - k)$ where $0 \le k \le n$. Count the number of such paths for each k and derive the formula

$$\binom{2n}{n} = \binom{n}{0}^2 + \binom{n}{1}^2 + \cdots + \binom{n}{n}^2.$$

17.   Prove for $2 \le k \le n - 2$ that

$$\binom{n}{k} = \binom{n-2}{k-2} + 2\binom{n-2}{k-1} + \binom{n-2}{k}.$$

18.   (cont.) Find a similar formula for $\binom{n}{k}$ in terms of the binomial coefficients $\binom{n-3}{j}$.

3. BINOMIAL THEOREM

The binomial theorem is a formula for the expanded product

$$(x + y)^n.$$

Let us compute a few cases:

$$(x + y)^1 = x + y$$
$$(x + y)^2 = x^2 + 2xy + y^2$$
$$(x + y)^3 = x^3 + 3x^2y + 3xy^2 + y^3$$
$$(x + y)^4 = x^4 + 4x^3y + 6x^2y^2 + 4xy^3 + y^4.$$

Obviously, the general formula should express $(x + y)^n$ as a sum of monomials $a_{n,k}x^{n-k}y^k$. The problem is, what are the correct coefficients? In multiplying out

$$(x + y)^n = (x + y)(x + y)(x + y) \cdots (x + y) \qquad (n \text{ factors}),$$

we obtain terms that are products of n factors of x's and y's. We obtain $x^{n-k}y^k$ each way we can choose y from k of the factors $x + y$ and x from the remaining $n - k$ factors. The number of such ways is the number of combinations of n things taken k at a time. Therefore, the coefficient of $x^{n-k}y^k$ is $\binom{n}{k}$.

Binomial Theorem   For each positive integer n,

$$(x + y)^n = x^n + \binom{n}{1}x^{n-1}y + \binom{n}{2}x^{n-2}y^2 + \binom{n}{3}x^{n-3}y^3$$

$$+ \cdots + \binom{n}{k}x^{n-k}y^k + \cdots + \binom{n}{n-1}xy^{n-1} + y^n.$$

■ *Example 3.1*

Expand   (a) $(x + 1)^5$ (b) $(2a - b)^6$.

SOLUTION (a)   Use the Binomial Theorem with $y = 1$ and $n = 5$. The coefficients needed are

$$\binom{5}{0} = 1, \quad \binom{5}{1} = 5, \quad \binom{5}{2} = 10, \quad \binom{5}{3} = 10, \quad \binom{5}{4} = 5, \quad \binom{5}{5} = 1.$$

By the formula,

$$(x + 1)^5 = x^5 + 5x^4 + 10x^3 + 10x^2 + 5x + 1.$$

(b)   Use the formula with $x = 2a$, $y = -b$, and $n = 6$. The coefficients are

$$\binom{6}{0} = \binom{6}{6} = 1, \quad \binom{6}{1} = \binom{6}{5} = 6, \quad \binom{6}{2} = \binom{6}{4} = 15, \quad \binom{6}{3} = 20.$$

The formula yields

$$(2a - b)^6 = (2a)^6 + 6(2a)^5(-b) + 15(2a)^4(-b)^2 + 20(2a)^3(-b)^3 + 15(2a)^2(-b)^4$$
$$+ 6(2a)(-b)^5 + (-b)^6$$
$$= 64a^6 - 192a^5b + 240a^4b^2 - 160a^3b^3 + 60a^2b^4 - 12ab^5 + b^6.$$

■ *Example 3.2*

Use the Binomial Theorem to prove that $(1.01)^{900} > 10$.

SOLUTION

$$(1.01)^{900} = (1 + 0.01)^{900} = 1 + \binom{900}{1}(0.01) + \cdots$$

$$> 1 + (900)(0.01) = 1 + 9 = 10.$$

Pascal's Triangle

The binomial coefficients $\binom{n}{k}$ have many interesting properties, some which appear in a display called **Pascal's Triangle.** We write the coefficients $\binom{n}{k}$ for $k = 0$, 1, 2, 3, $\cdots$ on the n-th line:

```
                    1
                 1     1
              1     2     1
           1     3     3     1
        1     4     6     4     1
     1     5    10    10     5     1
  1     6    15    20    15     6     1
```
·   ·   ·   ·   ·   ·   ·   ·   ·

Note that each interior entry is the sum of the two numbers above it. That is because

$$\binom{n+1}{k+1} = \binom{n}{k+1} + \binom{n}{k},$$

as proved in the last section. Note also that the sum of the numbers in the n-th row is 2^n. That is simply because

$$2^n = (1+1)^n = \binom{n}{0} + \binom{n}{1} + \cdots + \binom{n}{n}.$$

The alternating sum of each row is 0, that is

$$\binom{n}{0} - \binom{n}{1} + \binom{n}{2} - + \cdots + \binom{n}{n} = 0.$$

That is simply because $(1-1)^n = 0$.

EXERCISES

Expand:

1. $(x+2)^3$

2. $(x-y)^3$.

3. $(1-2x)^4$

4. $(2+x)^5$

5. $\left(x + \dfrac{1}{x}\right)^6$

6. $\left(xy + \dfrac{1}{x^2}\right)^7$.

7. Compute

$$3^5 - \binom{5}{1}3^4 \cdot 2 + \binom{5}{2}3^3 \cdot 2^2 - \binom{5}{3}3^2 \cdot 2^3 + \binom{5}{4}3 \cdot 2^4 - 2^5.$$

8. Find the coefficient of $x^{10}y^4$ in $(x^2 - 2y)^9$.

9. Find the coefficient of x^4y^{18} in $(3x - y^3)^{10}$.

10. Find the coefficient of x^8y^{12} in $(x^2 - y^3)^8$.

11. Find the coefficient of $1/x^3$ in $\left(2x + \dfrac{1}{x^2}\right)^6$.

12. Find the constant term in $\left(x - \dfrac{1}{2x}\right)^{10}$.

Compute by the binomial theorem:

13. $(102)^5$

14. 99^3.

Compute to six decimal accuracy:

15. $(0.99)^{10}$

16. $(1.01)^{12}$.

17. Look at lines 1–4 of Pascal's triangle. Notice that $11 = 11^1$, $121 = 11^2$, $1331 = 11^3$, and $14641 = 11^4$. Explain.

18. (cont.) How can you interpret line 5 as 11^5?

Prove:

19. $(1.002)^{10} > 1.02$

20. (cont.) $(1.002)^{10} > 1.02018$.

In Section 1, we computed the sums of arithmetic and geometric sequences. In this section, we discuss techniques for computing other sums. We begin by introducing some standard notation.

Summation Notation

The symbol

$$\sum_{j=1}^{n} a_j$$

is used to abbreviate the sum

$$a_1 + a_2 + \cdots + a_n.$$

The greek letter Σ stands for sum, a_j represents the typical term in the desired sum, and the **summation index** j labels the terms. For example, if $a_j = j^2$, the notation represents the sum of terms j^2, for $j = 1, 2, 3, \cdots, n$. Thus

$$\sum_{j=1}^{n} j^2 = 1^2 + 2^2 + \cdots + n^2.$$

There is nothing special about the letter j; any other letter will serve just as well for a summation index. For instance,

$$\sum_{j=1}^{n} j^2 = \sum_{k=1}^{n} k^2 = \sum_{i=1}^{n} i^2 = 1^2 + 2^2 + \cdots + n^2.$$

The index need not run from 1 to n. In general,

$$\sum_{j=s}^{t} a_j = a_s + a_{s+1} + \cdots + a_t.$$

Here are a few examples of summation notation in action:

$$\sum_{j=6}^{10} j(j+1) = 6 \cdot 7 + 7 \cdot 8 + 8 \cdot 9 + 9 \cdot 10 + 10 \cdot 11,$$

$$\sum_{j=0}^{n} a_j x^j = a_0 + a_1 x + a_2 x^2 + \cdots + a_n x^n,$$

$$\sum_{j=3}^{12} a_{j-1} x^{2j} = a_2 x^6 + a_3 x^8 + a_4 x^{10} + \cdots + a_{11} x^{24}.$$

A few more examples illustrate the versatility of the notation:

$$\sum_{j=1}^{n} a_{3j} = a_3 + a_6 + a_9 + \cdots + a_{3n},$$

$$\sum_{j=1}^{5} a_j b_{5+j} = a_1 b_6 + a_2 b_7 + a_3 b_8 + a_4 b_9 + a_5 b_{10},$$

$$\sum_{n=2}^{9} \frac{(-1)^n}{n!} x^{2n-1} = \frac{x^3}{1 \cdot 2} - \frac{x^5}{1 \cdot 2 \cdot 3} + \frac{x^7}{1 \cdot 2 \cdot 3 \cdot 4} - + \cdots - \frac{x^{17}}{1 \cdot 2 \cdot 3 \cdots 9}.$$

Let us express some previous results of this chapter in summation notation.

Arithmetic Progression:
$$\sum_{k=1}^{n} [a + (k-1)d] = \frac{n}{2}[2a + (n-1)d]$$

Geometric Progression:
$$\sum_{k=1}^{n} ar^{k-1} = a\frac{r^n - 1}{r - 1} \qquad (r \neq 1)$$

Binomial Theorem:
$$\sum_{k=0}^{n} \binom{n}{k} x^{n-k}y^k = (x + y)^n.$$

For later use, we note a special case of the first formula, which gives the sum of the first n positive integers. We just set $a = 1$ and $d = 1$:

$$1 + 2 + 3 + \cdots + n = \sum_{k=1}^{n} k = \tfrac{1}{2}n(n + 1).$$

Rules for Sums

There are two important rules concerning sums. The first is a generalized distributive law allowing you to take out a common factor; the second is a generalized commutative–associative law allowing you to rearrange certain sums in a natural way.

$$\sum_{j=s}^{t} ca_j = c \sum_{j=s}^{t} a_j$$

$$\sum_{j=s}^{t} (a_j + b_j) = \sum_{j=s}^{t} a_j + \sum_{j=s}^{t} b_j$$

Example:

$$\sum_{j=1}^{n} j(j+2) = \sum_{j=1}^{n} (j^2 + 2j) = \sum_{j=1}^{n} j^2 + 2\sum_{j=1}^{n} j.$$

In a typical application, we might have formulas for the two sums on the right. Then we could obtain, free of charge, a formula for the sum on the left.

One sum is so simple that it sometimes causes confusion:

$$\sum_{j=1}^{n} 1 = n.$$

The notation on the left represents the sum of *n* ones.

Changing the Index

We can often simplify computations involving summation notation by changing the summation index. A typical change of index is based on the following observation. The quantity $j + 1$ runs from 1 to n as j runs from 0 to $n - 1$, and the quantity $j - 1$ runs from 1 to n as j runs from 2 to $n + 1$. Therefore we can describe the same sum in three ways:

$$\sum_{j=1}^{n} a_j = \sum_{j=0}^{n-1} a_{j+1} = \sum_{j=2}^{n+1} a_{j-1}.$$

Example:

$$x^2 + 2x^4 + 3x^6 + 4x^8 = \sum_{j=1}^{4} jx^{2j} = \sum_{j=0}^{3} (j+1)x^{2(j+1)} = \sum_{j=2}^{5} (j-1)x^{2(j-1)}.$$

Another typical change: as j runs from 0 to n, the quantity $n - j$ runs from n to 0.

Example:

$$a_0 b_6 + a_1 b_5 + a_2 b_4 + \cdots + a_5 b_1 + a_6 b_0 = \sum_{j=0}^{6} a_j b_{6-j} = \sum_{j=0}^{6} a_{6-j} b_j.$$

The following example illustrates how changing the index can simplify expressions involving several sums.

■ *Example 4.1*

Simplify $$\sum_{j=2}^{10} \frac{1}{j} - \sum_{j=1}^{8} \frac{1}{j+2}.$$

SOLUTION Most terms are common to both sums and will cancel. In the second sum, let $k = j + 2$. Then as j runs from 1 to 8, the index k runs from 3 to 10. Therefore

$$\sum_{j=2}^{10} \frac{1}{j} - \sum_{j=1}^{8} \frac{1}{j+2} = \sum_{k=2}^{10} \frac{1}{k} - \sum_{k=3}^{10} \frac{1}{k} = \left(\frac{1}{2} + \sum_{k=3}^{10} \frac{1}{k}\right) - \sum_{k=3}^{10} \frac{1}{k} = \frac{1}{2}.$$

Answer: $\frac{1}{2}$.

· ·

Evaluation of Sums

Let us turn to the main business of this section, computing sums. We shall discuss a general principle that allows us to evaluate a wide variety of sums.

Suppose $b_1, b_2, \cdots, b_{n+1}$ is a sequence and $a_j = b_{j+1} - b_j$. Then

$$\sum_{j=1}^{n} a_j = \sum_{j=1}^{n} (b_{j+1} - b_j)$$

$$= (b_2 - b_1) + (b_3 - b_2) + \cdots + (b_{n+1} - b_n)$$

$$= -b_1 + (b_2 - b_2) + (b_3 - b_3) + \cdots + (b_n - b_n) + b_{n+1}$$

$$= b_{n+1} - b_1.$$

$$\boxed{\text{If } a_j = b_{j+1} - b_j, \text{ then } \sum_{j=1}^{n} a_j = b_{n+1} - b_1.}$$

This statement seems simple, yet in practice it can be very powerful. Suppose you want to compute $\sum_{1}^{n} a_j$. If you can find a sequence $b_1, b_2, \cdots$ such that $b_{j+1} - b_j = a_j$, then you can write the answer down immediately, $b_{n+1} - b_1$. For example, if $b_j = j^2$, then $b_{j+1} - b_j = 2j + 1$, hence

$$\sum_{j=1}^{n} (2j + 1) = b_{n+1} - b_1 = (n + 1)^2 - 1.$$

A corollary of this result is a familiar formula. Since

$$\sum_{j=1}^{n} (2j + 1) = 2 \sum_{j=1}^{n} j + \sum_{j=1}^{n} 1 = 2 \sum_{j=1}^{n} j + n,$$

we have

$$2 \sum_{j=1}^{n} j = (n + 1)^2 - 1 - n = n^2 + n = n(n + 1),$$

so we find again the sum of the first n positive integers

$$\sum_{j=1}^{n} j = \tfrac{1}{2}n(n + 1).$$

Example 4.2

Evaluate $\displaystyle\sum_{j=1}^{n} \frac{1}{j(j + 1)}.$

SOLUTION   By partial fractions,

$$\frac{1}{j(j + 1)} = \frac{1}{j} - \frac{1}{j + 1} = b_{j+1} - b_j,$$

where $b_j = -1/j$. Therefore,

$$\sum_{j=1}^{n} \frac{1}{j(j + 1)} = b_{n+1} - b_1 = -\frac{1}{n + 1} + 1 = \frac{n}{n + 1}.$$

Answer $n/(n + 1).$

Example 4.3

Evaluate $\displaystyle\sum_{j=1}^{n} j^2.$

SOLUTION   If $b_j = j^3$, then

$$b_{j+1} - b_j = (j + 1)^3 - j^3 = (j^3 + 3j^2 + 3j + 1) - j^3 = 3j^2 + 3j + 1.$$

Hence

$$\sum_{j=1}^{n} (3j^2 + 3j + 1) = b_{n+1} - b_1 = (n + 1)^3 - 1^3,$$

$$3 \sum_{j=1}^{n} j^2 + 3 \sum_{j=1}^{n} j + \sum_{j=1}^{n} 1 = (n + 1)^3 - 1,$$

$$3 \sum_{j=1}^{n} j^2 = (n + 1)^3 - 1 - 3 \sum_{j=1}^{n} j - \sum_{j=1}^{n} 1.$$

But we know the two sums on the right. The first is $\tfrac{1}{2}n(n + 1)$ and the second is n. Therefore

$$3 \sum_{j=1}^{n} j^2 = (n + 1)^3 - 1 - \tfrac{3}{2}n(n + 1) - n$$

$$= \tfrac{1}{2}[2(n + 1)^3 - 2 - 3(n^2 + n) - 2n]$$

$$= \tfrac{1}{2}[2(n + 1)^3 - 3n(n + 1) - 2(n + 1)] = \tfrac{1}{2}(n + 1)[2(n + 1)^2 - 3n - 2]$$

$$= \tfrac{1}{2}(n + 1)(2n^2 + n) = \tfrac{1}{2}n(n + 1)(2n + 1).$$

Answer $\displaystyle\sum_{j=1}^{n} j^2 = \tfrac{1}{6}n(n + 1)(2n + 1).$

· ·

Remark: In Example 4.3, we found $\displaystyle\sum_{1}^{n} j^2$ in terms of $\displaystyle\sum_{1}^{n} j$ and $\displaystyle\sum_{1}^{n} 1$, which were known. In a similar way, we can now find $\displaystyle\sum_{1}^{n} j^3$. See Exercise 19.

The method can also be applied to obtain certain estimates involving sums. In the following example, we are able to trap a complicated sum between successive integers. The solution, however, depends on a clever trick.

■ *Example 4.4*

Prove $198 < 1 + \dfrac{1}{\sqrt{2}} + \dfrac{1}{\sqrt{3}} + \cdots + \dfrac{1}{\sqrt{10000}} < 199.$

SOLUTION   Start with

$$\sqrt{j + 1} - \sqrt{j} = \frac{1}{\sqrt{j + 1} + \sqrt{j}}.$$

Clearly $2\sqrt{j} < \sqrt{j} + \sqrt{j + 1} < 2\sqrt{j + 1}$; hence

$$\frac{1}{2\sqrt{j + 1}} < \frac{1}{\sqrt{j} + \sqrt{j + 1}} < \frac{1}{2\sqrt{j}},$$

so

$$\frac{1}{2\sqrt{j + 1}} < \sqrt{j + 1} - \sqrt{j} < \frac{1}{2\sqrt{j}}.$$

Now sum from $j = 1$ to $j = 10^4 - 1 = 9999$:

$$\frac{1}{2} \sum_{j=1}^{9999} \frac{1}{\sqrt{j + 1}} < 100 - 1 < \frac{1}{2} \sum_{j=1}^{9999} \frac{1}{\sqrt{j}}.$$

The left inequality is equivalent to

$$\sum_{j=2}^{10000} \frac{1}{\sqrt{j}} < 2(99) = 198, \qquad \text{hence} \quad \sum_{j=1}^{10000} \frac{1}{\sqrt{j}} < 198 + 1 = 199.$$

The right inequality is equivalent to

$$198 < \sum_{j=1}^{9999} \frac{1}{\sqrt{j}}, \qquad \text{hence} \quad 198 < \sum_{j=1}^{10000} \frac{1}{\sqrt{j}}.$$

EXERCISES

Express in summation notation:

1. $x_3 + x_4 + x_5 + \cdots + x_{20}$
2. $x_3 + x_5 + x_7 + \cdots + x_{19}$
3. the sum of the first 50 odd integers
4. the sum of the first n powers of 2
5. $1^2 + 2^3 + 3^4 + \cdots + n^{n+1}$
6. $xy^{10} + x^2y^8 + x^3y^6 + x^4y^4 + x^5y^2 + x^6$.

7. Write $\displaystyle\sum_{j=1}^{n} jx^{j-1}$ as a sum of multiples of x^j.

8. (cont.) Do the same for $\displaystyle\sum_{j=2}^{n} j(j-1)x^{j-2}$.

9. Prove that $\displaystyle\frac{1}{2}\sum_{j=1}^{n} \frac{1}{j} + \sum_{j=0}^{n-1}\frac{1}{2j+1} = \sum_{j=1}^{2n}\frac{1}{j}$.

10. Show that $\displaystyle\sum_{j=1}^{n} a_j b_j = \frac{1}{4}\left[\sum_{j=1}^{n}(a_j + b_j)^2 - \sum_{j=1}^{n}(a_j - b_j)^2\right]$.

11. Show that $\displaystyle\sum_{j=0}^{n} a_j b_{n-j} = \sum_{j=0}^{n} a_{n-j}b_j$.

Find the sum:

12. $1\cdot 2 + 2\cdot 3 + 3\cdot 4 + \cdots + 99\cdot 100$

13. $\displaystyle\frac{1}{1\cdot 3} + \frac{1}{2\cdot 4} + \frac{1}{3\cdot 5} + \cdots + \frac{1}{49\cdot 51}$

14. $1^2 + 3^2 + 5^2 + 7^2 + \cdots + 99^2$.

15. Which is larger,

$$1 + 2 + 3 + 4 + \cdots + 100 \quad \text{or} \quad 1.1 + 1.1^2 + 1.1^3 + \cdots + 1.1^{99}?$$

16. Compute $\displaystyle\sum_{j=1}^{n} \frac{1}{j(j+2)}$ using partial fractions.

17. Compute $\sum_{j=1}^{n} j \cdot 2^j$. [Hint: Consider $b_{j+1} - b_j$, where $b_j = j \cdot 2^j$.]

18. Compute $\sum_{j=1}^{n} j(j+1)(j+2)$. [Hint: Consider $b_j - b_{j-1}$, where $b_j = j(j+1)(j+2)(j+3)$.]

19. Compute $(j+1)^4 - j^4$. Use the result to derive the formula

$$1^3 + 2^3 + 3^3 + \cdots + n^3 = \frac{n^2(n+1)^2}{4} = (1 + 2 + \cdots + n)^2.$$

20. Use the relation

$$\sum_{j=0}^{n} j^3 = \sum_{j=0}^{n} (n-j)^3$$

to obtain the formula for $\sum_{j=1}^{n} j^3$ in another way.

21. Prove that no matter how large n is,

$$1 + \frac{1}{2^2} + \frac{1}{3^2} + \cdots + \frac{1}{n^2} < 2.$$

$\left[\text{Hint: } \frac{1}{j^2} < \frac{1}{j(j-1)} \quad \text{for } j \geq 2. \right]$

22. Prove that no matter how large n is,

$$\frac{1}{1!} + \frac{1}{2!} + \cdots + \frac{1}{n!} < 2.$$

[Hint: Compare $1/j!$ with $1/2^{j-1}$.]

23*. Let $A_i = a_1 + a_2 + \cdots + a_i$ and $B_i = b_1 + b_2 + \cdots + b_i$ for $i = 1, \cdots, n$. Prove

$$\sum_{i=1}^{n} A_i b_i = A_n B_n - \sum_{i=1}^{n-1} a_{i+1} B_i.$$

24*. Find

$$\frac{3}{(1 \cdot 2)^2} + \frac{5}{(2 \cdot 3)^2} + \frac{7}{(3 \cdot 4)^2} + \cdots + \frac{2n+1}{[n(n+1)]^2}.$$

5. MATHEMATICAL INDUCTION

Often we deal in mathematics with statements that are really infinitely many statements at once. For example, the statement "if $a_1, a_2, \cdots, a_n$ are n positive numbers, then

$$\log a_1 a_2 \cdots a_n = \log a_1 + \log a_2 + \cdots + \log a_n,"$$

consists of infinitely many distinct statements:

$$n = 2, \quad \log a_1 a_2 = \log a_1 + \log a_2$$
$$n = 3, \quad \log a_1 a_2 a_3 = \log a_1 + \log a_2 + \log a_3$$
$$n = 4, \quad \log a_1 a_2 a_3 a_4 = \log a_1 + \log a_2 + \log a_3 + \log a_4$$

and so on. Other examples: "for each positive integer n, the inequality $2^n > n$ holds", and "for each positive integer n, the number $n^2 - n + 41$ is a prime number (has no divisors other than 1 and itself)".

Usually, statements of this type are suggested by experimental evidence. For instance, we observe that

$$2^1 > 1, \qquad 2^2 > 2, \qquad 2^3 > 3, \qquad 2^4 > 4;$$

we conjecture that $2^n > n$ for each positive integer n. We observe that the formula $n^2 - n + 41$ yields 41, 43, 47, 53, 61, all prime numbers, for $n = 1, 2, 3, 4, 5$, and conjecture that $n^2 - n + 41$ always yields prime numbers. But does experimental evidence assure that our conjecture is correct? Not necessarily. For example $n^2 - n + 41$ yields primes for $n = 1, 2, 3, \cdots, 40$ but breaks down for 41; the number $41^2 - 41 + 41$ is divisible by 41. The statement, "all integers n are less than 1,000,001" is false, yet the first 1,000,000 cases are correct!

We need a method to prove conjectures based on experimental evidence without treating each case separately (life is too short). This is what mathematical induction does.

To understand the idea, imagine a row of dominoes standing on end. If they are spaced close to each other, and the first one is knocked down, they will all fall. We know this for two reasons:

(1) The first domino *will* fall.

(2) *If* a domino falls, then it will knock down the next one.

Be careful of (2). We do not assert a given domino *will* fall, only that *if* it does, it will knock down the next.

This idea is stated formally in an axiom for the positive integers, called the **principle of mathematical induction:**

Let S be a set of positive integers such that

(1) the integer k belongs to S

(2) if an integer n belongs to S, then $n + 1$ also belongs to S.

Then S contains all integers $n \geq k$.

Here is how the axiom is used. Suppose T is a theorem (statement) about all integers $n \geq k$, i.e., T consists of a sequence of statements $T_k, T_{k+1}, T_{k+2}, \cdots$. To prove all the statements we do two steps:

(1) We prove somehow that T_k is true.

(2) We prove that if T_n *were* true for an integer n, then T_{n+1} would also be true.

Thus the set S of positive integers for which T_n is true contains k, and contains $n + 1$ whenever it contains n. By mathematical induction, S contains all integers $n \geq k$; the theorem is true for all $n \geq k$.

Remark: The critical step is going from n to $n + 1$. Remember you do not prove the theorem for n, you merely show if it *were* true for n, then it would be true for $n + 1$. Thus you prove the theorem for $n + 1$, *assuming* it is true for n, a boot-strap operation.

■ *Example 5.1*

Prove that $2^n > n$ for all integers $n \geq 1$.

SOLUTION We use a two-part procedure by mathematical induction.

(1)   For $n = 1$, the statement is $2^1 > 1$, which is true.
(2)   We show that if $2^n > n$, then it would follow that $2^{n+1} > n + 1$. Assuming $2^n > n$, multiply both sides of the inequality by 2:

$$2^{n+1} > 2n.$$

But $2n \geq n + 1$ for all positive integers. Hence $2^{n+1} \geq n + 1$. This completes the second step of the induction; the theorem is proved for all $n \geq 1$.

．．．．．．．．．．．．．．．．．．．．．．．．

■ *Example 5.2*

Prove that an n-piece jigsaw puzzle, $n \geq 2$, requires $n - 1$ joinings (fittings of a piece to another piece, or a block of pieces to another block).

SOLUTION (1)   For $n = 2$, the assertion is that a 2-piece puzzle requires 1 joining. This is obviously true.

(2)   Show that if an n-piece puzzle required $n - 1$ joinings, then an $(n + 1)$-piece puzzle would require n joinings. To begin an $(n + 1)$-piece puzzle, you make one joining connecting two pieces. But now you have an n-piece puzzle! If that requires $(n - 1)$ joinings, then the $(n + 1)$-piece puzzle requires $1 + (n - 1) = n$. This completes the proof by induction.

．．．．．．．．．．．．．．．．．．．．．．

Inductive Definitions

Suppose a sequence of numbers $a_1, a_2, a_3, \cdots$ is defined this way: $a_1 = 1$ and $a_{n+1} = 2a_n + 1$. The sequence is not given explicitly, only that it starts with 1 and that each number is one more than twice the preceding one. We say that the sequence is **defined inductively** or **recursively**. Another example is the sequence

$$a_1 = 0, \qquad a_2 = 1, \qquad a_{n+1} = \tfrac{1}{2}(a_n + a_{n-1});$$

the first two numbers are 0, 1, and after that, each number is the average of the two preceding ones.

The method of mathematical induction is tailor-made for inductively defined sequences. Usually, we conjecture properties of such sequences and prove our conjectures by induction.

■ *Example 5.3*

If $a_1 = 1$, and $a_{n+1} = 2a_n + 1$ for $n > 1$, find an explicit formula for a_n and prove it by mathematical induction.

SOLUTION Trying a few cases we see

$$a_1 = 1, \qquad a_2 = 3, \qquad a_3 = 7, \qquad a_4 = 15, \qquad a_5 = 31.$$

We observe that these numbers are 1 less than 2, 4, 8, 16, 32 or 2^1, 2^2, 2^3, 2^4, 2^5. We conjecture, therefore, that $a_n = 2^n - 1$ for $n > 1$.

We now prove the conjecture by mathematical induction. (1) We are given $a_1 = 1 = 2^1 - 1$; thus the conjecture is true for $n = 1$. (2) Now we assume $a_n = 2^n - 1$. If so, then

$$a_{n+1} = 2a_n + 1 = 2(2^n - 1) + 1 = 2^{n+1} - 2 + 1 = 2^{n+1} - 1.$$

Hence if the conjecture is true for n, it is true for $n + 1$. By mathematical induction, the conjecture is proved.

. . . .

Remark: Inductive definitions are important in a careful development of the rules of arithmetic for real and complex numbers. For example, sums and products are defined at first only for two numbers. Sums and products of n numbers, $n > 2$, are then defined inductively:

$$a_1 + a_2 + \cdots + a_{n+1} = (a_1 + \cdots + a_n) + a_{n+1}, \qquad a_1 a_2 \cdots a_{n+1} = (a_1 \cdots a_n)a_{n+1}.$$

EXERCISES

Prove by mathematical induction:

1. $\log(a_1 a_2 a_3 \cdots a_n) = \log a_1 + \log a_2 + \log a_3 + \cdots + \log a_n, \qquad n \geq 2$.
2. The sum of n even numbers is even.
3. The sum of $2n + 1$ odd numbers is odd.
4. To tie n pieces of string into one long string requires $n - 1$ knots.
5. $\dfrac{1}{1 \cdot 3} + \dfrac{1}{3 \cdot 5} + \cdots + \dfrac{1}{(2n - 1)(2n + 1)} = \dfrac{n}{2n + 1}, \qquad n \geq 1$.
6. $1^3 + 2^3 + 3^3 + \cdots + n^3 = \left[\dfrac{n(n + 1)}{2}\right]^2, \qquad n \geq 1$.
7. $1 \cdot 2 + 2 \cdot 3 + 3 \cdot 4 + \cdots + n(n + 1) = \frac{1}{3}n(n + 1)(n + 2), \qquad n \geq 1$.
8. $\dfrac{1}{2} + \dfrac{2}{2^2} + \dfrac{3}{2^3} + \cdots + \dfrac{n}{2^n} = 2 - (n + 2)2^{-n}, \qquad n \geq 1$.

9. Find a simple expression for
$$(1 - x)(1 + x)(1 + x^2)(1 + x^4) \cdots (1 + x^{2^n})$$
and prove your answer by induction.
10. If n people stand in line at a ticket counter, and if the first person in line is a woman and the last person is a man, prove that somewhere in the line there is a man directly behind a woman.
11. In the sequence of figures shown in Fig. 5.1 (on next page), the first is an equilateral triangle of side 1. At each step an equilateral triangle is constructed on each side of the preceding figure with length $\frac{1}{3}$ of the side. Find the number of sides of the n-th figure. Prove your answer.

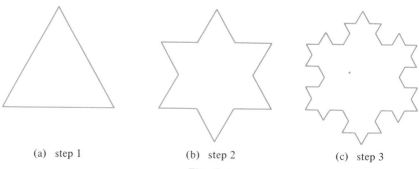

(a) step 1   (b) step 2   (c) step 3

Fig. 5.1

12. (cont.) Find the perimeter of the n-th figure.

13*. (cont.) Show that the area of the n-th figure is

$$\frac{\sqrt{3}}{20}\left[8 - 3\left(\frac{4}{9}\right)^{n-1}\right].$$

14. (cont.) Deduce that the area of the n-th figure approaches 8/5 of the area of the first figure as $n \longrightarrow \infty$.

Remark: Exercises 12 and 14 show that the area of the figures is bounded but their perimeters grow without limit as $n \longrightarrow \infty$. This is the same principle by which a fairly small state like Maine can have an extremely long coastline.

Establish the inequality by induction:

15. $(3/2)^n > n,$ $n \geq 1$ **16.** $3^n > 2^n + 10n,$ $n \geq 4$ **17.** $2^n > n^2 + 27,$ $n \geq 6$

18. $\dfrac{1}{n+1} + \dfrac{1}{n+2} + \dfrac{1}{n+3} + \cdots + \dfrac{1}{2n} > \dfrac{3}{5},$ $n \geq 3.$

19. Find and prove a formula for

$$\left(1 - \frac{1}{4}\right)\left(1 - \frac{1}{9}\right)\left(1 - \frac{1}{16}\right)\cdots\left(1 - \frac{1}{n^2}\right).$$

20. Find and prove a formula for the number of committees that can be formed from a group of n people (allowing committees of $1, 2, 3, \cdots, n$ members).

21. Define a sequence by $a_0 = 0$, $a_1 = 1$, and $a_{n+2} = 2a_{n+1} - a_n$. Find a formula for a_n and prove it by induction.

22. Define a sequence by $x_0 = 0$, $x_1 = 1$, and $x_{n+2} = x_{n+1} + x_n$. Prove that $x_3, x_6, x_9, \cdots$ are even integers and the others are odd.

23. (cont.) Compute $x_2, x_3, \cdots, x_{12}$. These numbers increase rapidly but it is not obvious at what rate. By induction prove the rough estimate

$$(\sqrt{2})^n < x_n < 2^n.$$

24. Define a sequence by $x_0 = 1$, $x_1 = 0$, and $x_{n+2} = 3x_{n+1} - 2x_n$. Prove that $x_n = 2 - 2^n$.

25. A pyramid of n decreasing rings is placed on one of three pegs as shown in Fig. 5.2. The game (tower of Hanoi puzzle) is to transfer the pyramid to peg 2 in as few moves as possible. There are two rules: (1) only one ring can be moved at a time (2) a ring may be moved from one peg to another but may not be placed on top of a ring smaller than itself. Find the least number of moves required and confirm your answer by induction.

Fig. 5.2

26. Consider the sequence
$$x_1 = \sqrt{2}, \qquad x_2 = \sqrt{2 + \sqrt{2}}, \qquad x_3 = \sqrt{2 + \sqrt{2 + \sqrt{2}}}, \qquad \text{etc.}$$
Prove that $x_n < 2$ for $n \geq 1$.

In each case give a formula for x_n and prove your assertion by induction:

27. $x_0 = 0$, $x_1 = 2$, and x_n is the average of $x_0, x_1, \cdots, x_{n-1}$ for $n \geq 2$.
28. $x_0 = 1$, $x_1 = 1$, and $x_n = x_0 + x_1 + \cdots + x_{n-1}$ for $n \geq 2$.
29. $x_0 = 1$, $x_1 = 2$, and $x_n = 2(x_0 + x_1 + \cdots + x_{n-1})$ for $n \geq 2$.
30. Define a sequence as follows:

$$a_0 = 1, \qquad a_{n+1} = \frac{1}{2}\left(a_n + \frac{2}{a_n}\right).$$

   Show that $1 \leq a_n \leq 2$ for $n \geq 0$.
31*. Define a sequence by $x_0 = 0$, $x_1 = 1$, and $x_{n+2} = \frac{1}{2}(x_n + x_{n+1})$. Thus each number is the average of the two preceding. Find a formula for x_n and a number that x_n approaches as n increases.
32*. Show that n lines in general position divide the plane into r_n regions, where $r_{n+1} = r_n + n + 1$. Conclude by induction that n lines divide the plane into $\frac{1}{2}(n^2 + n + 2)$ regions.
33*. Prove the division algorithm (p. 286) by mathematical induction.
34*. Initially flask A contains 1 liter of wine and flask B contains 1 liter of water. A **transfer operation** consists of pouring $\frac{1}{2}$ liter from A into B, mixing, pouring $\frac{1}{2}$ liter from B into A, and mixing. How much wine is in A and how much in B after n transfer operations?

Test 1

1. Express in summation notation:
   (a) The sum of all integers between 100 and 1000 that are divisible by 6.
   (b) $4 \cdot 45 + 7 \cdot 43 + 10 \cdot 41 + \cdots + 31 \cdot 27$.
2. Find

$$\sum_{j=10}^{100} \frac{1}{j(j-2)}.$$

(Do not simplify answer.)

3. In how many ways can two cards be drawn from a 52-card deck? In how many of these are both cards of the same suit?
4. Write out the expansion of $(-2x + 3y)^5$.
5. Find and prove a formula for the sequence defined by $a_1 = 2$, $a_{n+1} = 2a_n - 1$.

Test 2

1. Express in summation notation:
   (a) The sum of the squares of all odd integers less than 50.

   (b) $1 - \dfrac{x^2}{3!} + \dfrac{x^4}{5!} - + \cdots + \dfrac{x^{20}}{21!}$.

2. A Senate committee consists of 8 members of party A and 7 members of party B. How many 5-person subcommittees can be formed having a majority from party A?
3. Expand by the binomial theorem:
$$\left(\frac{1}{x} - \frac{x}{2} \right)^4.$$

4. A sequence is defined by
$$x_1 = 2, \qquad x_2 = 3, \qquad x_{n+1} = x_n x_{n-1} \cdots x_2 x_1 \qquad \text{for } n \geq 2.$$
   Find and prove a formula for x_n.
5. Compute $1 \cdot 2 + 2 \cdot 3 + 3 \cdot 4 + \cdots + 99 \cdot 100$.

CHAPTER 1

Section 2, p. 5

1. $x + \frac{1}{2}y$ **3.** $2(a + b)$ **5.** $0.03x$ **7.** $350/r$ **9.** $\frac{1}{6}b^2$

11. $\pi(r + 1)^2$ **13.** $(x + y)(x - y)$ **15.** $g = 17 + \frac{1}{2}b$ **17.** $1, -\frac{1}{7}$

19. $7, 7$ **21.** $10, -6$ **23.** $\frac{13}{3}, \frac{5}{3}$

25. $x = 1, y = 0$ yields 0 and 1, for example.

27. $x = y = 1$ yields 4 and 2, for example.

29. $x = y = 0$ yields 4 and 0, for example.

Section 3, p. 11

1. $(10 + 1)(1500 - 8) = 15000 + 1500 - 80 - 8 = 16500 - 88 = 16412$

3. $(20 - 1)(200 + 11) = 4000 - 200 + 220 - 11 = 4009$

5. $3 \cdot 4 \cdot 5(6 - 2) = 240$ **7.** $4(3x - 2y + 5)$ **9.** $2xy(z - 3)$

11. $3x^2(1 - 3y + 4x)$ **13.** $(x + y)(2x + 2y + 3v)$ **15.** $xy + x + y + 1$

17. $x^2y - 2xy + x - 2$ **19.** $2x^2 - x - 1$ **21.** $4xy - 10x - 10y + 25$

23. $3x^2 + 2xy - 8y^2$ **25.** $y^2 + 4y + 4$ **27.** $x^2 + xy + 2x + y + 1$

29. $4x^2 + 3xy + y^2 + 2xz + yz$ **31.** $au + av + bu + bv + u + v + 2a + 2b + 2$

33. $(-a) + a = 0$ and $(-a) + [-(-a)] = 0$, hence $-(-a) = a$.

35. $(a + b) + [(-a) + (-b)] = [a + (-a)] + [b + (-b)] = 0 + 0 = 0$, hence
$(-a) + (-b) = -(a + b)$.

37. $(a/b)(c/d) = (ab^{-1})(cd^{-1}) = (ac)(b^{-1}d^{-1}) = (ac)(bd)^{-1} = (ac)/(bd)$.

39. $ka/kb = (ka)(kb)^{-1} = kak^{-1}b^{-1} = (kk^{-1})(ab^{-1}) = ab^{-1} = a/b$.

41. $(x + y)(x^{-1} - y^{-1}) = 1 + yx^{-1} - xy^{-1} - 1 = yx^{-1} - xy^{-1}$.

43. $x^{-1}y^{-1}(x^2 + y^2) = (x^{-1}x)xy^{-1} + x^{-1}(y^{-1}y)y = xy^{-1} + x^{-1}y$.

Section 4, p. 17

1. $(2k + 1) + (2n + 1) = 2(k + n + 1)$, even
3. $(2k)(2n + 1) = 2(2nk + k)$, even
5. All three are odd.
7. You can't; any combination gives an even sum: $2m + 4n + 6p = 2(m + 2n + 3p)$.
9. $n(n + 1) - 20$ **11.** $3n^2$ **13.** $4n^2$ **15.** $\frac{5}{2}n(n + 1)$ **17.** $3/5$
19. $16/3$ **21.** $-34/5$ **23.** $47/15$ **25.** $2/5$ **27.** $5/36$ **29.** $7/12$
31. $-2/5$ **33.** $13/14$

Section 5, p. 23

1. 27 **3.** $1/36$ **5.** 4 **7.** $-32/243$ **9.** $1/72$ **11.** $1/2$ **13.** 2^4
15. 2^{-7} **17.** 2^3 **19.** 2^4 **21.** x^5y^4 **23.** $(a^2 + 1)/a$ **25.** $1/b^{16}$
27. $-1/16x^5y^2$ **29.** $1/2^{10}x^{15}y^{10}z^5$ **31.** $2y^2/x^2$ **33.** y^{12}/x^2
35. 4×10^{-1} **37.** 2.37 **39.** 7.6×10^{-5} **41.** 1.7×10 **43.** 1.24×10^4
45. 3.832×10^6 **47.** 6.48×10^5 **49.** 9.6×10^{-9} **51.** 6.336×10^6

Section 6, p. 27

1. 9 **3.** $1/3$ **5.** $3a^3$ **7.** $5\sqrt{2}$ **9.** $1/2\sqrt{3} = \sqrt{3}/6$
11. $3x\sqrt{2x}/(y + z)^2$ **13.** 10 **15.** $3/4$ **17.** $1/2$ **19.** $2\sqrt[4]{2}$ **21.** $2ab^3$
23. $2uv\sqrt[5]{2v}$ **25.** $\sqrt[5]{6}/6$ **27.** $(x + y + 2\sqrt{xy})/(x - y)$ **29.** $a^2 - 2b^2$
31. x^2 **33.** $3x^2\sqrt{5y}$ **35.** x^2
37. $2y\sqrt{5xz}$ **39.** $\sqrt{a^2 + x^2} = \sqrt{a^2\left(1 + \dfrac{x^2}{a^2}\right)} = a\sqrt{1 + \left(\dfrac{x}{a}\right)^2}$
41. $2/(\sqrt{3} + 1) - 1/(\sqrt{3} - 1) = [2(\sqrt{3} - 1) - (\sqrt{3} + 1)]/(\sqrt{3} - 1)(\sqrt{3} + 1) = (\sqrt{3} - 3)/2$
43. This is just the definition of the n-th root.
45. Let $\sqrt[n]{a} = r$ and $\sqrt[n]{b} = s$. Then $a = r^n$ and $b = s^n$, hence $ab = (rs)^n$. Therefore $\sqrt[n]{ab} = rs = \sqrt[n]{a}\sqrt[n]{b}$.
47. 2.38 **49.** 0.77 **51.** 5.24 **53.** 6.04
55. $(\sqrt{2} + \sqrt{3})^2 = (\sqrt{2})^2 + 2\sqrt{2}\sqrt{3} + (\sqrt{3})^2 = 5 + 2\sqrt{6}$. Now take square roots.
57. $(\sqrt{7} - \sqrt{2})^2 = (\sqrt{7})^2 - 2\sqrt{7}\sqrt{2} + (\sqrt{2})^2 = 9 - 2\sqrt{14}$. Now take square roots.

Section 7, p. 31

1. 9 **3.** $1/5$ **5.** $100{,}000$ **7.** $343/8$ **9.** $2^{7/3}$ **11.** $2^{7/2}$
13. $1/125x^6$ **15.** u^3/v^9 **17.** $x^{10}y^{15}/z^{20}$ **19.** x^4/y^2 **21.** $32x^2\sqrt{x}$
23. $2u - u^{1/6}$ **25.** $\sqrt[6]{32}$ **27.** 1 **29.** $3/2\sqrt[3]{6a^2}$ **31.** x^3
33. $1/x^{1/5}y^{2/5}$ **35.** $y^{9/20}/x^{3/2}$ **37.** 1

Section 8, p. 34

1. $6x + 1$ **3.** $8x^2 + 8x + 2$ **5.** $-6x^2 + 6x - 3$ **7.** $\frac{5}{3}x^2 - 2x - \frac{13}{2}$
9. $x^3 + 8x^2 - 16x - 4$ **11.** $-x^5 - x^4 - x^3 - x^2 + 8$
13. $x^8 + 7x^6 - 5x^5 + 4x^4 - 4x^3 + 2x^2$ **15.** $x^2 - x - 6$ **17.** $6x^2 + 17x + 5$

19. $x^3 + 7x^2 + 5x - 6$ **21.** $x^5 + 2x^3 + x$

23. $x^7 - x^6 - \frac{1}{2}x^5 + \frac{1}{2}x^4 - \frac{1}{2}x^3 - \frac{1}{4}x^2$ **25.** $x^4 + 2x^3 + 2x^2 + x$ **27.** $x^5 - 1$

29. $1 + x + x^2 + x^3$

31. $(1 + x)(1 + x^2) \cdots (1 + x^{2^n}) = 1 + x + x^2 + \cdots + x^{2^{n+1}-1}$

33. $1 - x^4$ **35.** $(1 - x)(1 + x)(1 + x^2) \cdots (1 + x^{2^n}) = 1 - x^{2^{n+1}}$

37. $x^4 + x^2 + 1$ **39.** $2x^3 + 4x^2 - x + 5$

41. $x^2 + (a + b)x + ab$ **43.** $x^4 + 2x^3 + 2x^2 + 2x + 1$

45. $abx^2 + (a^2 + b^2)x + ab$ **47.** 2   **49.** 7   **51.** -20 **53.** 1   **55.** 3

Section 9,   p. 38

1. $5x + y + 2$ **3.** $-5x + 6y + 5z - 2$ **5.** $2x^2 + 4y^2 - 3y$

7. $-2x^2 + yz + xz - x - y$ **9.** $2x^2 + 5xy + 3y^2$ **11.** $25x^2 - 4y^2$

13. $16u^2 - 24uv + 9v^2$ **15.** $x^3 + 3x^2y + 3xy^2 + y^3$

17. $2x^2 - y^2 + xy + 2xz - yz$ **19.** $x^3 + x^2y - xy^2 - y^3 + y^2 + xy - x$

21. $x^4 - 2x^2y^2 - 3y^4$ **23.** $x^2 + y^2 + z^2 + 2xy + 2xz + 2yz$

25. $rst + rs + rt + st + r + s + t + 1$ **27.** $a^3 + 6a^2b + 11ab^2 + 6b^3$

29. $x^4 - 4x^2y^2 + 8xy^3 - 4y^4$ **31.** $(x + 8)^2$ **33.** $(z^3 + 2)^2$

35. not a square   **37.** $(x^2 + 2)^2$ **39.** $(ab^4 - 2c^2)^2$

41. Start with x. Obtain $[(x + 1)^2 - (x - 1)^2]/x = 4x/x = 4$.   **43.** 16   **45.** 12

Section 10,   p. 42

1. $(x + 1)(x + 3)$ **3.** $(x + 3)(x - 2)$ **5.** $(x - 2)(x - 7)$ **7.** $(x - 3)^2$

9. $x^2(x + 5)(x - 4)$ **11.** $(x + 4)(2x + 3)$ **13.** $(2x + 1)^2$

15. $(x^2 - 2)(x^2 - 3)$ **17.** $(2x + 9)(2x - 9)$ **19.** $(x + y + z)(x + y - z)$

21. $(a^2 + 4b)(a^2 - 4b)$ **23.** $(v^2 + 9)(v^2 + v - 9)$ **25.** $(2c + 3)(4c^2 - 6c + 9)$

27. $(5x + 2y)(7x^2 + 2xy + 4y^2)$ **29.** $(x - 1)(x^4 + x^3 + x^2 + x + 1)$

31. $x^2(x - 1)(x^2 + x + 1)$ **33.** $(x - 1)(x + 1)(x^2 + 1)(x^4 + 1)$

35. $8(xy^2 + 5)(x^2y^4 - 5xy^2 + 25)$

37. $x^4 - 1 = (x - 1)[x^3 + x^2 + x + 1] = (x - 1)[(x^2 + 1)(x + 1)]$

39. $1027 = 10^3 + 3^3$, divisible by $10 + 3$.

Section 11,   p. 45

1. $1/(x + 4)$ **3.** $1/(x^2 - x + 1)$ **5.** $(x + y)/(x^2 + xy + y^2)$

7. $1/(x + y + 2z)$ **9.** $1/x(x - 1)$ **11.** $2/(2x + 3)(2x + 5)$

13. $(5x + 4)/(x^2 - 4)$ **15.** $(x^2 + y^2)/xy$

17. $(x^2 + 3xy + y^2)/(x + 2y)(x + 3y)$ **19.** $(x + 1)/x$ **21.** $(x^2 + x)/(y^2 + y)$

23. $1/xy$ **25.** $-3xy + 2xz + yz$ **27.** $(x + 3)/2x$ **29.** $(a - 1)/a$

31. $(u^2 - v^2)/uv$ **33.** $(u^{12} - v^{12})/u^6v^6$ **35.** $u^2 + v^2$

Section 12,   p. 47

1. $(x/5)^2 = x^2/25$ **3.** $\sqrt{4x + 4} = 2\sqrt{x + 1}$ **5.** $\sqrt{2^9} = 2^{9/2} = 16\sqrt{2}$

7. $\sqrt[3]{\sqrt[3]{x}} = \sqrt[9]{x}$ **9.** Leave as $\sqrt{x^2 + 1}$.

11. $2a(a^2 + 4a + 1) = 2a^3 + 8a^2 + 2a$ **13.** $a^3 + b^3 = (a + b)(a^2 - ab + b^2)$

15. $a^{-2}a^{-2} = a^{-4}$ **17.** $\sqrt{x^3 + 2x^2 + x} = (x + 1)\sqrt{x}$ **19.** $(25x)(4x) = 100x^2$

21. $(x + y)/(x + z) = x/(x + z) + y/(x + z)$

23. $(x + 1)(y + 1)(z + 1) = xyz + xy + xz + yz + x + y + z + 1$

25. $(4x^3 - 7x^2 + 1)/x = 4x^2 - 7x + 1/x$ **27.** $(x + y)^3 = x^3 + 3x^2y + 3xy^2 + y^3$

29. $(x^2 + 4x + 8)/(x^2 + 2x + 2) = 1 + (2x + 6)/(x^2 + 2x + 2)$, not constant.

31. Leave as -2^4. Note that $-2^4 = -16$ and $2^{-4} = 1/16$.

CHAPTER 2

Section 1, p. 52

1. C.E. **3.** I. **5.** C.E. **7.** I. **9.** C.E. **11.** 6

13. 1, -5 **15.** 0, 2, 7 **17.** 3 **19.** 2, -2, 3, -3

Section 2, p. 57

1. 6 **3.** 2 **5.** $-45/4$ **7.** 5/6 **9.** 1 **11.** -4

13. $(d - b)/(a - c)$ **15.** $x = 2 - \frac{3}{2}y$ **17.** $x = (y + 1)/(3y + 1)$

19. $x = (1 - 5y)/(3y - 2)$ **21.** $x = 1, y = 1$ **23.** $x = 4, y = 3$

25. $x = 1, y = 3$ **27.** $x = 10/29, y = 7/29$ **29.** no solutions

31. $x = 2, y = 1$ or 2, -1 or -2, 1 or -2, -1

Section 3, p. 62

1. $0, \frac{4}{3}$ **3.** $7, -2$ **5.** $\frac{1}{2}, \frac{1}{3}$ **7.** $\frac{3}{5}, -4$ **9.** $(x - 4)^2 - 16$

11. $(x + 3)^2 - 8$ **13.** $(x + k)^2 - k^2$ **15.** $4(x + \frac{3}{2})^2 - 9$

17. $2(x - \frac{1}{8})^2 - \frac{1}{32}$ **19.** $-(x - 2)^2 + 4$ **21.** $3(x - \frac{5}{6})^2 - \frac{13}{12}$

23. $-3(x - \frac{1}{6})^2 + \frac{61}{12}$ **25.** $-2 \pm \sqrt{2}$ **27.** $-1 \pm \frac{1}{2}\sqrt{2}$ **29.** no real solutions

31. $\frac{1}{6}(-5 \pm \sqrt{19})$ **33.** no real solutions **35.** $9, -1$ **37.** 1

39. $\frac{1}{5}(2 \pm \sqrt{39})$ **41.** none **43.** one **45.** two **47.** $-3y^2 \pm \sqrt{9y^4 + y}$

49. $\frac{1}{4}y(-5 \pm \sqrt{17})$ **51.** $\frac{1}{2}y, \frac{1}{3}y$

Section 4, p. 66

1. $0, -2, -3$ **3.** $-\frac{1}{2}, -\frac{1}{3}, 2$ **5.** $2, \frac{1}{2}(-3 \pm \sqrt{5})$ **7.** $\frac{1}{2}, -\frac{1}{2}, 3, -3$

9. -5 **11.** $1, -1$ **13.** no real solution **15.** $\pm\frac{1}{5}\sqrt{5}, \pm\frac{1}{2}\sqrt{2}$ **17.** 4, 9

19. $0, \frac{5}{2}, \pm\sqrt{\frac{1}{2}(\sqrt{29} - 1)}$ **21.** $\frac{1}{2}, -\frac{1}{2}, \pm\frac{1}{2}\sqrt{2}$ **23.** $\pm\sqrt{\frac{1}{2}(1 + \sqrt{37})}$

25. $x = 7, y = 9; x = 9, y = 7$ **27.** $x = 2, y = -1; x = -1, y = 2$

29. $x = \frac{1}{2}, y = 1; x = -\frac{1}{4}, y = -\frac{7}{2}$ **31.** $x = 3, y = 0; x = -5, y = 20/3$

Section 5, p. 71

1. $5\frac{1}{3}\%$ **3.** $2\frac{2}{3}\%$ **5.** 18/19 hr **7.** $2h_1h_2/(h_1 + h_2)$ hr **9.** 80

11. $66\frac{2}{3}$ g **13.** 375 g **15.** no solution **17.** \$3125 in X and \$6875 in B

19. $P \approx \$12.27$, $Q \approx 363600$ **21.** 4   **23.** $P = \$17500$, $Q = 3000$
25. no solution

Section 6,   p. 77

1. $4 < x < 9$ **3.** $x \geq 0$ **5.** $x < y \leq z$

7.

-2

9.

0 1

11.

-3 3

13. $5667 < 6093$, hence $5667 - 4128 < 6093 - 4128$.
15. $\pi < 4$, hence $1/\pi > 1/4$.   **17.** $9 \cdot 11 \cdot 13 = 99 \cdot 13 < 100 \cdot 13$
19. $\frac{1}{3}(8561 + 8774 + 8819) > \frac{1}{3}(8561 + 8561 + 8561) = 8561$
21. $\pi^2 < 10$, hence $\pi^6 = (\pi^2)^3 < 10^3$. [$\pi < 3.15$, so $\pi^2 < 9.9225$.]
23. $9 + 99 + 999 + 9999 < 10 + 100 + 1000 + 10000 = 11110$
25. $\dfrac{1}{\sqrt{5}} + \dfrac{1}{\sqrt{6}} + \dfrac{1}{\sqrt{7}} + \dfrac{1}{\sqrt{8}} < \dfrac{1}{\sqrt{4}} + \dfrac{1}{\sqrt{4}} + \dfrac{1}{\sqrt{4}} + \dfrac{1}{\sqrt{4}} = 2$

$\dfrac{1}{\sqrt{5}} + \dfrac{1}{\sqrt{6}} + \dfrac{1}{\sqrt{7}} + \dfrac{1}{\sqrt{8}} > \dfrac{1}{\sqrt{9}} + \dfrac{1}{\sqrt{9}} + \dfrac{1}{\sqrt{9}} + \dfrac{1}{\sqrt{9}} = \dfrac{4}{3}$

27. $b^2 - a^2 = (b - a)(b + a) > 0$ since each factor is positive.
29. $a < 0$ and $b > 0$. Hence $1/a < 0$ and $1/b > 0$, so $1/a < 1/b$.
31. $1/a - 1/b = (b - a)/ab > 0$ since both numerator and denominator are positive.
Hence $1/b < 1/a$.
33. $(A + B) - (a + b) = (A - a) + (B - b) > 0$ since $A - a > 0$ and
$B - b > 0$. Hence $a + b < A + B$.
35. $12 < ab < 13.65$ **37.** $60 < S < 65$
39. 68^2: $(67)(69) = (68 - 1)(68 + 1) = 68^2 - 1 < 68^2$.
41. 123^4: $(121)(122)(124)(125) = (123 - 2)(123 - 1)(123 + 1)(123 + 2)$
$= (123^2 - 4)(123^2 - 1) < (123)^2(123)^2 = (123)^4$.
43. best, large economy; worst, giant   **45.** best, mammoth; worst, super
47. decreases

Section 7,   p. 82

1. 4   **3.** 7   **5.** 5   **7.** 30   **9.** 4/9   **11.** 36   **13.** 0   **15.** ± 3
17. 3, 5   **19.** $-\frac{1}{2} \leq x \leq \frac{1}{2}$ **21.** $-4 < x < 4$ **23.** $-2 < x < 4$ **25.** $|x| = 2$
27. $|x - 1| < |x - 5|$ **29.** $|x - 7| < 3$ **31.** $-1, 2$
33. There is no point within 2 units of 1 and also within 3 units of 12.
35. If $a = b = c = 0$, then $|a| + |b| + |c| = 0$. Otherwise, at least one of the non-negative
numbers $|a|, |b|, |c|$ is positive, hence $|a| + |b| + |c| > 0$.
37. $|7x - 7a| = 7|x - a| < 7 \cdot 10^{-6} < 10^{-5}$
39. First we observe that $6.9 < x < 7.1$, hence $|x| < 8$. Next,

$$|xy - 35| = |x(y - 7) + 7(x - 5)| \leq |(x(y - 7)| + |7(x - 5)|$$
$$= |x| \, |y - 7| + 7|x - 5| < 8 \times 10^{-6} + 7 \times 10^{-6}$$
$$= 15 \times 10^{-6} = 1.5 \times 10^{-5}.$$

41. $|x^2 - 9| = |(x + 3)(x - 3)| = |x + 3| \, |x - 3| < 7 \times 10^{-6} < 10^{-5}$. (Since
$|x - 3| < 10^{-6} < 1$, $0 < x < 4$, and $0 < x + 3 < 7$.)

Section 8, p. 87

1. $x < \frac{2}{3}$ **3.** $x < -\frac{3}{2}$ **5.** $-\frac{13}{2} < x < \frac{7}{2}$ **7.** $x \geq -\frac{1}{4}$ **9.** $-5 < x < 3$
11. $x > 7$ or $x < -1$ **13.** all $x \neq 0$ **15.** all x **17.** $-\frac{3}{2} < x < \frac{5}{2}$
19. $-2 < x < 2$ **21.** $x > -\frac{5}{2}$ **23.** $-1 < x < 0$ or $x < -2$
25. $x > \frac{3}{8}$ or $x < 0$ **27.** $x > 2$ **29.** $5 < t < 15$
31. $x(1 - x) = -(x^2 - x + \frac{1}{4}) + \frac{1}{4} = -(x - \frac{1}{2})^2 + \frac{1}{4} \leq \frac{1}{4}.$ **33.** $c \geq 20$ or $c \leq -20$
35. The inequality is equivalent to $2x^2 + 2y^2 \geq x^2 + 2xy + y^2$ or
$x^2 - 2xy + y^2 = (x - y)^2 \geq 0$, which is true for all x, y.
37. Set $a = \frac{1}{2}(x + y)$ and $b = \frac{1}{2}(z + w)$. Then

$$\frac{x + y + z + w}{4} = \frac{a + b}{2} \geq \sqrt{ab} = \sqrt{\left(\frac{x + y}{2}\right)\left(\frac{z + w}{2}\right)} \geq \sqrt{\sqrt{xy}\,\sqrt{zw}} = \sqrt[4]{xyzw}.$$

39. $x^4 + x^2 - 2x + 3 = x^4 + (x - 1)^2 + 2 \geq 2 > 0$

CHAPTER 3

Section 2, p. 92

9.
$(-3, 0)$

11.

13.

15.

17.

19.

21.

23. $(\pm 1, \pm 1)$

25. $(9, 0), (0, 12)$ or $(12, 0), (0, 9)$

Section 3, p. 98

1. $5, 9, 6, \dfrac{2}{x} + 5 = \dfrac{2 + 5x}{x}, 2x - 1$

3.

5.

7.

9.

11.

13.

15.   **17.** all real x **19.** all real x

21. $x \neq \frac{3}{2}$ **23.** $x \neq \frac{5}{3}$ **25.** $x \geq 6$ **27.** $|x| \leq \frac{2}{3}$ **29.** $x \geq \frac{3}{2}$
31. $|x| \leq \frac{1}{2}$ **33.** all x such that $x \leq 1$ or $x \geq 4$

35.

Section 4, p. 100

1. $3x - 1, -6x - 2$ **3.** $x^2 - 2x + 1, -2x^3 + x^2$
5. No; their domains have no point in common.
7. $[f \circ g](x) = 3x - 5, [g \circ f](x) = 3x - 1$ **9.** $2x^2 + 4x + 2, -2x^2 - 1$
11. $-4x, -4x$ **13.** $9, 3$ **15.** $g(x)$ **17.** x **19.** $x + 1, 3$, etc.
21. No; $f(x)$ is defined only for $x \geq \frac{5}{2}$, but $g(x) \leq 1$.
23. Yes; $f[\frac{1}{2}(x_0 + x_1)] = \frac{1}{2}a(x_0 + x_1) + b$ and
$$\frac{1}{2}[f(x_0) + f(x_1)] = \frac{1}{2}[(ax_0 + b) + (ax_1 + b)]$$
$$= \frac{1}{2}a(x_0 + x_1) + b.$$
25. $f[\frac{1}{2}(x_0 + x_1)] = 2/(x_0 + x_1) = 2f(x_0 + x_1).$

Section 5, p. 108

1.

3.

5.

7.

9.

(different scales)

11.

13.

15.

17. $\frac{4}{3}$ **19.** 0

21. 1   **23.** $\frac{3}{2}$ **25.** 1   **27.** $y = x + 1, 1$ **29.** $y = 3, 3$ **31.** $y = \frac{1}{2}x - 3$
33. $y = 2x$ **35.** $y = \frac{4}{3}x + \frac{4}{3}$ **37.** $y = x + \frac{1}{2}$ **39.** $y = -5x + 3.5$ **41.** $x = 0$
43. $3, -7$ **45.** $-1, 7$ **47.** 2, 3; as given   **49.** $\frac{1}{2}, \frac{1}{3}; x/\frac{1}{2} + y/\frac{1}{3} = 1$

Section 6,   p. 115

1.

3.

5.

7.

9.

11.

13.

lowest: $(2, -3)$

15.

lowest: $(-\frac{1}{2}, \frac{3}{4})$

17.

highest: $(-1, 1)$

19.

highest: $(-2, 1)$

21.

$(\frac{3}{2}, -\frac{7}{2})$

23.

lowest: $(-2, -20)$

25.

highest: $(2, -2)$

27.

$(\frac{1}{8}, \frac{1}{16})$

29.

$(\frac{3}{4}, -\frac{9}{8})$

31.

$(-\frac{1}{2}, -\frac{17}{4})$

33.

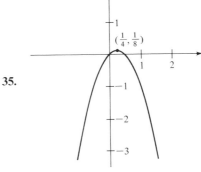

35.

37. When $x = 0$, then $y = a \cdot 0^2 + b \cdot 0 = 0$.   **39.** $b = 0$

41. $x(1 - x) = \frac{1}{4} - (x - \frac{1}{2})^2 \geq \frac{1}{4}$

43. $A^2 = (\frac{1}{2}ab)^2$ where $a^2 + b^2 = 16$, hence $4A^2 = a^2(16 - a^2) = 64 - (a^2 - 8)^2 \leq 64$. Therefore $A^2 \leq 16$, $A \leq 4$.   **45.**   800 ft/sec

Section 7,   p. 121

1. $x^2, x^4, 1/(x^2 + 1)$

3.

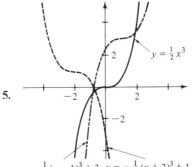

5.

7. $1, 3/(x^2 - 9), x^2/(1 - x^2)$ **9.** $g(-x) = \frac{1}{2}[f(-x) - f(x)] = -g(x)$

CHAPTER 4

Section 2,   p. 129

1.

3.

5.

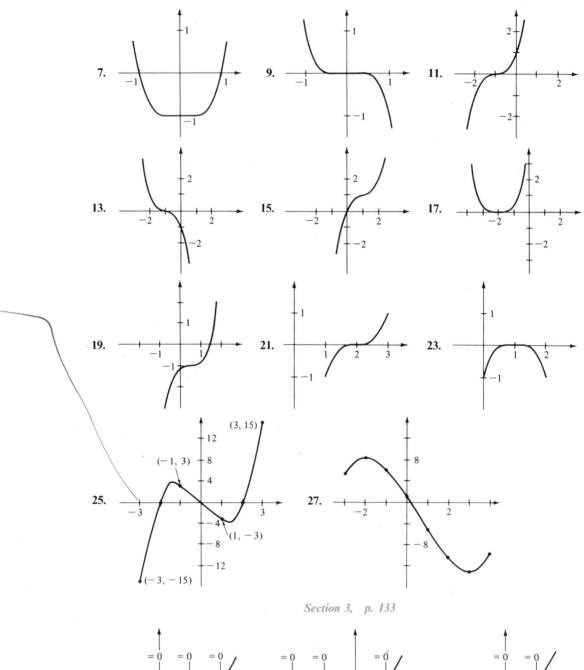

7.   9.   11.

13.   15.   17.

19.   21.   23.

25. (3, 15), (−1, 3), (1, −3), (−3, −15)   27.

Section 3, p. 133

1.   3.   5.

7.

9.

11.

13.

15.

17. $3 < x < 5$ or $x > 8$ **19.** $x < -2$, or $-1 < x < 1$, or $x > 2$

Section 4,   p. 136

1. $x \neq 1$ **3.** all x **5.** $x \neq 2$ **7.** $x \neq 0, 2$ **9.** $x \neq 1$ **11.** $x \neq \pm 1$
13. $x \neq \frac{1}{6}(5 \pm \sqrt{61})$ **15.** $24/5, -8/3, 0$

Section 5,   p. 142

1.

3.

5.

7.

9.

11. $r(x) \longrightarrow 0-$

1.   **3.**

5.   **7.**   **9.**

11.   **13.**   **15.**

17. By long division, $f(x) = g(x)(ax + b) + h(x)$, where $a \neq 0$ and $\deg h(x) < \deg g(x)$, so that $h(x)/g(x) \longrightarrow 0$ as $x \longrightarrow \infty$. Therefore $r(x) \approx ax + b$ as $x \longrightarrow \infty$.

CHAPTER 5

1.   **3.**   **5.**

7.

9.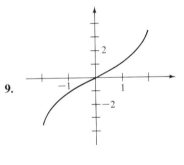

11. in 9 min   **13.** $n = 167$; since $2^{10} > 10^3$, $2^{170} > 10^{51}$, $2^{167} > \frac{1}{8}10^{51} > 10^{50}$.   **15.** a^x

17. no intersection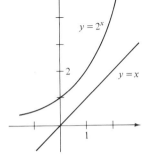

19. Set $a = 10^3$ and $b = 10^6$. Then $2^a = (2^{10})^{100} > (10^3)^{100} = a^{100}$ and
$2^b = (2^{10})^{100000} > 10^{300000} = 10^{600} \cdot 10^{300000-600} > b^{100} \cdot 10^{200000}$.
21. $(\frac{3}{2})^4 = \frac{81}{16} \approx 5$; $(\frac{3}{2})^{20} \approx 5^5 = 3125$.

Section 2, p. 159

1. 4   **3.** -2 **5.** $\frac{3}{2}$ **7.** 3   **9.** 10   **11.** -4 **13.** $-\frac{3}{2}$ **15.** 0.788
17. -0.250 **19.** 1.653   **21.** -1.632 **23.** -0.921 **25.** 17   **27.** 1   **29.** $\sqrt{ab}$
31. Set $a = \log x$ and $b = \log y$. Then $x = 10^a$, $y = 10^b$, $x/y = 10^{a-b}$, $\log(x/y) = a - b$.
33. $\log a + \log b = \log(ab)$
35. $c = \log_a b$ **37.** $x > 1$

39.

41.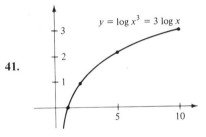

43. $\log_6 5 > \log_7 5$ because $7^{\log_6 5} > 6^{\log_6 5} = 5 = 7^{\log_7 5}$.

45.

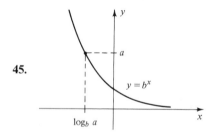

Section 3, p. 166

1.

3.

5.

7.

9.

11.

13.

15. $\dfrac{1}{\sqrt{7} + \sqrt{5}}$

17. $\dfrac{x - 1}{x(\sqrt{x} - 1)}$

19. $\dfrac{\sqrt{1 + x} - 1}{x} = \dfrac{\sqrt{1 + x} - 1}{x} \dfrac{\sqrt{1 + x} + 1}{\sqrt{1 + x} + 1} = \dfrac{1}{\sqrt{1 + x} + 1} \longrightarrow \dfrac{1}{2}$

21. $\sqrt{10001} - 100 = (\sqrt{10001} - 100) \dfrac{\sqrt{10001} + 100}{\sqrt{10001} + 100}$

$$= \dfrac{1}{\sqrt{10001} + 100} < \dfrac{1}{\sqrt{10000} + 100} = \dfrac{1}{200} = 0.005$$

23. $(b^{1/3} - a^{1/3})(b^{2/3} + b^{1/3}a^{1/3} + a^{2/3}) = (b^{1/3})^3 - (a^{1/3})^3 = b - a$

Section 4, p. 169

1. 0.44, 0.31, 0.11, 0.26   **3.** 0.000, 0.000, 16.244, 3.786   **5.** (a) 5.10   (b) 5.12
7. 1050, 55.5, 10.0   **9.** 0.40   **11.** 2.3

Section 5, p. 172

[There may be slight discrepancies in this and the following sections due to round-off errors.]
1. 0.1644   **3.** 1.5502   **5.** 1.0414   **7.** 1.7194   **9.** 2.4777   **11.** 0.1928
13. 1.2059   **15.** $0.8521 - 1$ **17.** 1.407   **19.** 1913   **21.** 4.774×10^5
23. 6.043×10^{-5} **25.** 0.01101

Section 6, p. 176

1. 19.20   **3.** 2764   **5.** 0.05679   **7.** 1.140   **9.** 64.89   **11.** 0.5639
13. 6.953×10^{-2} **15.** 0.5680   **17.** 2.646   **19.** 7.936×10^{-2} **21.** 4.672
23. 16.35   **25.** 238.0   **27.** 5058   **29.** 146   **31.** 1.328   **33.** 0.9951
35. 0.9817   **37.** 5.236×10^{-4} **39.** 5.37
41. $11^{12} \approx 3.14 \times 10^{12} > 12^{11} \approx 7.43 \times 10^{11}$
43. $400^{401} \approx 3 \times 10^{1043} > 401^{400} \approx 2 \times 10^{1041}$
45. $73 \cdots 78 \approx 1.85 \times 10^{11} > 75^6 \approx 1.78 \times 10^{11}$ **47.** 1.431   **49.** 9
51. $\log 5 \approx 0.6990$ **53.** $(1.01)^{2200} \approx 10^{9.46} > 10^9$
55. $\log(1.01)^{232} = 232 \log 1.01 > (232)(0.00432) = 1.00224 > 1$, hence $(1.01)^{232} > 10$.
57. 2.576   **59.** 2.352

Section 7, p. 182

1. 4.321 $\boxed{1/x}$ ≈ 0.2314

3. 4.23 $\boxed{\text{ENT} \uparrow}$ 3 $\boxed{+}$ 4.23 $\boxed{\text{ENT} \uparrow}$ 5 $\boxed{+}$ $\boxed{\times}$ ≈ 66.73

5. 7.3 $\boxed{\text{ENT} \uparrow}$ 1 $\boxed{+}$ $\boxed{x^2}$ $\boxed{1/x}$ ≈ 0.01452

7. 5.1 $\boxed{\text{ENT} \uparrow}$ 3.7 $\boxed{\sqrt{x}}$ $\boxed{\div}$ ≈ 2.651

9. 7.52 $\boxed{\sqrt{x}}$ 4.21 $\boxed{\sqrt{x}}$ $\boxed{-}$ $\boxed{1/x}$ ≈ 1.448

11. 6 $\boxed{\text{ENT} \uparrow}$ 0.03721 $\boxed{\times}$ 19 $\boxed{\text{ENT} \uparrow}$ 1.428 $\boxed{\times}$ $\boxed{+}$ 27 $\boxed{\text{ENT} \uparrow}$ 3.142 $\boxed{\times}$ $\boxed{-}$ ≈ -57.48

13. 4 $\boxed{\text{ENT} \uparrow}$ 1.721 $\boxed{x^2}$ $\boxed{\times}$ 7 $\boxed{\text{ENT} \uparrow}$ 3.998 $\boxed{x^2}$ $\boxed{\times}$ $\boxed{+}$ 5 $\boxed{\text{ENT} \uparrow}$ 1.072 $\boxed{x^2}$ $\boxed{\times}$
 $\boxed{+}$ 2 $\boxed{\text{ENT} \uparrow}$ 1.911 $\boxed{x^2}$ $\boxed{\times}$ $\boxed{+}$ 4 $\boxed{\text{ENT} \uparrow}$ 7 $\boxed{+}$ 5 $\boxed{+}$ 2 $\boxed{+}$ $\boxed{\div}$ $\boxed{\sqrt{x}}$ ≈ 2.757

15. 0.05 $\boxed{\text{ENT} \uparrow}$ 12 $\boxed{\div}$ $\boxed{\text{ENT} \uparrow}$ $\boxed{\text{ENT} \uparrow}$ 1 $\boxed{+}$ 120 $\boxed{\pm}$ $\boxed{x \leftrightharpoons y}$ $\boxed{x^y}$ $\boxed{\pm}$ 1 $\boxed{+}$
 $\boxed{\div}$ ≈ 0.01061

17. $\frac{1}{3}a^2 b$ **19.** $a + b + b^2$ **21.** 3 $\boxed{1/x}$ a $\boxed{x^2}$ b $\boxed{\times}$ $\boxed{x^y}$

23. 1.371   **25.** 10.9   **27.** 0.666

1. $\log 10 / \log 2 \approx 3.322$ **3.** 0 **5.** no solution **7.** 0
9. $\log 3 / \log 7 \approx 0.5646$ **11.** $\log 4 / \log 3 \approx 1.262$ hr
13. $(151 \times 10^6)(178/151)^5 \approx 344 \times 10^6$ **15.** $3.64 / \log 2 \approx 12.1$ days
17. $30(25/30)^5 \approx 12.1$ in.

1. $B: (1 + 0.07/4)^4 \approx 1.072 > (1 + 0.0675/365)^{365} \approx 1.070$
3. $(1 + j/12)^{12} = 1.25$, $j = 12[(1.25)^{1/12} - 1] \approx 22.52\%$
5. $10000 = P[1.005^{180} - 1]/0.005$, $P \approx \$34.39$
7. $2405.08 = 10[(1 + j/12)^{144} - 1]/(j/12)$, $j \approx 8.0\%$
9. $P = (28000)(0.085/12)/[1 - (1 + 0.085/12)^{-216}] \approx \$253.53$
11. In one year the investment of 28000 is worth

$$W = (28000)(1 + 0.085/12)^{12} \approx 30474.95.$$

In one year the monthly payments of 253.53 accumulate

$$A = (253.53)[(1 + 0.085/12)^{12} - 1]/(0.085/12) \approx 3163.73.$$

The balance due after one year is

$$B = W - A \approx 27311.22.$$

The reduction of the debt in one year is

$$R = 28000 - B \approx 688.78.$$

The total paid in one year is

$$T = (253.53)(12) = 3042.36.$$

Finally, the interest paid in the first year is

$$T - R \approx \$2353.58.$$

13. Balance due $= (3500)(1.02)^{72} - (90)(1.02^{72} - 1)/0.02 \approx 14563.99 - 14225.13 = \$338.86$.
15. We are given $1 + i = (1 + j/p)^p$, hence

$$1 + \frac{j}{p} = (1 + i)^{1/p}, \quad \frac{j}{p} = (1 + i)^{1/p} - 1, \quad j = p[(1 + i)^{1/p} - 1].$$

17. $208.34 = (2275)(j/12)/[1 - (1 + j/12)^{-12}]$, hence

$$\frac{j/12}{1 - (1 + j/12)^{-12}} \approx 0.00915780.$$

From $j = 17.7$ we get 915373 and from $j = 17.8$ we get 915843. By interpolation, $j \approx 17.87\%$.

CHAPTER 6

1. $\sqrt{37}$ **3.** 4 **5.** $\sqrt{74}$ **7.** $(\frac{1}{2}, \pm\frac{1}{2}\sqrt{3}), (\pm\frac{1}{2}\sqrt{3}, \frac{1}{2})$ **9.** $60°$
11. $216°$ **13.** $18°$ **15.** $40°$ **17.** $330°$ **19.** $2°$ **21.** $\frac{3}{4}\pi$ **23.** $\frac{1}{30}\pi$
25. $\frac{3}{8}\pi$ **27.** $\frac{1}{20}\pi$ **29.** $\frac{119}{60}\pi$ **31.** 0.0209 **33.** 1.17 **35.** $17.2°$
37. $60.2°$ **39.** $\frac{7}{6}\pi$ **41.** $\frac{1}{8}\pi$ **43.** π **45.** $\frac{3}{4}\pi$

Section 3,   p. 204

1. $-1, 0$ 3. $-\frac{1}{2}\sqrt{2}, \frac{1}{2}\sqrt{2}$ 5. $\frac{1}{2}\sqrt{3}, \frac{1}{2}$ 7. $\frac{1}{2}, -\frac{1}{2}\sqrt{3}$ 9. $-\frac{1}{2}\sqrt{3}, -\frac{1}{2}$
11. $0, -1$ 13. $\frac{1}{3}\pi \pm 2n\pi$ and $-\frac{1}{3}\pi \pm 2n\pi$ 15. $\frac{1}{2}\pi \pm n\pi$
17. $\frac{4}{3}\pi \pm 2n\pi$ and $\frac{5}{3}\pi \pm 2n\pi$ 19. $\frac{1}{4}\pi \pm n\pi$ 21. 1
23. $\sin\alpha = \sin(\frac{1}{2}\pi - \beta) = \cos\beta$ 25. 2π 27. π 29. $\frac{2}{5}\pi$ 31. π
33. A 35. $-A$ 37. A 39. $-\sqrt{1 - A^2}$

Section 4,   p. 208

1.

θ	0	$\frac{1}{6}\pi$	$\frac{1}{4}\pi$	$\frac{1}{3}\pi$	$\frac{1}{2}\pi$
$\tan\theta$	0	$\frac{1}{3}\sqrt{3}$	1	$\sqrt{3}$	undefined
$\cot\theta$	undefined	$\sqrt{3}$	1	$\frac{1}{3}\sqrt{3}$	0
$\sec\theta$	1	$\frac{2}{3}\sqrt{3}$	$\sqrt{2}$	2	undefined
$\csc\theta$	undefined	2	$\sqrt{2}$	$\frac{2}{3}\sqrt{3}$	1

3. $\cot(\theta + \frac{1}{2}\pi) = \dfrac{\cos(\theta + \frac{1}{2}\pi)}{\sin(\theta + \frac{1}{2}\pi)} = \dfrac{-\sin\theta}{\cos\theta} = -\tan\theta$

5. $1 + \tan^2\theta = 1 + (\sin^2\theta/\cos^2\theta) = (\cos^2\theta + \sin^2\theta)/\cos^2\theta = 1/\cos^2\theta = \sec^2\theta$

7. $\tan(\theta + \frac{1}{2}\pi) = \dfrac{\sin(\theta + \frac{1}{2}\pi)}{\cos(\theta + \frac{1}{2}\pi)} = \dfrac{\cos\theta}{-\sin\theta} = -\cot\theta$

9. $\cot(-\theta) = \cos(-\theta)/\sin(-\theta) = \cos\theta/(-\sin\theta) = -\cot\theta$
11. $\sec(-\theta) = 1/\cos(-\theta) = 1/\cos\theta = \sec\theta$
13. $\csc(-\theta) = 1/\sin(-\theta) = 1/(-\sin\theta) = -\csc\theta$
15. $\frac{4}{5}, \frac{3}{5}, \frac{4}{3}$ 17. $1, 0,$ undefined   19. $\frac{12}{13}, -\frac{5}{13}, -\frac{12}{5}$
21. $-\frac{3}{13}\sqrt{13}, -\frac{2}{13}\sqrt{13}, \frac{3}{2}$ 23. π 25. 2π 27. 2π 29. π 31. 1

Section 5,   p. 214

1.

3.

5.

7.

9.

11.

13.

15.

17.

19.

21.

23.

25.

27. $(x, y) = (3 \cos t, 3 \sin t)$

29. $(x, y) = (-3 \sin t, 3 \cos t)$
31. $(x, y) = (3 \cos 2\pi t, 3 \sin 2\pi t)$

33.

Section 6,   p. 221

1.

3.

5.

7.

9.

11. $\cos^2 \theta = 1 - \sin^2 \theta \approx 1 - \theta^2 = (1 - \frac{1}{2}\theta^2)^2 + \frac{1}{4}\theta^4 \approx (1 - \frac{1}{2}\theta^2)^2$, hence
$\cos \theta \approx 1 - \frac{1}{2}\theta^2$. We can ignore θ^4 because it is very small compared to θ^2.

13.   1.0   15.   2.0   17.   10.0   19.

21.

CHAPTER 7

Section 1,   p. 227

1. $\sin^2\theta - \cos^2\theta$ **3.** $\sin\theta$ **5.** $1/\sqrt{1 + \cot^2\theta}$ **7.** $\sqrt{1 + \cot^2\theta}$

9. $-\sqrt{1 + \cot^2\theta}/\cot\theta$ **11.** $\sin\theta/\sqrt{1 - \sin^2\theta}$ **13.** $1/\sqrt{1 - \cos^2\theta}$

15. $\sin\theta \sec\theta = \sin\theta(1/\cos\theta) = \sin\theta/\cos\theta = \tan\theta.$

17. $\sec\theta - \cos\theta = \dfrac{1}{\cos\theta} - \cos\theta = \dfrac{1 - \cos^2\theta}{\cos\theta} = \dfrac{\sin^2\theta}{\cos\theta} = \sin\theta\,\dfrac{\sin\theta}{\cos\theta} = \sin\theta\tan\theta.$

19. $\sec^2\theta - \csc^2\theta = (1 + \tan^2\theta) - (1 + \cot^2\theta) = \tan^2\theta - \cot^2\theta.$

21. $\sec^2\theta + \csc^2\theta = (1/\cos^2\theta) + (1/\sin^2\theta) = (\sin^2\theta + \cos^2\theta)/(\cos^2\theta \sin^2\theta)$
$= 1/(\cos^2\theta \sin^2\theta) = \sec^2\theta \csc^2\theta.$

23. $\sec^4\theta - \tan^4\theta = (1 + \tan^2\theta)^2 - \tan^4\theta = 1 + 2\tan^2\theta.$

25. $\cot^4\theta + \cot^2\theta = \cot^2\theta(\cot^2\theta + 1) = (\csc^2\theta - 1)\csc^2\theta = \csc^4\theta - \csc^2\theta.$

27. LHS $= (c - s)^2 + (c + s)^2 = (c^2 - 2cs + s^2) + (c^2 + 2cs + s^2) = 2(c^2 + s^2) = 2.$

29. LHS $= \dfrac{1 - s/c}{1 + s/c} = \dfrac{c - s}{c + s},$ RHS $= \dfrac{c/s - 1}{c/s + 1} = \dfrac{c - s}{c + s}.$

31. $\sec\theta(\sec\theta - \cos\theta) = \sec^2\theta - 1 = \tan^2\theta,$ hence $\tan\theta/(\sec\theta - \cos\theta) = \sec\theta/\tan\theta.$
Alternative solution: LHS $= (s/c)/[(1/c) - c] = s/(1 - c^2) = 1/s,$
RHS $= (1/c)/(s/c) = 1/s.$

33. RHS $= \left(\dfrac{s}{c} + \dfrac{c}{s}\right)(c + s) = \left(\dfrac{s^2 + c^2}{cs}\right)(c + s) = \dfrac{1}{cs}(c + s)$

$= \dfrac{\cdot\,1}{s} + \dfrac{1}{c} = $ LHS.

35. Equivalent relations:

$$(\cos \alpha - \sin \beta)(\cos \alpha + \sin \beta) = (\cos \beta + \sin \alpha)(\cos \beta - \sin \alpha),$$
$$\cos^2 \alpha - \sin^2 \beta = \cos^2 \beta - \sin^2 \alpha,$$
$$\cos^2 \alpha + \sin^2 \alpha = \cos^2 \beta + \sin^2 \beta.$$

The last is an identity, hence so is the given relation.

Section 2,   p. 230

1. $\cos \theta, \sin \theta$ **3.** $-\sin \theta, \cos \theta$ **5.** $\frac{1}{2}\sqrt{2}(\cos \theta - \sin \theta), \frac{1}{2}\sqrt{2}(\cos \theta + \sin \theta)$
7. $(\cot \alpha \cot \beta - 1)/(\cot \alpha + \cot \beta)$ **9.** $\frac{1}{4}(\sqrt{2} + \sqrt{6})$ **11.** $\frac{1}{4}(\sqrt{2} + \sqrt{6})$
13. $4 \cos^3 \theta - 3 \cos \theta$ **15.** $(\cot^3 \theta - 3 \cot \theta)/(3 \cot^2 \theta - 1)$
17. $\sin 4\theta = 2 \sin 2\theta \cos 2\theta = 2(2 \sin \theta \cos \theta)(1 - 2 \sin^2 \theta) = 4 \sin \theta \cos \theta - 8 \sin^3 \theta \cos \theta.$
19. $\cot 2\theta + \csc 2\theta = (1 + \cos 2\theta)/\sin 2\theta = (2 \cos^2 \theta)/(2 \sin \theta \cos \theta) = \cot \theta.$

21. $\text{LHS} = \dfrac{\sin \theta}{\cos \theta} + \dfrac{\cos \theta}{\sin \theta} = \dfrac{\sin^2 \theta + \cos^2 \theta}{\sin \theta \cos \theta} = \dfrac{1}{\sin \theta \cos \theta},$

$\text{RHS} = \dfrac{2}{\sin 2\theta} = \dfrac{2}{2 \sin \theta \cos \theta} = \text{LHS}.$

23. $\tan 2\theta = 2 \tan \theta/(1 + \tan^2 \theta) < 2 \tan \theta.$

25. $\cot 2\theta = \dfrac{\cos 2\theta}{\sin 2\theta} = \dfrac{\cos^2 \theta - \sin^2 \theta}{2 \sin \theta \cos \theta} = \dfrac{\cos^2 \theta/\sin^2 \theta - 1}{2 \cos \theta/\sin \theta} = \dfrac{\cot^2 \theta - 1}{2 \cot \theta}.$

27. $\cot(\alpha + \beta) = \dfrac{\cos(\alpha + \beta)}{\sin(\alpha + \beta)} = \dfrac{\cos \alpha \cos \beta - \sin \alpha \sin \beta}{\sin \alpha \cos \beta + \cos \alpha \sin \beta}$

$= \dfrac{\cos \alpha \cos \beta/\sin \alpha \sin \beta - 1}{\cos \beta/\sin \beta + \cos \alpha/\sin \alpha} = \dfrac{\cot \alpha \cot \beta - 1}{\cot \beta + \cot \alpha}.$

29. $1 + \cos 2\theta = 2 \cos^2 \theta$ and $1 - \cos 2\theta = 2 \sin^2 \theta,$
hence $(1 + \cos 2\theta)/(1 - \cos 2\theta) = \cos^2 \theta/\sin^2 \theta = \cot^2 \theta.$

Section 3,   p. 235

1. $\sin^2 22.5° = \sin^2 \frac{1}{2}(45°) = \frac{1}{2}(1 - \cos 45°) = \frac{1}{2}(1 - \frac{1}{2}\sqrt{2}) = \frac{1}{4}(2 - \sqrt{2}),$
$\sin 22.5° = \frac{1}{2}\sqrt{2 - \sqrt{2}}.$
3. $\tan 67.5° = \tan \frac{1}{2}(135°) = \sin 135°/(1 + \cos 135°) = \sqrt{2}/(2 - \sqrt{2}) = \sqrt{2} + 1.$
5. $\tan \frac{1}{2}\theta = \sin \theta/(1 + \cos \theta) = (\sin \theta)(1 - \cos \theta)/(1 - \cos^2 \theta)$
$= (\sin \theta)(1 - \cos \theta)/\sin^2 \theta = (1 - \cos \theta)/\sin \theta.$
7. $5 \cos(\theta - \theta_0), \theta_0 \approx 0.64$ **9.** $\sqrt{5} \cos(\theta + \theta_0), \theta_0 \approx 1.11$
11. $\sqrt{13} \cos(\theta - \theta_0), \theta_0 \approx 0.98$ **13.** $\sqrt{10} \cos(\theta + \theta_0), \theta_0 \approx 0.32$
15. $|\sin \theta + \cos \theta| = \sqrt{2}|\cos(\theta - \frac{1}{4}\pi)| \leq \sqrt{2}$ **17.** $\sqrt{10}$
19. $\frac{1}{2}(\cos \theta - \cos 3\theta)$ **21.** $\frac{1}{2}(\sin 7\theta + \sin \theta)$ **23.** $2 \sin \frac{3}{2}\theta \cos \frac{1}{2}\theta$
25. $2 \cos \frac{9}{2}\theta \sin \frac{1}{2}\theta$ **27.** $-2 \sin \frac{11}{2}\theta \sin \frac{1}{2}\theta$
29. $\tan \alpha + \tan \beta = (\sin \alpha/\cos \alpha) + (\sin \beta/\cos \beta)$
$= (\sin \alpha \cos \beta + \cos \alpha \sin \beta)/(\cos \alpha \cos \beta) = \sin(\alpha + \beta)/\cos \alpha \cos \beta.$
31. $\sin^4 \theta = \frac{1}{4}(1 - \cos 2\theta)^2 = \frac{1}{4}(1 - 2 \cos 2\theta + \cos^2 2\theta)$
$= \frac{1}{4}[1 - 2 \cos 2\theta + \frac{1}{2}(1 + \cos 4\theta)] = \frac{1}{8}(3 - 4 \cos 2\theta + \cos 4\theta).$
33. $\cos(\alpha + \beta) \cos(\alpha - \beta) = (\cos \alpha \cos \beta - \sin \alpha \sin \beta)(\cos \alpha \cos \beta + \sin \alpha \sin \beta)$
$= \cos^2 \alpha \cos^2 \beta - \sin^2 \alpha \sin^2 \beta$
$= \cos^2 \alpha(1 - \sin^2 \beta) - (1 - \cos^2 \alpha)\sin^2 \beta = \cos^2 \alpha - \sin^2 \beta.$

35. $\sin 2\theta = 2 \sin \theta \cos \theta = 2[2t/(1 + t^2)][(1 - t^2)/(1 + t^2)] = 4t(1 - t^2)/(1 + t^2)^2$.

37. $\dfrac{\sin(\beta - \alpha)}{\sin \alpha \sin \beta} = \dfrac{\sin \beta \cos \alpha - \sin \alpha \cos \beta}{\sin \alpha \sin \beta}$

$$= \frac{\cos \alpha}{\sin \alpha} - \frac{\cos \beta}{\sin \beta} = \cot \alpha - \cot \beta.$$

39. $\dfrac{\sin \alpha - \sin \beta}{\cos \alpha + \cos \beta} = \dfrac{2 \cos \frac{1}{2}(\alpha + \beta) \sin \frac{1}{2}(\alpha - \beta)}{2 \cos \frac{1}{2}(\alpha + \beta) \cos \frac{1}{2}(\alpha - \beta)} = \dfrac{\sin \frac{1}{2}(\alpha - \beta)}{\cos \frac{1}{2}(\alpha - \beta)} = \tan \frac{1}{2}(\alpha - \beta)$.

Section 4, p. 243

1. $\frac{1}{3}\pi$ **3.** $-\frac{1}{4}\pi$ **5.** 0.93   **7.** 0.90   **9.** $\frac{5}{6}\pi$ **11.** $\frac{1}{6}\pi$ **13.** -1.11

15. 1.26

17. Set $\theta = \arcsin x$ so $-\frac{1}{2}\pi \le \theta \le \frac{1}{2}\pi$. Then $-\frac{1}{2}\pi \le -\theta \le \frac{1}{2}\pi$ and $\sin(-\theta) = -\sin \theta$ $= -x$, hence $\arcsin(-x) = -\theta = -\arcsin x$.

19. Set $\theta = \text{arc cot } x$ so $0 < \theta < \pi$. Then $0 < \pi - \theta < \pi$ and $\cot(\pi - \theta) = -\cot \theta = -x$, hence $\text{arc cot}(-x) = \pi - \theta = \pi - \text{arc cot } x$.

21. Set $\theta = \arccos x$ so $0 \le \theta \le \frac{1}{2}\pi$. Then $2x^2 - 1 = 2 \cos^2 \theta - 1 = \cos 2\theta$ and $0 \le 2\theta \le \pi$, hence $\arccos(2x^2 - 1) = 2\theta = 2 \arccos x$.

23. Let $\alpha = \arctan \frac{1}{2}$ and $\beta = \arctan \frac{1}{3}$. Then $\tan \alpha = \frac{1}{2}$, $\tan \beta = \frac{1}{3}$, and $0 < \alpha + \beta < \frac{1}{4}\pi + \frac{1}{4}\pi = \frac{1}{2}\pi$.
Hence $\tan(\alpha + \beta) = (\frac{1}{2} + \frac{1}{3})/(1 - \frac{1}{2} \cdot \frac{1}{3}) = 1$, so $\alpha + \beta = \frac{1}{4}\pi$.

25.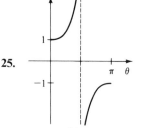

$x = \sec \theta$ increases

27.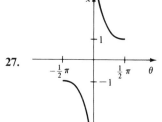

$x = \csc \theta$ decreases

29. If x is small, $\sin x \approx x$. Hence $x = \arcsin(\sin x) \approx \arcsin x$.

31. $\frac{1}{7}\pi$ **33.** 100

35. For $0 \le \theta \le \pi$ we have $-\frac{1}{2}\pi \le \frac{1}{2}\pi - \theta \le \frac{1}{2}\pi$ and $\sin(\frac{1}{2}\pi - \theta) = \cos \theta$, hence $\frac{1}{2}\pi - \theta = \arcsin(\cos \theta)$.

37. $\cot(\text{arc tan } x) = 1/\tan(\text{arc tan } x) = 1/x$.

39. $\tan(\text{arc sin } x) = \dfrac{\sin(\text{arc sin } x)}{\cos(\text{arc sin } x)} = \dfrac{x}{\sqrt{1 - \sin^2(\text{arc sin } x)}} = \dfrac{x}{\sqrt{1 - x^2}}$.

The sign is right because $-\frac{1}{2}\pi \le \text{arc sin } x \le \frac{1}{2}\pi$, so $\cos(\text{arc sin } x) \ge 0$.

Section 5, p. 249

1. $0, \frac{1}{6}\pi, \frac{5}{6}\pi, \pi$ **3.** $\frac{1}{4}\pi, \frac{3}{4}\pi, \frac{5}{4}\pi, \frac{7}{4}\pi$ **5.** $0, \pi$ **7.** $0, \frac{1}{4}\pi, \frac{3}{4}\pi, \pi, \frac{5}{4}\pi, \frac{7}{4}\pi$

9. $\frac{1}{4}\pi, \frac{3}{4}\pi, \frac{5}{4}\pi, \frac{7}{4}\pi$ **11.** $\frac{1}{8}\pi, \frac{5}{8}\pi, \frac{9}{8}\pi, \frac{13}{8}\pi$ **13.** $0, \frac{1}{4}\pi, \frac{1}{2}\pi, \frac{3}{4}\pi, \pi, \frac{5}{4}\pi, \frac{3}{2}\pi, \frac{7}{4}\pi$

15. $\frac{2}{3}\pi, \frac{4}{3}\pi$ **17.** $\frac{1}{3}\pi, \frac{2}{3}\pi, \frac{4}{3}\pi, \frac{5}{3}\pi$ **19.** $18.88°, 71.12°$

21. $\sin \theta = 3 - \sin^2 \theta \geq 2$, impossible. Alternatively: $|\sin \theta| = |\frac{1}{2}(-1 \pm \sqrt{13})| > 1$,
impossible.

23. $s = 2a \arctan(y/a)$ **25.** $(0, 1)$ **27.** $(\frac{1}{2}\sqrt{3}, -\frac{1}{2})$ **29.** $(-\sqrt{2}, -\sqrt{2})$

31. $\{\sqrt{2}, \frac{1}{4}\pi\}$ **33.** $\{\sqrt{2}, \frac{3}{4}\pi\}$ **35.** $\{2, -\frac{1}{6}\pi\}$ **37.** $\theta = \frac{1}{4}\pi$

39. $r(\cos \theta + \sin \theta) = 1$

CHAPTER 8

[There may be slight discrepancies in the answers due to round-off errors.]

Section 1,   p. 254

1. $67.4°, 22.6°$ **3.** $37.7°, 52.3°$ **5.** $47.0°, 43.0°$ **7.** $46.7°, 43.3°$

9. $38.7°, 51.3°$ **11.** $71.0, 19.0$ **13.** $a \approx 7.512, b \approx 29.04$

15. $b \approx 31.31, c \approx 34.71$ **17.** $a \approx 194.8, b \approx 132.3$ **19.** $a \approx 58.22, c \approx 80.79$

21. 77.75 ft   **23.** 14.10 ft   **25.** $23.6°$ **27.** $8 \cdot 2 \cdot 10 \sin 22.5° \approx 61.23$

29. 0.3925 mi

31. The diagonal d satisfies $d^2 = 8^2 + 12^2 + 15^2 = 433$;
$\cos \alpha = 8/d, \cos \beta = 12/d, \cos \gamma = 15/d. \, \alpha \approx 67.4°, \beta \approx 54.8°, \gamma \approx 43.9°$.

33. 287.6 ft/sec ≈ 196.1 mph   **35.** $\cos \theta$ **37.** $\tan \theta$ **39.** $\csc \theta$

41. $a \csc \theta + [(c - a \cot \theta)^2 + b^2]^{1/2}$

43. $A = \frac{1}{2}ab = \frac{1}{2}(c \sin \alpha)(c \cos \alpha) = \frac{1}{4}c^2(2 \sin \alpha \cos \alpha) = \frac{1}{4}c^2 \sin 2\alpha$

45. $30°$ **47.** $\theta = \frac{1}{2} \arcsin[3(\sqrt{109} + 3)/50] \approx 26.87°$

Section 3,   p. 265

1. $\gamma = 110°, b \approx 13.44, c \approx 16.49$ **3.** $\gamma = 113.5°, a \approx 7.298, b \approx 10.51$

5. $\gamma = 28°, b \approx 15.29, c \approx 10.73$ **7.** $\alpha \approx 56.4°, \beta \approx 87.6°, c \approx 7.06$

9. no solution

11. $\beta \approx 25.0°, \gamma \approx 137.4°, c \approx 308.3$ or $\beta \approx 155.0°, \gamma \approx 7.4°, c \approx 58.66$

13. $\alpha \approx 22.3°, \beta \approx 27.1°, \gamma \approx 130.6°$ **15.** $\alpha \approx 45.7°, \beta \approx 113.3°, c \approx 6.009$

17. $\beta \approx 11.4°, \gamma \approx 21.4°, a \approx 54.89$ **19.** 137.2 ft   **21.** 305.2 **23.** 4.111 mi

25. If $c > b$, one solution,
otherwise no solution

27. 283.6 ft

29. $\beta \approx 77.83°, \gamma \approx 49.74°, a \approx 23.21$ **31.** $\beta \approx 51.52°, \gamma \approx 102.90°, a \approx 4.126$

33. $\alpha \approx 22.26°, \beta \approx 33.59°, c \approx 47.32$ **35.** $\alpha \approx 27.59°, \gamma \approx 121.29°, b \approx 2.424$

37. $a^2 = b^2 + c^2 - 2bc \cos \alpha \approx 819.1 + 499.1 - 779.7 \approx 538.5$,
$a \approx 23.21, \sin \beta = b(\sin \alpha)/a \approx 0.9774$ ($\log \sin \beta \approx 0.9901 - 1$), $\beta \approx 77.81°$, etc.

39. To find b: 6 | ENT ↑ | 50 | SIN | × | 20 | SIN | ÷ | ≈ 13.44. Obviously $\gamma = 110°$. To
find c: 6 | ENT ↑ | 110 | SIN | × | 20 | SIN | ÷ | ≈ 16.48.

41. To find c: 10 $\boxed{x^2}$ 12 $\boxed{x^2}$ $\boxed{+}$ 20 $\boxed{\text{ENT} \uparrow}$ 12 $\boxed{\times}$ 36 $\boxed{\text{COS}}$ $\boxed{\times}$ $\boxed{-}$

$\boxed{\sqrt{x}}$ ≈ 7.059. To find α: $10\,\boxed{\text{ENT} \uparrow}\,36$ $\boxed{\text{SIN}}$ $\boxed{\times}$ 7.059 $\boxed{\div}$ $\boxed{\text{ARC}}$ $\boxed{\text{SIN}}$ $\approx 56.37°$, etc.

43. To find β: 15.81 $\boxed{\text{ENT} \uparrow}$ 35.6 $\boxed{\text{SIN}}$ $\boxed{\times}$ 4.53 $\boxed{\div}$ $\boxed{\text{ARC}}$ $\boxed{\text{SIN}}$. The calculation fails (blinking display) because arc sin 2.032 is impossible. No solution.

45. To find γ: 4.1 $\boxed{x^2}$ 9.5 $\boxed{x^2}$ $\boxed{+}$ 10.6 $\boxed{x^2}$ $\boxed{-}$ 2 $\boxed{\div}$ 4.1 $\boxed{\div}$ 9.5 $\boxed{\div}$ $\boxed{\text{ARC}}$

$\boxed{\text{COS}}$ $\approx 93.90°$. Similarly, $\beta \approx 63.40°$, $\alpha \approx 22.70°$.

Section 4, p. 271

1. $\dfrac{b - c}{a} = \dfrac{b \sin \alpha - c \sin \alpha}{a \sin \alpha} = \dfrac{a \sin \beta - a \sin \gamma}{a \sin \alpha}$

$= \dfrac{\sin \beta - \sin \gamma}{\sin \alpha} = \dfrac{2 \cos \frac{1}{2}(\beta + \gamma) \sin \frac{1}{2}(\beta - \gamma)}{2 \sin \frac{1}{2}\alpha \cos \frac{1}{2}\alpha}.$

But $\cos \frac{1}{2}(\beta + \gamma) = \sin \frac{1}{2}\alpha$, so the formula follows.

3. $\sin \alpha + \sin \beta + \sin \gamma = \dfrac{a}{2R} + \dfrac{b}{2R} + \dfrac{c}{2R} = \frac{1}{2}(a + b + c)/R = s/R.$

5. $A = \frac{1}{2}a(b) \sin \gamma = \frac{1}{2}a(a \sin \beta/\sin \alpha) \sin \gamma = \frac{1}{2}a^2 \sin \beta \sin \gamma/\sin \alpha.$

7. In Fig. 4.4b, let h denote the hypotenuse.

Then $h^2 = x^2 + r^2 = (s - a)^2 + (s - a)(s - b)(s - c)/s$

$= (s - a)[s(s - a) + (s - b)(s - c)]/s.$

But $s(s - a) + (s - b)(s - c) = 2s^2 - s(a + b + c) + bc = bc$, hence $h^2 = (s - a)bc/s$. Therefore $\cos \frac{1}{2}\alpha = x/h = (s - a)/\sqrt{(s - a)bc/s} = \sqrt{s(s - a)/bc}.$

9. $R = abc/4\sqrt{s(s - a)(s - b)(s - c)}$

11. $\alpha \approx 25.50°$, $\beta \approx 94.75°$, $\gamma \approx 59.75°$

13. $\alpha \approx 49.38°$, $\beta \approx 86.32°$, $\gamma \approx 44.30°$

Section 5, p. 277

1. $(0, 3)$ **3.** $(2, -2)$ **5.** $(2, 12)$ **7.** $(3, 0)$ **9.** $(0, 0)$ **11.** $(1, 1)$

13. $(10, 4)$ **15.** $(18, 10)$ **17.** $(7, 3)$ **19.** $(-4, 4)$ **21.** $(1, -10)$

23. $a = -2, b = 5$

Section 6, p. 283

1. 6 **3.** 0 **5.** 3 **7.** 24

9. $\|\mathbf{v} + \mathbf{w}\|^2 = (\mathbf{v} + \mathbf{w}) \cdot (\mathbf{v} + \mathbf{w}) = \mathbf{v} \cdot \mathbf{v} + \mathbf{v} \cdot \mathbf{w} + \mathbf{w} \cdot \mathbf{v} + \mathbf{w} \cdot \mathbf{w}$

$= \|\mathbf{v}\|^2 + 2\mathbf{v} \cdot \mathbf{w} + \|\mathbf{w}\|^2$

11. $\|\mathbf{v} + \mathbf{w}\|^2 - \|\mathbf{v} - \mathbf{w}\|^2 = (\|\mathbf{v}\|^2 + 2\mathbf{v} \cdot \mathbf{w} + \|\mathbf{w}\|^2) - (\|\mathbf{v}\|^2 - 2\mathbf{v} \cdot \mathbf{w} + \|\mathbf{w}\|^2) = 4\mathbf{v} \cdot \mathbf{w}.$

13. $\frac{1}{2}\pi$ **15.** arc cos $0.6 \approx 0.93$ **17.** arc cos $0.6 \approx 0.93$

19. If one vertex is $\mathbf{0}$, the others are $\mathbf{u}, \mathbf{v}, \mathbf{u} + \mathbf{v}$, where $\|\mathbf{u}\| = \|\mathbf{v}\|$. The diagonals are $\mathbf{u} + \mathbf{v}$ and a segment parallel to $\mathbf{u} - \mathbf{v}$. But $(\mathbf{u} + \mathbf{v}) \cdot (\mathbf{u} - \mathbf{v}) = \|\mathbf{u}\|^2 - \|\mathbf{v}\|^2 = 0.$

CHAPTER 9

Section 1,   p. 290

1. $1; 4$ 3. $\frac{1}{2}; \frac{3}{2}$ 5. $x + 1; 5$ 7. $2; -2$ 9. $2x^2 + 2x + 4; 5$
11. $x - 1; x$ 13. $x + 1; 2x + 2$ 15. $\frac{1}{2}; 2x - \frac{1}{2}$ 17. $x^2 - x; x + 1$
19. $x^3 - x; x$ 21. 2 23. 6 25. 1 27. 5 29. 94 31. 119
33. no; $f(-4) = 140$ 35. yes   37. $1^{61} - 1^{12} - 1^4 + 1 = 0$

Section 2,   p. 298

1. $a(x - 3), a \neq 0$ 3. $-2(x + 1)(x - 3)$ 5. $ax^2 - 5ax + c, a \neq 0$
7. The graph of $y = ax + b$, for $a \neq 0$, intersects the graph of $y = k$ where $ax + b = k$.
There is only one x satisfying this condition: $x = (k - b)/a$.
9. The equation $ax + b = cx^2 + dx + e$ is quadratic, so it has at most two real roots.
11. $D = b^2 - 4ac \geq -4ac > 0$, hence there are two real roots.
13. If $r < 0$, then $ar^2 + br + c > 0$.
15. $-\frac{1}{2}$ 17. $-2, -2, 2, 2$ 19. $-2, -1, -1, 1, 1, 1$
21. The cubic equation $ax^3 + bx^2 + cx + d = k$ has at most three real roots.
23. $hx + k = ax^3 + bx^2 + cx + d$ has at most three real roots.
25. $(x - 1)^2(x - 2)^2(x - 3)^2$
27. Yes, because $x - x^5 - 1000$ has odd degree, hence has a real zero.
29. $\frac{17}{100}x(x - 5)(x - 8)$ 31. -1 33. $-1, 2$ 35. $\frac{2}{3}$ 37. -1
39. $2 < r < 3$
41. $f(1) = -5 < 0, f(2) = 29 > 0$. If $x > 2$, then $f(x) = x^5 + x^2 - 7 > 2^5 + 2^2 - 7$
$= f(2) = 29 > 0$.   43. $f(x) > 0$ for $x > 0$.
45. $x^n + a_1 x^{n-1} + \cdots + a_n = (x - r_1)(x - r_2) \cdots (x - r_n)$.
The coefficient of x^{n-1} on the RHS is $-(r_1 + r_2 + \cdots + r_n)$.

Section 3,   p. 304

1. $\dfrac{\frac{1}{2}}{x} - \dfrac{\frac{1}{2}}{x + 4}$ 3. $\dfrac{1}{x + 1} - \dfrac{2}{x + 2}$ 5. $\dfrac{-1}{x} + \dfrac{2}{x - 1}$

7. $\dfrac{1}{x - 1} + \dfrac{1}{x + 2}$ 9. $\dfrac{-7}{x + 2} + \dfrac{9}{x + 3}$ 11. $\dfrac{\frac{1}{2}}{2x - 1} - \dfrac{\frac{1}{2}}{2x + 1}$

13. $1 + \dfrac{1}{x - 2} - \dfrac{1}{x + 2}$ 15. $\dfrac{2}{x - 2} - \dfrac{1}{x - 1}$ 17. $\dfrac{\frac{1}{2}}{x - 1} - \dfrac{4}{x - 2} + \dfrac{\frac{9}{2}}{x - 3}$

19. $\dfrac{\frac{3}{2}}{x - 2} - \dfrac{4}{x - 3} + \dfrac{\frac{5}{2}}{x - 4}$ 21. $\dfrac{1}{x} + \dfrac{\frac{9}{4}}{x - 2} - \dfrac{\frac{5}{4}}{x + 2}$

23. $\dfrac{1}{x + 1} - \dfrac{1}{(x + 1)^2}$ 25. $\dfrac{2}{(x - 1)^4} + \dfrac{2}{(x - 1)^3} + \dfrac{1}{(x - 1)^2}$

27. $\dfrac{\frac{9}{10}}{x + 3} + \dfrac{\frac{1}{10}x - \frac{3}{10}}{x^2 + 1}$ 29. $\dfrac{-\frac{1}{2}}{x + 2} + \dfrac{\frac{1}{2}x}{x^2 + 4}$

31. $\dfrac{-2x}{(x^2 + 2)^2} + \dfrac{x}{x^2 + 2}$ 33. $\dfrac{-2}{(x^2 + 2)^2} + \dfrac{1}{x^2 + 2}$

Section 4, p. 311

1. $(-\frac{8}{19}, \frac{1}{19}, \frac{7}{19})$ 3. $(0, 0, 0)$ 5. $(\frac{3}{14}, -\frac{1}{4}, -\frac{3}{28})$ 7. $(\frac{12}{7}, 0, -\frac{1}{7})$

9. $(-\frac{5}{2}, -\frac{3}{2}, \frac{9}{2})$ 11. $\frac{3}{2}x^2 + \frac{1}{2}x$ 13. $\frac{1}{2}x^2 + \frac{1}{2}$

15. Add the first and third equations: $0 = 1$.

17. $(t, -\frac{1}{2}t - \frac{1}{2}, -\frac{5}{2}t + \frac{1}{2})$, where t is arbitrary.

19. $(5, 6, 0, -1)$

Section 5, p. 318

1. 7   3. -17 5. a^2 7. 1

9. $(-3)(-2)(-1) + 0 \cdot 1 \cdot 5 + 3 \cdot 4 \cdot 2 - (-3) \cdot 1 \cdot 2 - 0 \cdot 4 \cdot (-1) - 3 \cdot (-2) \cdot 5 = 54$

11. $3[4 \cdot 2 - (-2) \cdot 5] - [(-3) \cdot 2 - 0 \cdot 5] - [(-3)(-2) - 0 \cdot 4] = 54$

13. $-4[0 \cdot (-1) - 2 \cdot 3] + (-2)[(-3)(-1) - 5 \cdot 3] - [(-3) \cdot 2 - 0 \cdot 5] = 54$

15. -12 17. 16   19. -65 21. -130 23. -6 25. -17 27. 5

29. $x^3 + 2x$ 31. 0   33. $(-\frac{1}{3}, \frac{2}{3})$ 35. $(-1, 1)$ 37. $(\frac{1}{19}, \frac{7}{19})$ 39. $(\frac{5}{2}, 2, \frac{3}{2})$

41. $(-\frac{8}{15}, -\frac{13}{15}, \frac{22}{15})$ 43. $(\frac{13}{3}, \frac{5}{3}, -3)$ 45. $(\frac{3}{14}, -\frac{1}{4}, -\frac{3}{28})$

CHAPTER 10

Section 1, p. 325

1. $9 + i$ 3. $-4 + 5i$ 5. $3 + 2i$ 7. $7 - i$ 9. $7 + 22i$ 11. $2i$

13. $\frac{3}{5} - \frac{4}{5}i$ 15. $1 + i$ 17. $22/533 - 7i/533$ 19. $-37/145 - 9i/145$

21. $82/85 + 39i/85$ 23. $3 - 3i$ 25. $8 - i$ 27. $-4 - 2i$ 29. 5

31. $\sqrt{10}$ 33. $\sqrt{5} + \sqrt{17}$ 35. $\frac{1}{5}\sqrt{5}$ 37. 1

39. $(a + bi) + (c + di) = (a + c) + (b + d)i = (c + a) + (d + b)i = (c + di) + (a + bi)$

41. $(a + bi)[(c + di)(e + fi)] = (a + bi)[(ce - df) + (cf + de)i]$
$= [a(ce - df) - b(cf + de)] + [b(ce - df) + a(cf + de)]i$
$= [(ac - bd)e - (ad + bc)f] + [(ac - bd)f + (ad + bc)e]i$
$= [(ac - bd) + (ad + bc)i](e + fi) = [(a + bi)(c + di)](e + fi).$

43. $(a + bi)[(c + di) + (e + fi)] = (a + bi)[(c + e) + (d + f)i]$
$= [a(c + e) - b(d + f)] + [a(d + f) + b(c + e)]i$
$= [(ac - bd) + (ae - bf)] + [(ad + bc) + (af + be)]i$
$= [(ac - bd) + (ad + bc)i] + [(ae - bf) + (af + be)i]$
$= (a + bi)(c + di) + (a + bi)(e + fi).$

45. $|a - bi|^2 = a^2 + b^2 = |a + bi|^2.$

47. $|\alpha\beta|^2 = (\alpha\beta)\overline{(\alpha\beta)} = \alpha\beta\bar{\alpha}\bar{\beta} = (\alpha\bar{\alpha})(\beta\bar{\beta}) = |\alpha|^2|\beta|^2,$
hence $|\alpha\beta| = |\alpha|\,|\beta|.$ 49. -1

Section 2, p. 332

1. $3(\cos\frac{3}{2}\pi + i\sin\frac{3}{2}\pi)$ 3. $\sqrt{2}(\cos\frac{7}{4}\pi + i\sin\frac{7}{4}\pi)$

5. $2(\cos\frac{7}{6}\pi + i\sin\frac{7}{6}\pi)$ 7. $6(\cos\frac{1}{6}\pi + i\sin\frac{1}{6}\pi)$

9. $\sqrt{2}(\cos\frac{1}{4}\pi + i\sin\frac{1}{4}\pi), \sqrt{2}(\cos\frac{7}{4}\pi + i\sin\frac{7}{4}\pi); 2(\cos 0 + i\sin 0)$

11. $2(\cos\frac{1}{3}\pi + i\sin\frac{1}{3}\pi); 4(\cos\frac{2}{3}\pi + i\sin\frac{2}{3}\pi)$

13. $2(\cos \frac{11}{6}\pi + i \sin \frac{11}{6}\pi)$, $\sqrt{2}(\cos \frac{1}{4}\pi + i \sin \frac{1}{4}\pi)$; $\sqrt{2}(\cos \frac{19}{12}\pi + i \sin \frac{19}{12}\pi)$.
$(\cos \frac{19}{12}\pi = \frac{1}{2}\sqrt{3} - \frac{1}{2}$, etc.)

15. $\sqrt{2}(\cos \frac{7}{4}\pi + i \sin \frac{7}{4}\pi)$, $4(\cos \frac{1}{3}\pi + i \sin \frac{1}{3}\pi)$; $\frac{1}{4}\sqrt{2}(\cos \frac{17}{12}\pi + i \sin \frac{17}{12}\pi)$

17. $\sqrt{3}, 3, \frac{1}{3}\pi$ **19.** $\frac{1}{2}(\sqrt{3}-1)$, $-\frac{1}{2}(\sqrt{3}+1)$, $-\frac{5}{12}\pi = \arctan(-2-\sqrt{3})$

21. $3\sqrt{2}$ **23.** $\sqrt{65}$ **25.** $2\sqrt{5}$ **27.** $\frac{1}{2}\sqrt{2}$ **29.** $\frac{2}{5}\sqrt{10}$

31. $\frac{1}{29}\sqrt{290}$ **33.** $\pm\frac{1}{2}(\sqrt{2} - i\sqrt{2})$

35. $\pm 2^{1/4}(\cos \frac{1}{8}\pi + i \sin \frac{1}{8}\pi) = \pm\frac{1}{2}2^{1/4}[(2 + 2^{1/2})^{1/2} + i(2 - 2^{1/2})^{1/2}]$

37. $\pm\sqrt{2}(\cos \frac{11}{12}\pi + i \sin \frac{11}{12}\pi) = \pm\frac{1}{2}\sqrt{2}[-(2 + 3^{1/2})^{1/2} + i(2 - 3^{1/2})^{1/2}]$

39. $\pm 8^{1/4}(\cos \frac{3}{8}\pi + i \sin \frac{3}{8}\pi) = \pm\frac{1}{2}8^{1/4}[(2 - 2^{1/2})^{1/2} + i(2 + 2^{1/2})^{1/2}]$

41. Use $|\alpha\beta|^2 = |\alpha|^2|\beta|^2$ with $\alpha = x + yi$ and $\beta = u + vi$.

43.

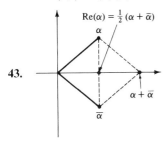

Section 3, p. 336

1. $1 \pm 2i$ **3.** $-\frac{1}{2} \pm \frac{1}{2}\sqrt{23}$ **5.** $\pm i, \pm 2i$ **7.** $1, -\frac{1}{2} \pm \frac{1}{2}i\sqrt{3}$

9. $\pm 1, \pm i$ **11.** $(x + 1)(x^2 + x + 1)$

13. $z + 1$ is not a factor of $g(z) = z^4 - z^3 + z^2 - z + 1$ because $g(-1) = 5 \neq 0$.

15. Non-real zeros occur in pairs.   **17.** $a(x^2 - 4)(x^2 - 2x + 2)$, $a \neq 0$.

19. $a(x^2 + b_1 x + c_1)(x^2 + b_2 x + c_2)$, $a \neq 0$, $b_1^2 - 4c_1 < 0$, $b_2^2 - 4c_2 < 0$.

Section 4, p. 343

1. $-32 - 32i$ **3.** -64 **5.** $-\frac{1}{2} + \frac{1}{2}i\sqrt{3}$ **7.** $\pm 1, \pm i$

9. $\pm 1, \frac{1}{2}(-1 \pm i\sqrt{3}), \frac{1}{2}(1 \pm i\sqrt{3})$ **11.** $i, \frac{1}{2}(\pm\sqrt{3} - i)$

13. $\pm\alpha, \pm i\alpha$, where $\alpha = \cos\frac{1}{8}\pi + i \sin\frac{1}{8}\pi = \frac{1}{2}[(2 + 2^{1/2})^{1/2} + i(2 - 2^{1/2})^{1/2}]$

15. $\pm\alpha, \pm i\alpha$, where $\alpha = 2^{1/8}(\cos\frac{3}{16}\pi + i \sin\frac{3}{16}\pi)$ **17.** $\frac{1}{2}(-1 + i\sqrt{3}), \frac{1}{2}(-1 - i\sqrt{3})$

19. $0 = \alpha^n - 1 = (\alpha - 1)(\alpha^{n-1} + \cdots + \alpha + 1)$ and $\alpha - 1 \neq 0$.

21. $\alpha^5 = 1$ and $\alpha \neq 1$, hence $\alpha^4 + \alpha^3 + \alpha^2 + \alpha + 1 = 0$,
$\beta^2 + \beta - 1 = \alpha^2 + \alpha + 1 + \alpha^{-1} + \alpha^{-2} = 0$.

23. $\alpha + \alpha^{-1} = \beta$, $\alpha\beta = \alpha(\alpha + \alpha^{-1}) = \alpha^2 + 1$.   **25.** $\pm i$ **27.** $\frac{1}{2} \pm \frac{1}{2}i\sqrt{3}$

29. $\frac{1}{2}\sqrt{2}(1 \pm i), \frac{1}{2}\sqrt{2}(-1 \pm i)$ **31.** $\cos\frac{1}{10}\pi k + i \sin\frac{1}{10}\pi k$, where $k = 1, 3, 7, 9$

33. $z^6 - 1 = (z^3 - 1)(z + 1)(z^2 - z + 1)$. The zeros of $z^3 - 1$ are the cube roots of 1; the
zero of $z + 1$ is a square root of 1.

35. $z^9 - 1 = (z^3 - 1)(z^6 + z^3 + 1)$, etc.

37. $1 = \alpha^7 = (\alpha^2)^3\alpha = (\alpha^3)^2\alpha = \alpha/(\alpha^4)^2 = \alpha/(\alpha^5)^3 = (\alpha^6)\alpha$.
Hence $\alpha^2 \neq 1$, $\alpha^3 \neq 1$, $\cdots$, $\alpha^6 \neq 1$.

39. $\alpha^2 = 3 - 4i$, $\alpha^4 = -7 - 24i$. Then $(-\alpha)^4 = (-1)^4\alpha^4 = \alpha^4$ and
$(\pm i\alpha)^4 = (\pm i)^4\alpha^4 = \alpha^4$.

41. $z^n - 1 = (z - 1)(z - z_2) \cdots (z - z_n)$ and
$z^n - 1 = (z - 1)(z^{n-1} + z^{n-2} + \cdots + z + 1)$; compare.

CHAPTER 11

Section 1, p. 349

1. $81, 4 + 7(n - 1) = 7n - 3$ 3. $256, 2^{n-4}$ 5. $\frac{1}{2}n(n + 1)$ 7. 575
9. -55 11. $10x + 90$ 13. $\frac{1}{2}n[4a + 3b(n + 1)]$ 15. $3(2^{10} - 1) = 3069$
17. $255/16$ 19. 2032 21. $(x^6 - 1)/x^6(x - 1)$
23. $ab^2c^5(b^7c^{14} - 1)/(bc^2 - 1)$ 25. $a_1 = a_2 = a_3$
27. If $a_j = a_1 + (j - 1)d$ and $b_j = b_1 + (j - 1)e$, then
    $a_j + 3b_j = (a_1 + 3b_1) + (j - 1)(d + 3e)$, an arithmetic progression with common
    difference $d + 3e$.
29. $1 + (1 + 2 + \cdots + 2^{j-1}) = 1 + (2^j - 1)/(2 - 1) = 2^j$ 31. $29\frac{27}{32}$ ft
33. $18!/13!$ 35. $2^6 \cdot 6!$ 37. 72 39. 252 41. 1365

Section 2, p. 353

1. $8! = 40,320$ 3. $10! = 3,628,800; 9 \times 9! = 3,265,920$ 5. 32
7. 1260 9. $(8!)^2$ 11. $2,598,960$ 13. 3744
15. $2n$ moves are required, n moves up and n moves right, and these may occur in any
    order. A path is completely determined by choosing the n up moves out of the $2n$
    possible moves, hence the given number of paths.
17. $\binom{n}{k} = \binom{n-1}{k} + \binom{n-1}{k-1} = \left[\binom{n-2}{k} + \binom{n-2}{k-1}\right] + \left[\binom{n-2}{k-1} + \binom{n-2}{k-2}\right]$.

Section 3, p. 356

1. $(x + 2)^3 = x^3 + 6x^2 + 12x + 8$ 3. $1 - 8x + 24x^2 - 32x^3 + 16x^4$
5. $x^6 + 6x^4 + 15x^2 + 20 + 15x^{-2} + 6x^{-4} + x^{-6}$
7. $(3 - 2)^5 = 1$ 9. $\binom{10}{4} \times 3^4 = 17,010$ 11. $\binom{6}{3} \times 2^3 = 160$
13. $11,040,808,032$ 15. 0.904382
17. $11^2 = (10 + 1)^2 = 10^2 + 2 \cdot 10 + 1$,
    $11^3 = (10 + 1)^3 = 10^3 + 3 \cdot 10^2 + 3 \cdot 10 + 1$,
    $11^4 = (10 + 1)^4 = 10^4 + 4 \cdot 10^3 + 6 \cdot 10^2 + 4 \cdot 10 + 1$.
19. $(1.002)^{10} = (1 + 0.002)^{10} = 1 + 10(0.002) + \cdots > 1 + 0.02 = 1.02$.

Section 4, p. 363

1. $\displaystyle\sum_{j=3}^{20} x_j$ 3. $\displaystyle\sum_{j=1}^{50} (2j - 1)$ 5. $\displaystyle\sum_{j=1}^{n} j^{j+1}$ 7. $\displaystyle\sum_{j=0}^{n-1} (j + 1)x^j$

9. $\displaystyle\sum_{j=1}^{2n} \frac{1}{j} = \sum_{j=1}^{n} \frac{1}{2j} + \sum_{j=0}^{n-1} \frac{1}{2j + 1} = \frac{1}{2}\sum_{j=1}^{n} \frac{1}{j} + \sum_{j=0}^{n-1} \frac{1}{2j + 1}$

11. Set $k = n - j$ so k goes over $n, n - 1, \cdots, 1, 0$ as j goes from 0 to n. Then

$$\sum_{j=0}^{n} a_j b_{n-j} = \sum_{k=0}^{n} a_{n-k} b_k = \sum_{j=0}^{n} a_{n-j} b_j.$$

13. $\frac{1}{2}(1 + \frac{1}{2} - \frac{1}{50} - \frac{1}{51})$

15. $1 + 2 + \cdots + 100 = 5050 < 1.1 + 1.1^2 + \cdots + 1.1^{99}$
$= (1.1^{100} - 1)/(1.1 - 1) - 1 \approx 137800.$

17. $(n - 1)2^{n+1} + 2$

19. $(n + 1)^4 - 1 = \sum_{j=1}^{n} [(j + 1)^4 - j^4] = 4 \sum_{j=1}^{n} j^3 + 6 \sum_{j=1}^{n} j^2 + 4 \sum_{j=1}^{n} j + \sum_{j=1}^{n} 1$

$$= 4 \sum_{j=1}^{n} j^3 + n(n + 1)(2n + 1) + 2n(n + 1) + n.$$

Solve: $\sum_{j=1}^{n} j^3 = \frac{1}{4} n^2 (n + 1)^2.$

21. $1 + \sum_{j=2}^{n} \frac{1}{j^2} < 1 + \sum_{j=2}^{n} \frac{1}{j(j - 1)} = 1 + \sum_{j=2}^{n} \left(\frac{1}{j - 1} - \frac{1}{j} \right) = 2 - \frac{1}{n + 1} < 2.$

23. Set $A_0 = B_0 = 0$. Then $\sum_{i=1}^{n} A_i b_i = \sum_{i=1}^{n} A_i (B_i - B_{i-1}) = \sum_{i=1}^{n} A_i B_i - \sum_{i=1}^{n} A_i B_{i-1}$

$$= \left(A_n B_n + \sum_{i=1}^{n-1} A_i B_i \right) - \left(\sum_{i=1}^{n-1} A_{i+1} B_i + A_1 B_0 \right) = A_n B_n + \sum_{i=1}^{n-1} (A_i - A_{i+1}) B_i$$

$$= A_n B_n - \sum_{i=1}^{n-1} a_{i+1} B_i.$$

Section 5,   p. 367

1. $\log(a_1 \cdots a_{n+1}) = \log[(a_1 \cdots a_n)a_{n+1}] = \log(a_1 \cdots a_n) + \log a_{n+1}$, etc.

3. $(2b_1 + 1) + \cdots + (2b_{2(n+1)+1} + 1)$
$= (2b_1 + 1) + \cdots + (2b_{2n+1} + 1) + (2b_{2n+2} + 1) + (2b_{2n+3} + 1)$
$= (\text{odd}) + 2(b_{2n+2} + b_{2n+3} + 1) = (\text{odd}) + (\text{even}) = (\text{odd}).$

5. $\dfrac{n}{2n + 1} + \dfrac{1}{(2n + 1)(2n + 3)} = \dfrac{n + 1}{2n + 3}$, etc.

7. $\frac{1}{3} n(n + 1)(n + 2) + (n + 1)(n + 2) = \frac{1}{3}(n + 1)(n + 2)(n + 3)$, etc.

9. $1 - x^k, k = 2^{n+1}$ **11.** $3 \times 4^{n-1}$

13. Let A_n be the area of the n-th figure. Then $A_1 = \frac{1}{4} \sqrt{3}$. To the n-th figure are added $3 \times 4^{n-1}$ equilateral triangles of side length $1/3^n$ (proved by induction). Hence

$$A_{n+1} = A_n + (3)(4^{n-1}) \left(\frac{\sqrt{3}}{(4)(3^{2n})} \right) = A_n + \frac{3}{16} \left(\frac{4}{9} \right)^n \sqrt{3}.$$

Now set $B_n = \frac{1}{20} \sqrt{3}[8 - 3(\frac{4}{9})^{n-1}]$. Then $B_1 = \frac{1}{4} \sqrt{3} = A_1$ and
$B_{n+1} - B_n = \frac{3}{16}(\frac{4}{9})^n \sqrt{3} = A_{n+1} - A_n$. By induction, $A_n = B_n$.

15. $\frac{3}{2} > 1, (\frac{3}{2})^2 = \frac{9}{4} > 2, (\frac{3}{2})^{n+1} = (\frac{3}{2})(\frac{3}{2})^n > \frac{3}{2} n \geq n + 1$ for $n \geq 2$, etc.

17. $2^6 > 6^2 + 27, 2^{n+1} = 2 \cdot 2^n > 2n^2 + 54$. But
$(2n^2 + 54) - [(n + 1)^2 + 27] = n^2 - 2n + 26 = (n - 1)^2 + 25 > 0$, etc.

19. $(n + 1)/2n$ **21.** $a_n = n$

23. $1, 2, 3, 5, 8, 13, 21, 34, 55, 89, 144.$
Use $(\sqrt{2})^{n-1} + (\sqrt{2})^n > (\sqrt{2})^{n+1}$ (because $1 + \sqrt{2} > 2$) and $2^{n-1} + 2^n < 2^{n+1}$.

25. $2^n - 1$ **27.** $x_n = 1$ **29.** $x_n = 2 \cdot 3^{n-1}$

31. $x_n = \frac{2}{3}[1 - (-\frac{1}{2})^n] \longrightarrow \frac{2}{3}$ as $n \longrightarrow \infty$.

33. If $\deg g(x) = 0$, the result is obvious, so we may assume $m = \deg g(x) > 0$. We argue by induction on $n = \deg f(x)$. If $f(x) = 0$ or $\deg f(x) < \deg g(x)$, we have $f(x) = g(x) \times 0 + f(x)$. If the result is true for $\deg f(x) \leq n$ and we are given $f(x)$ with $\deg f(x) = n + 1 \geq \deg g(x)$, then choose a constant c so that $f_1(x) = f(x) - cx^{n+1-m}$ has $\deg f_1(x) \leq n$ or $f_1(x) = 0$. By the induction hypothesis, $f_1(x) = g(x)q_1(x) + r_1(x)$, where $r_1(x) = 0$ or $\deg r_1(x) < m$. Then

$$f(x) = g(x)[cx^{n+1-m} + q_1(x)] + r_1(x),$$

etc.

TABLES

Table 1   *4-place logarithm*

N	0	1	2	3	4	5	6	7	8	9	Proportional Parts 1 2 3 4 5 6 7 8 9
10	0000	0043	0086	0128	0170	0212	0253	0294	0334	0374	4 8 12 17 21 25 29 33 37
11	0414	0453	0492	0531	0569	0607	0645	0682	0719	0755	4 8 11 15 19 23 26 30 34
12	0792	0828	0864	0899	0934	0969	1004	1038	1072	1106	3 7 10 14 17 21 24 28 31
13	1139	1173	1206	1239	1271	1303	1335	1367	1399	1430	3 6 10 13 16 19 23 26 29
14	1461	1492	1523	1553	1584	1614	1644	1673	1703	1732	3 6 9 12 15 18 21 24 27
15	1761	1790	1818	1847	1875	1903	1931	1959	1987	2014	3 6 8 11 14 17 20 22 25
16	2041	2068	2095	2122	2148	2175	2201	2227	2253	2279	3 5 8 11 13 16 18 21 24
17	2304	2330	2355	2380	2405	2430	2455	2480	2504	2529	2 5 7 10 12 15 17 20 22
18	2553	2577	2601	2625	2648	2672	2695	2718	2742	2765	2 5 7 9 12 14 16 19 21
19	2788	2810	2833	2856	2878	2900	2923	2945	2967	2989	2 4 7 9 11 13 16 18 20
20	3010	3032	3054	3075	3096	3118	3139	3160	3181	3201	2 4 6 8 11 13 15 17 19
21	3222	3243	3263	3284	3304	3324	3345	3365	3385	3404	2 4 6 8 10 12 14 16 18
22	3424	3444	3464	3483	3502	3522	3541	3560	3579	3598	2 4 6 8 10 12 14 15 17
23	3617	3636	3655	3674	3692	3711	3729	3747	3766	3784	2 4 6 7 9 11 13 15 17
24	3802	3820	3838	3856	3874	3892	3909	3927	3945	3962	2 4 5 7 9 11 12 14 16
25	3979	3997	4014	4031	4048	4065	4082	4099	4116	4133	2 3 5 7 9 10 12 14 15
26	4150	4166	4183	4200	4216	4232	4249	4265	4281	4298	2 3 5 7 8 10 11 13 15
27	4314	4330	4346	4362	4378	4393	4409	4425	4440	4456	2 3 5 6 8 9 11 13 14
28	4472	4487	4502	4518	4533	4548	4564	4579	4594	4609	2 3 5 6 8 9 11 12 14
29	4624	4639	4654	4669	4683	4698	4713	4728	4742	4757	1 3 4 6 7 9 10 12 13
30	4771	4786	4800	4814	4829	4843	4857	4871	4886	4900	1 3 4 6 7 9 10 11 13
31	4914	4928	4942	4955	4969	4983	4997	5011	5024	5038	1 3 4 6 7 8 10 11 12
32	5051	5065	5079	5092	5105	5119	5132	5145	5159	5172	1 3 4 5 7 8 9 11 12
33	5185	5198	5211	5224	5237	5250	5263	5276	5289	5302	1 3 4 5 6 8 9 10 12
34	5315	5328	5340	5353	5366	5378	5391	5403	5416	5428	1 3 4 5 6 8 9 10 11
35	5441	5453	5465	5478	5490	5502	5514	5527	5539	5551	1 2 4 5 6 7 9 10 11
36	5563	5575	5587	5599	5611	5623	5635	5647	5658	5670	1 2 4 5 6 7 8 10 11
37	5682	5694	5705	5717	5729	5740	5752	5763	5775	5786	1 2 3 5 6 7 8 9 10
38	5798	5809	5821	5832	5843	5855	5866	5877	5888	5899	1 2 3 5 6 7 8 9 10
39	5911	5922	5933	5944	5955	5966	5977	5988	5999	6010	1 2 3 4 5 7 8 9 10
40	6021	6031	6042	6053	6064	6075	6085	6096	6107	6117	1 2 3 4 5 6 8 9 10
41	6128	6138	6149	6160	6170	6180	6191	6201	6212	6222	1 2 3 4 5 6 7 8 9
42	6232	6243	6253	6263	6274	6284	6294	6304	6314	6325	1 2 3 4 5 6 7 8 9
43	6335	6345	6355	6365	6375	6385	6395	6405	6415	6425	1 2 3 4 5 6 7 8 9
44	6435	6444	6454	6464	6474	6484	6493	6503	6513	6522	1 2 3 4 5 6 7 8 9
45	6532	6542	6551	6561	6571	6580	6590	6599	6609	6618	1 2 3 4 5 6 7 8 9
46	6628	6637	6646	6656	6665	6675	6684	6693	6702	6712	1 2 3 4 5 6 7 7 8
47	6721	6730	6739	6749	6758	6767	6776	6785	6794	6803	1 2 3 4 5 5 6 7 8
48	6812	6821	6830	6839	6848	6857	6866	6875	6884	6893	1 2 3 4 4 5 6 7 8
49	6902	6911	6920	6928	6937	6946	6955	6964	6972	6981	1 2 3 4 4 5 6 7 8
50	6990	6998	7007	7016	7024	7033	7042	7050	7059	7067	1 2 3 3 4 5 6 7 8
51	7076	7084	7093	7101	7110	7118	7126	7135	7143	7152	1 2 3 3 4 5 6 7 8
52	7160	7168	7177	7185	7193	7202	7210	7218	7226	7235	1 2 2 3 4 5 6 7 7
53	7243	7251	7259	7267	7275	7284	7292	7300	7308	7316	1 2 2 3 4 5 6 6 7
54	7324	7332	7340	7348	7356	7364	7372	7380	7388	7396	1 2 2 3 4 5 6 6 7

N	0	1	2	3	4	5	6	7	8	9	1 2 3 4 5 6 7 8 9

Tables 1–5 and 7 are from the "Handbook of Tables for Mathematics," 3rd Edition (Robert C. Weast and Samuel M. Selby, eds.), The Chemical Rubber Co., Cleveland, Ohio, 1967, and are used by permission.

Table 1 4-place logarithm (continued)

N	0	1	2	3	4	5	6	7	8	9	Proportional Parts								
											1	2	3	4	5	6	7	8	9
55	7404	7412	7419	7427	7435	7443	7451	7459	7466	7474	1	2	2	3	4	5	5	6	7
56	7482	7490	7497	7505	7513	7520	7528	7536	7543	7551	1	2	2	3	4	5	5	6	7
57	7559	7566	7574	7582	7589	7597	7604	7612	7619	7627	1	2	2	3	4	5	5	6	7
58	7634	7642	7649	7657	7664	7672	7679	7686	7694	7701	1	1	2	3	4	4	5	6	7
59	7709	7716	7723	7731	7738	7745	7752	7760	7767	7774	1	1	2	3	4	4	5	6	7
60	7782	7789	7796	7803	7810	7818	7825	7832	7839	7846	1	1	2	3	4	4	5	6	6
61	7853	7860	7868	7875	7882	7889	7896	7903	7910	7917	1	1	2	3	4	4	5	6	6
62	7924	7931	7938	7945	7952	7959	7966	7973	7980	7987	1	1	2	3	3	4	5	6	6
63	7993	8000	8007	8014	8021	8028	8035	8041	8048	8055	1	1	2	3	3	4	5	5	6
64	8062	8069	8075	8082	8089	8096	8102	8109	8116	8122	1	1	2	3	3	4	5	5	6
65	8129	8136	8142	8149	8156	8162	8169	8176	8182	8189	1	1	2	3	3	4	5	5	6
66	8195	8202	8209	8215	8222	8228	8235	8241	8248	8254	1	1	2	3	3	4	5	5	6
67	8261	8267	8274	8280	8287	8293	8299	8306	8312	8319	1	1	2	3	3	4	5	5	6
68	8325	8331	8338	8344	8351	8357	8363	8370	8376	8382	1	1	2	3	3	4	4	5	6
69	8388	8395	8401	8407	8414	8420	8426	8432	8439	8445	1	1	2	2	3	4	4	5	6
70	8451	8457	8463	8470	8476	8482	8488	8494	8500	8506	1	1	2	2	3	4	4	5	6
71	8513	8519	8525	8531	8537	8543	8549	8555	8561	8567	1	1	2	2	3	4	4	5	5
72	8573	8579	8585	8591	8597	8603	8609	8615	8621	8627	1	1	2	2	3	4	4	5	5
73	8633	8639	8645	8651	8657	8663	8669	8675	8681	8686	1	1	2	2	3	4	4	5	5
74	8692	8698	8704	8710	8716	8722	8727	8733	8739	8745	1	1	2	2	3	4	4	5	5
75	8751	8756	8762	8768	8774	8779	8785	8791	8797	8802	1	1	2	2	3	3	4	5	5
76	8808	8814	8820	8825	8831	8837	8842	8848	8854	8859	1	1	2	2	3	3	4	5	5
77	8865	8871	8876	8882	8887	8893	8899	8904	8910	8915	1	1	2	2	3	3	4	4	5
78	8921	8927	8932	8938	8943	8949	8954	8960	8965	8971	1	1	2	2	3	3	4	4	5
79	8976	8982	8987	8993	8998	9004	9009	9015	9020	9025	1	1	2	2	3	3	4	4	5
80	9031	9036	9042	9047	9053	9058	9063	9069	9074	9079	1	1	2	2	3	3	4	4	5
81	9085	9090	9096	9101	9106	9112	9117	9122	9128	9133	1	1	2	2	3	3	4	4	5
82	9138	9143	9149	9154	9159	9165	9170	9175	9180	9186	1	1	2	2	3	3	4	4	5
83	9191	9196	9201	9206	9212	9217	9222	9227	9232	9238	1	1	2	2	3	3	4	4	5
84	9243	9248	9253	9258	9263	9269	9274	9279	9284	9289	1	1	2	2	3	3	4	4	5
85	9294	9299	9304	9309	9315	9320	9325	9330	9335	9340	1	1	2	2	3	3	4	4	5
86	9345	9350	9355	9360	9365	9370	9375	9380	9385	9390	1	1	2	2	3	3	4	4	5
87	9395	9400	9405	9410	9415	9420	9425	9430	9435	9440	0	1	1	2	2	3	3	4	4
88	9445	9450	9455	9460	9465	9469	9474	9479	9484	9489	0	1	1	2	2	3	3	4	4
89	9494	9499	9504	9509	9513	9518	9523	9528	9533	9538	0	1	1	2	2	3	3	4	4
90	9542	9547	9552	9557	9562	9566	9571	9576	9581	9586	0	1	1	2	2	3	3	4	4
91	9590	9595	9600	9605	9609	9614	9619	9624	9628	9633	0	1	1	2	2	3	3	4	4
92	9638	9643	9647	9652	9657	9661	9666	9671	9675	9680	0	1	1	2	2	3	3	4	4
93	9685	9689	9694	9699	9703	9708	9713	9717	9722	9727	0	1	1	2	2	3	3	4	4
94	9731	9736	9741	9745	9750	9754	9759	9763	9768	9773	0	1	1	2	2	3	3	4	4
95	9777	9782	9786	9791	9795	9800	9805	9809	9814	9818	0	1	1	2	2	3	3	4	4
96	9823	9827	9832	9836	9841	9845	9850	9854	9859	9863	0	1	1	2	2	3	3	4	4
97	9868	9872	9877	9881	9886	9890	9894	9899	9903	9908	0	1	1	2	2	3	3	4	4
98	9912	9917	9921	9926	9930	9934	9939	9943	9948	9952	0	1	1	2	2	3	3	4	4
99	9956	9961	9965	9969	9974	9978	9983	9987	9991	9996	0	1	1	2	2	3	3	3	4
N	0	1	2	3	4	5	6	7	8	9	1	2	3	4	5	6	7	8	9

Table 2 4-place antilogarithm

	0	1	2	3	4	5	6	7	8	9	Proportional Parts 1	2	3	4	5	6	7	8	9
.00	1000	1002	1005	1007	1009	1012	1014	1016	1019	1021	0	0	1	1	1	1	2	2	2
.01	1023	1026	1028	1030	1033	1035	1038	1040	1042	1045	0	0	1	1	1	1	2	2	2
.02	1047	1050	1052	1054	1057	1059	1062	1064	1067	1069	0	0	1	1	1	1	2	2	2
.03	1072	1074	1076	1079	1081	1084	1086	1089	1091	1094	0	0	1	1	1	1	2	2	2
.04	1096	1099	1102	1104	1107	1109	1112	1114	1117	1119	0	1	1	1	1	2	2	2	2
.05	1122	1125	1127	1130	1132	1135	1138	1140	1143	1146	0	1	1	1	1	2	2	2	2
.06	1148	1151	1153	1156	1159	1161	1164	1167	1169	1172	0	1	1	1	1	2	2	2	2
.07	1175	1178	1180	1183	1186	1189	1191	1194	1197	1199	0	1	1	1	1	2	2	2	2
.08	1202	1205	1208	1211	1213	1216	1219	1222	1225	1227	0	1	1	1	1	2	2	2	3
.09	1230	1233	1236	1239	1242	1245	1247	1250	1253	1256	0	1	1	1	1	2	2	2	3
.10	1259	1262	1265	1268	1271	1274	1276	1279	1282	1285	0	1	1	1	1	2	2	2	3
.11	1288	1291	1294	1297	1300	1303	1306	1309	1312	1315	0	1	1	1	2	2	2	3	3
.12	1318	1321	1324	1327	1330	1334	1337	1340	1343	1346	0	1	1	1	2	2	2	3	3
.13	1349	1352	1355	1358	1361	1365	1368	1371	1374	1377	0	1	1	1	2	2	2	3	3
.14	1380	1384	1387	1390	1393	1396	1400	1403	1406	1409	0	1	1	1	2	2	2	3	3
.15	1413	1416	1419	1422	1426	1429	1432	1435	1439	1442	0	1	1	1	2	2	2	3	3
.16	1445	1449	1452	1455	1459	1462	1466	1469	1472	1476	0	1	1	1	2	2	2	3	3
.17	1479	1483	1486	1489	1493	1496	1500	1503	1507	1510	0	1	1	1	2	2	2	3	3
.18	1514	1517	1521	1524	1528	1531	1535	1538	1542	1545	0	1	1	1	2	2	2	3	3
.19	1549	1552	1556	1560	1563	1567	1570	1574	1578	1581	0	1	1	1	2	2	3	3	3
.20	1585	1589	1592	1596	1600	1603	1607	1611	1614	1618	0	1	1	1	2	2	3	3	3
.21	1622	1626	1629	1633	1637	1641	1644	1648	1652	1656	0	1	1	2	2	2	3	3	3
.22	1660	1663	1667	1671	1675	1679	1683	1687	1690	1694	0	1	1	2	2	2	3	3	3
.23	1698	1702	1706	1710	1714	1718	1722	1726	1730	1734	0	1	1	2	2	2	3	3	4
.24	1738	1742	1746	1750	1754	1758	1762	1766	1770	1774	0	1	1	2	2	2	3	3	4
.25	1778	1782	1786	1791	1795	1799	1803	1807	1811	1816	0	1	1	2	2	2	3	3	4
.26	1820	1824	1828	1832	1837	1841	1845	1849	1854	1858	0	1	1	2	2	3	3	3	4
.27	1862	1866	1871	1875	1879	1884	1888	1892	1897	1901	0	1	1	2	2	3	3	3	4
.28	1905	1910	1914	1919	1923	1928	1932	1936	1941	1945	0	1	1	2	2	3	3	4	4
.29	1950	1954	1959	1963	1968	1972	1977	1982	1986	1991	0	1	1	2	2	3	3	4	4
.30	1995	2000	2004	2009	2014	2018	2023	2028	2032	2037	0	1	1	2	2	3	3	4	4
.31	2042	2046	2051	2056	2061	2065	2070	2075	2080	2084	0	1	1	2	2	3	3	4	4
.32	2089	2094	2099	2104	2109	2113	2118	2123	2128	2133	0	1	1	2	2	3	3	4	4
.33	2138	2143	2148	2153	2158	2163	2168	2173	2178	2183	0	1	1	2	2	3	3	4	4
.34	2188	2193	2198	2203	2208	2213	2218	2223	2228	2234	1	1	2	2	3	3	4	4	5
.35	2239	2244	2249	2254	2259	2265	2270	2275	2280	2286	1	1	2	2	3	3	4	4	5
.36	2291	2296	2301	2307	2312	2317	2323	2328	2333	2339	1	1	2	2	3	3	4	4	5
.37	2344	2350	2355	2360	2366	2371	2377	2382	2388	2393	1	1	2	2	3	3	4	4	5
.38	2399	2404	2410	2415	2421	2427	2432	2438	2443	2449	1	1	2	2	3	3	4	4	5
.39	2455	2460	2466	2472	2477	2483	2489	2495	2500	2506	1	1	2	2	3	3	4	5	5
.40	2512	2518	2523	2529	2535	2541	2547	2553	2559	2564	1	1	2	2	3	4	4	5	5
.41	2570	2576	2582	2588	2594	2600	2606	2612	2618	2624	1	1	2	2	3	4	4	5	5
.42	2630	2636	2642	2649	2655	2661	2667	2673	2679	2685	1	1	2	2	3	4	4	5	6
.43	2692	2698	2704	2710	2716	2723	2729	2735	2742	2748	1	1	2	3	3	4	4	5	6
.44	2754	2761	2767	2773	2780	2786	2793	2799	2805	2812	1	1	2	3	3	4	4	5	6
.45	2818	2825	2831	2838	2844	2851	2858	2864	2871	2877	1	1	2	3	3	4	5	5	6
.46	2884	2891	2897	2904	2911	2917	2924	2931	2938	2944	1	1	2	3	3	4	5	5	6
.47	2951	2958	2965	2972	2979	2985	2992	2999	3006	3013	1	1	2	3	3	4	5	5	6
.48	3020	3027	3034	3041	3048	3055	3062	3069	3076	3083	1	1	2	3	4	4	5	6	6
.49	3090	3097	3105	3112	3119	3126	3133	3141	3148	3155	1	1	2	3	4	4	5	6	6
	0	1	2	3	4	5	6	7	8	9	1	2	3	4	5	6	7	8	9

	0	1	2	3	4	5	6	7	8	9	Proportional Parts 1 2 3 4 5 6 7 8 9
.50	3162	3170	3177	3184	3192	3199	3206	3214	3221	3228	1 1 2 3 4 4 5 6 7
.51	3236	3243	3251	3258	3266	3273	3281	3289	3296	3304	1 2 2 3 4 5 5 6 7
.52	3311	3319	3327	3334	3342	3350	3357	3365	3373	3381	1 2 2 3 4 5 5 6 7
.53	3388	3396	3404	3412	3420	3428	3436	3443	3451	3459	1 2 2 3 4 5 6 6 7
.54	3467	3475	3483	3491	3499	3508	3516	3524	3532	3540	1 2 2 3 4 5 6 6 7
.55	3548	3556	3565	3573	3581	3589	3597	3606	3614	3622	1 2 2 3 4 5 6 7 7
.56	3631	3639	3648	3656	3664	3673	3681	3690	3698	3707	1 2 3 3 4 5 6 7 8
.57	3715	3724	3733	3741	3750	3758	3767	3776	3784	3793	1 2 3 4 4 5 6 7 8
.58	3802	3811	3819	3828	3837	3846	3855	3864	3873	3882	1 2 3 4 4 5 6 7 8
.59	3890	3899	3908	3917	3926	3936	3945	3954	3963	3972	1 2 3 4 5 5 6 7 8
.60	3981	3990	3999	4009	4018	4027	4036	4046	4055	4064	1 2 3 4 5 6 6 7 8
.61	4074	4083	4093	4102	4111	4121	4130	4140	4150	4159	1 2 3 4 5 6 7 8 9
.62	4169	4178	4188	4198	4207	4217	4227	4236	4246	4256	1 2 3 4 5 6 7 8 9
.63	4266	4276	4285	4295	4305	4315	4325	4335	4345	4355	1 2 3 4 5 6 7 8 9
.64	4365	4375	4385	4395	4406	4416	4426	4436	4446	4457	1 2 3 4 5 6 7 8 9
.65	4467	4477	4487	4498	4508	4519	4529	4539	4550	4560	1 2 3 4 5 6 7 8 9
.66	4571	4581	4592	4603	4613	4624	4634	4645	4656	4667	1 2 3 4 5 6 7 9 10
.67	4677	4688	4699	4710	4721	4732	4742	4753	4764	4775	1 2 3 4 5 7 8 9 10
.68	4786	4797	4808	4819	4831	4842	4853	4864	4875	4887	1 2 3 4 6 7 8 9 10
.69	4898	4909	4920	4932	4943	4955	4966	4977	4989	5000	1 2 3 5 6 7 8 9 10
.70	5012	5023	5035	5047	5058	5070	5082	5093	5105	5117	1 2 4 5 6 7 8 9 11
.71	5129	5140	5152	5164	5176	5188	5200	5212	5224	5236	1 2 4 5 6 7 8 10 11
.72	5248	5260	5272	5284	5297	5309	5321	5333	5346	5358	1 2 4 5 6 7 9 10 11
.73	5370	5383	5395	5408	5420	5433	5445	5458	5470	5483	1 3 4 5 6 8 9 10 11
.74	5495	5508	5521	5534	5546	5559	5572	5585	5598	5610	1 3 4 5 6 8 9 10 12
.75	5623	5636	5649	5662	5675	5689	5702	5715	5728	5741	1 3 4 5 7 8 9 10 12
.76	5754	5768	5781	5794	5808	5821	5834	5848	5861	5875	1 3 4 5 7 8 9 11 12
.77	5888	5902	5916	5929	5943	5957	5970	5984	5998	6012	1 3 4 5 7 8 10 11 12
.78	6026	6039	6053	6067	6081	6095	6109	6124	6138	6152	1 3 4 6 7 8 10 11 13
.79	6166	6180	6194	6209	6223	6237	6252	6266	6281	6295	1 3 4 6 7 9 10 11 13
.80	6310	6324	6339	6353	6368	6383	6397	6412	6427	6442	1 3 4 6 7 9 10 12 13
.81	6457	6471	6486	6501	6516	6531	6546	6561	6577	6592	2 3 5 6 8 9 11 12 14
.82	6607	6622	6637	6653	6668	6683	6699	6714	6730	6745	2 3 5 6 8 9 11 12 14
.83	6761	6776	6792	6808	6823	6839	6855	6871	6887	6902	2 3 5 6 8 9 11 13 14
.84	6918	6934	6950	6966	6982	6998	7015	7031	7047	7063	2 3 5 6 8 10 11 13 15
.85	7079	7096	7112	7129	7145	7161	7178	7194	7211	7228	2 3 5 7 8 10 12 13 15
.86	7244	7261	7278	7295	7311	7328	7345	7362	7379	7396	2 3 5 7 8 10 12 13 15
.87	7413	7430	7447	7464	7482	7499	7516	7534	7551	7568	2 3 5 7 9 10 12 14 16
.88	7586	7603	7621	7638	7656	7674	7691	7709	7727	7745	2 4 5 7 9 11 12 14 16
.89	7762	7780	7798	7816	7834	7852	7870	7889	7907	7925	2 4 5 7 9 11 13 14 16
.90	7943	7962	7980	7998	8017	8035	8054	8072	8091	8110	2 4 6 7 9 11 13 15 17
.91	8128	8147	8166	8185	8204	8222	8241	8260	8279	8299	2 4 6 8 9 11 13 15 17
.92	8318	8337	8356	8375	8395	8414	8433	8453	8472	8492	2 4 6 8 10 12 14 15 17
.93	8511	8531	8551	8570	8590	8610	8630	8650	8670	8690	2 4 6 8 10 12 14 16 18
.94	8710	8730	8750	8770	8790	8810	8831	8851	8872	8892	2 4 6 8 10 12 14 16 18
.95	8913	8933	8954	8974	8995	9016	9036	9057	9078	9099	2 4 6 8 10 12 15 17 19
.96	9120	9141	9162	9183	9204	9226	9247	9268	9290	9311	2 4 6 8 11 13 15 17 19
.97	9333	9354	9376	9397	9419	9441	9462	9484	9506	9528	2 4 7 9 11 13 15 17 20
.98	9550	9572	9594	9616	9638	9661	9683	9705	9727	9750	2 4 7 9 11 13 16 18 20
.99	9772	9795	9817	9840	9863	9886	9908	9931	9954	9977	2 5 7 9 11 14 16 18 20
	0	1	2	3	4	5	6	7	8	9	1 2 3 4 5 6 7 8 9

Table 3 *Powers and roots*

n	n²	√n	√10n	n³	∛n	∛10n	∛100n
1	1	1.000 000	3.162 278	1	1.000 000	2.154 435	4.641 589
2	4	1.414 214	4.472 136	8	1.259 921	2.714 418	5.848 035
3	9	1.732 051	5.477 226	27	1.442 250	3.107 233	6.694 330
4	16	2.000 000	6.324 555	64	1.587 401	3.419 952	7.368 063
5	25	2.236 068	7.071 068	125	1.709 976	3.684 031	7.937 005
6	36	2.449 490	7.745 967	216	1.817 121	3.914 868	8.434 327
7	49	2.645 751	8.366 600	343	1.912 931	4.121 285	8.879 040
8	64	2.828 427	8.944 272	512	2.000 000	4.308 869	9.283 178
9	81	3.000 000	9.486 833	729	2.080 084	4.481 405	9.654 894
10	100	3.162 278	10.00000	1 000	2.154 435	4.641 589	10.00000
11	121	3.316 625	10.48809	1 331	2.223 980	4.791 420	10.32280
12	144	3.464 102	10.95445	1 728	2.289 428	4.932 424	10.62659
13	169	3.605 551	11.40175	2 197	2.351 335	5.065 797	10.91393
14	196	3.741 657	11.83216	2 744	2.410 142	5.192 494	11.18689
15	225	3.872 983	12.24745	3 375	2.466 212	5.313 293	11.44714
16	256	4.000 000	12.64911	4 096	2.519 842	5.428 835	11.69607
17	289	4.123 106	13.03840	4 913	2.571 282	5.539 658	11.93483
18	324	4.242 641	13.41641	5 832	2.620 741	5.646 216	12.16440
19	361	4.358 899	13.78405	6 859	2.668 402	5.748 897	12.38562
20	400	4.472 136	14.14214	8 000	2.714 418	5.848 035	12.59921
21	441	4.582 576	14.49138	9 261	2.758 924	5.943 922	12.80579
22	484	4.690 416	14.83240	10 648	2.802 039	6.036 811	13.00591
23	529	4.795 832	15.16575	12 167	2.843 867	6.126 926	13.20006
24	576	4.898 979	15.49193	13 824	2.884 499	6.214 465	13.38866
25	625	5.000 000	15.81139	15 625	2.924 018	6.299 605	13.57209
26	676	5.099 020	16.12452	17 576	2.962 496	6.382 504	13.75069
27	729	5.196 152	16.43168	19 683	3.000 000	6.463 304	13.92477
28	784	5.291 503	16.73320	21 952	3.036 589	6.542 133	14.09460
29	841	5.385 165	17.02939	24 389	3.072 317	6.619 106	14.26043
30	900	5.477 226	17.32051	27 000	3.107 233	6.694 330	14.42250
31	961	5.567 764	17.60682	29 791	3.141 381	6.767 899	14.58100
32	1 024	5.656 854	17.88854	32 768	3.174 802	6.839 904	14.73613
33	1 089	5.744 563	18.16590	35 937	3.207 534	6.910 423	14.88806
34	1 156	5.830 952	18.43909	39 304	3.239 612	6.979 532	15.03695
35	1 225	5.916 080	18.70829	42 875	3.271 066	7.047 299	15.18294
36	1 296	6.000 000	18.97367	46 656	3.301 927	7.113 787	15.32619
37	1 369	6.082 763	19.23538	50 653	3.332 222	7.179 054	15.46680
38	1 444	6.164 414	19.49359	54 872	3.361 975	7.243 156	15.60491
39	1 521	6.244 998	19.74842	59 319	3.391 211	7.306 144	15.74061
40	1 600	6.324 555	20.00000	64 000	3.419 952	7.368 063	15.87401
41	1 681	6.403 124	20.24846	68 921	3.448 217	7.428 959	16.00521
42	1 764	6.480 741	20.49390	74 088	3.476 027	7.488 872	16.13429
43	1 849	6.557 439	20.73644	79 507	3.503 398	7.547 842	16.26133
44	1 936	6.633 250	20.97618	85 184	3.530 348	7.605 905	16.38643
45	2 025	6.708 204	21.21320	91 125	3.556 893	7.663 094	16.50964
46	2 116	6.782 330	21.44761	97 336	3.583 048	7.719 443	16.63103
47	2 209	6.855 655	21.67948	103 823	3.608 826	7.774 980	16.75069
48	2 304	6.928 203	21.90890	110 592	3.634 241	7.829 735	16.86865
49	2 401	7.000 000	22.13594	117 649	3.659 306	7.883 735	16.98499
50	2 500	7.071 068	22.36068	125 000	3.684 031	7.937 005	17.09976

Table 3 *Powers and roots* (*continued*) **409**

n	n²	√n	√10n	n³	∛n	∛10n	∛100n
50	2 500	7.071 068	22.36068	125 000	3.684 031	7.937 005	17.09976
51	2 601	7.141 428	22.58318	132 651	3.708 430	7.989 570	17.21301
52	2 704	7.211 103	22.80351	140 608	3.732 511	8.041 452	17.32478
53	2 809	7.280 110	23.02173	148 877	3.756 286	8.092 672	17.43513
54	2 916	7.348 469	23.23790	157 464	3.779 763	8.143 253	17.54411
55	3 025	7.416 198	23.45208	166 375	3.802 952	8.193 213	17.65174
56	3 136	7.483 315	23.66432	175 616	3.825 862	8.242 571	17.75808
57	3 249	7.549 834	23.87467	185 193	3.848 501	8.291 344	17.86316
58	3 364	7.615 773	24.08319	195 112	3.870 877	8.339 551	17.96702
59	3 481	7.681 146	24.28992	205 379	3.892 996	8.387 207	18.06969
60	3 600	7.745 967	24.49490	216 000	3.914 868	8.434 327	18.17121
61	3 721	7.810 250	24.69818	226 981	3.936 497	8.480 926	18.27160
62	3 844	7.874 008	24.89980	238 328	3.957 892	8.527 019	18.37091
63	3 969	7.937 254	25.09980	250 047	3.979 057	8.572 619	18.46915
64	4 096	8.000 000	25.29822	262 144	4.000 000	8.617 739	18.56636
65	4 225	8.062 258	25.49510	274 625	4.020 726	8.662 391	18.66256
66	4 356	8.124 038	25.69047	287 496	4.041 240	8.706 588	18.75777
67	4 489	8.185 353	25.88436	300 763	4.061 548	8.750 340	18.85204
68	4 624	8.246 211	26.07681	314 432	4.081 655	8.793 659	18.94536
69	4 761	8.306 624	26.26785	328 509	4.101 566	8.836 556	19.03778
70	4 900	8.366 600	26.45751	343 000	4.121 285	8.879 040	19.12931
71	5 041	8.426 150	26.64583	357 911	4.140 818	8.921 121	19.21997
72	5 184	8.485 281	26.83282	373 248	4.160 168	8.962 809	19.30979
73	5 329	8.544 004	27.01851	389 017	4.179 339	9.004 113	19.39877
74	5 476	8.602 325	27.20294	405 224	4.198 336	9.045 042	19.48695
75	5 625	8.660 254	27.38613	421 875	4.217 163	9.085 603	19.57434
76	5 776	8.717 798	27.56810	438 976	4.235 824	9.125 805	19.66095
77	5 929	8.774 964	27.74887	456 533	4.254 321	9.165 656	19.74681
78	6 084	8.831 761	27.92848	474 552	4.272 659	9.205 164	19.83192
79	6 241	8.888 194	28.10694	493 039	4.290 840	9.244 335	19.91632
80	6 400	8.944 272	28.28427	512 000	4.308 869	9.283 178	20.00000
81	6 561	9.000 000	28.46050	531 441	4.326 749	9.321 698	20.08299
82	6 724	9.055 385	28.63564	551 368	4.344 481	9.359 902	20.16530
83	6 889	9.110 434	28.80972	571 787	4.362 071	9.397 796	20.24694
84	7 056	9.165 151	28.98275	592 704	4.379 519	9.435 388	20.32793
85	7 225	9.219 544	29.15476	614 125	4.396 830	9.472 682	20.40828
86	7 396	9.273 618	29.32576	636 056	4.414 005	9.509 685	20.48800
87	7 569	9.327 379	29.49576	658 503	4.431 048	9.546 403	20.56710
88	7 744	9.380 832	29.66479	681 472	4.447 960	9.582 840	20.64560
89	7 921	9.433 981	29.83287	704 969	4.464 745	9.619 002	20.72351
90	8 100	9.486 833	30.00000	729 000	4.481 405	9.654 894	20.80084
91	8 281	9.539 392	30.16621	753 571	4.497 941	9.690 521	20.87759
92	8 464	9.591 663	30.33150	778 688	4.514 357	9.725 888	20.95379
93	8 649	9.643 651	30.49590	804 357	4.530 655	9.761 000	21.02944
94	8 836	9.695 360	30.65942	830 584	4.546 836	9.795 861	21.10454
95	9 025	9.746 794	30.82207	857 375	4.562 903	9.830 476	21.17912
96	9 216	9.797 959	30.98387	884 736	4.578 857	9.864 848	21.25317
97	9 409	9.848 858	31.14482	912 673	4.594 701	9.898 983	21.32671
98	9 604	9.899 495	31.30495	941 192	4.610 436	9.932 884	21.39975
99	9 801	9.949 874	31.46427	970 299	4.626 065	9.966 555	21.47229
100	10 000	10.00000	31.62278	1 000 000	4.641 589	10.00000	21.54435

Table 4   Trigonometric (degrees)

Deg.	Sin	Tan	*Cot	Cos	
0.0	0.00000	0.00000	∞	1.0000	**90.0**
.1	.00175	.00175	573.0	1.0000	89.9
.2	.00349	.00349	286.5	1.0000	.8
.3	.00524	.00524	191.0	1.0000	.7
.4	.00698	.00698	143.24	1.0000	.6
.5	.00873	.00873	114.59	1.0000	.5
.6	.01047	.01047	95.49	0.9999	.4
.7	.01222	.01222	81.85	.9999	.3
.8	.01396	.01396	71.62	.9999	.2
.9	.01571	.01571	63.66	.9999	89.1
1.0	0.01745	0.01746	57.29	0.9998	**89.0**
.1	.01920	.01920	52.08	.9998	88.9
.2	.02094	.02095	47.74	.9998	.8
.3	.02269	.02269	44.07	.9997	.7
.4	.02443	.02444	40.92	.9997	.6
.5	.02618	.02619	38.19	.9997	.5
.6	.02792	.02793	35.80	.9996	.4
.7	.02967	.02968	33.69	.9996	.3
.8	.03141	.03143	31.82	.9995	.2
.9	.03316	.03317	30.14	.9995	88.1
2.0	0.03490	0.03492	28.64	0.9994	**88.0**
.1	.03664	.03667	27.27	.9993	87.9
.2	.03839	.03842	26.03	.9993	.8
.3	.04013	.04016	24.90	.9992	.7
.4	.04188	.04191	23.86	.9991	.6
.5	.04362	.04366	22.90	.9990	.5
.6	.04536	.04541	22.02	.9990	.4
.7	.04711	.04716	21.20	.9989	.3
.8	.04885	.04891	20.45	.9988	.2
.9	.05059	.05066	19.74	.9987	87.1
3.0	0.05234	0.05241	19.081	0.9986	**87.0**
.1	.05408	.05416	18.464	.9985	86.9
.2	.05582	.05591	17.886	.9984	.8
.3	.05756	.05766	17.343	.9983	.7
.4	.05931	.05941	16.832	.9982	.6
.5	.06105	.06116	16.350	.9981	.5
.6	.06279	.06291	15.895	.9980	.4
.7	.06453	.06467	15.464	.9979	.3
.8	.06627	.06642	15.056	.9978	.2
.9	.06802	.06817	14.669	.9977	86.1
4.0	0.06976	0.06993	14.301	0.9976	**86.0**
.1	.07150	.07168	13.951	.9974	85.9
.2	.07324	.07344	13.617	.9973	.8
.3	.07498	.07519	13.300	.9972	.7
.4	.07672	.07695	12.996	.9971	.6
.5	.07846	.07870	12.706	.9969	.5
.6	.08020	.08046	12.429	.9968	.4
.7	.08194	.08221	12.163	.9966	.3
.8	.08368	.08397	11.909	.9965	.2
.9	.08542	.08573	11.664	.9963	85.1
5.0	0.08716	0.08749	11.430	0.9962	**85.0**
.1	.08889	.08925	11.205	.9960	84.9
.2	.09063	.09101	10.988	.9959	.8
.3	.09237	.09277	10.780	.9957	.7
.4	.09411	.09453	10.579	.9956	.6
.5	.09585	.09629	10.385	.9954	.5
.6	.09758	.09805	10.199	.9952	.4
.7	.09932	.09981	10.019	.9951	.3
.8	.10106	.10158	9.845	.9949	.2
.9	.10279	.10334	9.677	.9947	84.1
6.0	0.10453	0.10510	9.514	0.9945	**84.0**
	Cos	Cot	*Tan	Sin	Deg.

Deg.	Sin	Tan	Cot	Cos	
6.0	0.10453	0.10510	9.514	0.9945	**84.0**
.1	.10626	.10687	9.357	.9943	83.9
.2	.10800	.10863	9.205	.9942	.8
.3	.10973	.11040	9.058	.9940	.7
.4	.11147	.11217	8.915	.9938	.6
.5	.11320	.11394	8.777	.9936	.5
.6	.11494	.11570	8.643	.9934	.4
.7	.11667	.11747	8.513	.9932	.3
.8	.11840	.11924	8.386	.9930	.2
.9	.12014	.12101	8.264	.9928	83.1
7.0	0.12187	0.12278	8.144	0.9925	**83.0**
.1	.12360	.12456	8.028	.9923	82.9
.2	.12533	.12633	7.916	.9921	.8
.3	.12706	.12810	7.806	.9919	.7
.4	.12880	.12988	7.700	.9917	.6
.5	.13053	.13165	7.596	.9914	.5
.6	.13226	.13343	7.495	.9912	.4
.7	.13399	.13521	7.396	.9910	.3
.8	.13572	.13698	7.300	.9907	.2
.9	.13744	.13876	7.207	.9905	82.1
8.0	0.13917	0.14054	7.115	0.9903	**82.0**
.1	.14090	.14232	7.026	.9900	81.9
.2	.14263	.14410	6.940	.9898	.8
.3	.14436	.14588	6.855	.9895	.7
.4	.14608	.14767	6.772	.9893	.6
.5	.14781	.14945	6.691	.9890	.5
.6	.14954	.15124	6.612	.9888	.4
.7	.15126	.15302	6.535	.9885	.3
.8	.15299	.15481	6.460	.9882	.2
.9	.15471	.15660	6.386	.9880	81.1
9.0	0.15643	0.15838	6.314	0.9877	**81.0**
.1	.15816	.16017	6.243	.9874	80.9
.2	.15988	.16196	6.174	.9871	.8
.3	.16160	.16376	6.107	.9869	.7
.4	.16333	.16555	6.041	.9866	.6
.5	.16505	.16734	5.976	.9863	.5
.6	.16677	.16914	5.912	.9860	.4
.7	.16849	.17093	5.850	.9857	.3
.8	.17021	.17273	5.789	.9854	.2
.9	.17193	.17453	5.730	.9851	80.1
10.0	0.1736	0.1763	5.671	0.9848	**80.0**
.1	.1754	.1781	5.614	.9845	79.9
.2	.1771	.1799	5.558	.9842	.8
.3	.1788	.1817	5.503	.9839	.7
.4	.1805	.1835	5.449	.9836	.6
.5	.1822	.1853	5.396	.9833	.5
.6	.1840	.1871	5.343	.9829	.4
.7	.1857	.1890	5.292	.9826	.3
.8	.1874	.1908	5.242	.9823	.2
.9	.1891	.1926	5.193	.9820	79.1
11.0	0.1908	0.1944	5.145	0.9816	**79.0**
.1	.1925	.1962	5.097	.9813	78.9
.2	.1942	.1980	5.050	.9810	.8
.3	.1959	.1998	5.005	.9806	.7
.4	.1977	.2016	4.959	.9803	.6
.5	.1994	.2035	4.915	.9799	.5
.6	.2011	.2053	4.872	.9796	.4
.7	.2028	.2071	4.829	.9792	.3
.8	.2045	.2089	4.787	.9789	.2
.9	.2062	.2107	4.745	.9785	78.1
12.0	0.2079	0.2126	4.705	0.9781	**78.0**
	Cos	Cot	Tan	Sin	Deg.

*Interpolation in this section of the table is inaccurate.

Table 4 *Trigonometric* (*degrees*) (*continued*)   **411**

Deg.	Sin	Tan	Cot	Cos		Deg.	Sin	Tan	Cot	Cos	
12.0	0.2079	0.2126	4.705	0.9781	**78.0**	**18.0**	0.3090	0.3249	3.078	0.9511	**72.0**
.1	.2096	.2144	4.665	.9778	77.9	.1	.3107	.3269	3.060	.9505	71.9
.2	.2113	.2162	4.625	.9774	.8	.2	.3123	.3288	3.042	.9500	.8
.3	.2130	.2180	4.586	.9770	.7	.3	.3140	.3307	3.024	.9494	.7
.4	.2147	.2199	4.548	.9767	.6	.4	.3156	.3327	3.006	.9489	.6
.5	.2164	.2217	4.511	.9763	.5	.5	.3173	.3346	2.989	.9483	.5
.6	.2181	.2235	4.474	.9759	.4	.6	.3190	.3365	2.971	.9478	.4
.7	.2198	.2254	4.437	.9755	.3	.7	.3206	.3385	2.954	.9472	.3
.8	.2215	.2272	4.402	.9751	.2	.8	.3223	.3404	2.937	.9466	.2
.9	.2233	.2290	4.366	.9748	77.1	.9	.3239	.3424	2.921	.9461	71.1
13.0	0.2250	0.2309	4.331	0.9744	**77.0**	**19.0**	0.3256	0.3443	2.904	0.9455	**71.0**
.1	.2267	.2327	4.297	.9740	76.9	.1	.3272	.3463	2.888	.9449	70.9
.2	.2284	.2345	4.264	.9736	.8	.2	.3289	.3482	2.872	.9444	.8
.3	.2300	.2364	4.230	.9732	.7	.3	.3305	.3502	2.856	.9438	.7
.4	.2317	.2382	4.198	.9728	.6	.4	.3322	.3522	2.840	.9432	.6
.5	.2334	.2401	4.165	.9724	.5	.5	.3338	.3541	2.824	.9426	.5
.6	.2351	.2419	4.134	.9720	.4	.6	.3355	.3561	2.808	.9421	.4
.7	.2368	.2438	4.102	.9715	.3	.7	.3371	.3581	2.793	.9415	.3
.8	.2385	.2456	4.071	.9711	.2	.8	.3387	.3600	2.778	.9409	.2
.9	.2402	.2475	4.041	.9707	76.1	.9	.3404	.3620	2.762	.9403	70.1
14.0	0.2419	0.2493	4.011	0.9703	**76.0**	**20.0**	0.3420	0.3640	2.747	0.9397	**70.0**
.1	.2436	.2512	3.981	.9699	75.9	.1	.3437	.3659	2.733	.9391	69.9
.2	.2453	.2530	3.952	.9694	.8	.2	.3453	.3679	2.718	.9385	.8
.3	.2470	.2549	3.923	.9690	.7	.3	.3469	.3699	2.703	.9379	.7
.4	.2487	.2568	3.895	.9686	.6	.4	.3486	.3719	2.689	.9373	.6
.5	.2504	.2586	3.867	.9681	.5	.5	.3502	.3739	2.675	.9367	.5
.6	.2521	.2605	3.839	.9677	.4	.6	.3518	.3759	2.660	.9361	.4
.7	.2538	.2623	3.812	.9673	.3	.7	.3535	.3779	2.646	.9354	.3
.8	.2554	.2642	3.785	.9668	.2	.8	.3551	.3799	2.633	.9348	.2
.9	.2571	.2661	3.758	.9664	75.1	.9	.3567	3819	2.619	.9342	69.1
15.0	0.2588	0.2679	3.732	0.9659	**75.0**	**21.0**	0.3584	0.3839	2.605	0.9336	**69.0**
.1	.2605	.2698	3.706	.9655	74.9	.1	.3600	.3859	2.592	.9330	68.9
.2	.2622	.2717	3.681	.9650	.8	.2	.3616	.3879	2.578	.9323	.8
.3	.2639	.2736	3.655	.9646	.7	.3	.3633	.3899	2.565	.9317	.7
.4	.2656	.2754	3.630	.9641	.6	.4	.3649	.3919	2.552	.9311	.6
.5	.2672	.2773	3.606	.9636	.5	.5	.3665	.3939	2.539	.9304	.5
.6	.2689	.2792	3.582	.9632	.4	.6	.3681	.3959	2.526	.9298	.4
.7	.2706	.2811	3.558	.9627	.3	.7	.3697	.3979	2.513	.9291	.3
.8	.2723	.2830	3.534	.9622	.2	.8	.3714	.4000	2.500	.9285	.2
.9	.2740	.2849	3.511	.9617	74.1	.9	.3730	.4020	2.488	.9278	68.1
16.0	0.2756	0.2867	3.487	0.9613	**74.0**	**22.0**	0.3746	0.4046	2.475	0.9272	**68.0**
.1	.2773	.2886	3.465	.9608	73.9	.1	.3762	.4061	2.463	.9265	67.9
.2	.2790	.2905	3.442	.9603	.8	.2	.3778	.4081	2.450	.9259	.8
.3	.2807	.2924	3.420	.9598	.7	.3	.3795	.4101	2.438	.9252	.7
.4	.2823	.2943	3.398	.9593	.6	.4	.3811	.4122	2.426	.9245	.6
.5	.2840	.2962	3.376	.9588	.5	.5	.3827	.4142	2.414	.9239	.5
.6	.2857	.2981	3.354	.9583	.4	.6	.3843	.4163	2.402	.9232	.4
.7	.2874	.3000	3.333	.9578	.3	.7	.3859	.4183	2.391	.9225	.3
.8	.2890	.3019	3.312	.9573	.2	.8	.3875	.4204	2.379	.9219	.2
.9	.2907	.3038	3.291	.9568	73.1	.9	.3891	.4224	2.367	.9212	67.1
17.0	0.2924	0.3057	3.271	0.9563	**73.0**	**23.0**	0.3907	0.4245	2.356	0.9205	**67.0**
.1	.2940	.3076	3.251	.9558	72.9	.1	.3923	.4265	2.344	.9198	66.9
.2	.2957	.3096	3.230	.9553	.8	.2	.3939	.4286	2.333	.9191	.8
.3	.2974	.3115	3.211	.9548	.7	.3	.3955	.4307	2.322	.9184	.7
.4	.2990	.3134	3.191	.9542	.6	.4	.3971	.4327	2.311	.9178	.6
.5	.3007	.3153	3.172	.9537	.5	.5	.3987	.4348	2.300	.9171	.5
.6	.3024	.3172	3.152	.9532	.4	.6	.4003	.4369	2.289	.9164	.4
.7	.3040	.3191	3.133	.9527	.3	.7	.4019	.4390	2.278	.9157	.3
.8	.3057	.3211	3.115	.9521	.2	.8	.4035	.4411	2.267	.9150	.2
.9	.3074	.3230	3.096	.9516	72.1	.9	.4051	.4431	2.257	.9143	66.1
18.0	0.3090	0.3249	3.078	0.9511	**72.0**	**24.0**	0.4067	0.4452	2.246	0.9135	**66.0**
	Cos	Cot	Tan	Sin	Deg.		Cos	Cot	Tan	Sin	Deg.

Table 4 *Trigonometric (degrees) (continued)*

Deg.	Sin	Tan	Cot	Cos		Deg.	Sin	Tan	Cot	Cos	
24.0	0.4067	0.4452	2.246	0.9135	**66.0**	**30.0**	0.5000	0.5774	1.7321	0.8660	**60.0**
.1	.4083	.4473	2.236	.9128	65.9	.1	.5015	.5797	1.7251	.8652	59.9
.2	.4099	.4494	2.225	.9121	.8	.2	.5030	.5820	1.7182	.8643	.8
.3	.4115	.4515	2.215	.9114	.7	.3	.5045	.5844	1.7113	.8634	.7
.4	.4131	.4536	2.204	.9107	.6	.4	.5060	.5867	1.7045	.8625	.6
.5	.4147	.4557	2.194	.9100	.5	.5	.5075	.5890	1.6977	.8616	.5
.6	.4163	.4578	2.184	.9092	.4	.6	.5090	.5914	1.6909	.8607	.4
.7	.4179	.4599	2.174	.9085	.3	.7	.5105	.5938	1.6842	.8599	.3
.8	.4195	.4621	2.164	.9078	.2	.8	.5120	.5961	1.6775	.8590	.2
.9	.4210	.4642	2.154	.9070	65.1	.9	.5135	.5985	1.6709	.8581	59.1
25.0	0.4226	0.4663	2.145	0.9063	**65.0**	**31.0**	0.5150	0.6009	1.6643	0.8572	**59.0**
.1	.4242	.4684	2.135	.9056	64.9	.1	.5165	.6032	1.6577	.8563	58.9
.2	.4258	.4706	2.125	.9048	.8	.2	.5180	.6056	1.6512	.8554	.8
.3	.4274	.4727	2.116	.9041	.7	.3	.5195	.6080	1.6447	.8545	.7
.4	.4289	.4748	2.106	.9033	.6	.4	.5210	.6104	1.6383	.8536	.6
.5	.4305	.4770	2.097	.9026	.5	.5	.5225	.6128	1.6319	.8526	.5
.6	.4321	.4791	2.087	.9018	.4	.6	.5240	.6152	1.6255	.8517	.4
.7	.4337	.4813	2.078	.9011	.3	.7	.5255	.6176	1.6191	.8508	.3
.8	.4352	.4834	2.069	.9003	.2	.8	.5270	.6200	1.6128	.8499	.2
.9	.4368	.4856	2.059	.8996	64.1	.9	.5284	.6224	1.6066	.8490	58.1
26.0	0.4384	0.4877	2.050	0.8988	**64.0**	**32.0**	0.5299	0.6249	1.6003	0.8480	**58.0**
.1	.4399	.4899	2.041	.8980	63.9	.1	.5314	.6273	1.5941	.8471	57.9
.2	.4415	.4921	2.032	.8973	.8	.2	.5329	.6297	1.5880	.8462	.8
.3	.4431	.4942	2.023	.8965	.7	.3	.5344	.6322	1.5818	.8453	.7
.4	.4446	.4964	2.014	.8957	.6	.4	.5358	.6346	1.5757	.8443	.6
.5	.4462	.4986	2.006	.8949	.5	.5	.5373	.6371	1.5697	.8434	.5
.6	.4478	.5008	1.997	.8942	.4	.6	.5388	.6395	1.5637	.8425	.4
.7	.4493	5029	1.988	.8934	.3	.7	.5402	.6420	1.5577	.8415	.3
.8	.4509	.5051	1.980	.8926	.2	.8	.5417	.6445	1.5517	.8406	.2
.9	.4524	.5073	1.971	.8918	63.1	.9	.5432	.6469	1.5458	.8396	57.1
27.0	0.4540	0.5095	1.963	0.8910	**63.0**	**33.0**	0.5446	0.6494	1.5399	0.8387	**57.0**
.1	.4555	.5117	1.954	.8902	62.9	.1	.5461	.6519	1.5340	.8377	56.9
.2	.4571	.5139	1.946	.8894	.8	.2	.5476	.6544	1.5282	.8368	.8
.3	.4586	.5161	1.937	.8886	.7	.3	.5490	.6569	1.5224	.8358	.7
.4	.4602	.5184	1.929	.8878	.6	.4	.5505	.6594	1.5166	.8348	.6
.5	.4617	.5206	1 921	.8870	.5	.5	.5519	.6619	1.5108	.8339	.5
.6	.4633	.5228	1.913	.8862	.4	.6	.5534	.6644	1.5051	.8329	.4
.7	.4648	.5250	1.905	.8854	.3	.7	.5548	.6669	1.4994	.8320	.3
.8	.4664	.5272	1.897	.8846	.2	.8	.5563	.6694	1.4938	.8310	.2
.9	.4679	.5295	1.889	.8838	62.1	.9	.5577	.6720	1.4882	.8300	56.1
28.0	0.4695	0.5317	1.881	0.8829	**62.0**	**34.0**	0.5592	0.6745	1.4826	0.8290	**56.0**
.1	.4710	.5340	1.873	.8821	61.9	.1	.5606	.6771	1.4770	.8281	55.9
.2	.4726	.5362	1.865	.8813	.8	.2	.5621	.6796	1.4715	.8271	.8
.3	.4741	.5384	1.857	.8805	.7	.3	.5635	.6822	1.4659	.8261	.7
.4	.4756	.5407	1.849	.8796	.6	.4	.5650	.6847	1.4605	.8251	.6
.5	.4772	.5430	1.842	.8788	.5	.5	.5664	.6873	1.4550	.8241	.5
.6	.4787	.5452	1.834	.8780	.4	.6	.5678	.6899	1.4496	.8231	.4
.7	.4802	.5475	1.827	.8771	.3	.7	.5693	.6924	1.4442	.8221	.3
.8	.4818	.5498	1.819	.8763	.2	.8	.5707	.6950	1.4388	.8211	.2
.9	.4833	.5520	1.811	.8755	61.1	.9	.5721	.6976	1.4335	.8202	55.1
29.0	0.4848	0.5543	1.804	0.8746	**61.0**	**35.0**	0.5736	0.7002	1.4281	0.8192	**55.0**
.1	.4863	.5566	1.797	.8738	60.9	.1	.5750	.7028	1.4229	.8181	54.9
.2	.4879	.5589	1.789	.8729	.8	.2	.5764	.7054	1.4176	.8171	.8
.3	.4894	.5612	1.782	.8721	.7	.3	.5779	.7080	1.4124	.8161	.7
.4	.4909	.5635	1.775	.8712	.6	.4	.5793	.7107	1.4071	.8151	.6
.5	.4924	.5658	1.767	.8704	.5	.5	.5807	.7133	1.4019	.8141	.5
.6	.4939	.5681	1.760	.8695	.4	.6	.5821	.7159	1.3968	.8131	.4
.7	.4955	.5704	1.753	.8686	.3	.7	.5835	.7186	1.3916	.8121	.3
.8	.4970	.5727	1.746	.8678	.2	.8	.5850	.7212	1.3865	.8111	.2
.9	.4985	.5750	1.739	.8669	60.1	.9	.5864	.7239	1.3814	.8100	54.1
30.0	0.5000	0.5774	1.732	0.8660	**60.0**	**36.0**	0.5878	0.7265	1.3764	0.8090	**54.0**
	Cos	Cot	Tan	Sin	Deg.		Cos	Cot	Tan	Sin	Deg.

Table 4 Trigonometric (degrees) (continued)   **413**

Deg.	Sin	Tan	Cot	Cos	
36.0	0.5878	0.7265	1.3764	0.8090	**54.0**
.1	.5892	.7292	1.3713	.8080	53.9
.2	.5906	.7319	1.3663	.8070	.8
.3	.5920	.7346	1.3613	.8059	.7
.4	.5934	.7373	1.3564	.8049	.6
.5	.5948	.7400	1.3514	.8039	.5
.6	.5962	.7427	1.3465	.8028	.4
.7	.5976	.7454	1.3416	.8018	.3
.8	.5990	.7481	1.3367	.8007	.2
.9	.6004	.7508	1.3319	.7997	53.1
37.0	0.6018	0.7536	1.3270	0.7986	**53.0**
.1	.6032	.7563	1.3222	.7976	52.9
.2	.6046	.7590	1.3175	.7965	.8
.3	.6060	.7618	1.3127	.7955	.7
.4	.6074	.7646	1.3079	.7944	.6
.5	.6088	.7673	1.3032	.7934	.5
.6	.6101	.7701	1.2985	.7923	.4
.7	.6115	.7729	1.2938	.7912	.3
.8	.6129	.7757	1.2892	.7902	.2
.9	.6143	.7785	1.2846	.7891	52.1
38.0	0.6157	0.7813	1.2799	0.7880	**52.0**
.1	.6170	.7841	1.2753	.7869	51.9
.2	.6184	.7869	1.2708	.7859	.8
.3	.6198	.7898	1.2662	.7848	.7
.4	.6211	.7926	1.2617	.7837	.6
.5	.6225	.7954	1.2572	.7826	.5
.6	.6239	.7983	1.2527	.7815	.4
.7	.6252	.8012	1.2482	.7804	.3
.8	.6266	.8040	1.2437	.7793	.2
.9	.6280	.8069	1.2393	.7782	51.1
39.0	0.6293	0.8098	1.2349	0.7771	**51.0**
.1	.6307	.8127	1.2305	.7760	50.9
.2	.6320	.8156	1.2261	.7749	.8
.3	.6334	.8185	1.2218	.7738	.7
.4	.6347	.8214	1.2174	.7727	.6
.5	.6361	.8243	1.2131	.7716	.5
.6	.6374	.8273	1.2088	.7705	.4
.7	.6388	.8302	1.2045	.7694	.3
.8	.6401	.8332	1.2002	.7683	.2
.9	.6414	.8361	1.1960	.7672	50.1
40.0	0.6428	0.8391	1.1918	0.7660	**50.0**
.1	.6441	.8421	1.1875	.7649	49.9
.2	.6455	.8451	1.1833	.7638	.8
.3	.6468	.8481	1.1792	.7627	.7
.4	.6481	.8511	1.1750	.7615	.6
40.5	0.6494	0.8541	1.1708	0.7604	**49.5**
	Cos	Cot	Tan	Sin	Deg.

Deg.	Sin	Tan	Cot	Cos	
40.5	0.6494	0.8541	1.1708	0.7604	**49.5**
.6	.6508	.8571	1.1667	.7593	.4
.7	.6521	.8601	1.1626	.7581	.3
.8	.6534	.8632	1.1585	.7570	.2
.9	.6547	.8662	1.1544	.7559	49.1
41.0	0.6561	0.8693	1.1504	0.7547	**49.0**
.1	.6574	.8724	1.1463	.7536	48.9
.2	.6587	.8754	1.1423	.7524	.8
.3	.6600	.8785	1.1383	.7513	.7
.4	.6613	.8816	1.1343	.7501	.6
.5	.6626	.8847	1.1303	.7490	.5
.6	.6639	.8878	1.1263	.7478	.4
.7	.6652	.8910	1.1224	.7466	.3
.8	.6665	.8941	1.1184	.7455	.2
.9	.6678	.8972	1.1145	.7443	48.1
42.0	0.6691	0.9004	1.1106	0.7431	**48.0**
.1	.6704	.9036	1.1067	.7420	47.9
.2	.6717	.9067	1.1028	.7408	.8
.3	.6730	.9099	1.0990	.7396	.7
.4	.6743	.9131	1.0951	.7385	.6
.5	.6756	.9163	1.0913	.7373	.5
.6	.6769	.9195	1.0875	.7361	.4
.7	.6782	.9228	1.0837	.7349	.3
.8	.6794	.9260	1.0799	.7337	.2
.9	.6807	.9293	1.0761	.7325	47.1
43.0	0.6820	0.9325	1.0724	0.7314	**47.0**
.1	.6833	.9358	1.0686	.7302	46.9
.2	.6845	.9391	1.0649	.7290	.8
.3	.6858	.9424	1.0612	.7278	.7
.4	.6871	.9457	1.0575	.7266	.6
.5	.6884	.9490	1.0538	.7254	.5
.6	.6896	.9523	1.0501	.7242	.4
.7	.6909	.9556	1.0464	.7230	.3
.8	.6921	.9590	1.0428	.7218	.2
.9	.6934	.9623	1.0392	.7206	46.1
44.0	0.6947	0.9657	1.0355	0.7193	**46.0**
.1	.6959	.9691	1.0319	.7181	45.9
.2	.6972	.9725	1.0283	.7169	**.8**
.3	.6984	.9759	1.0247	.7157	**.7**
.4	.6997	.9793	1.0212	.7145	.6
.5	.7009	.9827	1.0176	.7133	.5
.6	.7022	.9861	1.0141	.7120	.4
.7	.7034	.9896	1.0105	.7108	.3
.8	.7046	.9930	1.0070	.7096	.2
.9	.7059	.9965	1.0035	.7083	45.1
45.0	0.7071	1.0000	1.0000	0.7071	**45.0**
	Cos	Cot	Tan	Sin	Deg.

Table 5 Trigonometric (radians)

Rad.	Sin	Tan	Cot	Cos	Rad.	Sin	Tan	Cot	Cos
.00	.00000	.00000	∞	1.00000	**.50**	.47943	.54630	1.8305	.87758
.01	.01000	.01000	99.997	0.99995	.51	.48818	.55936	1.7878	.87274
.02	.02000	.02000	49.993	.99980	.52	.49688	.57256	1.7465	.86782
.03	.03000	.03001	33.323	.99955	.53	.50553	.58592	1.7067	.86281
.04	.03999	.04002	24.987	.99920	.54	.51414	.59943	1.6683	.85771
.05	.04998	.05004	19.983	.99875	.55	.52269	.61311	1.6310	.85252
.06	.05996	.06007	16.647	.99820	.56	.53119	.62695	1.5950	.84726
.07	.06994	.07011	14.262	.99755	.57	.53963	.64097	1.5601	.84190
.08	.07991	.08017	12.473	.99680	.58	.54802	.65517	1.5263	.83646
.09	.08988	.09024	11.081	.99595	.59	.55636	.66956	1.4935	.83094
.10	.09983	.10033	9.9666	.99500	**.60**	.56464	.68414	1.4617	.82534
.11	.10978	.11045	9.0542	.99396	.61	.57287	.69892	1.4308	.81965
.12	.11971	.12058	8.2933	.99281	.62	.58104	.71391	1.4007	.81388
.13	.12963	.13074	7.6489	.99156	.63	.58914	.72911	1.3715	.80803
.14	.13954	.14092	7.0961	.99022	.64	.59720	.74454	1.3431	.80210
.15	.14944	.15114	6.6166	.98877	.65	.60519	.76020	1.3154	.79608
.16	.15932	.16138	6.1966	.98723	.66	.61312	.77610	1.2885	.78999
.17	.16918	.17166	5.8256	.98558	.67	.62099	.79225	1.2622	.78382
.18	.17903	.18197	5.4954	.98384	.68	.62879	.80866	1.2366	.77757
.19	.18886	.19232	5.1997	.98200	.69	.63654	.82534	1.2116	.77125
.20	.19867	.20271	4.9332	**.98007**	**.70**	.64422	.84229	1.1872	.76484
.21	.20846	.21314	4.6917	.97803	.71	.65183	.85953	1.1634	.75836
.22	.21823	.22362	4.4719	.97590	.72	.65938	.87707	1.1402	.75181
.23	.22798	.23414	4.2709	.97367	.73	.66687	.89492	1.1174	.74517
.24	.23770	.24472	4.0864	.97134	.74	.67429	.91309	1.0952	.73847
.25	.24740	.25534	3.9163	.96891	.75	.68164	.93160	1.0734	.73169
.26	.25708	.26602	3.7591	.96639	.76	.68892	.95045	1.0521	.72484
.27	.26673	.27676	3.6133	.96377	.77	.69614	.96967	1.0313	.71791
.28	.27636	.28755	3.4776	.96106	.78	.70328	.98926	1.0109	.71091
.29	.28595	.29841	3.3511	.95824	.79	.71035	1.0092	.99084	.70385
.30	.29552	.30934	3.2327	.95534	**.80**	.71736	1.0296	.97121	.69671
.31	.30506	.32033	3.1218	.95233	.81	.72429	1.0505	.95197	.68950
.32	.31457	.33139	3.0176	.94924	.82	.73115	1.0717	.93309	.68222
.33	.32404	.34252	2.9195	.94604	.83	.73793	1.0934	.91455	.67488
.34	.33349	.35374	2.8270	.94275	.84	.74464	1.1156	.89635	.66746
.35	.34290	.36503	2.7395	.93937	.85	.75128	1.1383	.87848	.65998
.36	.35227	.37640	2.6567	.93590	.86	.75784	1.1616	.86091	.65244
.37	.36162	.38786	2.5782	.93233	.87	.76433	1.1853	.84365	.64483
.38	.37092	.39941	2.5037	.92866	.88	.77074	1.2097	.82668	.63715
.39	.38019	.41105	2.4328	.92491	.89	.77707	1.2346	.80998	.62941
.40	.38942	.42279	2.3652	.92106	**.90**	.78333	1.2602	.79355	.62161
.41	.39861	.43463	2.3008	.91712	.91	.78950	1.2864	.77738	.61375
.42	.40776	.44657	2.2393	.91309	.92	.79560	1.3133	.76146	.60582
.43	.41687	.45862	2.1804	.90897	.93	.80162	1.3409	.74578	.59783
.44	.42594	.47078	2.1241	.90475	.94	.80756	1.3692	.73034	.58979
.45	.43497	.48306	2.0702	.90045	.95	.81342	1.3984	.71511	.58168
.46	.44395	.49545	2.0184	.89605	.96	.81919	1.4284	.70010	.57352
.47	.45289	.50797	1.9686	.89157	.97	.82489	1.4592	.68531	.56530
.48	.46178	.52061	1.9208	.88699	.98	.83050	1.4910	.67071	.55702
.49	.47063	.53339	1.8748	.88233	.99	.83603	1.5237	.65631	.54869
.50	.47943	.54630	1.8305	.87758	**1.00**	.84147	1.5574	.64209	.54030

Rad.	Sin	Tan	Cot	Cos	Rad.	Sin	Tan	Cot	Cos

Table 5 *Trigonometric* (*radians*) (*continued*)   **415**

Rad.	Sin	Tan	Cot	Cos	Rad.	Sin	Tan	Cot	Cos
1.00	.84147	1.5574	.64209	.54030	**1.50**	.99749	14.101	.07091	.07074
1.01	.84683	1.5922	.62806	.53186	1.51	.99815	16.428	.06087	.06076
1.02	.85211	1.6281	.61420	.52337	1.52	.99871	19.670	.05084	.05077
1.03	.85730	1.6652	.60051	.51482	1.53	.99917	24.498	.04082	.04079
1.04	.86240	1.7036	.58699	.50622	1.54	.99953	32.461	.03081	.03079
1.05	.86742	1.7433	.57362	.49757	1.55	.99978	48.078	.02080	.02079
1.06	.87236	1.7844	.56040	.48887	1.56	.99994	92.621	.01080	.01080
1.07	.87720	1.8270	.54734	.48012	1.57	1.00000	1255.8	.00080	.00080
1.08	.88196	1.8712	.53441	.47133	1.58	.99996	−108.65	−.00920	−.00920
1.09	.88663	1.9171	.52162	.46249	1.59	.99982	−52.067	−.01921	−.01920
1.10	.89121	1.9648	.50897	.45360	**1.60**	.99957	−34.233	−.02921	−.02920
1.11	.89570	2.0143	.49644	.44466	1.61	.99923	−25.495	−.03922	−.03919
1.12	.90010	2.0660	.48404	.43568	1.62	.99879	−20.307	−.04924	−.04918
1.13	.90441	2.1198	.47175	.42666	1.63	.99825	−16.871	−.05927	−.05917
1.14	.90863	2.1759	.45959	.41759	1.64	.99761	−14.427	−.06931	−.06915
1.15	.91276	2.2345	.44753	.40849	1.65	.99687	−12.599	−.07937	−.07912
1.16	.91680	2.2958	.43558	.39934	1.66	.99602	−11.181	−.08944	−.08909
1.17	.92075	2.3600	.42373	.39015	1.67	.99508	−10.047	−.09953	−.09904
1.18	.92461	2.4273	.41199	.38092	1.68	.99404	− 9.1208	−.10964	−.10899
1.19	.92837	2.4979	.40034	.37166	1.69	.99290	− 8.3492	−.11977	−.11892
1.20	.93204	2.5722	.38878	.36236	**1.70**	.99166	− 7.6966	−.12993	−.12884
1.21	.93562	2.6503	.37731	.35302	1.71	.99033	− 7.1373	−.14011	−.13875
1.22	.93910	2.7328	.36593	.34365	1.72	.98889	− 6.6524	−.15032	−.14865
1.23	.94249	2.8198	.35463	.33424	1.73	.98735	− 6.2281	−.16056	−.15853
1.24	.94578	2.9119	.34341	.32480	1.74	.98572	− 5.8535	−.17084	−.16840
1.25	.94898	3.0096	.33227	.31532	1.75	.98399	− 5.5204	−.18115	−.17825
1.26	.95209	3.1133	.32121	.30582	1.76	.98215	− 5.2221	−.19149	−.18808
1.27	.95510	3.2236	.31021	.29628	1.77	.98022	− 4.9534	−.20188	−.19789
1.28	.95802	3.3413	.29928	.28672	1.78	.97820	− 4.7101	−.21231	−.20768
1.29	.96084	3.4672	.28842	.27712	1.79	.97607	− 4.4887	−.22278	−.21745
1.30	.96356	3.6021	.27762	.26750	**1.80**	.97385	− 4.2863	−.23330	−.22720
1.31	.96618	3.7471	.26687	.25785	1.81	.97153	− 4.1005	−.24387	−.23693
1.32	.96872	3.9033	.25619	.24818	1.82	.96911	− 3.9294	−.25449	−.24663
1.33	.97115	4.0723	.24556	.23848	1.83	.96659	− 3.7712	−.26517	−.25631
1.34	.97348	4.2556	.23498	.22875	1.84	.96398	− 3.6245	−.27590	−.26596
1.35	.97572	4.4552	.22446	.21901	1.85	.96128	− 3.4881	−.28669	−.27559
1.36	.97786	4.6734	.21398	.20924	1.86	.95847	− 3.3608	−.29755	−.28519
1.37	.97991	4.9131	.20354	.19945	1.87	.95557	− 2.2419	−.30846	−.29476
1.38	.98185	5.1774	.19315	.18964	1.88	.95258	− 3.1304	−.31945	−.30430
1.39	.98370	5.4707	.18279	.17981	1.89	.94949	− 3.0257	−.33051	−.31381
1.40	.98545	5.7979	.17248	.16997	**1.90**	.94630	− 2.9271	−.34164	−.32329
1.41	.98710	6.1654	.16220	.16010	1.91	.94302	− 2.8341	−.35284	−.33274
1.42	.98865	6.5811	.15195	.15023	1.92	.93965	− 2.7463	−.36413	−.34215
1.43	.99010	7.0555	.14173	.14033	1.93	.93618	− 2.6632	−.37549	−.35153
1.44	.99146	7.6018	.13155	.13042	1.94	.93262	− 2.5843	−.38695	−.36087
1.45	.99271	8.2381	.12139	.12050	1.95	.92896	− 2.5095	−.39849	−.37018
1.46	.99387	8.9886	.11125	.11057	1.96	.92521	− 2.4383	−.41012	−.37945
1.47	.99492	9.8874	.10114	.10063	1.97	.92137	− 2.3705	−.42185	−.38868
1.48	.99588	10.983	.09105	.09067	1.98	.91744	− 2.3058	−.43368	−.39788
1.49	.99674	12.350	.08097	.08071	1.99	.91341	− 2.2441	−.44562	−.40703
1.50	.99749	14.101	.07091	.07074	**2.00**	.90930	− 2.1850	−.45766	−.41615
Rad.	Sin	Tan	Cot	Cos	Rad.	Sin	Tan	Cot	Cos

Deg.	L. Sin	L. Tan	L. Cot	L. Cos	
0.0	− ∞	− ∞	∞	0.0000	**90.0**
.1	7.2419	7.2419	2.7581	0.0000	.9
.2	7.5429	7.5429	2.4571	0.0000	.8
.3	7.7190	7.7190	2.2810	0.0000	.7
.4	7.8439	7.8439	2.1561	0.0000	.6
.5	7.9408	7.9409	2.0591	0.0000	.5
.6	8.0200	8.0200	1.9800	0.0000	.4
.7	8.0870	8.0870	1.9130	0.0000	.3
.8	8.1450	8.1450	1.8550	0.0000	.2
.9	8.1961	8.1962	1.8038	9.9999	.1
1.0	8.2419	8.2419	1.7581	9.9999	**89.0**
.1	8.2832	8.2833	1.7167	9.9999	.9
.2	8.3210	8.3211	1.6789	9.9999	.8
.3	8.3558	8.3559	1.6441	9.9999	.7
.4	8.3880	8.3881	1.6119	9.9999	.6
.5	8.4179	8.4181	1.5819	9.9999	.5
.6	8.4459	8.4461	1.5539	9.9998	.4
.7	8.4723	8.4725	1.5275	9.9998	.3
.8	8.4971	8.4973	1.5027	9.9998	.2
.9	8.5206	8.5208	1.4792	9.9998	.1
2.0	8.5428	8.5431	1.4569	9.9997	**88.0**
.1	8.5640	8.5643	1.4357	9.9997	.9
.2	8.5842	8.5845	1.4155	9.9997	.8
.3	8.6035	8.6038	1.3962	9.9996	.7
.4	8.6220	8.6223	1.3777	9.9996	.6
.5	8.6397	8.6401	1.3599	9.9996	.5
.6	8.6567	8.6571	1.3429	9.9996	.4
.7	8.6731	8.6736	1.3264	9.9995	.3
.8	8.6889	8.6894	1.3106	9.9995	.2
.9	8.7041	8.7046	1.2954	9.9994	.1
3.0	8.7188	8.7194	1.2806	9.9994	**87.0**
.1	8.7330	8.7337	1.2663	9.9994	.9
.2	8.7468	8.7475	1.2525	9.9993	.8
.3	8.7602	8.7609	1.2391	9.9993	.7
.4	8.7731	8.7739	1.2261	9.9992	.6
.5	8.7857	8.7865	1.2135	9.9992	.5
.6	8.7979	8.7988	1.2012	9.9991	.4
.7	8.8098	8.8107	1.1893	9.9991	.3
.8	8.8213	8.8223	1.1777	9.9990	.2
.9	8.8326	8.8336	1.1664	9.9990	.1
4.0	8.8436	8.8446	1.1554	9.9989	**86.0**
.1	8.8543	8.8554	1.1446	9.9989	.9
.2	8.8647	8.8659	1.1341	9.9988	.8
.3	8.8749	8.8762	1.1238	9.9988	.7
.4	8.8849	8.8862	1.1138	9.9987	.6
.5	8.8946	8.8960	1.1040	9.9987	.5
.6	8.9042	8.9056	1.0944	9.9986	.4
.7	8.9135	8.9150	1.0850	9.9985	.3
.8	8.9226	8.9241	1.0759	9.9985	.2
.9	8.9315	8.9331	1.0669	9.9984	.1
5.0	8.9403	8.9420	1.0580	9.9983	**85.0**
.1	8.9489	8.9506	1.0494	9.9983	.9
.2	8.9573	8.9591	1.0409	9.9982	.8
.3	8.9655	8.9674	1.0326	9.9981	.7
.4	8.9736	8.9756	1.0244	9.9981	.6
.5	8.9816	8.9836	1.0164	9.9980	.5
.6	8.9894	8.9915	1.0085	9.9979	.4
.7	8.9970	8.9992	1.0008	9.9978	.3
.8	9.0046	9.0068	0.9932	9.9978	.2
.9	9.0120	9.0143	0.9857	9.9977	.1
6.0	9.0192	9.0216	0.9784	9.9976	**84.0**
	L. Cos	L. Cot	L. Tan	L. Sin	Deg.

Deg.	L. Sin	L. Tan	L. Cot	L. Cos	
6.0	9.0192	9.0216	0.9784	9.9976	**84.0**
.1	9.0264	9.0289	0.9711	9.9975	.9
.2	9.0334	9.0360	0.9640	9.9975	.8
.3	9.0403	9.0430	0.9570	9.9974	.7
.4	9.0472	9.0499	0.9501	9.9973	.6
.5	9.0539	9.0567	0.9433	9.9972	.5
.6	9.0605	9.0633	0.9367	9.9971	.4
.7	9.0670	9.0699	0.9301	9.9970	.3
.8	9.0734	9.0764	0.9236	9.9969	.2
.9	9.0797	9.0828	0.9172	9.9968	.1
7.0	9.0859	9.0891	0.9109	9.9968	**83.0**
.1	9.0920	9.0954	0.9046	9.9967	.9
.2	9.0981	9.1015	0.8985	9.9966	.8
.3	9.1040	9.1076	0.8924	9.9965	.7
.4	9.1099	9.1135	0.8865	9.9964	.6
.5	9.1157	9.1194	0.8806	9.9963	.5
.6	9.1214	9.1252	0.8748	9.9962	.4
.7	9.1271	9.1310	0.8690	9.9961	.3
.8	9.1326	9.1367	0.8633	9.9960	.2
.9	9.1381	9.1423	0.8577	9.9959	.1
8.0	9.1436	9.1478	0.8522	9.9958	**82.0**
.1	9.1489	9.1533	0.8467	9.9956	.9
.2	9.1542	9.1587	0.8413	9.9955	.8
.3	9.1594	9.1640	0.8360	9.9954	.7
.4	9.1646	9.1693	0.8307	9.9953	.6
.5	9.1697	9.1745	0.8255	9.9952	.5
.6	9.1747	9.1797	0.8203	9.9951	.4
.7	9.1797	9.1848	0.8152	9.9950	.3
.8	9.1847	9.1898	0.8102	9.9949	.2
.9	9.1895	9.1948	0.8052	9.9947	.1
9.0	9.1943	9.1997	0.8003	9.9946	**81.0**
.1	9.1991	9.2046	0.7954	9.9945	.9
.2	9.2038	9.2094	0.7906	9.9944	.8
.3	9.2085	9.2142	0.7858	9.9943	.7
.4	9.2131	9.2189	0.7811	9.9941	.6
.5	9.2176	9.2236	0.7764	9.9940	.5
.6	9.2221	9.2282	0.7718	9.9939	.4
.7	9.2266	9.2328	0.7672	9.9937	.3
.8	9.2310	9.2374	0.7626	9.9936	.2
.9	9.2353	9.2419	0.7581	9.9935	.1
10.0	9.2397	9.2463	0.7537	9.9934	**80.0**
.1	9.2439	9.2507	0.7493	9.9932	.9
.2	9.2482	9.2551	0.7449	9.9931	.8
.3	9.2524	9.2594	0.7406	9.9929	.7
.4	9.2565	9.2637	0.7363	9.9928	.6
.5	9.2606	9.2680	0.7320	9.9927	.5
.6	9.2647	9.2722	0.7278	9.9925	.4
.7	9.2687	9.2764	0.7236	9.9924	.3
.8	9.2727	9.2805	0.7195	9.9922	.2
.9	9.2767	9.2846	0.7154	9.9921	.1
11.0	9.2806	9.2887	0.7113	9.9919	**79.0**
.1	9.2845	9.2927	0.7073	9.9918	.9
.2	9.2883	9.2967	0.7033	9.9916	.8
.3	9.2921	9.3006	0.6994	9.9915	.7
.4	9.2959	9.3046	0.6954	9.9913	.6
.5	9.2997	9.3085	0.6915	9.9912	.5
.6	9.3034	9.3123	0.6877	9.9910	.4
.7	9.3070	9.3162	0.6838	9.9909	.3
.8	9.3107	9.3200	0.6800	9.9907	.2
.9	9.3143	9.3237	0.6763	9.9906	.1
12.0	9.3179	9.3275	0.6725	9.9904	**78.0**
	L. Cos	L. Cot	L. Tan	L. Sin	Deg.

Note: Entries in the L. Sin column are $10 + \log \sin \theta$. The columns for L. Tan and L. Cos are read similarly except L. Cos for $0.0 \leq \theta \leq 0.8$.

Table 6 *Log-trig (degrees) (continued)* **417**

Deg.	L. Sin	L. Tan	L. Cot	L. Cos	
12.0	9.3179	9.3275	0.6725	9.9904	**78.0**
.1	9.3214	9.3312	0.6688	9.9902	.9
.2	9.3250	9.3349	0.6651	9.9901	.8
.3	9.3284	9.3385	0.6615	9.9899	.7
.4	9.3319	9.3422	0.6578	9.9897	.6
.5	9.3353	9.3458	0.6542	9.9896	.5
.6	9.3387	9.3493	0.6507	9.9894	.4
.7	9.3421	9.3529	0.6471	9.9892	.3
.8	9.3455	9.3564	0.6436	9.9891	.2
.9	9.3488	9.3599	0.6401	9.9889	.1
13.0	9.3521	9.3634	0.6366	9.9887	**77.0**
.1	9.3554	9.3668	0.6332	9.9885	.9
.2	9.3586	9.3702	0.6298	9.9884	.8
.3	9.3618	9.3736	0.6264	9.9882	.7
.4	9.3650	9.3770	0.6230	9.9880	.6
.5	9.3682	9.3804	0.6196	9.9878	.5
.6	9.3713	9.3837	0.6163	9.9876	.4
.7	9.3745	9.3870	0.6130	9.9875	.3
.8	9.3775	9.3903	0.6097	9.9873	.2
.9	9.3806	9.3935	0.6065	9.9871	.1
14.0	9.3837	9.3968	0.6032	9.9869	**76.0**
.1	9.3867	9.4000	0.6000	9.9867	.9
.2	9.3897	9.4032	0.5968	9.9865	.8
.3	9.3927	9.4064	0.5936	9.9863	.7
.4	9.3957	9.4095	0.5905	9.9861	.6
.5	9.3986	9.4127	0.5873	9.9859	.5
.6	9.4015	9.4158	0.5842	9.9857	.4
.7	9.4044	9.4189	0.5811	9.9855	.3
.8	9.4073	9.4220	0.5780	9.9853	.2
.9	9.4102	9.4250	0.5750	9.9851	.1
15.0	9.4130	9.4281	0.5719	9.9849	**75.0**
.1	9.4158	9.4311	0.5689	9.9847	.9
.2	9.4186	9.4341	0.5659	9.9845	.8
.3	9.4214	9.4371	0.5629	9.9843	.7
.4	9.4242	9.4400	0.5600	9.9841	.6
.5	9.4269	9.4430	0.5570	9.9839	.5
.6	9.4296	9.4459	0.5541	9.9837	.4
.7	9.4323	9.4488	0.5512	9.9835	.3
.8	9.4350	9.4517	0.5483	9.9833	.2
.9	9.4377	9.4546	0.5454	9.9831	.1
16.0	9.4403	9.4575	0.5425	9.9828	**74.0**
.1	9.4430	9.4603	0.5397	9.9826	.9
.2	9.4456	9.4632	0.5368	9.9824	.8
.3	9.4482	9.4660	0.5340	9.9822	.7
.4	9.4508	9.4688	0.5312	9.9820	.6
.5	9.4533	9.4716	0.5284	9.9817	.5
.6	9.4559	9.4744	0.5256	9.9815	.4
.7	9.4584	9.4771	0.5229	9.9813	.3
.8	9.4609	9.4799	0.5201	9.9811	.2
.9	9.4634	9.4826	0.5174	9.9808	.1
17.0	9.4659	9.4853	0.5147	9.9806	**73.0**
.1	9.4684	9.4880	0.5120	9.9804	.9
.2	9.4709	9.4907	0.5093	9.9801	.8
.3	9.4733	9.4934	0.5066	9.9799	.7
.4	9.4757	9.4961	0.5039	9.9797	.6
.5	9.4781	9.4987	0.5013	9.9794	.5
.6	9.4805	9.5014	0.4986	9.9792	.4
.7	9.4829	9.5040	0.4960	9.9789	.3
.8	9.4853	9.5066	0.4934	9.9787	.2
.9	9.4876	9.5092	0.4908	9.9785	.1
18.0	9.4900	9.5118	0.4882	9.9782	**72.0**
	L. Cos	L. Cot	L. Tan	L. Sin	Deg.

Deg.	L. Sin	L. Tan	L. Cot	L. Cos	
18.0	9.4900	9.5118	0.4882	9.9782	**72.0**
.1	9.4923	9.5143	0.4857	9.9780	.9
.2	9.4946	9.5169	0.4831	9.9777	.8
.3	9.4969	9.5195	0.4805	9.9775	.7
.4	9.4992	9.5220	0.4780	9.9772	.6
.5	9.5015	9.5245	0.4755	9.9770	.5
.6	9.5037	9.5270	0.4730	9.9767	.4
.7	9.5060	9.5295	0.4705	9.9764	.3
.8	9.5082	9.5320	0.4680	9.9762	.2
.9	9.5104	9.5345	0.4655	9.9759	.1
19.0	9.5126	9.5370	0.4630	9.9757	**71.0**
.1	9.5148	9.5394	0.4606	9.9754	.9
.2	9.5170	9.5419	0.4581	9.9751	.8
.3	9.5192	9.5443	0.4557	9.9749	.7
.4	9.5213	9.5467	0.4533	9.9746	.6
.5	9.5235	9.5491	0.4509	9.9744	.5
.6	9.5256	9.5516	0.4484	9.9741	.4
.7	9.5278	9.5539	0.4461	9.9738	.3
.8	9.5299	9.5563	0.4437	9.9735	.2
.9	9.5320	9.5587	0.4413	9.9733	.1
20.0	9.5341	9.5611	0.4389	9.9730	**70.0**
.1	9.5361	9.5634	0.4366	9.9727	.9
.2	9.5382	9.5658	0.4342	9.9724	.8
.3	9.5402	9.5681	0.4319	9.9722	.7
.4	9.5423	9.5704	0.4296	9.9719	.6
.5	9.5443	9.5727	0.4273	9.9716	.5
.6	9.5463	9.5750	0.4250	9.9713	.4
.7	9.5484	9.5773	0.4227	9.9710	.3
.8	9.5504	9.5796	0.4204	9.9707	.2
.9	9.5523	9.5819	0.4181	9.9704	.1
21.0	9.5543	9.5842	0.4158	9.9702	**69.0**
.1	9.5563	9.5864	0.4136	9.9699	.9
.2	9.5583	9.5887	0.4113	9.9696	.8
.3	9.5602	9.5909	0.4091	9.9693	.7
.4	9.5621	9.5932	0.4068	9.9690	.6
.5	9.5641	9.5954	0.4046	9.9687	.5
.6	9.5660	9.5976	0.4024	9.9684	.4
.7	9.5679	9.5998	0.4002	9.9681	.3
.8	9.5698	9.6020	0.3980	9.9678	.2
.9	9.5717	9.6042	0.3958	9.9675	.1
22.0	9.5736	9.6064	0.3936	9.9672	**68.0**
.1	9.5754	9.6086	0.3914	9.9669	.9
.2	9.5773	9.6108	0.3892	9.9666	.8
.3	9.5792	9.6129	0.3871	9.9662	.7
.4	9.5810	9.6151	0.3849	9.9659	.6
.5	9.5828	9.6172	0.3828	9.9656	.5
.6	9.5847	9.6194	0.3806	9.9653	.4
.7	9.5865	9.6215	0.3785	9.9650	.3
.8	9.5883	9.6236	0.3764	9.9647	.2
.9	9.5901	9.6257	0.3743	9.9643	.1
23.0	9.5919	9.6279	0.3721	9.9640	**67.0**
.1	9.5937	9.6300	0.3700	9.9637	.9
.2	9.5954	9.6321	0.3679	9.9634	.8
.3	9.5972	9.6341	0.3659	9.9631	.7
.4	9.5990	9.6362	0.3638	9.9627	.6
.5	9.6007	9.6383	0.3617	9.9624	.5
.6	9.6024	9.6404	0.3596	9.9621	.4
.7	9.6042	9.6424	0.3576	9.9617	.3
.8	9.6059	9.6445	0.3555	9.9614	.2
.9	9.6076	9.6465	0.3535	9.9611	.1
24.0	9.6093	9.6486	0.3514	9.9607	**66.0**
	L. Cos	L. Cot	L. Tan	L. Sin	Deg.

Deg.	L. Sin	L. Tan	L. Cot	L. Cos	
24.0	9.6093	9.6486	0.3514	9.9607	**66.0**
.1	9.6110	9.6506	0.3494	9.9604	.9
.2	9.6127	9.6527	0.3473	9.9601	.8
.3	9.6144	9.6547	0.3453	9.9597	.7
.4	9.6161	9.6567	0.3433	9.9594	.6
.5	9.6177	9.6587	0.3413	9.9590	.5
.6	9.6194	9.6607	0.3393	9.9587	.4
.7	9.6210	9.6627	0.3373	9.9583	.3
.8	9.6227	9.6647	0.3353	9.9580	.2
.9	9.6243	9.6667	0.3333	9.9576	.1
25.0	9.6259	9.6687	0.3313	9.9573	**65.0**
.1	9.6276	9.6706	0.3294	9.9569	.9
.2	9.6292	9.6726	0.3274	9.9566	.8
.3	9.6308	9.6746	0.3254	9.9562	.7
.4	9.6324	9.6765	0.3235	9.9558	.6
.5	9.6340	9.6785	0.3215	9.9555	.5
.6	9.6356	9.6804	0.3196	9.9551	.4
.7	9.6371	9.6824	0.3176	9.9548	.3
.8	9.6387	9.6843	0.3157	9.9544	.2
.9	9.6403	9.6863	0.3137	9.9540	.1
26.0	9.6418	9.6882	0.3118	9.9537	**64.0**
.1	9.6434	9.6901	0.3099	9.9533	.9
.2	9.6449	9.6920	0.3080	9.9529	.8
.3	9.6465	9.6939	0.3061	9.9525	.7
.4	9.6480	9.6958	0.3042	9.9522	.6
.5	9.6495	9.6977	0.3023	9.9518	.5
.6	9.6510	9.6996	0.3004	9.9514	.4
.7	9.6526	9.7015	0.2985	9.9510	.3
.8	9.6541	9.7034	0.2966	9.9506	.2
.9	9.6556	0.7053	0.2947	9.9503	.1
27.0	9.6570	9.7072	0.2928	9.9499	**63.0**
.1	9.6585	9.7090	0.2910	9.9495	.9
.2	9.6600	9.7109	0.2891	9.9491	.8
.3	9.6615	9.7128	0.2872	9.9487	.7
.4	9.6629	9.7146	0.2854	9.9483	.6
.5	9.6644	9.7165	0.2835	9.9479	.5
.6	9.6659	9.7183	0.2817	9.9475	.4
.7	9.6673	9.7202	0.2798	9.9471	.3
.8	9.6687	9.7220	0.2780	9.9467	.2
.9	9.6702	9.7238	0.2762	9.9463	.1
28.0	9.6716	9.7257	0.2743	9.9459	**62.0**
.1	9.6730	9.7275	0.2725	9.9455	.9
.2	9.6745	9.7293	0.2707	9.9451	.8
.3	9.6759	9.7311	0.2689	9.9447	.7
.4	9.6773	9.7330	0.2670	9.9443	.6
.5	9.6787	9.7348	0.2652	9.9439	.5
.6	9.6801	9.7366	0.2634	9.9435	.4
.7	9.6814	9.7384	0.2616	9.9431	.3
.8	9.6828	9.7402	0.2598	9.9427	.2
.9	9.6842	9.7420	0.2580	9.9422	.1
29.0	9.6856	9.7438	0.2562	9.9418	**61.0**
.1	9.6869	9.7455	0.2545	9.9414	.9
.2	9.6883	9.7473	0.2527	9.9410	.8
.3	9.6896	9.7491	0.2509	9.9406	.7
.4	9.6910	9.7509	0.2491	9.9401	.6
.5	9.6923	9.7526	0.2474	9.9397	.5
.6	9.6937	9.7544	0.2456	9.9393	.4
.7	9.6950	9.7562	0.2438	9.9388	.3
.8	9.6963	9.7579	0.2421	9.9384	.2
.9	9.6977	9.7597	0.2403	9.9380	.1
30.0	9.6990	9.7614	0.2386	9.9375	**60.0**
	L. Cos	L. Cot	L. Tan	L. Sin	Deg.

Deg.	L. Sin	L. Tan	L. Cot	L. Cos	
30.0	9.6990	9.7614	0.2386	9.9375	**60.0**
.1	9.7003	9.7632	0.2368	9.9371	.9
.2	9.7016	9.7649	0.2351	9.9367	.8
.3	9.7029	9.7667	0.2333	9.9362	.7
.4	9.7042	9.7684	0.2316	9.9358	.6
.5	9.7055	9.7701	0.2299	9.9353	.5
.6	9.7068	9.7719	0.2281	9.9349	.4
.7	9.7080	9.7736	0.2264	9.9344	.3
.8	9.7093	9.7753	0.2247	9.9340	.2
.9	9.7106	9.7771	0.2229	9.9335	.1
31.0	9.7118	9.7788	0.2212	9.9331	**59.0**
.1	9.7131	9.7805	0.2195	9.9326	.9
.2	9.7144	9.7822	0.2178	9.9322	.8
.3	9.7156	9.7839	0.2161	9.9317	.7
.4	9.7168	9.7856	0.2144	9.9312	.6
.5	9.7181	9.7873	0.2127	9.9308	.5
.6	9.7193	9.7890	0.2110	9.9303	.4
.7	9.7205	9.7907	0.2093	9.9298	.3
.8	9.7218	9.7924	0.2076	9.9294	.2
.9	9.7230	9.7941	0.2059	9.9289	.1
32.0	9.7242	9.7958	0.2042	9.9284	**58.0**
.1	9.7254	9.7975	0.2025	9.9279	.9
.2	9.7266	9.7992	0.2008	9.9275	.8
.3	9.7278	9.8008	0.1992	9.9270	.7
.4	9.7290	9.8025	0.1975	9.9265	.6
.5	9.7302	9.8042	0.1958	9.9260	.5
.6	9.7314	9.8059	0.1941	9.9255	.4
.7	9.7326	9.8075	0.1925	9.9251	.3
.8	9.7338	9.8092	0.1908	9.9246	.2
.9	9.7349	9.8109	0.1891	9.9241	.1
33.0	9.7361	9.8125	0.1875	9.9236	**57.0**
.1	9.7373	9.8142	0.1858	9.9231	.9
.2	9.7384	9.8158	0.1842	9.9226	.8
.3	9.7396	9.8175	0.1825	9.9221	.7
.4	9.7407	9.8191	0.1809	9.9216	.6
.5	9.7419	9.8208	0.1792	9.9211	.5
.6	9.7430	9.8224	0.1776	9.9206	.4
.7	9.7442	9.8241	0.1759	9.9201	.3
.8	9.7453	9.8257	0.1743	9.9196	.2
.9	9.7464	9.8274	0.1726	9.9191	.1
34.0	9.7476	9.8290	0.1710	9.9186	**56.0**
.1	9.7487	9.8306	0.1694	9.9181	.9
.2	9.7498	9.8323	0.1677	9.9175	.8
.3	9.7509	9.8339	0.1661	9.9170	.7
.4	9.7520	9.8355	0.1645	9.9165	.6
.5	9.7531	9.8371	0.1629	9.9160	.5
.6	9.7542	9.8388	0.1612	9.9155	.4
.7	9.7553	9.8404	0.1596	9.9149	.3
.8	9.7564	9.8420	0.1580	9.9144	.2
.9	9.7575	9.8436	0.1564	9.9139	.1
35.0	9.7586	9.8452	0.1548	9.9134	**55.0**
.1	9.7597	9.8468	0.1532	9.9128	.9
.2	9.7607	9.8484	0.1516	9.9123	.8
.3	9.7618	9.8501	0.1499	9.9118	.7
.4	9.7629	9.8517	0.1483	9.9112	.6
.5	9.7640	9.8533	0.1467	9.9107	.5
.6	9.7650	9.8549	0.1451	9.9101	.4
.7	9.7661	9.8565	0.1435	9.9096	.3
.8	9.7671	9.8581	0.1419	9.9091	.2
.9	9.7682	9.8597	0.1403	9.9085	.1
36.0	9.7692	9.8613	0.1387	9.9080	**54.0**
	L. Cos	L. Cot	L. Tan	L. Sin	Deg.

Table 6 *Log-trig (degrees)* *(continued)* **419**

Deg.	L. Sin	L. Tan	L. Cot	L. Cos	
36.0	9.7692	9.8613	0.1387	9.9080	**54.0**
.1	9.7703	9.8629	0.1371	9.9074	.9
.2	9.7713	9.8644	0.1356	9.9069	.8
.3	9.7723	9.8660	0.1340	9.9063	.7
.4	9.7734	9.8676	0.1324	9.9057	.3
.5	9.7744	9.8692	0.1308	9.9052	.5
.6	9.7754	9.8708	0.1292	9.9046	.4
.7	9.7764	9.8724	0.1276	9.9041	.3
.8	9.7774	9.8740	0.1260	9.9035	.2
.9	9.7785	9.8755	0.1245	9.9029	.1
37.0	9.7795	9.8771	0.1229	9.9023	**53.0**
.1	9.7805	9.8787	0.1213	9.9018	.9
.2	9.7815	9.8803	0.1197	9.9012	.8
.3	9.7825	9.8818	0.1182	9.9006	.7
.4	9.7835	9.8834	0.1166	9.9000	.6
.5	9.7844	9.8850	0.1150	9.8995	.5
.6	9.7854	9.8865	0.1135	9.8989	.4
.7	9.7864	9.8881	0.1119	9.8983	.3
.8	9.7874	9.8897	0.1103	9.8977	.2
.9	9.7884	9.8912	0.1088	9.8971	.1
38.0	9.7893	9.8928	0.1072	9.8965	**52.0**
.1	9.7903	9.8944	0.1056	9.8959	.9
.2	9.7913	9.8959	0.1041	9.8953	.8
.3	9.7922	9.8975	0.1025	9.8947	.7
.4	9.7932	9.8990	0.1010	9.8941	.6
.5	9.7941	9.9006	0.0994	9.8935	.5
.6	9.7951	9.9022	0.0978	9.8929	.4
.7	9.7960	9.9037	0.0963	9.8923	.3
.8	9.7970	9.9053	0.0947	9.8917	.2
.9	9.7979	9.9068	0.0932	9.8911	.1
39.0	9.7989	9.9084	0.0916	9.8905	**51.0**
.1	9.7998	9.9099	0.0901	9.8899	.9
.2	9.8007	9.9115	0.0885	9.8893	.8
.3	9.8017	9.9130	0.0870	9.8887	.7
.4	9.8026	9.9146	0.0854	9.8880	.6
.5	9.8035	9.9161	0.0839	9.8874	.5
.6	9.8044	9.9176	0.0824	9.8868	.4
.7	9.8053	9.9192	0.0808	9.8862	.3
.8	9.8063	9.9207	0.0793	9.8855	.2
.9	9.8072	9.9223	0.0777	9.8849	.1
40.0	9.8081	9.9238	0.0762	9.8843	**50.0**
.1	9.8090	9.9254	0.0746	9.8836	.9
.2	9.8099	9.9269	0.0731	9.8830	.8
.3	9.8108	9.9284	0.0716	9.8823	.7
.4	9.8117	9.9300	0.0700	9.8817	.6
.5	9.8125	9.9315	0.0685	9.8810	.5
.6	9.8134	9.9330	0.0670	9.8804	.4
.7	9.8143	9.9346	0.0654	9.8797	.3
.8	9.8152	9.9361	0.0639	9.8791	.2
.9	9.8161	9.9376	0.0624	9.8784	.1
41.0	9.8169	9.9392	0.0608	9.8778	**49.0**
	L. Cos	L. Cot	L. Tan	L. Sin	Deg.

Deg.	L. Sin	L. Tan	L. Cot	L. Cos	
41.0	9.8169	9.9392	0.0608	9.8778	**49.0**
.1	9.8178	9.9407	0.0593	9.8771	.9
.2	9.8187	9.9422	0.0578	9.8765	.8
.3	9.8195	9.9438	0.0562	9.8758	.7
.4	9.8204	9.9453	0.0547	9.8751	.6
.5	9.8213	9.9468	0.0532	9.8745	.5
.6	9.8221	9.9483	0.0517	9.8738	.4
.7	9.8230	9.9499	0.0501	9.8731	.3
.8	9.8238	9.9514	0.0486	9.8724	.2
.9	9.8247	9.9529	0.0471	9.8718	.1
42.0	9.8255	9.9544	0.0456	9.8711	**48.0**
.1	9.8264	9.9560	0.0440	9.8704	.9
.2	9.8272	9.9575	0.0425	9.8697	.8
.3	9.8280	9.9590	0.0410	9.8690	.7
.4	9.8289	9.9605	0.0395	9.8683	.6
.5	9.8297	9.9621	0.0379	9.8676	.5
.6	9.8305	9.9636	0.0364	9.8669	.4
.7	9.8313	9.9651	0.0349	9.8662	.3
.8	9.8322	9.9666	0.0334	9.8655	.2
.9	9.8330	9.9681	0.0319	9.8648	.1
43.0	9.8338	9.9697	0.0303	9.8641	**47.0**
.1	9.8346	9.9712	0.0288	9.8634	.9
.2	9.8354	9.9727	0.0273	9.8627	.8
.3	9.8362	9.9742	0.0258	9.8620	.7
.4	9.8370	9.9757	0.0243	9.8613	.6
.5	9.8378	9.9772	0.0228	9.8606	.5
.6	9.8386	9.9788	0.0212	9.8598	.4
.7	9.8394	9.9803	0.0197	9.8591	.3
.8	9.8402	9.9818	0.0182	9.8584	.2
.9	9.8410	9.9833	0.0167	9.8577	.1
44.0	9.8418	9.9848	0.0152	9.8569	**46.0**
.1	9.8426	9.9864	0.0136	9.8562	.9
.2	9.8433	9.9879	0.0121	9.8555	.8
.3	9.8441	9.9894	0.0106	9.8547	.7
.4	9.8449	9.9909	0.0091	9.8540	.6
.5	9.8457	9.9924	0.0076	9.8532	.5
.6	9.8464	9.9939	0.0061	9.8525	.4
.7	9.8472	9.9955	0.0045	9.8517	.3
.8	9.8480	9.9970	0.0030	9.8510	.2
.9	9.8487	9.9985	0.0015	9.8502	.1
45.0	9.8495	0.0000	0.0000	9.8495	**45.0**
	L. Cos	L. Cot	L. Tan	L. Sin	Deg.

Table 7   *Trigonometric (π radians)*

$x\pi$		Sin (πx)	Tan (πx)	Cot (πx)	Cos (πx)
$x =$					
.00 or 1.00		.00000	.00000	inf	1.00000
.01	.99	.03143	.03143	31.821	.99951
.02	.98	.06279	.06291	15.895	.99803
.03	.97	.09411	.09453	10.579	.99556
.04	.96	.12533	.12633	7.9158	.99211
.05	.95	.15643	.15838	6.3138	.98769
.06	.94	.18738	.19076	5.2422	.98229
.07	.93	.21814	.22353	4.4737	.97592
.08	.92	.24869	.25676	3.8947	.96858
.09	.91	.27899	.29053	3.4420	.96029
.10	.90	.30902	.32492	3.0777	.95106
.11	.89	.33874	.36002	2.7776	.94088
.12	.88	.36812	.39593	2.5257	.92978
.13	.87	.39715	.43274	2.3109	.91775
.14	.86	.42578	.47056	2.1251	.90483
.15	.85	.45399	.50953	1.9626	.89101
.16	.84	.48175	.54975	1.8190	.87631
.17	.83	.50904	.59140	1.6909	.86074
.18	.82	.53583	.63462	1.5757	.84433
.19	.81	.56208	.67960	1.4715	.82708
.20	.80	.58779	.72654	1.3764	.80902
.21	.79	.61291	.77568	1.2892	.79016
.22	.78	.63742	.82727	1.2088	.77051
.23	.77	.66131	.88162	1.1343	.75011
.24	.76	.68455	.93906	1.0649	.72897
.25	.75	.70711	1.0000	1.0000	.70711
.26	.74	.72897	1.0649	.93906	.68455
.27	.73	.75011	1.1343	.88162	.66131
.28	.72	.77051	1.2088	.82727	.63742
.29	.71	.79016	1.2892	.77568	.61291
.30	.70	.80902	1.3764	.72654	.58779
.31	.69	.82708	1.4715	.67960	.56208
.32	.68	.84433	1.5757	.63462	.53583
.33	.67	.86074	1.6909	.59140	.50904
.34	.66	.87631	1.8190	.54975	.48175
.35	.65	.89101	1.9626	.50953	.45399
.36	.64	.90483	2.1251	.47056	.42578
.37	.63	.91775	2.3109	.43274	.39715
.38	.62	.92978	2.5257	.39593	.36812
.39	.61	.94088	2.7776	.36002	.33874
.40	.60	.95106	3.0777	.32492	.30902
.41	.59	.96029	3.4420	.29053	.27899
.42	.58	.96858	3.8947	.25676	.24869
.43	.57	.97592	4.4737	.22353	.21814
.44	.56	.98229	5.2422	.19076	.18738
.45	.55	.98769	6.3138	.15838	.15643
.46	.54	.99211	7.9158	.12633	.12533
.47	.53	.99556	10.579	.09453	.09411
.48	.52	.99803	15.895	.06291	.06279
.49	.51	.99951	31.821	.03143	.03141
.50	.50	1.0000	inf	.00000	.00000

INDEX

A

Abscissa, 91
Absolute
  inequality, 83
  value, 78, 80, 324
Addition
  of functions, 99
  law, 228
  of vectors, 275
Algebraic notation, 3
Angle, 193, 196
Annuity, 187
Antilogarithm, 171
Arc cos, 239
Arc cot, 242
Arc csc, 243 (Exs. 27, 28)
Arc sec, 243 (Exs. 25, 26)
Arc sin, 164
Arc tan, 242
Arithmetic–geometric mean inequality, 87
Arithmetic progression, 345
Associative law, 7, 323
Asymptote, 147
  horizontal, 147
  oblique, 144 (Fig. 6.2), 147 (Ex. 17)
  vertical, 146

B

Bacteria growth, 184
Base of exponential function, 151
Binomial coefficient, 352, 354
Binomial Theorem, 354
Business applications, 186

C

Cartesian coordinate system, 91
Characteristic, 169
Circumradius, 267
Closed system, 12
Coefficient
  of linear system, 305
  of polynomial, 32
Combination, 352
Common
  denominator, 16, 43
  difference, 346
  errors in algebra, 46
  logarithm, 157
  ratio, 347
Commutative law, 6, 323
Completing square, 37, 58
Complex
  arithmetic, 322, 323
  conjugate, 324
  number, 321
  plane, 326
Composition of functions, 99
Compound interest, 186
Computation
  with logarithms, 172
  with pocket calculator, 179, 265
Conditional
  equation, 49
  inequality, 83
Coordinate, 90
Cosecant, 206
Cosine, 200

Cotangent, 208
Cramer's rule, 317
Cubic function, 127
Curve fitting, 309

D

Degree, 32, 35
De Moivre's Theorem, 337
Dependent variable, 93
Determinant, 312
Discriminant, 61, 292
Distance, 80, 194
Distributive law, 7, 323
Division
    algorithm, 286
    of polynomials, 285
    shortcuts, 287
Domain, 93, 94, 120, 135
Dot product, 279
Double-angle formula, 229

E

Elimination, 305
Equation(s), 49
    exponential, 183
    linear, 52
    polynomial, 63
    quadratic, 57, 321
    quadratic type, 64
    simultaneous, 55, 65, 305
    trigonometric, 242
Equilibrium price, 70
Estimates with inequalities, 76
Evaluation of sums, 360
Even
    function, 117
    integer, 12
Expansion by minors, 314
Exponent, 18
Exponential
    equation, 183
    function, 149
    rate of growth, 152
Extent, 120

F

Factor Theorem, 290, 334
Factored
    polynomial, 130
    rational function, 142
Factorial, 349
Factoring, 38

Formula, 49
Fraction, 16
Function(s), 89, 93
    composition of, 99
    construction of, 99
    even and odd, 117
    exponential, 149
    inverse, 236
    logarithm, 153
    nearest integer, 97
    not given by formula, 98
    periodic, 201, 220
    rational, 123, 134, 285
    trigonometric, 193
Fundamental Theorem of Algebra, 334

G

Gauss, K. F., 326, 334
Gaussian plane, 326
Geometric progression, 347
Graph, 89, 95
    of exponential functions, 150, 151, 158
    of factored polynomial, 130
    of factored rational function, 142
    of inverse trig function, 236
    of linear function, 101
    of logarithm function, 154
    of polynomial, 123
    of quadratic function, 111
    of rational function, 137
    of sine and cosine, 209
    of tangent, secant, etc., 215

H

Half-angle formula, 231, 270
Heron's formula, 269
Horizontal
    asymptote, 147
    shift, 117
    tangent, 219

I

Identity, 49, 51
    law, 9
    proving trig identities, 225
    trigonometric, 223
Imaginary part, 326
Inconsistent system, 56, 309
Independent variable, 93
Induction, 365
Inductive definition, 366
Inner product, 279

Inradius, 269
Integer, 12
Integral exponent, 18
Intercept, 106, 107
Interpolation, 170
Inverse
　function, 236
　law, 10
　trig function, 238
Irrational number, 17
Irreducible factor, 336

L

Law
　of cosines, 256
　of sines, 257
　of tangents, 259
Leading
　coefficient, 32
　term, 32
Linear
　equation, 101
　function, 52
　system, 305
Location of zeros, 296
Logarithm(s), 153
　to base b, 157
　common, 157
　computation with, 172
　function, 153
Lowest terms
　fraction in, 16
　rational expression in, 42, 134

M

Mantissa, 169
Mathematical induction, 365
Matrix, 308
Maximum problems, 114
Memory register, 180
Minor, 314
Mixture problems, 69
Modulus, 324
Mollweide's formula, 267
Monomial, 35
Mortgage, 190
Multiple zero, 295
Multiplication
　of functions, 99
　of vector by scalar, 273
Multiplicity, 132, 295, 335

N

n-th root, 26
Negative number, 73
Newton's formula, 267
Numerical solution of triangles, 260

O

Oblique triangle, 256
Odd
　function, 117
　integer, 12
Order, 72
Ordinate, 91
Origin, 91, 272

P

Parallelogram
　of forces, 275
　rule, 327
Parentheses, 4
Parity, 17
Partial fractions, 299, 361
Pascal's triangle, 355
Perfect square, 36
Period, 201, 220
Periodic
　function, 201, 220
　payment, 190
Permutation, 351
Perpendicular vectors, 282
Pocket calculator, 177, 265
Point–slope form, 108
Polar
　coordinates, 246
　equation of line, 248
　form of complex number, 328
Polynomial, 31
　algebra, 32
　equation, 63
　function, 123
　in several variables, 35
Positive number, 73
Power
　function, 160
　integer, 18
　rational, 28
　real, 160
Present value, 190
Primitive n-th root, 343
Product of sines and cosines, 234

Q

Quadratic
  equation, 57, 321
  formula, 61, 333
  function, 109
  polynomial, 292
  type, 64
Quotient
  of polynomials, 287
  of real numbers, 10

R

Radian, 196
Radical, 23, 26
Radioactive decay, 185
Rate problems, 68
Rational
  exponent, 28, 162
  expression, 42
  function, 123, 134, 285
  number, 16
  zero, 297
Rationalizing
  denominator, 26
  numerator, 164
Real
  number, 5
  part, 326
Reciprocal, 10, 323
Rectangular
  coordinate system, 91
  form of complex number, 328
Reflecting in axis, 118
Remainder
  in polynomial division, 287
  Theorem, 289
Right triangle, 251
  ratio of sides of, 208, 251
Root(s)
  of equation, 49, 201
  of unity, 338
Round-off, 167

S

Saw-tooth function, 97
Scientific notation, 21
Secant, 208
Segment, 73
Sequence, 345
Signed projection, 280
Significant figure, 167

Simple
  harmonic motion, 210
  zero, 295, 335
Simultaneous
  linear equations, 55, 305
  linear and quadratic equations, 65
Sine, 200
Sinking fund, 187
Slope, 105
  –intercept form, 108
  of sine and cosine curves, 219
Solution
  of equation, 49
  by factoring, 58, 63
  of linear system, 55
Square root
  of complex number, 332
  functions involving, 164
  of real number, 23, 26
Stretching, 118
Subscript, 13
Summation
  index, 357, 359
  notation, 357
Sum(s)
  of arithmetic progression, 346
  evaluation of, 360
  of geometric progression, 348
  involving integers, 14
  of sines, cosines, 235
Supply and demand, 70
Symmetry, 116
  in origin, 117
  in y-axis, 117
Synthetic division, 288
System of linear equations, 55, 305
  problems leading to, 69

T

Tables, use of, 169, 182
Tangent, 205
Tips on graphing, 116
Triangle inequality
  real numbers, 78
  vectors, 278
Trigonometric
  equation, 243
  function, 193
  function defined by point in plane, 207
  function in terms of other trig functions, 224
  identity, 223
Two
  -intercept form, 108
  -point form, 108

U

Underdetermined system, 57, 310
Unit circle, 195

V

Vector(s), 272
    addition of, 275
    inner product of, 279
    length of, 278

Vertical

Vertical
    asymptote, 146, 147 (Ex. 18)
    shift, 117
    stack, 180

Z

Zero
    existence of, 295, 334
    of function, 291
    of polynomial, 291, 334

5
B   6
C   7
D   8
E   9
F   0
G   1
H   2
I   3
J   4